HYPERSPECTRAL DATA EXPLOITATION

The Wiley Bicentennial–Knowledge for Generations

Each generation has its unique needs and aspirations. When Charles Wiley first opened his small printing shop in lower Manhattan in 1807, it was a generation of boundless potential searching for an identity. And we were there, helping to define a new American literary tradition. Over half a century later, in the midst of the Second Industrial Revolution, it was a generation focused on building the future. Once again, we were there, supplying the critical scientific, technical, and engineering knowledge that helped frame the world. Throughout the 20th Century, and into the new millennium, nations began to reach out beyond their own borders and a new international community was born. Wiley was there, expanding its operations around the world to enable a global exchange of ideas, opinions, and know-how.

For 200 years, Wiley has been an integral part of each generation's journey, enabling the flow of information and understanding necessary to meet their needs and fulfill their aspirations. Today, bold new technologies are changing the way we live and learn. Wiley will be there, providing you the must-have knowledge you need to imagine new worlds, new possibilities, and new opportunities.

Generations come and go, but you can always count on Wiley to provide you the knowledge you need, when and where you need it!

William J. Pesce
President and Chief Executive Officer

Peter Booth Wiley
Chairman of the Board

HYPERSPECTRAL DATA EXPLOITATION
THEORY AND APPLICATIONS

Edited by

CHEIN-I CHANG, PhD
University of Maryland—Baltimore County
Baltimore, MD

WILEY-INTERSCIENCE
A JOHN WILEY & SONS, INC., PUBLICATION

Published by John Wiley & Sons, Inc., Hoboken, New Jersey.
Published simultaneously in Canada.

Wiley Bicentennial Logo: Richard J. Pacifico

Library of Congress Cataloging-in-Publication Data:

Hyperspectral data exploitation : theory and applications / edited by Chein-I Chang.
p. cm.
Includes index.
ISBN: 978-0-471-74697-3 (cloth)
1. Remote sensing. 2. Multispectral photography. 3. Image processing–Digital techniques.
I. Chang, Chein-I.
G70.4.H97 2007
526.9′82–dc22 2006032486

10 9 8 7 6 5 4 3 2 1

MIX
Paper from responsible sources
FSC
www.fsc.org
FSC® C013604

CONTENTS

PREFACE

Hyperspectral imaging has become one of most promising and emerging techniques in remote sensing. It has made great advance in recent years due to introduction of new techniques from versatile disciplines, particularly statistical signal processing from engineering aspects. Such clear evidence can be witnessed by hundreds of articles published in journals and conference proceedings every year as well as many annual conferences held in various venues. The rapid growth in this subject has made many researchers difficult to keep up with new developments and advances in its technology. Despite the fact that many books have been published in the area of remote sensing image processing, most have been focused on multispectral image processing rather than hyperspectral signal/image processing. Until recently, only a few appeared as book forms in this particular field. One example is my first book, *Hyperspectral Imaging: Spectral Techniques for Detection and Classification*, published in 2003 by Kluwer/Plenum Academic Publishers (now part of Springer-Verlag Publishers); this was primarily written for subpixel detection and mixed pixel classification that were designed and developed in my laboratory. Unfortunately, many other topics are also of interest, but were not covered in this particular book. In order to address this need, I have made a significant effort to invite experts in hyperspectral imaging from academia and industries to write chapters in their expertise and share their research works with readers. This book is essentially a result of their contributions. A total of 13 chapters (Chapters 2 to 14) are included in this book and cover a wide spectrum of topics in hyperspectral data exploitation including imaging systems, data modeling, data representation, band selection and partition, and classification to data compression. Each chapter has been contributed by an expert in his/her specialty or by experts in their specialties. Also included is Chapter 1, an overview written by me which provides readers with a discussion on design philosophy in developing hyperspectral imaging techniques from a hyperspectral imagery point of view as well as brief reviews of each of the 13 chapters including coherent connections among different chapters. Therefore, this chapter can serve as a guide to direct readers to particular topics in which they are interested.

The ultimate goal of this book is to offer readers a peek at the cutting-edge research in hyperspectral data exploitation. In particular, this book can be found very useful for practitioners and engineers who are interested in this area. It is hoped that the chapters presented in this book have just done that.

Last but not least, I would like to thank all contributors for their participation in this book project. I owe them a great debt of gratitude for their efforts which make this book possible. This book would not have been possible without their contributions.

CHEIN-I CHANG

University of Maryland, Baltimore County
December 2006

CONTRIBUTORS

JON ATLI BENEDIKTSSON, Department of Electrical and Computer Engineering, University of Iceland, 107 Reykjavik, Iceland

JEFFERY H. BOWLES, Remote Sensing Division, Naval Research Laboratory, Washington, DC 20375

LORENZO BRUZZONE, Department of Information and Communication Technology, University of Trento, I-38050 Trento, Italy

ANDREA F. CATTONI, Department of Biophysical and Electronic Engineering, University of Genoa, I-16145 Genoa, Italy

CHEIN-I CHANG, Remote Sensing Signal and Image Processing Laboratory, Department of Computer Science and Electrical Engineering, University of Maryland—Baltimore Country, Baltimore, MD 21250

JOCELYN CHANUSSOT, Laboratoire des Images et des Signaux, 38402 Saint Martin d'Heres, France

MINGMIN CHI, Department of Information and Communication Technology, University of Trento, I-38050 Trento, Italy

JOSÉ M. B. DIAS, Instituto de Telecomunicações, Lisbon 1049-001, Portugal

MICHAEL T. EISMANN, AFRL's Sensors Directorate, Electro Optical Technology Division, Electro Optical Targeting Branch, Wright-Patterson AFB, OH 45433

MATHIEU FAUVEL, Laboratoire des Images et des Signaux, 38402 Saint Martin d'Heres, France; and Department of Electrical and Computer Engineering, University of Iceland, 107 Reykjavik, Iceland

JAMES E. FOWLER, Department of Electrical and Computer Engineering, GeoResources Institute, Mississippi State University, Mississippi State, MS 39762

DAVID B. GILLIS, Remote Sensing Division, Naval Research Laboratory, Washington, DC 20375

Xiuping. JIA, School of Information Technology and Electrical Engineering, University College, The University of New South Wales, Australian Defense Force Academy, Campbell ACT 2600, Australia

JOHN P. KEREKES, Chester F. Carlson Center for Imaging Science, Rochester Institute of Technology, Rochester, NY 14623

MATTIA MARCONCINI, Department of Information and Communication Technology, University of Trento, I-38050 Trento, Italy

GABRIELE MOSER, Department of Biophysical and Electronic Engineering, University of Genoa, I-16145 Genoa, Italy

JOSÉ M. P. NASCIMENTO, Instituto Superior de Engenharia de Lisboa, Lisbon 1049-001, Portugal

ANTONIO J. PLAZA, Department of Computer Science, University of Extremadura, E-10071 Caceres, Spain

JOHN A. RICHARDS, College of Engineering and Computer Science, The Australian National University, Canberra ACT 0200, Australia

JUSTIN T. RUCKER, Department of Electrical and Computer Engineering, Geo-Resources Institute, Mississippi State University, Mississippi State, MS 39762

JOHN R. SCHOTT, Chester F. Carlson Center for Imaging Science, Rochester Institute of Technology, Rochester, NY 14623

SEBASTIANO B. SERPICO, Department of Biophysical and Electronic Engineering, University of Genoa, I-16145 Genoa, Italy

SYLVIA S. SHEN, The Aerospace Corporation, Chantilly, VA, USA

DAVID W. J. STEIN, MIT Lincoln Laboratory, Lexington, MA 02421

MICHAEL E. WINTER, Hawaii Institute of Geophysics and Planetology, University of Hawaii, Honolulu, HI 96822

CHAPTER 1

OVERVIEW

CHEIN-I CHANG
Remote Sensing Signal and Image Processing Laboratory, Department of Computer Science and Electrical Engineering, University of Maryland—Baltimore County, Baltimore, MD 21250

1.1. INTRODUCTION

Hyperspectral imaging has become a fast growing technique in remote sensing image processing due to recent advances of hyperspectral imaging technology. It makes use of as many as hundreds of contiguous spectral bands to expand the capability of multispectral sensors that use tens of discrete spectral bands. As a result, with such high spectral resolution many subtle objects and materials can now be uncovered and extracted by hyperspectral imaging sensors with very narrow diagnostic spectral bands for detection, discrimination, classification, identification, recognition, and quantification. Many of its applications are yet to be explored. It has been common sense to think of hyperspectral imaging as a natural extension of multispectral imaging with band expansion. Accordingly, all techniques developed for multispectral imagery are considered to be readily applicable to hyperspectral imagery. Unfortunately, this intuitive interpretation may be somewhat misleading. To understand the fundamental difference between multispectral and hyperspectral images from a data processing perspective, we use a good example in mathematics for illustration, which is the difference between real analysis and complex analysis where the variables considered are real variables in real analysis as opposed to complex variables in complex analysis. Since real variables can be considered as real parts of complex variables, this may lead many to a belief that real analysis is a special case of complex analysis, which is certainly not true. One piece of clear evidence is derivatives. When a derivative is considered in real analysis, it has only two directions along the real line: left limit and right limit. However, in complex analysis, the direction of a derivative can be any curve in the complex plane. As a result, only partial derivatives in complex analysis can be considered as a natural extension of derivatives in real analysis. When a complex variable is

Hyperspectral Data Exploitation: Theory and Applications, Edited by Chein-I Chang

differentiable in the complex plane, it is usually called total differentiable or analytic because it must satisfy the so-called Cauchy–Riemann equation. This simple example provides a similar interpretation to explain the key difference between multispectral and hyperspectral images. In the early days, multispectral imagery was used in remote sensing mainly for land cover/use classification in agriculture applications, disaster assessment and management, ecology, environmental monitoring, geology, geographical information system (GIS), and so on. In these cases, low spectral resolution multispectral imagery may provide sufficient information for data analysis, and the techniques developed for multispectral image processing are primarily derived from the traditional two-dimensional spatial domain-based image processing that takes advantage of spatial correlation to perform various tasks. Compared to multispectral imagery, hyperspectral imagery utilizes hundreds of spectral bands for data acquisition and collection with two prominent improvements, very fine spectral resolution, and hundreds of spectral bands. It is these differences that distinguish hyperspectral imagery from multispectral imagery in their utility in many applications as demonstrated by the chapters presented in this book.

1.2. ISSUES OF MIXED PIXELS AND SUBPIXELS

Due to its low spectral resolution, a multispectral image pixel may not have information that is as rich as that of a hyperspectral image pixel. In this case, it must rely on its surrounding image pixels to provide spatial correlation and information to help to make up insufficient spectral information provided by multiple discrete spectral bands. Because of that, this may be one of main reasons that early development of multsipectral image processing has been focused on spatial domain-based techniques. The issues of subpixels and mixed pixels usually arise from very high spectral resolution produced by hyperspectral imagery and have become crucial but may not be critical to multispectral imagery. First of all, targets or objects of interest are different. In multispectral imagery, land covers or patterns are often of major interest. Therefore, the techniques developed for multispectral image analysis generally perform pattern classification and recognition. As a complete opposite, the objects of interest in hyperspectral imagery usually appear either in a form mixed by a number of material substances or at subpixel level with targets embedded in a single pixel due to their sizes smaller than the ground sampling distance (GSD). In both cases, these objects may not be identified *a priori* or by visual inspection. Therefore, they are generally considered as insignificant targets but are indeed of major interest from an intelligence or information point of view. More specifically, in hyperspectral data exploitation the objects of particular interest are those targets which have their small spatial presence and low probability existence in either form of a mixed pixel or a subpixel. Such targets may include special spices in agriculture and ecology, toxic wastes in environmental monitoring, rare minerals in geology, drug/smuggler trafficking in law enforcement, military vehicles and landmines in battlefields, chemical/biological agents in bioterrorism, and weapon concealment and mass graves in intelligence gathering. Under such circumstances, they can only be detected at mixed or subpixel level, and the traditional spatial domain (i.e., literal)-based image processing techniques may not be suitable and may also not be effective even if they can be applied. So, a great challenge in extraction of such targets is that these targets provide very limited spatial information and are generally difficult to be visualized in data. Therefore, the techniques developed for hyperspectral image analysis generally perform *target*-based detection, discrimination, classification, identification, recognition, and quantification as opposed to *pattern*-based multispectral imaging techniques. Consequently, a direct extension of multispectral imaging techniques to hyperspectral imagery may not be applicable in hyperspectral data exploitation. In order to address this issue, an approach directly from a hyperspectral imagery point of view is highly desirable and may offer insights into design and development of hyperspectral imaging algorithms because a single hyperspectral image pixel alone may already provide a wealth of spectral information for data processing without appealing to its spatial correlation with other sample pixels due to its limited spatial information.

1.3. PIGEON-HOLE PRINCIPLE

The advent of hyperspectral imagery has changed the way we think of multispectral imagery because we now have hundreds of spectral bands available for our use. Thus, one major issue is how to effectively use and take advantage of spectral information provided by these hundreds spectral bands to perform target detection, discrimination, classification and identification. This interesting issue can be addressed by the following well-known pigeon-hole principle in discrete mathematics [1].

Suppose that there are 13 pigeons flying into a dozen pigeon holes (nests). According to the pigeon-hole principle, there exists at least one pigeon hole that must accommodate at least two pigeons. Now, assume that L is the total number of spectral bands and p is the number of target classes to be classified. A hyperspectral image pixel is actually an L-dimensional column vector. By virtue of the pigeon-hole principle, we interpret a pigeon hole as a spectral band while a pigeon is considered as a target (or an object) so that we can actually use a spectral band to detect, discriminate, and classify a distinct target. With this interpretation, L spectral bands can be used to classify L different targets. Since there are hundreds of spectral bands available from hyperspectral imagery, technically speaking, hundreds of spectrally distinct targets can be also classified and discriminated by these spectral bands. In order to make this idea work, three issues need to be addressed. One is that the number of spectral bands must be greater than or equal to the number of targets to be classified; that is, $L \geq p$, which always seems true for hyperspectral imagery, but not valid for multispectral imagery, in which $L < p$, such as three-band SPOT data that may have more than three target substances present in the data. Furthermore, the first issue also gives rise to a second issue that is a well-known curse of dimensionality [2]—that is, to determine the value of p if $L \geq p$. This has been a most difficult and challenging issue for any hyperspectral image analyst to resolve, since it is nearly impossible to know the exact value of p in real-world problems and it may not be reliable even if the value of p is provided by prior knowledge. In multivariate data analysis, the value of p can be estimated by so-called intrinsic dimensionality (ID) [3], which is defined as the minimum number of parameters used to specify the data. However, this concept is only of theoretic interest, and no method has been proposed for this purpose in the literature regarding how to find it. A common strategy is on a trial-and-error basis. A similar problem is also encountered in passive array processing where the number of signal sources arriving at an array of sensors is of major interest and a key issue. In order to estimate this number, two criteria—an information criterion (AIC) suggested by Akaike and minimum description length developed by Schwarz and Rissanen [4]—have been shown successfully in such estimation. Unfortunately, a key assumption made on these criteria is that the noise must be independent identically distributed, which is usually not a valid assumption in hyperspectral images as shown in Chang [5] and in Chang and Du [6]. In order to cope with this dilemma, a new concept coined and suggested by Chang [5], called virtual dimensionality (VD), was recently proposed to estimate the number of spectrally distinct signatures in hyperspectral imagery. Its applications to hyperspectral data exploitation such as linear spectral unmixing (Chapters 4–6 in this book), dimensionality reduction (Chapter 8 in this book), band selection (Chapters 9 and 10 in this book), and so on, are also reported in Chang [7, 8]. Finally, the third and last issue is that once a spectral band is being used to accommodate one target, it cannot be used again to accommodate another distinct target. How do we make sure that this will not happen? One way to do so is to perform orthogonal subspace projection (OSP) developed in Harsanyi and Chang [9] on the hyperspectral imagery so that no two or more distinct targets will be accommodated by a single spectral band. This implies that no two pigeons will be allowed to fly into a single pigeon hole (nest) in terms of the pigeon-hole principle. Once these three issues—that is, (1) $L \geq p$, (2) determination of p, and (3) no two distinct target signatures to be accommodated by a single spectral band—are addressed, the idea of using the pigeon-hole principle for hyperspectral data exploitation can be realized and becomes feasible. Most importantly, it provides an alternative approach that uses spectral bands as a means to perform detection, and discrimination, classification, and identification without counting on spatial information or correlation. This is particularly important for targets that are small or insignificant due to their limited spatial presence and cannot be captured by spatial correlation or information. As a result, hyperspectral imaging techniques developed from this aspect are generally carried out on a pixel-by-pixel basis rather than on a spatial domain basis.

1.4. ORGANIZATION OF CHAPTERS IN THE BOOK

This book has 13 chapters contributed by researchers from various disciplinary areas whose expertise is in hyperspectral data exploitation. Each of these chapters addresses different problems caused by the above-mentioned issues. In particular, these 13 chapters are organized into three categories, Part I: Tutorials, Part II: Theory, and Part III: Applications.

1.4.1. Part I: Tutorials

The tutorials part consists of two tutorial chapters that review some basics of hyperspectral data exploitation, hyperspectral imaging systems, and algorithm design rationale for target detection and classification. Chapter 2 by Kerekes and Schott offers an excellent introduction of hyperspectral imaging systems including two popular airborne hyperspectral imagers, known as Airborne Visible/InfraRed Imaging Spectrometer (AVIRIS) and Hyperspectral Digital Image Collection Experiment (HYDICE), and a satellite-operated HYPERION. It is then followed by Chapter 3 by Chang, which is a review of matched filter-based target detection and classification algorithms.

1.4.2. Part II: Theory

The theory part is comprised of eight chapters that essentially address key issues in data modeling and representation by various approaches: linear mixing model (LMM) with deterministic endmembers (Chapter 4) and random endmembers (Chapters 5 and 6), endmember extraction (Chapter 7), dimensionality reduction (Chapter 8), band selection (Chapter 9), band partition (Chapter 10), and semisupervised support vector machines (Chapter 11).

Chapter 4 by Bowles and Gillis describes an optical real-time adaptive spectral identification system developed by the Naval Research Laboratory, known as ORASIS, which is a collection of algorithms to perform a series of tasks in sequence, an exemplar set selection, basis selection, endmember selection, and spectral unmixing. While the endmembers considered in Chapter 4 for spectral unmixing are deterministic, Chapter 5 by Eismann and Stein develops a stochastic mixing model (SMM) to describe statistical representation of hyperspectral data where the endmembers used in the model are considered as random vectors with probability density functions described by finite Gaussian mixtures. As an alternative to the stochastic mixing model discussed in Chapter 5, Chapter 6 by Nascimento and Dias presents Independent Component Analysis (ICA) and Independent Factor Analysis (IFA) for spectral unmixing where the abundance fractions of endmembers used in the linear mixing model for the ICA/IFA are described by a mixture of Dirichlet densities as opposed to a mixture of Gaussian densities assumed in the SMM in Chapter 5. Two common and key issues shared by Chapters 4–6 are (1) finding an appropriate set of endmembers to be used to form a linear mixing model and (2) performing data dimensionality reduction to reduce computational complexity. To address the first issue, Chapter 7 by Winter revisits his well-known endmember extraction algorithm, N-finder algorithm (N-FINDR), and further develops a new improved version of the N-FINDR, called maximum volume transform (MVT). Chapter 8 by Jia and Richards addresses the second issue by investigating data representation of hyperspectral data to cope with the so-called curse of dimensionality where feature extraction becomes a powerful and effective means to resolve this issue, such as variance used by the PCA, Fisher's ratio, or Rayleigh quotient used by Fisher's linear discriminant analysis (FLDA). Another approach to address the issue of data dimensionality reduction is band selection. Chapter 9 by Shen develops an entropy-based genetic algorithm to select optimal band sets for spectral imaging systems including five existing multispectral imaging systems and further substantiates the utility of optimal band selection in target detection and material identification. As an alternative to band selection, Chapter 10 by Serpico et al. proposes an approach to band partition which is based on feature extraction/selection for a specific classification application. Finally, Chapter 11 by Bruzzone et al. improves a well-known supervised classifier, support vector machines (SVMs), by introducing semisupervised SVMs for classification of hyperspectral remote sensing images.

1.4.3. Part III: Applications

The applications part consists of three chapters that address various data exploitation issues by different approaches using classification as an application. Chapter 12 by Benediktsson and co-workers proposes a generic framework to fuse decisions of multiple classifiers for hyperspectral classification including morphology-based classifier, neural network classifier, and SVMs. Chapter 13 by Plaza develops a morphology-based classification approach and its potential in parallel computing. Finally, this book concludes with one of the most important applications in hyperspectral data exploitation, namely, hyperspectral data compression. Chapter 14 by Fowler and Rucker which overviews 3-D wavelet-based hyperspectral data compression with classification as an application.

1.5. BRIEF DESCRIPTIONS OF CHAPTERS IN THE BOOK

In order to provide a quick glimpse of all the chapters presented in the book, this section intends to help the reader walk through each of these chapters by briefly summarizing their works and suggesting coherent connections among different chapters as follows.

Part I: Tutorials

Chapter 2. Hyperspectral Imaging Systems

John P. Kerekes and John R. Schott
Chester F. Carlson Center for Imaging Science
Rochester Institute of Technology, Rochester, NY, USA

This chapter offers an excellent overview of some currently used hyperspectral imaging systems: JPL/NASA developed the 224-band Airborne Visible InfraRed Imaging Spectrometer in 1987, Hughes/NRL developed the 210-band HYperspectral Digital Image Collection Experiment (HYDICE) in 1994, and TRW/NASA developed the 220-band HYPERION in 2000. In addition, two sensor models are also introduced for simulation in development and application of sensor technology: (1) Digital Imaging and Remote Sensing Image Generation (DIRSIG) developed by the Rochester Institute of Technology (RIT) and (2) Forecasting and Analysis of Spectroradiometric System Performance (FASSP) developed by Massachusetts Institute of Technology (MIT) Lincoln Laboratory. This chapter provides a good tutorial introduction of hyperspectral sensor design and technology to researchers working in the hyperspectral imaging area.

Chapter 3. Information-Processed Matched Filters for Hyperspectral Target Detection and Classification

Chein-I Chang
Remote Sensing Signal and Image Processing Laboratory
Department of Computer Science and Electrical Engineering
University of Maryland—Baltimore County, Baltimore, MD, USA

This chapter reviews hyperspectral target detection and classification algorithms from a matched filter perspective. Since most such algorithms share the same design principles of using a matched filter as a framework, this chapter presents an information-processed matched-filter approach to unifying these algorithms. It interprets a hyperspectral target detection and classification algorithm using two sequential filter operations. The first filter operation is an information-processed filter that processes *a priori* or *a posteriori* target information to suppress unwanted interference and noise effects. The follow-up second filter operation is a matched filter that extracts targets of interest for detection and classification. Three well-known specific techniques—Orthogonal Subspace Projection (OSP), Constrained Energy Mimimization (CEM), and Reed–Yu's RX-anomaly detection—are selected for this interpretation, each of which represents a particular category of algorithms that process a different level of information to enhance performance of the follow-up matched filter. While the OSP requires a complete prior knowledge, the RX-anomaly detection relies only on the *a posteriori* information provided by data samples. The CEM is somewhere in between, which requires *a priori* information of the desired targets used in the matched filter with *a posteriori* information obtained from data samples to suppress interfering effects while performing target extraction. The relationship among these three types of techniques shows how *a priori* target knowledge is approximated by *a posteriori* information as well as how a matched filter is affected by the information used in its matched signal.

Part II: Theory

Chapter 4. An Optical Real-Time Adaptive Spectral Identification System (ORASIS)

Jeffery H. Bowles and David B. Gillis
Remote Sensing Division
Naval Research Laboratory, Washington, DC, USA

This chapter presents a popular system, called the Optical Real-Time Adaptive Spectral Identification System (ORASIS), developed by the authors with their colleagues in the Naval Research Laboratory. It is a collection of a number of algorithms that are designed to perform various tasks in sequence. In its first-stage process, it develops a prescreener that finds an exemplar set and uses the found exemplar set as a code book to encode all image spectral signatures. This is followed by a second-stage process, which is basis selection that projects the exemplar set into a low-dimensional space spanned by an appropriate set of bases. This process is similar to dimensionality reduction that is commonly accomplished by the Principal Components Analysis (PCA). With this reduced data space the third-stage process performs a simplex-based endmember extraction to select a desired set of endmembers that are used to form a linear mixing model for least-squares error-based spectral unmixing that is carried out in the fourth and final state process to exploit three applications: automatic target recognition, terrain categorization, and compression.

Chapter 5. Stochastic Mixture Modeling

Michael T. Eismann[1] and David W. J. Stein[2]

[1]AFRL's Sensors Directorate, Electro Optical Technology Division
Electro Optical Targeting Branch, Wright-Patterson AFB, OH, USA

[2]MIT Lincoln Laboratory, Lexington, MA, USA

This chapter develops a stochastic mixing model (SMM) to address limitations of the commonly used linear mixture model (LMM) by capturing data variation that cannot be well described by linear mixing. Unlike the LMM which considers image endmembers as deterministic signatures, the SMM treats image endmembers used in a linear mixture model as random signatures. More specifically, a data sample is described by a linear mixture of a finite set of random endmembers that can be modeled by mixtures of Gaussian distributions. Two approaches are developed to estimate mixture density functions: (1) discrete SMM, which imposes physical abundance constraints, and (2) normal composition model (NCM), which is a continuous version of the SMM with no constraints imposed on abundance fractions. As a result, the NCM does not make assumption of existence of pure pixels as does in the discrete SMM. In order to estimate mixture density functions used to describe both models, the well-known Expectation-Maximization (EM) algorithm is used for this purpose. Interestingly, a similar approach using linear mixtures of random endmembers can be also found in Chapter 6 where two models, mixtures of Gaussian distrivutions and mixtures of Dirichlet distributions are introduced as counterparts of the discrete SMM and NCM dealing with the issue of presence of pure pixels in the data. The readers are strongly recommended to read this chapter along with Chapter 6 to have maximum benefits in gaining insights into linear mixtures of random endmembers.

Chapter 6. Unmixing Hyperspectral Data: Independent and Dependent Component Analysis

Jose M. P. Nascimento[1] and Jose M. B. Dias[2]
[1]Instituto Superior De Engenharia de Lisboa, Lisbon, Portugal
[2]Instituto de Telecomunicações, Lisbon, Portugal

This chapter presents approaches using independent component analysis (ICA) and independent factor analysis (IFA) to unmix hyperspectral data, and it further addresses issues of limitations on data independency and dependency due to constraints imposed on abundance fractions in the unmxing processing. The criterion used for finding an unmixing matrix for the ICA and IFA is the minimization of mutual information based on the calculation of a finite mixture of Gaussian distributions via the expectation–maximization (EM) algorithm to estimate mixture density functions where the resulting unmixng matrix is generally far from the true one if there are no pure pixels present in the data. In order to mitigate this problem, it introduces a new blind separation source unmixing technique where abundance fractions are modeled by mixtures of Dirichlet sources which enforce two physical constraints, namely, non-negativity and sum-to-one abundance fraction constraints. Once again, the EM algorithm is also used to estimate mixture density functions. Interestingly, the work in this chapter follows a very similar approach to the work in Chapter 5, where a data sample is also described by a finite mixture of Gaussian random endmembers whose mixture density functions are estimated by the EM algorithm. It will be very beneficial to the readers if both Chapter 5 and Chapter 6 are read together to gain their ideas developed for the models.

Chapter 7. Maximum Volume Transform for Endmember Spectra Determination

Michael E. Winter
Hawaii Institute of Geophysics and Planetology
University of Hawaii, Honolulu, HI, USA

This chapter revisits the well-known endmember extraction algorithm, called the N-finder algorithm (N-FINDR), which was developed by the author and further presents a new development of the N-FINDR, called the N-FINDR-based maximum volume transform (MVT). Endmember extraction has been a fundamental issue arising in hyperspectral data exploitations (as indicated in Chapters 4–6), where endmembers form a base of a linear mixing model. The N-FINDR is probably one of most widely used endmember extraction algorithms available in the literature. The work presented in this chapter offers a good review of the N-FINDR which should interest researchers working in automatic exploitation of hyperspectral imagery.

Chapter 8. Hyperspectral Data Representation

Xiuping. Jia[1] and John A. Richards[2]
[1]Australian Defense Force Academy, Australia
[2]The Australia National University, Australia

This chapter investigates hyperspectral data representation to explore the issue of the curse of dimensionality. In doing so, several selected supervised classification methods including standard maximum likelihood classification (MLC) with its variants—block-wise MLC, regularized MLC, and nonparametric weighted feature extraction (NWFE)—are used to reduce data dimensionality. In order to conduct a comparative analysis among these four algorithms, two sets of hyperspectral image data, Hyperion data, and Purdue's Indiana Indian Pine AVIRIS data are used for performance evaluation.

Chapter 9. Optimal Band Selection and Utility Evaluation for Spectral Systems

Sylvia S. Shen
The Aerospace Corporation, Chantilly, VA, USA

This chapter considers optimal band selection and utility evaluation for spectral imaging systems. For a given number of bands, it develops an information theoretic criterion-based genetic algorithm to find an optimal band set that yields the highest possible material separability. One of interesting findings in this chapter is to use 612 adjusted spectra obtained from a combined data base to conduct a comparative study of various optimal band sets with their respective five different existing spectral imaging systems: Landsat-7 ETM+, Multispectral Thermal Imager (MTI), Advanced Land Imager (ALI), Daedalus AADS 1268, and M7. Additionally, in order to assess utility of optimal band sets, two applications of anomaly detection by spectral unmixing and material identification by spectral matching are investigated for performance evaluation where two HYDICE data cubes are used for experiments to perform qualitative and quantitative study. The results demonstrate that a judicious selection of a band subset from original bands (e.g., as few as nine bands) can perform very effectively in separating man-made objects from natural background. This useful information provide insights into the development and optimization of multiband spectral sensors and algorithms using an exploitation-based optimal band selection to reduce data transmission and storage while retaining features used for target detection and material identification.

Chapter 10. Feature Reduction for Classification Purpose

Sebastiano B. Serpico, Gabriele Moser, and Andrea F. Cattoni
Department of Biophysical and Electronic Engineering
University of Genoa, Genoa, Italy

This chapter investigates approaches to feature extraction-based band partition where four band partition algorithms, called sequential forward band partitioning (SFBP), steepest ascent band partitioning (SABP), fast constrained band partitioning (FCBP), and convergent constrained band partitioning (CCBP), are developed with the Jeffries–Matusita distance used as the criterion for band partition from a classification point of view. It is interesting to compare the work in this chapter to that in Chapter 9, where the former performs a classification-based band partition, whereas the latter proposes a genetic algorithm-based band selection with its utility substantiated by anomaly detection and material identification.

Chapter 11. Semisupervised Support Vector Machines for Classification of Hyperspectral Remote Sensing Images

Lorenzo Bruzzone, Mingmin Chi, and Mattia Marconcini
Department of Information, and Communication Technology
University of Trento, Trento, Italy

This chapter presents an approach based on semisupervised support vector machines (SVMs) which combine advantages of semisupervised classification approaches with the advantages of distribution-free kernel-based methods based on SVMs so as to achieve better classification. Two such semisupervised SVM techniques are developed. One is a transductive SVM based on an iterative self-labeling procedure implemented in the dual formulation of the optimization problem related to the learning of the classifier. The other is a transductive SVM based on the cluster assumption implemented in the primal formulation of the optimization problem associated with the learning of the classification algorithm. A comparative analysis between these two techniques along with a standard inductive SVM is conducted by using a real hypersepctral data set for experiments. Experimental results demonstrate that the proposed semisupervised support vector machines perform effectively and increase the classification accuracy compared to standard inductive SVMs.

Part III: Applications

Chapter 12. Decision Fusion for Hyperspectral Classification

Mathieu Fauvel[1,2], Jocelyn Chanussot[1], and Jon Atli Benediktsson[2]
[1] Laboratoire des Images et des Signaux, Saint Martin d'Heres, France
[2] Department of Electrical and Computer Engineering
University of Iceland, Reykjavik, Iceland

This chapter presents a generic framework where the redundant or complementary results provided by multiple classifiers can actually be aggregated. Taking advantage of the specificities of each classifier, the decision fusion thus increases the overall classification performances. The proposed fusion approach is in two steps. In a first step, data are processed by each classifier separately and the algorithms provide for each pixel membership degrees for the considered classes. Then in a second step, a fuzzy decision rule is used to aggregate the results provided by the algorithms according to the classifiers' capabilities. The general framework proposed for combining information from several individual classifiers in multiclass classification is based on the definition of two measures of accuracy. The first one is a pointwise measure that estimates for each pixel the reliability of the information provided by each classifier. By modeling the output of a classifier as a fuzzy set, this pointwise reliability is defined as the degree of uncertainty of the fuzzy set. The second measure estimates the global accuracy of each classifier. It is defined *a priori* by the user. Finally, the results are aggregated with an adaptive fuzzy fusion ruled by these two accuracy measures. The method is illustrated by considering the classification of hyperspectral remote sensing images from urban areas. It is tested and validated with two classifiers on a ROSIS image from Pavia, Italy. The proposed method improves the classification results when compared with the separate use of the different classifiers.

Chapter 13. Morphological Hyperspectral Image Classification: A Parallel Processing Perspective

Antonio J. Plaza
Computer Science Department
University of Extremadura, Caceres, Spain

This chapter provides a detailed overview of recently developed approaches to morphological analysis of remotely sensed data. It first explores vector ordering strategies for the generalization of concepts from mathematical morphology to multichannel image data and further develops new, physically meaningful distance-based organization schemes to define morphological vector operations by extension. The problem of ties resulting from partial vector ordering is also addressed. Then, two new morphological algorithms for hyperspectral image classification are developed, which are (1) a supervised mixed pixel classification algorithm which integrates spatial and spectral information in simultaneous fashion and (2) an unsupervised morphological watershed-based image segmentation algorithm that first analyzes the data using spectral information and then refines the result using spatial context. While such integrated spatial/spectral approaches hold great promise in several applications, they also introduce new processing challenges. Several applications exist, however, where having the desired information calculated in (near) real time is highly desirable. For that purpose, this chapter also develops efficient parallel implementations of the morphological techniques addressed above. Three parallel computing platform used in experiments is a massively parallel Beowulf cluster called Thunderhead, made up of 256 processors and located at NASA's Goddard Space Flight Center in Maryland.

Chapter 14. Three-Dimensional Wavelet-Based Compression of Hyperspectral Imagery

James E. Fowler and Justin T. Rucker
Department of Electrical and Computer Engineering
GeoResources Institute
Mississippi State University, Mississippi State, MS USA

This chapter overviews 3D embedded wavelet-based algorithms with their applications to hyperspectral data compression. Six JPEG2000-based compression algorithms, (1) JPEG2000-band-independent fixed-rate (BIFR), (2) 2D JPEG2000-band-independent fixed-rate (BIFR), (3) JPEG2000-band-independent rate allocation (BIRA), (4) 2D JPEG2000-band-independent rate allocation (BIRA), (5) JPEG2000 multicomponent (JPEG2000-MC), (6) 2D JPEG2000 multicomponent (JPEG2000-MC), are studied for compression of hyperspectral image data. It is well known that the commonly used compression criteria mean-squared error (MSE) and signal-to-noise ratio (SNR) are not appropriate measures to evaluate hyperspectral data compression. In order to address this issue, this chapter introduces an application specific measure, called preservation of classification (POC), as a compression criterion where an unsupervised classifier, ISODATA, is used for evaluation of classification performance. Three hyperspectral AVIRIS data—Moffett, Jasper Ridge, and Cuprite—are then used to conduct a comparative analysis among the six considered compression algorithms using three different compression criteria, MSE, SNR, and POC. The experimental results have demonstarted that JPEG2000 can always benefit from a 1D spectral wavelet transform.

Finally, in order to provide a guide for what topics and techniques are discussed in each of the chapters, Table 1.1 summarizes the major tasks accomplished in each of chapters with acronyms defined as follows for reference. However, it should be noted that since Chapter 2 is completely devoted to design and development of hyperspectral imaging systems, it is not included in Table 1.1.

ACRONYMS

DR	Dimensionality reduction
EM	Expectation–maximization algorithm
FE	Feature extraction
GA	Genetic algorithm
ICA	Independent component analysis
IFA	Independent factor analysis
LMM	Linear–mixing model
LSE	Least–squares error
MNF	Maximum noise fraction
MLE	Maximum likelihood estimation

TABLE 1.1. Techniques Used to Perform Various Functionalities in Chapters

Chapters	Data Model and Representation	Endmember Extraction	Spectral Unmixing	Applications
Chapter 3	OSP-DR, LMM		OSP	Detection, classification
Chapter 4	Basis-DR, LMM	Simplex	LSE	Detection, classification, compression
Chapter 5	PCA-DR, SMM/NCM	N-FINDR	MLE	
Chapter 6	PCA-DR, LMM	Mutual information	ICA/IFA	
Chapter 7	MNF-DR	N-FINDR		
Chapter 8	FE-DR		MLE	Classification
Chapter 9	GA-based Band selection		Unspecified	Spectral matching, detection, identification
Chapter 10	Band partition			SVM/classification
Chapter 11				SVM/classification
Chapter 12				Morphology-NN SVM/classification
Chapter 13	PCA/MNF-DR			Morphology classification
Chapter 14	3D wavelet compression			ISODATA/ classification

NCM	Normal composition model
NN	Neural network
NWFE	Nonparametric weighted feature extraction
OSP	Orthogonal subspace projection
PCA	Principal components analysis
SMM	Stochastic mixing model
SVM	Support vector machine

Additionally, Table 1.2 also provides information about the types of image data that are used in Chapters 2–14, where a check symbol "√" indicates that an image scene is not specified in a particular chapter.

1.6. CONCLUSIONS

Hyperspectral imaging offers an effective means of detecting, discriminating, classifying, quantifying, and identifying targets via their spectral characteristics captured by high spectral-resolution sensors without accounting for their spatial information. The processing techniques that only make use of spectral properties

TABLE 1.2. Data Used in Various Chapters

Chapters	AVIRIS	HYDICE	HYPERION	Other Images
Chapter 2	√	√	√	DIRSIG
Chapter 3	Lab data			
Chapter 4	Cuprite	Forest		PHILLS
Chapter 5	Cuprite	Forest	√	
Chapter 6	Indian Pine			
Chapter 7	Cuprite			HyMap (Cuprite)
Chapter 8	Indian Pine		√	
Chapter 9		√		Landsat, ALI, MTI, Daedalus, M7
Chapter 10	Indian Pine			
Chapter 11			√	
Chapter 12				ROSIS
Chapter 13	Salinas Valley			
Chapter 14	Moffett, Cuprite, Jasper Ridge			

without taking into account spatial information are generally referred to as nonliteral (spectral) processing techniques as opposed to literal techniques referred to as traditional spatial domain-based image processing techniques. Over the past years, significant research efforts have been devoted to design and development of such nonliteral processing techniques with applications in hypespectral data exploitation. Many results have been published in various journals and presented in different conference meetings. Despite the fact that several books have recently been published [5,10–13], the subjects covered in these books are somewhat selective. The chapters presented in this book provide the most recent advances of many techniques which are not available in these books. In particular, it addresses many important key issues that should serve as a nice guide for researchers who are interested in exploitation of hyperspectral data.

REFERENCES

1. S. S. Epp, *Discrete Mathematics with Applications*, 2nd edition, Brooks/Cole, Pacific Grove, CA, 1995.
2. R. O. Duda and P. E. Hart, *Pattern Classification and Scene Analysis*, John Wiley & Sons, New York, 1973.
3. K. Fukunaga, *Statistical Pattern Recognition*, 2nd edition, Academic Press, New York, 1990.
4. M. Wax and T. Kailath, Detection of signals by information criteria, *IEEE Transactions on Acoustic, Speech, and Signal Processes*, vol. ASSP-33, no. 2, pp. 387–392, 1985
5. C.-I Chang, *Hyperspectral Imaging: Techniques for Spectral Detection and Classification*, Kluwer Academic/Plenum Publishers, New York, 2003.

6. C.-I Chang and Q. Du, Estimation of number of spectrally distinct signal sources in hyperspectral imagery, *IEEE Transactions on Geoscience and Remote Sensing*, vol. 42, no. 3, pp. 608–619, 2004.
7. C.-I Chang, Exploration of virtual dimensionality in hyperspectral image analysis, *Algorithms and Technologies for Multispectral, Hyperspectral, and Ultraspectral Imagery XII*, SPIE Defense and Security Symposium, Orlando, Florida, April 17–21, 2006.
8. C.-I Chang, Utility of Virtual Dimensionality in Hyperspectral Signal/Image Processing, Chapter 1, *Recent Advances in Hyperspectral Signal and Image Processing*, edited by C.-I Chang, Research Signpost, Trivandrum, Kerala, India, 2006.
9. J. C. Harsanyi and C.-I Chang, Hyperspectral image classification and dimensionality reduction: An orthogonal subspace projection approach, *IEEE Transactions on Geoscience and Remote Sensing*, vol. 32, no. 4, pp. 779–785, 1994.
10. P. K. Varshney and M. K. Arora (Ed.), *Advanced Image Processing Techniques for Remotely Sensed Hyperspectral Data*, Springer-Verlag, Berlin, 2004.
11. C.-I Chang (Ed.), *Recent Advances in Hyperspectral Signal and Image Processing*, Research Signpost, Transworld Research Network, Trivandrum, Kerala, India, 2006.
12. A. J. Plaza and C.-I Chang (Ed.), *High Performance Computing in Remote Sensing*, CRC Press, Boca Raton, FL, 2007.
13. C.-I Chang, *Hyperspectral Imaging: Signal Processing Algorithm Design and Analysis*, John Wiley & Sons, Hoboken, NJ, 2007.

PART I

TUTORIALS

CHAPTER 2

HYPERSPECTRAL IMAGING SYSTEMS

JOHN P. KEREKES AND JOHN R. SCHOTT
Chester F. Carlson Center for Imaging Science, Rochester Institute of Technology, Rochester, NY 14623

2.1 INTRODUCTION

Spectral imaging refers to the collection of optical images taken in multiple wavelength bands that are spatially aligned such that at each pixel there is a vector representing the response to the same spatial location for all wavelengths. A simple spectral imaging system is a digital color camera that records an intensity image in red, green, and blue spectral bands, which in composite create a color image. Hyperspectral imaging (HSI) systems are distinguished from color and multispectral imaging (MSI) systems in three main characteristics. First, color and MSI systems typically image the scene in just three to ten spectral bands, while HSI systems image in hundreds of co-registered bands. Second, MSI systems typically have spectral resolution (center wavelength divided by the width of the spectral band, $\lambda/\Delta\lambda$) on the order of 10, while HSI systems typically have spectral resolution on the order of 100. Third, while MSI systems often have their spectral bands widely and irregularly spaced, HSI systems have spectral bands that are contiguous and regularly spaced, leading to a continuous spectrum measured for each pixel. Figure 2.1 shows the hyperspectral imaging concept.

This figure also illustrates the notion of a "hypercube" by showing the hyperspectral data as a cube with the visible grayscale image on the face and representations of the spectra along the sides, indicating the presence of a complete spectrum of measurements for each pixel.

HSI systems are a technology enabled primarily by advances in optical detector array fabrication that occurred starting in the late 1970s. Through the development of linear and two-dimensional detector arrays, the collection of hundreds of spectrally contiguous and spatially co-registered images became feasible.

One of the first systems to demonstrate hyperspectral imaging in the context of remote sensing of the earth from aircraft was the Airborne Imaging Spectrometer

Hyperspectral Data Exploitation: Theory and Applications, Edited by Chein-I Chang

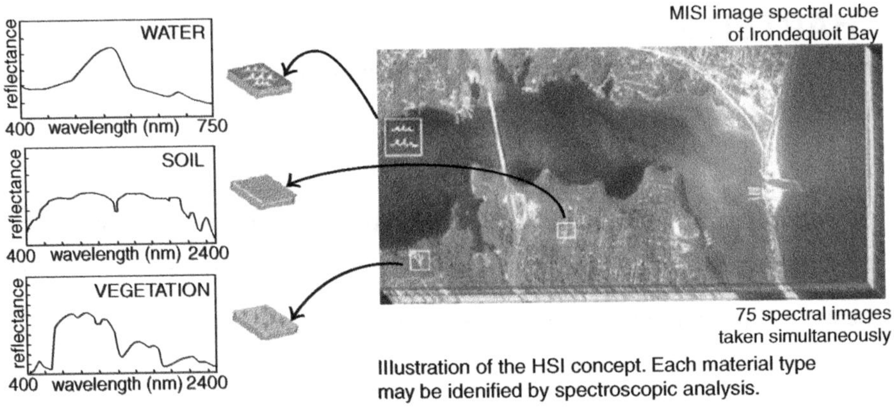

Figure 2.1. Hyperspectral imaging concept.

(AIS) built at NASA's Jet Propulsion Laboratory and first flown in November 1982 [1]. In the paper by Goetz et al. [1], the terms *imaging spectrometry* and *hyperspectral imaging* were introduced to the earth remote sensing community. Today, while the term hyperspectral imaging is sometimes broadly applied to any imaging system with hundreds to thousands of spectral channels, the original and most prominent use of the term refers to high-resolution (1- to 30-meter ground pixel size) imaging of the earth's surface for environmental and military applications. Table 2.1 provides a summary of several existing HSI sensor systems used for research and development (R&D), demonstration, commercial, and operational applications.

The intrinsic value of the data collected by these hyperspectral imaging systems lies in the intersection between the phenomenology of the surface spectral and spatial characteristics and the ability of the system to capture these characteristics. Laboratory and field studies have shown that spectral bandwidths on the order of 5 to 20 nm (corresponding to spectral resolution of $\sim$100 over the range 400 to 2500 nm, the range of the optical spectrum where the sun provides illumination) resolve most features of interest in the reflectance spectra of solid and liquids visible on the earth's surface. These features arise as vibrational or electronic resonances in the material structure or from the three-dimensional microgeometry in the top surface of a material. Thus, hyperspectral spectra contain sufficient spectral resolution for distinguishing or even sometimes identifying materials remotely.

Spatially, the earth's surface (including man-made objects) has relevant features of interest over a very wide range of sizes, from millimeters to kilometers and more. It is generally naïve to say that "with only enough spatial resolution, we can image objects such that the pixels are spectrally pure, or whole, with but one unique material present." It only takes a modest amount of reflection to realize that whether we image with pixels that are either 10 km or 1 mm across, we nearly always will be facing a "mixed pixel" problem where the measurement represents a response from

TABLE 2.1. Example Hyperspectral Imaging Systems

Sensor	Objective	Typical Altitude	Spectral Range	Number of Bands	Ground Pixel Size	Ground Swath
AHI [2]	University R&D	3 km	7.9–11.5 μm	210	3 m	0.7 km
AIS [1]	Science R&D	4 km	1.2–2.4 μm	128	8 m	0.3 km
ARCHER [3]	Civil Air Patrol	2 km	0.5–1.1 μm	512	5 m	1.3 km
AVIRIS [4]	Science R&D	20 km	0.4–2.5 μm	224	20 m	11 km
CASI [5]	Commercial operational	2 km	0.4–1.1 μm	288	1 m	1.4 km
COMPASS [6]	Military demonstration	3 km	0.4–2.5 μm	256	1 m	1.6 km
HYDICE [7]	Military R&D	6 km	0.4–2.5 μm	210	3 m	1 km
HyMAP [8]	Commercial operational	2 km	0.45–2.5 μm	126	5 m	2.3 km
Hyperion [9]	Space demonstration	705 km	0.4–2.5 μm	200	30 m	7.5 km
SEBASS [10]	Military R&D	3 km	2–5 and 8–14 μm	128 and 128	3 m	0.4 km
TRWIS III [11]	Commercial R&D	3 km	0.3–2.5 μm	384	3 m	0.7 km

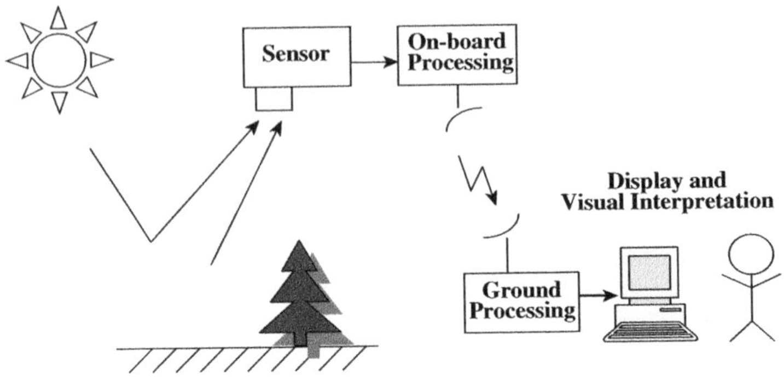

Figure 2.2. Hyperspectral imaging process.

a composite of multiple materials. For example, a field of "grass" contains blades typically of different species blended together with various weeds, plus contributions from the underlying soil that can be made up of various organic compounds and soil types. There is generally no optimal or minimal spatial resolution necessary for the vast majority of remote sensing applications. Therefore, the spatial resolution of most HSI systems cited above (1 to 30 m) represents a compromise between spatial resolution, coverage area, detector performance, data rate, and data volume considerations.

The above discussion and the rest of this chapter are designed to provide the reader with a context for the other chapters in this text which primarily deal with the application of mathematical algorithms for the extraction of information from hyperspectral imagery. To better understand how to select and apply those algorithms, it is useful to appreciate the process by which the data were collected.

Figure 2.2 provides a schematic of the hyperspectral imaging process as a system. The process begins with a source of illumination, which may be the sun or the thermal radiation of the surface itself. This is followed by the effects of the transmission media, or atmosphere, to create the optical radiance field incident upon the sensor's aperture. This field is sampled spatially, spectrally, temporally, and radiometrically to create the digital hypercube collected by the sensor. This cube is typically then processed to remove instrument artifacts and to convert recorded signal levels to calibrated scientific units. Additional processing is then applied before final interpretation by a human analyst. The process is viewed as a system in that the quality and utility of the final product depends on effects that occur at each stage of the process.

In this chapter we provide an introductory overview of hyperspectral imaging systems, citing appropriate references for the interested reader to pursue. We pay particular attention to point out where errors are introduced by the measurement and processing process, so the reader can better appreciate characteristics of real data. We conclude with a discussion of modeling and analysis approaches that can lead to enhanced designs and improved analysis of hyperspectral imaging systems and their data.

2.2. PHYSICS OF SCENE SPECTRAL RADIANCE

In order to understand the characteristics of the data collected by hyperspectral imaging systems, it is important to discuss the physics behind the scene radiance field incident on the imaging system. This involves the sources and paths of optical energy, surface reflectance and emittance characteristics, the effects of the atmosphere and adjacent materials, and the impact of finite spatial resolution and the "mixing" that occurs in every pixel.

2.2.1. Spectral Radiance

We begin by defining the fundamental physical quantity that describes the transfer of optical energy in remote sensing systems. *Spectral radiance* is the transfer of optical energy in a specific direction [12]. In particular, it describes the optical power per unit of solid angle—incident at, through, or exiting a surface of unit projected area—per unit of wavelength. It is generally measured in units of watts/(meter^2-steradian-μm). This quantity conveniently captures the spectral and radiometric characteristics of the incident optical flux at a sensor's aperture and is the usual quantity of calibrated data from a hyperspectral sensor.

2.2.2. Sources and Paths

As mentioned above, the spectral radiance measured by a remote sensing instrument can originate from the sun or from the scene itself. The text by Schott [12] describes eight different paths depending on the geometry of the scene and the sensing process.

While the relative magnitudes of the different paths will depend upon the scene characteristics, a few of the paths typically dominate when viewing a flat surface in the open. Equation (2.1) describes these dominant paths.

$$L_{\text{At-sensor}} = L_{\text{DN}}\rho + L_{\text{PR}} + L_{\text{BS}}(1 - \rho) + L_U \tag{2.1}$$

where $L_{\text{At-sensor}}$ is the radiance incident at sensor aperture, L_{DN} is the total down-welling radiance from the sun and thermally emitted by the atmosphere, L_{PR} is the path reflected radiance from sun and scattered by the atmosphere into the sensor's field of view, L_{BS} is the thermally emitted radiance from surface (assuming it is a blackbody), L_{U} is the total atmospherically emitted upwelling thermal radiance, and ρ is the surface reflectance factor.

Figure 2.3 presents typical spectral radiance values of the four terms shown in Eq. (2.1). For the parameters listed in the figure, one can see that in the visible through mid-wave infrared (0.4 to 4 μm), the ground-reflected down-welling dominates the total at-sensor radiance, but at longer wavelengths the thermally self-emitted radiance dominates. Atmospherically scattered path radiance is significant at wavelengths less than 0.7 μm, but falls off rapidly beyond that. The atmospheric thermally self-emitted radiance is the dominant source in the

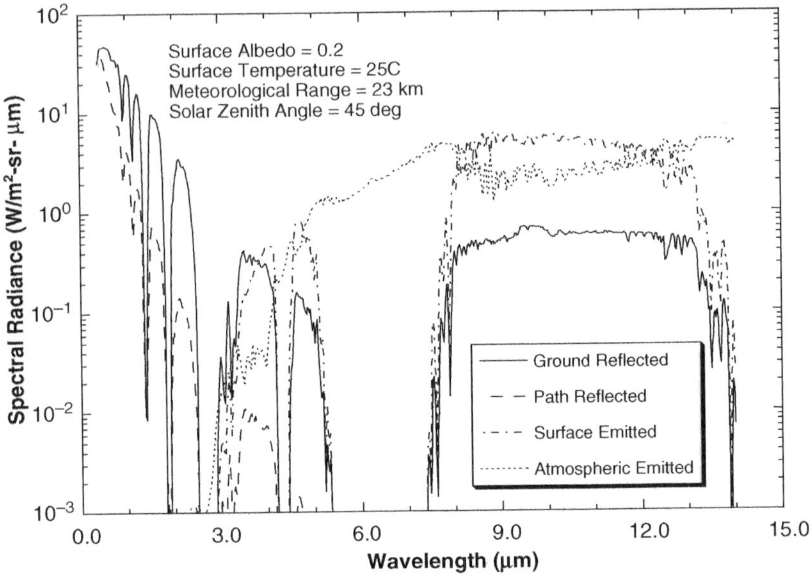

Figure 2.3. Example spectral radiance components [13].

atmospheric absorption region 5.0 to 7.5 μm and can be still significant in the thermal infrared atmospheric window region of 7.5 to 13.5 μm.

The overall envelopes of the radiance components show the general shape of the solar illumination from 0.4 to 4 μm, along with the thermal self-emission of the earth and its atmosphere from 4 to 14 μm. The narrow absorption features of the atmosphere can be seen as deviations from this envelope.

The figure also illustrates which regions of the spectrum are likely to contain information about the surface, which is the topic of this text. The regions where the scattering or emission by the atmosphere dominates are usually avoided by algorithms seeking surface characteristics, but they are often helpful in characterizing the atmosphere as part of an atmospheric compensation process.

2.2.3. Surface Reflectance

A primary inherent parameter of interest that can be retrieved and used in land and water remote sensing is the spectral reflectance of the surface. Information about the differences among materials and their condition (i.e., health of vegetation) is contained in this quantity. The causes of the differences in spectral reflectance can be traced back to the chemical and physiological makeup of the materials, and they are dependent upon the electronic, vibrational, and rotational resonances as well as the microgeometrical structure. For example, vegetation has the characteristic spectral reflectance shape shown in Figure 2.4 due to (a) the absorption of chlorophyll near 0.4 and 0.6 μm and (b) the cell microstructure leading to the high reflectance from 0.7 to 1.0 μm.

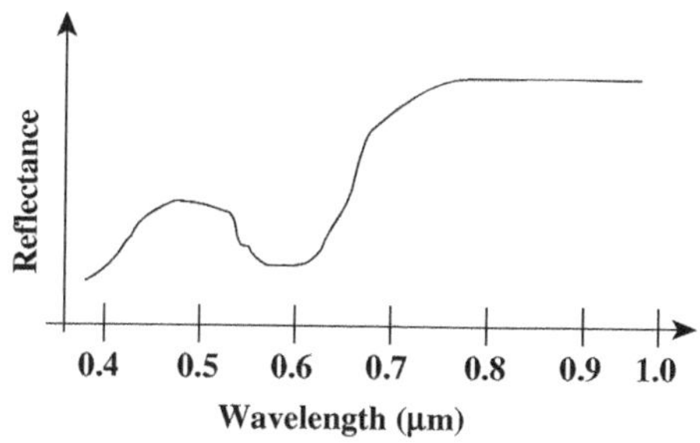

Figure 2.4. Typical spectral reflectance of vegetation.

An important aspect of surface reflectance in remote sensing is the fallacious concept of a spectral "signature" being uniquely associated with a given material. While it is true there are spectral characteristics that are common among materials, it is naïve to assume that a material will have a consistent, unique spectral shape when observed under a variety of conditions. The origins of this concept can be traced back to the laboratory environment where pure materials (solids, liquids, or gases) can be isolated and shown to have characteristic absorption features due to their chemical makeup.

However, in the remote sensing environment there are numerous effects that lead to variations or even masking of these spectral characteristics, to such a great degree that the concept of a spectral signature can be misleading. For example, materials often have a varying reflectance depending on the angles of illumination and view relative to the surface normal. The bi-directional reflectance distribution function (BRDF) [12] is a mathematical description of this variability, but at the typical scales of remote sensing, it is difficult to know the precise orientation of a surface, and thus the dependence on illumination and view angles becomes a source of variability in the material reflectance. Aging and contamination from exposure to the environment can also lead to variation for man-made objects.

Another aspect of variability is related to the specificity of naming the material. As mentioned in the earlier example, the spectral reflectance of a grass field is actually a composite of the reflectance from individual blades of grass and weeds, along with the soil, decaying organic matter, and even water. The proportions of all these materials in a remotely sensed pixel will vary spatially and lead to variability in a grass reflectance measurement. In addition, there is variability even among the "pure" blades of grass just due to species variation.

While these effects lead to variability in material reflectance, fortunately, the variations are often spectrally correlated and modest in magnitude, thus providing the opportunity for further processing to discriminate and associate measurements with particular materials. Indeed the other chapters in this text offer many ways to unravel these complexities, and the examples presented show the power of remote sensing in material mapping. But it is important that the reader appreciate the difficulties and not be blindly led by the spectral signature concept.

2.2.4. Atmospheric Effects

The radiance from a surface as measured by a remote sensing hyperspectral sensor must pass through the atmosphere, which by its nature will modify the spectral shape and magnitude according to the scattering, emission, and absorption of particles and gases present. Often, HSI data are *atmospherically compensated*, which is a processing step that attempts to convert the measured radiance to apparent surface reflectance by "removing" the effects of solar illumination and atmospheric transmittance (and temperature in the thermal infrared). Of course, there are portions of the spectrum where the atmosphere is opaque and no amount of processing will compensate for the lack of optical energy reaching the sensor. These regions can be identified from the plots in Figure 2.3 where the ground reflected and surface emitted curves go to zero.

In addition to the reduction in magnitude and change in shape of the surface-leaving radiance due to the atmosphere, the scattering and emission of the atmosphere add to the signal measured by the sensor, thereby providing an offset to the measured radiance. The path reflected and atmospheric emitted curves in Figure 2.3 give an example of the magnitude and spectral regions where these effects can dominate.

2.2.5. Adjacency Effect

An atmospheric effect of particular note arises due to the scattering from aerosols and molecules in the atmosphere and the spatial variation of surface reflectance across a scene being imaged. In the reflective part of the spectrum, where the sun's energy is the dominant source, the radiance measured in a given pixel (or sensor instantaneous field of view—IFOV) not only includes the radiance reflected from the region of the surface defined by the geometric projection of the sensor's IFOV, but can also include radiance reflected from the ground in adjacent areas and scattered by the atmosphere into the IFOV. This *adjacency effect* can lead to radiance reflected from surface areas hundreds of meters away from the area on the ground being imaged in a given pixel [14]. This effect is most significant in hazy atmospheres and when a dark area is surrounded by bright regions. An example is the presence of the vegetation red-edge and high near-infrared intensity for a pixel imaging a black asphalt road surrounded by a lush green field.

2.2.6. Mixing

Another dominant effect in hyperspectral remote sensing is the mixing of radiance contributions from the variety of materials present in a given pixel. As mentioned earlier, at the range of typical spatial resolutions of hyperspectral imagery, there are nearly always many distinct materials present on the surface within a given pixel. This can arise when the spatial resolution is modest and there are clearly different objects imaged within the pixel, or when the surface is a composite of many different materials mixed together so any practical imaging system will see only the mixture.

It is most common to assume that the radiance reflected by the different materials combine in a linear additive manner, even when that may not be the case. For many situations, this is a reasonable assumption and with appropriate processing can lead to the consistent extraction of the various components (often termed *endmembers*) and their relative *abundances* (subpixel fractions). However, there are cases where the mixing process leads to nonlinear combinations of the radiance and where the linear assumption fails to hold. This can occur where there is significant three-dimensional structure within a given pixel and where the optical energy makes multiple bounces between objects before exiting in the direction of the sensor. Without detailed knowledge of that three-dimensional structure, it can be very difficult to "un-mix" the contributions of the various materials.

2.3. SENSOR TECHNOLOGY

The central part of hyperspectral imaging systems is the sensor itself. This includes the hardware and processes that transform the optical radiance field incident at the imaging system aperture to the array of digital numbers that form the hypercube.

2.3.1. Optical Imaging

The key process of the imaging aspect of hyperspectral sensing is the focusing of light from a small region on the earth's surface into a given detector element forming a *pixel* in the resulting image. There are two main elements in determining the spatial resolution of an image. One is the size of the optical entrance aperture relative to the wavelength of light observed, and the other is the size of the detector element relative to the optical prescription of the imaging system. The detailed description of optical imaging systems is beyond the scope of this text, but the following provides a top-level view of these determining factors. The interested reader is referred to texts on the subject for more details [12, 15].

In most airborne or satellite remote sensing systems, the dominant system parameter is the size of the entrance aperture defined by the primary lens or mirror. The size of this element most often determines the volume and weight of the entire sensor package, which, in turn, drives the requirements for the size of the aircraft or satellite bus and ultimately the cost of manufacture and operation. Since it is usually desirable to have the highest spatial resolution possible within the volume and weight constraints dictated by the platform, remote sensing optical systems are typically designed to be diffraction limited in spatial resolution. That is, the size of the optical aperture defines the spatial resolution, and all other optical elements are designed from that starting point.

The achievable spatial resolution of a diffraction limited sensor system can be related to the size of the entrance aperture through Eq. (2.2), known as the *Rayleigh* criterion. The closest distance, d, between two point sources on the ground that can be distinguished is determined by the size of the aperture, D, the wavelength of light, λ, and the height of the sensor, H (relative to the ground), through Eq. (2.2).

$$d = 1.22\frac{\lambda}{D}H \tag{2.2}$$

For example, a sensor with an aperture of $D = 1$ m, operating at a wavelength of $\lambda = 1\,\mu\text{m}$, from an altitude of $H = 250$ km, can resolve point sources no closer than $d = 0.3$ m. The reader can explore for himself how large an aperture would be necessary for a spy satellite to read license plates!

2.3.2. Spatial Scanning

While the previous section discussed the topic of spatial resolution and focusing light into a pixel, an image is created by arranging these pixels into a two-dimensional array. Modern airborne and satellite imaging systems do this with a variety of techniques. There are four basic ways these sensors form images, each placing different requirements on focal plane complexity and platform stability. The following discussion assumes a single spectral band system; measurement of the spectrum for each pixel is addressed in the next section.

Line Scanners. This was the form of most early sensors since it can be used with individual large detectors and does not require an array. The across-track pixels, or lines of an image, are formed by sweeping the projected view of the detector across the ground with a rotating mirror. The down-track pixels, or columns, are formed by the forward movement of the sensor platform. The simplicity of the focal plane for these types of systems is traded-off with platform stability since changes in the platform's orientation in relation to the earth will lead to nonuniform spacing of the pixels. For example, it is quite common for aircraft to be buffeted by winds and to roll slightly relative to the flightline. With a mirror rotating at a constant velocity, the pixels then correspond to nonuniform sampling of the ground.

Whiskbroom Scanners. These types of systems operate in a manner very similar to that of the line scanners except that there are multiple detectors in the focal plane. The detectors are commonly oriented in a line along the platform direction. Thus, as the mirror sweeps across the field of view, the detectors collect several lines (rows) of the image simultaneously. The advantage to this mode is that the detectors can dwell longer for each sample and increase the signal-to-noise ratio. The disadvantage is the additional cost of the detectors and the additional processing necessary to compensate for variations in detector responsivity. Whiskbroom sensors also suffer from the same geometric distortion problems from platform roll, pitch, and yaw, as do the line scanners.

Pushbroom Scanners. These types of systems generally do not require the use of a moving mirror because the focal plane consists of a linear array of detectors that image a line across the entire field of view. Thus, one row of an image is collected all at once, while the platform motion provides the down-track sampling or columns of the image. An advantage here is the significantly longer dwell time for the detectors and the resultant dramatic improvement possible in signal-to-noise ratio. Another advantage is that the across-track sampling is done by the fixed spacing of the detectors in the array, thereby eliminating the nonuniform sampling possible with scanning systems. The disadvantages include the cost of the arrays, the limited field of view since the optics must image the full swath at once, and the added processing and calibration to compensate for detector nonuniformity.

Framing Cameras. This last type of system does not require any mirror or platform motion to create an image, because it relies on the use of two-dimensional detector arrays in the focal plane to image the entire swath and the down-track columns all at once. The advantage here is the enormous gain in available dwell time, as well as the ability to image from a stationary platform such as an aerostat or geosynchronous satellite. Also, since an entire image is collected at once, platform instability is much less an issue and the images have high geometric fidelity. The disadvantages include the potentially high cost of the array and the processing associated with nonuniformity correction as with the other detector arrays.

An important aspect of the spatial collection of remotely sensed imagery is the geometric fidelity and knowledge. An entire filed of study, *photogrammetry*, is concerned with this topic and the reader is referred to other texts for details [16]. While originally of most concern in land surveying and military targeting applications, the precise knowledge of the geographic location of image pixels is growing in importance, particularly because remotely sensed data of different modalities and from different platforms are being fused or integrated together as is possible in geographic information systems. Fortunately, technologies such as inertial navigation systems and Global Positioning System (GPS) units are providing improved accuracies and making this task easier.

2.3.3. Spectral Selection Techniques

The key characteristic that distinguishes hyperspectral imagery from other types of imagery is the collection of a contiguous high-resolution spectrum for each pixel in the image. There are several techniques used in existing systems for accomplishing the spectral measurement. Figure 2.5 provides pictorial descriptions of five different techniques. They can be grouped into two categories based on the type of focal plane and spatial scanning technique with which they can be used. The prism and grating methods spread the spectrum out spatially and require a linear (or two-dimensional) array and can be used with line scanner, whiskbroom, or push-broom scanning systems. The filter wheel, interferometer, or tunable filter systems collect the spectrum over time, and thus they can be used with either (a) single detector elements in an across-track scanning mode or (b) two-dimensional arrays in a framing camera mode. The following text provides additional details on each technique.

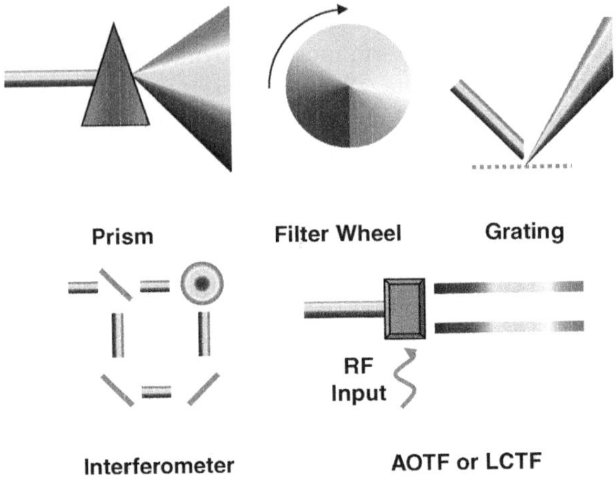

Figure 2.5. Spectral selection techniques [17].

Prism. As indicated in Figure 2.5, the incident light is diffracted in different directions, depending upon the wavelength. With a linear detector array placed at the appropriate distance from the prism, the various wavelengths of light are sampled by the detectors, collecting a contiguous spectrum. In hyperspectral systems, it is quite common to use this type of dispersive system with a two-dimensional detector array with one dimension sampling the spectrum and the other dimension sampling the scene in a pushbroom scanning mode.

Grating. These types of systems function in a way very similar to that of prisms by dispersing the light spatially, except that the light is reflected rather than refracted through the spectral selection element. While gratings can be preferred in space environments since they can be made from radiation-hard materials, they suffer from less efficiency, polarization effects, and the need to use order-sorting filters when covering a broad spectral range.

Filter Wheel. This approach is more typically used in multispectral systems where interference filters of arbitrary passbands are arranged around the edge of a wheel which then rotates in front of a broadband detector and collects data at the various wavelengths. Recently, circular variable filters (CVF) and linear variable filters (LVF) have emerged to enable the high-resolution, contiguous nature of hyperspectral sensors with this type of spectral selection technique.

Interferometer. These types of systems collect an interferogram which then must be digitally processed to obtain the optical spectrum. The most common interferometer is a *Michelson Fourier transform spectrometer* (FTS) that is configured to collect the Fourier transform of the spectrum by moving a mirror to change the optical path difference between two mirrors and causing constructive and destructive interference for the incident light [18]. This interferogram, measured

TABLE 2.2. Example Detector Materials

Material	Spectral Range	Quantum Efficiency	Operating Temperature
Si	0.3–1.0 μm	70–95%	300 K
InGaAs	0.8–1.7 μm	70–95%	220 K
InSb	0.3–5.5 μm	70–95%	77 K
HgCdTe	0.7–15 μm	50–80%	77 K
Si:As	2.5–25 μm	20–60%	10 K

over time, is then processed through an inverse Fourier transform to obtain the desired spectrum. Another type of interferometer is known as the *Sagnac*, in which the interferogram is spatially distributed across an array of detectors as opposed to the temporal sampling of the FTS.

Acousto-Optical Tunable Filter (AOTF) or Liquid Crystal Tunable Filter (LCTF). These devices work on the principal of selective transmission of light through a material in which acoustic waves are passed (AOTF) or a varying voltage is applied (LCTF). These types of systems have the advantage of being able to rapidly select the wavelength of light sensed and thus enable selective spectral sensing, but generally suffer from low throughput and limited angular fields of view.

2.3.4. Detectors

The detection and conversion of optical energy into an electrical signal is at the heart of hyperspectral imaging systems and is done with a variety of materials depending on the wavelength band [19]. Table 2.2 provides a list of commonly used detector materials and their characteristics.

Of note in this table is the trend that for longer wavelength sensing, the materials are less efficient overall and require cooling to be functional. The added weight and complexity due to cooling requirements leads to higher cost for systems operating at these longer wavelengths. Also, while the large commercial market for visible cameras has lowered the cost for silicon detector arrays, no such volume market has emerged for arrays operating at longer wavelengths, resulting in higher detector costs as well. Thus, the commercial airborne HSI systems that are becoming available are usually limited to the silicon spectral range, and the longer-wavelength, thermal infrared systems are limited to research or major governmental programs.

2.3.5. Electronics, Calibration, and Processing

After the optical energy is converted to an electrical signal by the detector, the first step is usually an analog-to-digital conversion. Once the signal is in digital counts, it can be moved around and operated on without additional noise corrupting the measurement.

The digital data are then either (a) further processed on-board the platform for real-time analyses or (b) saved to storage media for processing on the ground. The

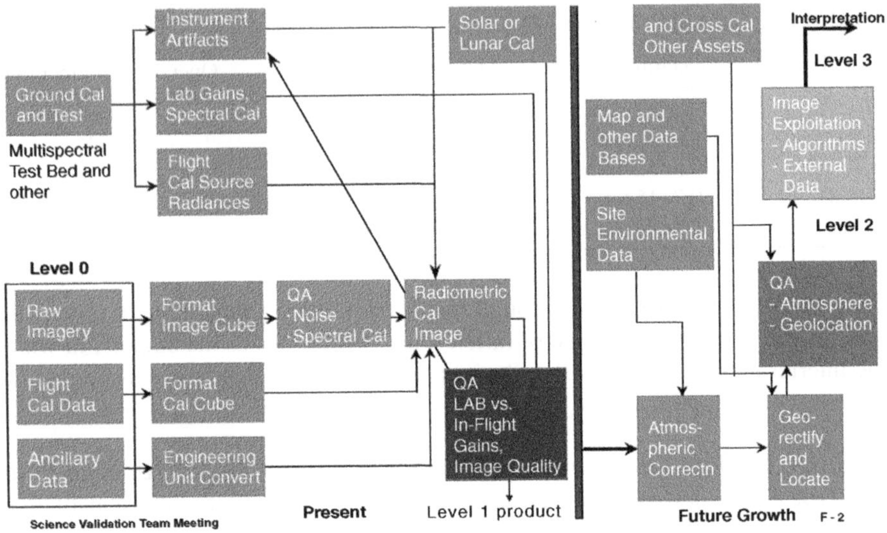

Figure 2.6. Processing flow for the EO-1 Hyperion instrument [20].

details for on-board processing are system specific, but the steps described below are generally completed for all non-real-time systems.

Figure 2.6 presents an example of the ground processing for the spaceborne EO-1 Hyperion sensor. The diagram shows the flow from the raw data to radiometrically and spectrally calibrated image cubes.

The figure uses a common nomenclature for the processing and analysis of hyperspectral image data. The following describes the characteristics of the data at each level of processing.

Level 0. This refers to the raw imagery and ancillary data that are produced by a sensor before any processing has been applied. Often, the image data are not even arranged in a row/column format, but retain the sequence as read directly from the readout electronics and A/D converters.

Level 1. This level of processing refers to the steps where the data are reformatted into an image, instrument artifacts are mitigated, the bands are spectrally calibrated, and the digital counts are calibrated to physical radiometric units. Usually, ancillary data files are associated at this level to provide the user with information about the quality of the data such as bad pixel locations, sensor noise levels, and estimates of the spectral and radiometric calibration accuracy.

Level 2. Generally, Level 2 refers to imagery that has been geometrically corrected and located on the earth. (Note that for some sensors this step is referred to as Level 1G, with the radiometric calibration described above as Level 1R.) Also, some programs perform atmospheric compensation at this stage and provide surface reflectance, emittance, and temperature as Level 2 products.

Level 3. Level 3 generally refers to hyperspectral image derived products such as classification, detection, or material identification maps.

As mentioned above, the exact definitions of the levels depend on the specific sensor program, but all systems perform these steps one way or another.

2.4. SENSOR PERFORMANCE METRICS

The utility of hyperspectral imagery depends very much on the quality of the data. This quality can be characterized by a number of performance metrics that address the various aspects of the data. This section describes commonly accepted metrics that characterize sensor performance. Further details can be found in other texts [21].

In each section below, we also briefly address the impact on exploitation and analysis of HSI data from the various sensor characteristics described.

2.4.1. Image Resolution

A primary measure of performance of any imaging system is the spatial resolution achieved. There are a number of ways that resolution can be characterized, and they are often used interchangeably. Earlier in this chapter we defined the *Rayleigh* criterion for determining resolution based on the ability to distinguish two point sources as determined by the optical aperture, range to the surface, and the wavelength being sensed. However, the achieved resolution will depend on the complete system including ground processing.

One metric that is commonly quoted as resolution is the *ground sample distance (GSD)*. GSD refers to the distance on the ground between the centers of the pixels in the image. While this can often serve as a reasonable surrogate for resolution, there are systems that may be spatially oversampled providing closely spaced, but blurry, pixels, leading to this metric being an optimistic value for the resolution.

A more accurate measure of the spatial resolution is often termed *ground resolved distance (GRD)*, which is the geometric projection on the ground of the sensor's *instantaneous field of view (IFOV)* for a given pixel. The IFOV can be defined as the angular interval between specified cutoff levels of the optical *point spread function (PSF)*. Equivalently, the resolution can be derived from the *modulation transfer function (MTF)*, which is the Fourier transform of the PSF.

Another fact to consider is that imaging systems may have different resolution along track versus cross track. In this case, the data are often resampled to provide equal sample spacing in each direction. Thus when printed or displayed on a computer monitor, the image will have the correct aspect ratio.

Impact on Exploitation. Clearly, spatial resolution has a large effect on the detection and identification of man-made objects in particular, because they often appear as subpixel, or unresolved, in typical systems. While HSI data have been demonstrated to allow subpixel object detection [22], it is also well known that such detection is easier when the object occupies a greater fraction of a pixel [23]. However, smaller pixels have been seen to lead to lower accuracy in land cover classification applications [24]. This results from the averaging of within-class variability that is more significant with larger pixels and reduces class variability, leading to greater separation among classes.

2.4.2. Spectral Metrics

Spectral performance can be measured in several ways. First, there is *spectral resolution*, which, like spatial resolution, can be described by both sample spacing and actual resolution. For hyperspectral imagers, it is quite common that the width of the spectral bandpass exceeds the sample spacing and the spectral channels overlap.

Next, there is *spectral calibration accuracy*, which is the accuracy of knowing the central wavelength (as well as the bandwidth) of each spectral channel. For some systems, particularly airborne ones, the true central wavelength of the channels can vary rapidly, even from line to line. This effect is known as *spectral jitter* and can lead to requirements that certain processing algorithms be applied separately to each line of the image.

Another effect is *spectral misregistration*, which is the result when the same pixel index in a hypercube represents radiance from slightly different areas on the ground from spectral channel to channel.

A spectral metric that is particular to dispersive spectral imaging systems using a two-dimensional focal plane is *spectral smile*. This refers to an artifact in which the central wavelength of a given spectral channel varies across the spatial dimension of the detector array. The name arises from a plot of constant center wavelength which can trace out an arc, or a smile, across the array. This can necessitate algorithms be applied differently across the columns of an image.

Impact on Exploitation. With the large number of spectral channels in hyperspectral data, exploitation algorithms can be relatively tolerant of small errors in the knowledge and registration of the spectral data. However, these effects can be critical for physics-based algorithms that rely on precise knowledge for their application. Physics-based algorithms for atmospheric compensation, such as FLAASH [25], can be very sensitive to accurate spectral knowledge since they rely upon matching the spectral measurements to tabulated spectra and models. Land cover classification has also been shown to be sensitive to spectral misregistration, with noticeable effects on accuracy at misregistrations greater than one-tenth of a pixel [26]. Target detection can be sensitive to spectral errors, most often in cases of small subpixel fractions or low contrast targets.

2.4.3. Radiometric

As was mentioned earlier, except for real-time systems, it is quite common for hyperspectral data to be calibrated to physical quantities, usually spectral radiance (W/m^2-sr-μm). The accuracy to which this can be accomplished is referred to as the *absolute radiometric accuracy*. This is usually measured as a long-term average difference between the reported and the true radiance for a measurement of a calibration source and can generally be thought of as a deterministic, or constant, error source.

All sensors suffer from random errors as well, of which the predominant one is *sensor noise*. This noise comes both from mechanisms in the detector itself, termed *photon noise*, and from the readout and electronics circuitry. The level of photon noise is proportional to the square root of total radiance incident on the detector, but the other noise sources are generally fixed in magnitude and thus can be lumped into a term known as *fixed noise*. It is convention to characterize the level of total noise by the standard deviation of the output signal for a constant radiance input. The various noise sources are usually statistically uncorrelated and add in quadrature. The primary metric for these random error sources is the signal-to-noise ratio (SNR) defined on a spectral channel by channel basis as

$$\mathrm{SNR}_\lambda = \frac{\bar{S}_\lambda}{\sigma_\lambda} \tag{2.3}$$

where $\bar{S}_\lambda$ is the measured mean signal, and σ_λ is the measured standard deviation for that signal level, each at spectral wavelength λ. Note that SNR is generally a function of the input signal level (and other sensor parameters) so that it is only meaningful when quoted for a given input signal.

Figure 2.7 presents an example input radiance and an estimated SNR for the NASA EO-1 Hyperion instrument. As can be seen, the effects of low atmospheric transmittance leading to regions across the spectrum with low input signal result in those regions having a low SNR.

Another random error source that can occur is residual error from focal plane nonuniformity corrections, or sometimes referred to as *pattern noise*. All detector arrays suffer from a variation in responsivity from detector to detector. Usually a correction map is determined from measurements in the laboratory to compensate for these variations and used to perform a *flat-fielding* of the array. However, these corrections often don't entirely eliminate these responsivity variations. This can happen because the responsivity of the detector elements may change from the laboratory to the field due to temperature sensitivity or bias voltage variations.

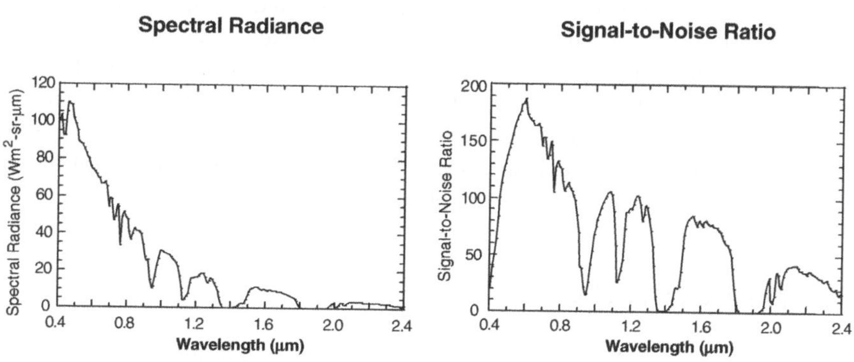

Figure 2.7. Example radiance input and resulting SNR for Hyperion.

Since one does not know *a priori* where on the focal plane a given object or location on the earth will be imaged, or what the exact magnitude of the error will be, this error source can be considered to be random in nature.

There are also noise effects that may be correlated across the spectral channels. For example, there may be cross-talk across the detectors which would lead to the photon noise being correlated across spectral channels. These effects are very instrument specific, but the user should be aware that they do exist.

Impact on Exploitation. In many hyperspectral systems, the actual sensor noise (photon + fixed) is quite low and is generally not a limiting factor on exploitation. Often, it is other sources of error that dominate performance including residual pattern noise, or even radiometric artifacts introduced by nonideal spatial or spectral responses. For systems where noise is significant, preprocessing algorithms may be applied to reduce the impact of noise on the final image product [27] Absolute calibration error (both spectral and radiometric) also can have significant effects on algorithms that use physics-based modeling as part of the process [28].

2.4.4. Geometric

Geometric accuracy refers to both (a) the fidelity of reproducing the spatial relationships on the ground accurately in the image and (b) the knowledge of the corresponding geographic location (e.g., latitude/longitude) of each pixel in the image. The spatial fidelity is often addressed during the Level 2 processing of the imagery with resampling to a uniform grid. The geographic accuracy also has two parts: the *absolute geopositioning accuracy* and the *random uncertainty* in the geopositioning accuracy in the reported locations.

Impact on Exploitation. As discussed earlier, accurate knowledge as to the location of the pixels within an image can be critical when attempting to process the data in conjunction with other data sources. In the military context, clearly this knowledge is very important when using the data for targeting purposes! More commonly, the process of resampling the data to a uniform grid, or ground projection, can in some cases modify the spectral integrity and introduce artifacts in the data. This can arise when the spatial sampling frequency determined by the pixel spacing is too low compared to the scene spatial frequencies admitted by the optical system. Since many HSI processing algorithms attempt to extract surface features and objects from the spectral data alone, these artifacts can mask their presence. Thus, it is often recommended that spectral processing be performed on the data before any geometric resampling is applied.

2.5. EXAMPLE SYSTEMS

The following brief descriptions of several hyperspectral imaging systems are provided for the reader as reference and as examples of typical systems. These

TABLE 2.3. HYDICE System Parameters [7]

Parameter	Description
Initial year of operation	1994
Size	30 cm × 30 cm × 30 cm
Weight	200 kg
Power	500 W
Aperture	2.54 cm
Spatial scanning technique	Pushbroom
Spectral selection technique	Grating
Focal plane technology	InSb two-dimensional array
Field of view	0.15 rad
Instantaneous field of view	0.5 mrad
Number of spectral channels	210
Spectral range	400–2500 nm
Spectral channel bandwidth	3–15 nm

particular systems were selected since their data have been widely distributed and have served as sources for many algorithm and analysis studies.

2.5.1. HYDICE

The Hyperspectral Digital Imagery Collection Experiment (HYDICE) instrument was built by Hughes Danbury Optical Systems (now part of Goodrich Corp.) under contract from the Naval Research Laboratory. The instrument was developed under dual-use funding and made available to researchers from the civilian as well as military R&D communities. Table 2.3 provides a summary of its system parameters.

HYDICE was one of the first airborne hyperspectral instruments to be operated from a relatively low altitude thereby achieving very high spatial resolution (1 m from 2-km altitude). The sensor was thus able to resolve man-made objects such as buildings, roads, and vehicles and has provided excellent data for algorithm development in urban, rural, coastal, and agricultural regions.

Of particular note is a series of data collections performed with HYDICE in the 1990s termed the *Radiance* experiments. In these experiments a number of man-made objects were deployed on the ground in fixed configurations with their locations accurately determined. Together with field measured reflectance spectra of these objects and their adjacent backgrounds, these experiments provided some of the best ground-truthed hyperspectral data sets around and have been used extensively by researchers developing and testing hyperspectral algorithms.

2.5.2. AVIRIS

The airborne visible/infrared imaging spectrometer (AVIRIS) was designed and constructed by NASA's Jet Propulsion Laboratory and has been operated as a

TABLE 2.4. AVIRIS System Parameters [4]

Parameter	Description
Initial year of operation	1987
Size	165 cm × 90 cm × 130 cm
Weight	340 kg
Power	800 W
Aperture	200-mm diameter
Spatial scanning technique	Line scanner
Spectral selection technique	Grating
Focal plane technology	Si, InGaAs, and InSb linear arrays
Field of view	0.6 rad
Instantaneous field of view	1 mrad
Number of spectral channels	224
Spectral range	360–2510 nm
Spectral channel bandwidth	10 nm

facility instrument for NASA and associated earth science researchers. Built as a follow-on to one of the first hyperspectral imagers, AIS [1], it has been a workhorse in collecting data for the scientific community. Over the years, NASA/JPL has continued to improve the instrument such that the data quality is extremely high and is regarded as one of the lowest noise hyperspectral imagers around. Table 2.4 provides a summary of its system parameters.

Deployed on NASA's high-altitude ER-2 or WB-57 flying up to 20 km above ground, AVIRIS achieves 20-m ground resolution over an approximate 11-km swath. With this broad coverage, AVIRIS has proven to be very useful for natural resource, land cover, and mineral mapping applications. It has also been found to be useful for the study of small-scale atmospheric phenomena such as the details of thin cirrus clouds and the mapping of water vapor horizontal structure.

In the last few years, AVIRIS has also been deployed on lower-altitude platforms and collected data with approximately 4-m resolution. These data have been useful for studying urban regions and natural features occurring at the smaller spatial scales.

2.5.3. Hyperion

Hyperion was built by the TRW Space and Electronics Group (now part of Northrup Grumman Space Technology), under contract from NASA Goddard Space Flight Center. It was developed under NASA's New Millenium Program, which was created to demonstrate new sensor and spacecraft technologies with improved performance and lower cost. Hyperion was constructed in under a year using parts left over from NASA's Lewis instrument (which suffered a launch failure in 1993). As such, the instrument was built with a "best performance" goal, as opposed to meeting stringent specifications. Despite a moderately high noise level,

TABLE 2.5. Hyperion System Parameters [9]

Parameter	Description
Initial year of operation	2000
Size	39 cm × 75 cm × 66 cm
Weight	49 kg
Power	126 W
Aperture	12 cm
Spatial scanning technique	Pushbroom
Spectral selection technique	Grating
Focal plane technology	Si and HgCdTe
Field of view	0.01 rad
Instantaneous field of view	40 μrad
Number of spectral channels	220
Spectral range	400–2500 nm
Spectral channel bandwidth	10 nm

the sensor has provided extremely useful data and has led to demonstrations of many spaceborne hyperspectral applications. Table 2.5 provides a summary of its system parameters.

Hyperion and the multispectral Advanced Land Imager (ALI) are onboard the EO-1 satellite, which orbits directly behind Landsat 7 acquiring nearly simultaneous images. This has allowed researchers to explore the added capability of the hyperspectral data relative to the ALI and the Landsat 7 Enhanced Thematic Mapper (ETM).

2.6. MODELING HYPERSPECTRAL IMAGING SYSTEMS

The modeling of hyperspectral imaging systems plays a number of roles in the development and application of the technology. One primary role is that by constructing and validating models, we demonstrate our understanding of the phenomenology and processes of hyperspectral imaging. Another major role is to create accurate simulations of hyperspectral images which can be used as test imagery for algorithm development and evaluation with known image truth. A third role is to optimize the design and operation of the imaging systems by allowing trade-off studies to characterize the impact of system parameter choices. The following sections describe a few of the existing models used for hyperspectral imaging systems.

2.6.1. First Principles Image Simulation—DIRSIG

The Digital Imaging and Remote Sensing Image Generation (DIRSIG) model has been developed at the Rochester Institute of Technology to produce broad-band, multispectral, hyperspectral, and lidar imagery through the integration of a suite of first-principles-based radiation propagation submodels [29]. The submodels are responsible for tasks ranging from the bi-directional reflectance distribution function (BRDF) predictions for a surface to the dynamic scanning geometry of a line scanning imaging instrument. In addition to these DIRS-developed submodels, the code uses several modeling tools used by the multi- and hyperspectral community including MODTRAN [30] and FASCODE [31]. All modeled components are combined using a spectral representation, and spectral radiance images can be produced for an arbitrary number of user-defined bandpasses spanning the visible though longwave infrared (0.4 to 20 μm).

The model uses 3D scene geometry, material, and thermodynamic properties with a ray-tracing approach that allows a virtual camera to be placed anywhere within the scene. The model tracks photons directly transmitted and scattered by the atmosphere from the sun, as well as those emitted by surfaces and the atmosphere. To accurately model land and material surfaces, techniques have been incorporated that introduce spatially and spectrally correlated reflectance variations producing typical texture variations observed in remotely sensed scenes. The model also can handle transmissive materials allowing the model to predict the solar load on objects beneath scene elements including vegetation. This allows the tool to model the absorption by transmissive volumes including clouds and man-made gas plumes. Geometric sensor modeling is another capability that allows the model to produce imagery that contains the geometric distortions that would be produced by scanning imaging systems such as line and pushbroom scanners. The optical modulation transfer function of the sensor is modeled in postprocessing of the sensor reaching radiance field.

Many scenes of natural and urban areas have been simulated with DIRSIG. Figure 2.8 presents one example. This project, dubbed MegaScene, involved the simulation of an area northeast of Rochester, New York, bordering on Lake Ontario. The scene was constructed in five tiles, each covering an area approximately 1.6 km^2. As indicated in the figure, there are over 25,000 discrete objects (houses, buildings, trees, etc.) in the scene with over 5.5×10^9 facets.

2.6.2. Analytical System Modeling—FASSP

An alternative to the physics-based image simulation and modeling technique is an analytical approach. The Forecasting and Analysis of Spectroradiometric System Performance (FASSP) model, developed at MIT Lincoln Laboratory, is based on such an approach [23]. FASSP is an end-to-end spectral imaging system model that includes significant effects of the complete remote sensing process, including those from the information extraction algorithms. This modeling approach uses statistical representations of various land cover classes and objects, and it analytically propagates them through the remote sensing system.

FASSP does not produce an image, but rather represents the characteristics of the scene classes by statistical models and computes expected performance through the use of analytical equations. This approach offers the advantages of reduced manual scene definition effort and computational time required, allowing trade-off and

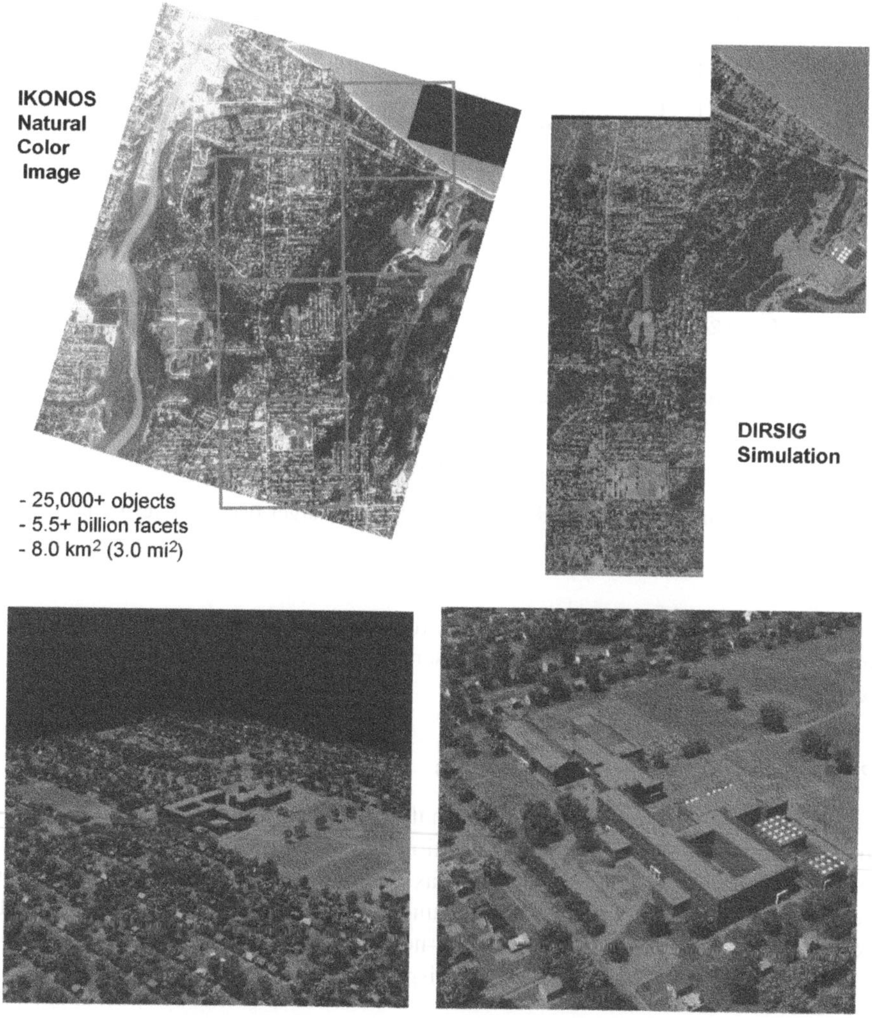

Figure 2.8. Example imagery simulated with DIRSIG.

sensitivity analyses to be conducted quickly, but with the disadvantages of not being tractable for certain situations involving nonlinear effects nor producing an actual image.

Figure 2.9 provides an example of the output from FASSP. Here, the receiver operating characteristic (ROC) curve showing the trade-off in detection versus false alarm rate is shown for a particular target, background, sensor, and imaging scenario and for two cases of target subpixel fraction.

The FASSP model has been found to be particularly useful for parameter sensitivity studies due to its quick execution time. One such study used the model to explore the impact of sensor noise characteristics on subpixel object detection

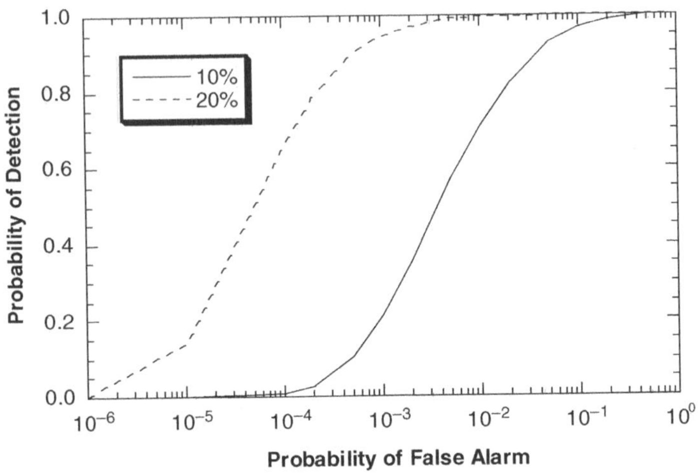

Figure 2.9. Example FASSP modeling result showing target detection versus false alarm rate for two target subpixel fill fractions.

applications [32]. Recent extensions to the longwave infrared have enabled the model to study parameter impacts for sensors operating at those wavelengths as well [13].

2.6.3. Other Modeling Approaches

Other modeling efforts have also contributed to the community modeling goals of image simulation and design optimization. In particular, a modeling effort led by Photon Research Associates [33] has resulted in more accurate simulations for specified scenes, while an effort undertaken at the former Environmental Research Institute of Michigan (ERIM—now part of General Dynamics) contributed to the design specification of a multispectral imaging system [34].

2.7. SUMMARY

This chapter has provided an overview of hyperspectral imaging systems, including their components and important features relevant to the analysis and interpretation of their data. Metrics for characterizing the performance of the systems were presented as well as approaches for their modeling.

During the discussion on sensor performance metrics, we briefly addressed the topic of how sensor characteristics affect the data exploitation process. This topic is one aspect of a broader spectral quality and utility question. That is, can we capture the essential components of the hyperspectral imaging system (including the scene being imaged and the exploitation process) in a way that would allow us to quantitatively measure or predict the quality and utility of data collected in a specific scenario? If such a metric were available, it would be very useful in sensor design and operation situations as well as for rapidly indexing and searching spectral image databases.

There have been some initial efforts at addressing this spectral quality metric question in the context of target detection [35], but the general problem remains an area for further research. However, it is clear that the modeling of spectral imaging systems will play a significant role in gaining the understanding and in the predictive component of such a spectral quality metric. It is also clear that only after hyperspectral imagery becomes widely available from operational platforms will we be in a position to fully develop and demonstrate the efficacy of a scientific community accepted quantitative metric for spectral quality.

REFERENCES

1. A. Goetz, G. Vane, J. Solomon, and B. Rock, Imaging spectrometry for earth remote sensing, *Science*, vol. 228, no. 4704, pp. 1147–1153, 1985.
2. P. Lucey, T. Williams, and M. Winter, Recent results from AHI, an LWIR hyperspectral imager, *Proceedings of Imaging Spectrometry IX*, SPIE, Vol. 5159, pp. 361–369, 2003.
3. B. Stevenson, R. O'Connor, W. Kendall, A. Stocker. W. Schaff, R. Holasek. D. Even, D. Alexa, J. Salvador, M. Eismann, R. Mack, P. Kee, S. Harris. B. Karch, and J. Kershenstein, The Civil Air Patrol ARCHER Hyperspectral Sensor System, *Proceedings of Airborne Intelligence, Surveillance, reconnaissance (ISR) Systems and Applications II*, SPIE Vol. 5787, pp. 17–28, 2005.
4. R. Green, M. Eastwood, C. Sarture, T. Chrien, M. Aronsson, B. Chippendale, J. Faust, B. Pavri, C. Chouit, M. Solis, M. Olah, and O. Williams, Imaging spectroscopy and the airborne visible/infrared imaging spectrometer (AVIRIS), *Remote Sensing of Environment*, vol. 65, pp. 227–248, 1998.
5. Itres website: www.itres.com, 2005.
6. C. Simi, E. Winter, M. Williams, and D. Driscoll, Compact Airborne Spectral Sensor (COMPASS), *Proceedings of Algorithms for Multispectral, Hyperspectral, and Ultraspectral Imagery VII*, SPIE, Vol. 4381, pp. 129–136, 2001.
7. L. Rickard, R. Basedow, E. Zalewski, P. Silverglate, and M. Landers, HYDICE: An Airborne System for Hyperspectral Imaging, *Proceedings of Imaging Spectrometry of the Terrestrial Environment*, SPIE, Vol. 1937, pp. 173–179, 1993.
8. T. Cocks, R. Jenssen, A. Stewart, I. Wilson, and T. Shields, The HyMap airborne hyperspectral sensor: The system, calibration, and performance, *Proceedings of First EARSEL Workshop on Imaging Spectroscopy,* Zurich, October 1998.
9. J. Pearlman, C. Segal, L. Liao, S. Carman, M. Folkman, B. Browne, L. Ong, and S. Ungar, Development and Operations of the EO-1 Hyperion Imaging Spectrometer, *Proceedings of Earth Observing Systems V*, SPIE, Vol. 4135, pp. 243–253, 2000.
10. J. Hackwell, D. Warren, R. Bongiovi, S. Hansel, T. Hayhurst, D. Mabry, M. Sivjee, and J. Skinner, LWIR/MWIR imaging hyperspectral sensor for airborne and ground-based remote sensing, *Proceedings of Imaging Spectrometry II*, SPIE, Vol. 2819, pp.102–107, 1996.

11. R. DeLong, T. Romesser, J. Marmo, and M. Folkman, Airborne and satellite imaging spectrometer development at TRW, *Proceedings of Imaging Spectrometry*, SPIE, Vol. 2480, pp. 287–294, 1995.
12. J. Schott, *Remote Sensing: The Image Chain Approach*, 2nd edition, Oxford University Press, New York, 2006.
13. J. Kerekes, and J. E. Baum, Full spectrum spectral imaging system analytical model, *IEEE Transactions on Geoscience and Remote Sensing*, vol. 43, no. 3, pp. 571–580, 2005.
14. Y. Kaufman, Atmospheric effect on spatial resolution of surface imagery, *Applied Optics*, vol. 23, no. 19, pp. 3400–3408, 1984.
15. C. Wyatt, *Radiometric System Design*, Macmillan, New York, 1987.
16. C. McGlone, (Ed.), *Manual of Photogrammetry*, American Society of Photogrammetry and Remote Sensing, Bethesda, MD, 2004.
17. T. Opar, MIT Lincoln Laboratory, personal communication, 2005.
18. R. Beer, *Remote Sensing by Fourier Transform Spectrometry*, John Wiley & Sons, Hoboken, NJ, 1992.
19. P. Norton, Detector focal plane array technology, in *Encyclopedia of Optical Engineering*, edited by R. Driggers, pp. 320–348, Marcel Dekker, New York, 2003.
20. EO-1 Science Team Presentation, 2000.
21. G. Holst, *Electro-Optical Imaging System Performance*, SPIE Press, Bellingham, Washington, 2005.
22. S. Subramanian and N. Gat, Subpixel object detection using hyperspectral imaging for search and rescue operations, *Proceedings of Automatic Target Recognition VIII*, SPIE, Vol. 3371, pp. 216–225, 1998.
23. J. Kerekes, and J. E. Baum, Spectral imaging system analytical model for subpixel object detection, *IEEE Transactions on Geoscience and Remote Sensing*, vol. 40, no. 5, pp. 1088–1101, 2002.
24. J. Kerekes, and S. M. Hsu, Spectral quality metrics for terrain classification, *Proceedings of Imaging Spectrometry X*, SPIE, Vol. 5546, pp. 382–389, 2004.
25. G. Anderson, G. Felde, M. Hoke, A. Ratkowski, T. Cooley, J. Chetwynd, J. Gardner, S. Adler-Golden, M. Matthew, A. Berk, L. Bernstein, P. Acharya, D. Miller, P. Lewis, MODTRAN4-based atmospheric correction algorithm: FLAASH (fast line-of-sight atmospheric analysis of spectral hypercubes), *Proceedings of Algorithms and Technologies for Multispectral, Hyperspectral, and Ultraspectral Imagery VIII*, SPIE, Vol. 4725, pp. 65–71, 2002.
26. P. Swain, V. Vanderbilt, and C. Jobusch, A quantitative applications-oriented evaluation of thematic mapper design specifications, *IEEE Transactions on Geoscience and Remote Sensing*, vol. 20, no. 3, pp. 370–377, 1982.
27. J. Lee, A. Woodyatt, and M. Berman, Enhancement of high spectral resolution remote-sensing data by a noise-adjusted principal components transform, *IEEE Transactions on Geoscience and Remote Sensing*, vol. 28, no. 3, pp. 295–304, 1990.
28. B. Thai, and G. Healey, Invariant subpixel material detection in hyperspectral imagery, *IEEE Transactions on Geoscience and Remote Sensing*, vol. 40, no. 3, pp. 599–608, 2002.
29. J. R. Schott, S. D. Brown, R. V. Raqueño, H. N. Gross, and G. Robinson, An advanced synthetic image generation model and its application to multi/hyperspectral algorithm development, *Canadian Journal of Remote Sensing*, vol. 25, no. 2, 1999.

30. A. Berk, L. S. Bernstein, G. P. Anderson, P. K. Acharya, D. C. Robertson, J. H. Chetwynd, and S. M. Adler-Golden, MODTRAN cloud and multiple scattering upgrades with application to AVIRIS, *Remote Sensing of Environment,* vol. 65, pp. 367–375, 1998.
31. H. D. Smith, Dube, M. Gardner, S. Clough, F. Kneizys, and L. Rothman, FASCODE- Fast Atmospheric Signature Code (Spectral Transmittance and Radiance), Air Force Geophysics Laboratory Technical Report AFGL-TR-78-0081, Hanscom AFB, MA, 1978.
32. M. Nischan, J. Kerekes, J. Baum, and R. Basedow, Analysis of HYDICE noise characteristics and their impact on subpixel object detection, *Proceedings of Imaging Spectrometry V,* SPIE, Vol. 3753, pp.112–123, 1999.
33. B. Shetler, D. Mergens, C. Chang, F. Mertz, J. Schott, S. Brown, R. Strunce, F. Maher, S. Kubica, R. de Jonckheere, and B. Tousley, Comprehensive hyperspectral system simulation: I. Integrated sensor scene modeling and the simulation architecture, *Algorithms for Multispectral, Hyperspectral, and Ultraspectral Imagery VI,* SPIE, Vol. 4049, 2000.
34. C. Schwartz, A. C. Kenton, W. F. Pont, and B. J. Thelen, Statistical parametric signature/sensor/detection model for multispectral mine target detection, *Proceedings of Detection Technologies for Mines and Minelike Targets*, SPIE, Vol. 2496, pp. 222–238, 1995.
35. J. Kerekes, A. Cisz, and R. Simmons, A comparative evaluation of spectral quality metrics for hyperspectral imagery, *Proceedings of Algorithms and Technologies for Multispectral, Hyperspectral, and Ultraspectral Imagery XI*, SPIE, Vol. 5806, pp. 469–480, 2005.

CHAPTER 3

INFORMATION-PROCESSED MATCHED FILTERS FOR HYPERSPECTRAL TARGET DETECTION AND CLASSIFICATION

CHEIN-I CHANG
Remote Sensing Signal and Image Processing Laboratory, Department of Computer Science and Electrical Engineering, University of Maryland—Baltimore County, Baltimore, MD 21250

3.1. INTRODUCTION

How to effectively use information in data analysis is an important subject. In hyperspectral image analysis the information provided by hundreds of contiguous spectral channels may be overwhelming, but may not necessarily be useful in all applications. In particular, some information resulting from unknown signal sources may contaminate and distort the information that we try to extract. Interestingly, this problem has not received much attention in the past. In this chapter, we investigate the issue of how information plays its role in target detection and classification in hyperspectral imagery from a signal processing viewpoint and study the effects of different levels of information on performance. Two types of information are considered: *a priori* information and *a posteriori* information. The former is generally referred to as the knowledge provided *a priori* before data processing takes place, whereas the latter is the knowledge obtained directly from the data during data processing. So, in terms of pattern classification they can be thought of as supervised information and unsupervised information respectively.

There have been many algorithms developed for detection and classification in hyperspectral imagery over the past years—in particular, linear spectral unmixing. Due to use of different degrees of target knowledge, these algorithms appear in different forms. Nevertheless, they do share some common ground. Most noticeable is the design principle, such as the concept of matched filter. This chapter is devoted

Hyperspectral Data Exploitation: Theory and Applications, Edited by Chein-I Chang

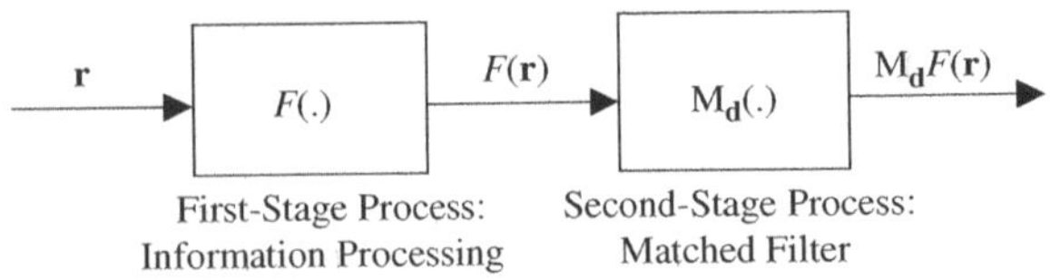

Figure 3.1. A block diagram of an IPMF interpretation.

to a study of hyperspectral target detection and classification algorithms from a viewpoint of target knowledge to be used in the algorithms. Specifically, an information-processed matched-filter (IPMF) interpretation is presented, which allows us to decompose an algorithm in two filter operations that can be carried out in two stages sequentially as depicted in Figure 3.1. The filter operation $F(.)$ in the first-stage process is an information-processed filter, which processes either *a priori* or *a posteriori* information to suppress interfering effects caused by unknown and unwanted target sources in the data. This process is generally referred to as a "whitening" process in communication and signal processing. This is then followed by a filter operation $M_d(.)$ in the second-stage process that is a matched filter to extract targets of interest for detection and classification. The performance of a matched filter is determined by the matched signal used in the filter as well as the information extracted in the information-processed filter.

In order to illustrate how information is used for data analysis, three types of techniques that utilize different levels of target information are considered in this chapter. The first type of technique is the orthogonal subspace projection (OSP)-based approach. It is a linear spectral mixture analysis method, which requires complete target knowledge *a priori*. When the information of target signatures is provided by prior knowledge and processed for target detection, such OSP is referred to as *a priori* OSP and is the one originally developed by Harsanyi and Chang [1]. If the information of target signatures is provided *a priori* and processed for estimation of target abundance fractions from the image data, it is called *a posteriori* OSP, which was studied by Chang and co-workers [2, 3]. A second type of technique is a linearly constrained minimum variance (LCMV)-based approach, which only requires prior knowledge of targets of interest without knowing image background. It was developed in Chang [4] with two special versions: constrained energy minimization (CEM) filter [5] and target-constrained interference-minimized filter (TCIMF) [6]. The difference between the OSP and LCMV approaches is that the former requires a linear mixture model that is not required for the latter. The CEM filter linearly constrains a desired target that is provided *a priori*, while taking advantaging of the sample correlation matrix to obtain *a posteriori* information that will be used for interference minimization. The TCIMF divides the targets of interest into desired targets and undesired targets separately. Like the CEM, the TCIMF also utilizes the prior target knowledge to extract the desired targets and annihilate the undesired targets, while using the *a posteriori* information obtained from the sample correlation matrix to minimize interference. A third type of techniques is anomaly detection, which requires no prior knowledge at all. Its performance is completely determined by the *a posteriori* information obtained by the sample covariance/correlation matrix. Of particular interest are (a) the RX algorithm developed by Reed and Yu [7], which uses the sample covariance matrix to generate *a posteriori* information for anomaly detection, and (b) the low probability detection (LPD) proposed in Harsanyi's dissertation [5], which uses the sample correlation matrix to detect targets that occur with low probabilities in the image data. So, generally speaking, the OSP requires the complete prior target knowledge as opposed to anomaly detection, which requires no prior knowledge with its performance completely determined by the *a posteriori* information generated from the data. The LCMV is somewhere in between, which requires partial *a priori* target knowledge as well as *a posteriori* information. These three types of techniques apply different levels of *a priori* or *a posteriori* target information, which results in various performances in detection and classification. A detailed study and analysis was conducted in Chang [8]. Interestingly, the relationship among these three seemingly different techniques can be well illustrated and interpreted by the proposed IPMF approach. More specifically, in the light of the IPMF approach, the OSP, LCMV, and anomaly detection implement an information-processed filter in the first stage in Figure 3.1 to extract *a priori* or *a posteriori* target information from data, and then they use a follow-up matched filter in the second stage in Figure 3.1 to extract desired targets of interest. They all operate the same functional form of a matched filter with the matched signal determined by the information derived in the information-processed filter. In other words, the OSP, LCMV, and anomaly detection essentially perform the same matched filter in different forms that reflect different levels of target information used in their filter designs. In order to explore the roles of *a priori* information and *a posteriori* information play in these three types of techniques, a set of experiments is conducted for demonstration.

3.2. TECHNIQUES USING DIFFERENT LEVELS OF TARGET INFORMATION

Three types of techniques—OSP, LCMV, and anomaly detection—are selected for evaluation, each of which utilizes different levels of target information to accomplish specific applications.

3.2.1. OSP-Based Techniques (Complete Prior Target Knowledge Is Required)

The OSP technique was originally developed for hyperspectral image classification [1]. It takes advantage of a linear mixture model to unmix the targets of interest. Suppose that L is the number of spectral bands. Let $\mathbf{r}$ be an $L \times 1$ column pixel vector in a hyperspectral image where boldface is used for vectors. Assume that $\{\mathbf{t}_1, \mathbf{t}_2, \ldots, \mathbf{t}_p\}$ is a set of k targets of interest present in the image and $\mathbf{m}_1$, $\mathbf{m}_2, \ldots, \mathbf{m}_p$ are their corresponding spectral signatures. Let $\mathbf{M}$ be an $L \times p$ target signature matrix denoted by $\lfloor \mathbf{m}_1 \mathbf{m}_2 \cdots \mathbf{m}_p \rfloor$, where $\mathbf{m}_j$ is an $L \times 1$ column vector represented by the signature of the jth target spectral resident in the image scene and p is the number of targets in the image scene. Let $\boldsymbol{\alpha} = (\alpha_1, \alpha_2, \cdots, \alpha_p)^T$ be a $p \times 1$ abundance column vector associated with $\mathbf{r}$, where $\boldsymbol{\alpha}_j$ denotes the fraction of the jth signature $\mathbf{m}_j$ present in the pixel vector $\mathbf{r}$. Then the spectral signature of $\mathbf{r}$ can be represented by the following linear mixture model:

$$\mathbf{r} = \mathbf{M}\boldsymbol{\alpha} + \mathbf{n} \tag{3.1}$$

where $\mathbf{n}$ is noise or can be interpreted as a measurement or model error.

3.2.2. *A Priori* OSP

Equation (3.1) assumes that the target knowledge $\mathbf{M}$ must be given *a priori*. Without loss of generality we further assume that $\mathbf{d} = \mathbf{m}_p$ is the desired target signature to be classified and $\mathbf{U} = \lfloor \mathbf{m}_1 \mathbf{m}_2 \cdots \mathbf{m}_{p-1} \rfloor$ is the undesired target signature matrix made up of the remaining $p-1$ undesired target signatures in $\mathbf{M}$. Then, we rewrite Eq. (3.1) as

$$\mathbf{r} = \mathbf{d}\alpha_p + \mathbf{U}\boldsymbol{\gamma} + \mathbf{n} \tag{3.2}$$

where $\boldsymbol{\gamma}$ is the abundance vector associated with $\mathbf{U}$. Equation (3.2) separates the desired target signature $\mathbf{d}$ from the undesired target signatures in $\mathbf{U}$. This allows us to design the following orthogonal subspace projector, denoted by $P_{\mathbf{U}}^{\perp}$, to annihilate $\mathbf{U}$ from $\mathbf{r}$ prior to classification:

$$P_{\mathbf{U}}^{\perp} = \mathbf{I} - \mathbf{U}\mathbf{U}^{\#} \tag{3.3}$$

where $\mathbf{U}^{\#} = (\mathbf{U}^T\mathbf{U})^{-1}\mathbf{U}^T$ is the pseudo-inverse of $\mathbf{U}$.

Applying $P_{\mathbf{U}}^{\perp}$ in (3.3) to (3.2) results in

$$P_{\mathbf{U}}^{\perp}\mathbf{r} = P_{\mathbf{U}}^{\perp}\mathbf{d}\alpha_p + P_{\mathbf{U}}^{\perp}\mathbf{n} \tag{3.4}$$

Equation (3.4) represents a standard signal detection problem. If the signal-to-noise ratio (SNR) is chosen as a criterion for optimality, the optimal solution to (3.4) is given by a matched filter $\mathbf{M_d}$ defined by

$$\mathbf{M_d}(P_{\mathbf{U}}^{\perp}\mathbf{r}) = \kappa\mathbf{d}^T P_{\mathbf{U}}^{\perp}\mathbf{r} \tag{3.5}$$

where the matched signal is specified by $\mathbf{d}$ and κ is a constant [9].

Setting $\kappa = 1$ in (3.5) yields the following OSP detector $\delta^{\mathrm{OSP}}(\mathbf{r})$ derived by Harsanyi and Chang [5]:

$$\delta^{\mathrm{OSP}}(\mathbf{r}) = \mathbf{d}^T P_{\mathbf{U}}^{\perp}\mathbf{r} = (\mathbf{d}^T P_{\mathbf{U}}^{\perp}\mathbf{d})\alpha_p + \mathbf{d}^T P_{\mathbf{U}}^{\perp}\mathbf{n} \tag{3.6}$$

The OSP detector given by Eq. (3.6) used the target knowledge $\mathbf{M}$ to detect targets whose signatures were specified by the desired signature $\mathbf{d}$. Its primary task was

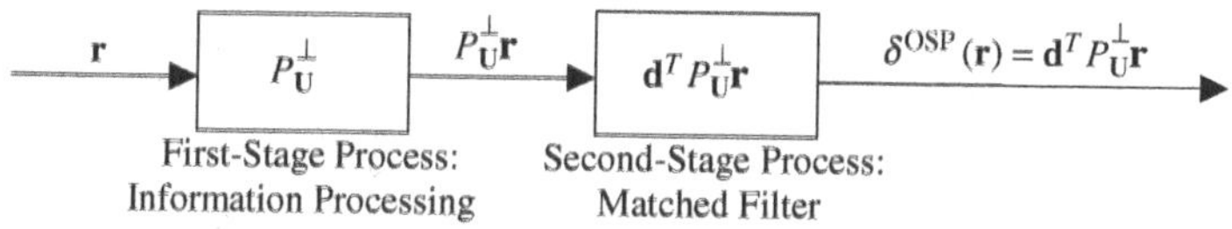

Figure 3.2. A block diagram of *a priori* OSP, $\delta^{OSP}(\mathbf{r})$.

designed for target detection and was not designed to estimate the target abundance fraction of **d** present in the pixel **r**. In other words, the Harsanyi–Chang developed OSP intended to detect all the targets specified by the desired spectral signature **d**, but did not attempt to estimate the abundance fractions of its detected targets. Accordingly, such an OSP is referred to as an *a priori* OSP whose block diagram given in Figure 3.2 can be obtained by replacing the function $F(.)$ and $M_d(.)$ in Figure 3.1 with $P_U^\perp$ specified by Eq. (3.3) and $M_d(P_U^\perp \mathbf{r})$ specified by Eq. (3.5), respectively, as follows.

3.2.3. *A Posteriori* OSP

As noted in previous section, *a priori* OSP was not designed to estimate target abundance fractions. However, in reality, $\boldsymbol{\alpha}$ is generally not known and needs to be estimated. In order to address this issue, several techniques have been developed by Chang and co-workers [3, 10] for estimation of $\boldsymbol{\alpha} = (\alpha_1, \alpha_2, \ldots, \alpha_p)^T$ and are based on *a posteriori* information obtained from the image data to be processed. Three *a posteriori* OSP detectors are of interest and can be also considered as OSP abundance estimators, signature subspace classifier (SSC) denoted by $\delta^{SSC}(\mathbf{r})$, oblique subspace classifier (OBC) denoted by $\delta^{OBC}(\mathbf{r})$, and Gaussian maximum likelihood classifier (GMLC) denoted by $\delta^{GMLC}(\mathbf{r})$, all of which are derived by Chang and co-workers [3, 10] and given as follows:

$$\delta^{SSC}(\mathbf{r}) = (\mathbf{d}^T P_U^\perp \mathbf{d})^{-1} \mathbf{d}^T P_U^\perp P_\mathbf{M} \mathbf{r} = \alpha_p + (\mathbf{d}^T P_U^\perp \mathbf{d})^{-1} \mathbf{d}^T P_U^\perp P_\mathbf{M} \mathbf{n} \tag{3.7}$$

where $P_\mathbf{M} = \mathbf{M}(\mathbf{M}^T\mathbf{M})^{-1}\mathbf{M}^T$ and

$$\delta^{OBC}(\mathbf{r}) = \delta^{GMLC}(\mathbf{r}) = (\mathbf{d}^T P_U^\perp \mathbf{d})^{-1} \delta^{OSP}(\mathbf{r}) = \alpha_p + (\mathbf{d}^T P_U^\perp \mathbf{d})^{-1} \mathbf{d}^T P_U^\perp \mathbf{n} \tag{3.8}$$

where the GMLC is identical to the OBC, provided that the noise **n** in Eq. (3.2) is assumed to be Gaussian.

Comparing Eqs. (3.7) and (3.8) to (3.6), there is a constant $(\mathbf{d}^T P_U^\perp \mathbf{d})^{-1}$ that appears in Eqs. (3.7) and (3.8) but is absent in the second equality of Eq. (3.6). This constant is a result of using *a posteriori* information extracted from the pixel vector **r** based on the least-squares error criterion. As shown in references 10–12, the constant $(\mathbf{d}^T P_U^\perp \mathbf{d})^{-1}$ was closely related to the estimation accuracy of $\boldsymbol{\alpha}$. To reflect this fact, the block diagram in Figure 3.3 can be further obtained by including the constant $(\mathbf{d}^T P_U^\perp \mathbf{d})^{-1}$ in Figure 3.2 to yield the following block diagram for *a posteriori* OSP.

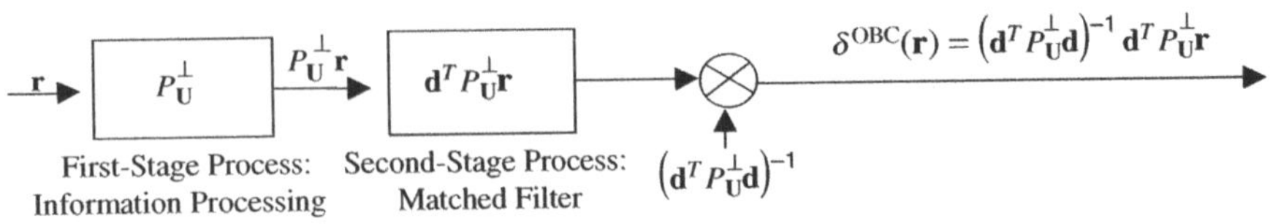

Figure 3.3. A block diagram of *a posteriori* OSP, $\delta^{OBC}(\mathbf{r})$.

3.3. LINEARLY CONSTRAINED MINIMUM VARIANCE (LCMV)

Unlike the OSP-based techniques which require a linear mixture model and the complete prior target knowledge $\mathbf{M}$, the LCMV-based techniques only require the prior knowledge of targets of interest. It makes use of the sample correlation matrix to minimize the interfering effects caused by unknown signal sources, which may include unknown image background signatures.

Assume that $\{\mathbf{r}_1, \mathbf{r}_2, \ldots, \mathbf{r}_N\}$ is the set of all image pixels in a remotely sensed image where $\mathbf{r}_i = (r_{i1}, r_{i2}, \ldots, r_{iL})^T$ for $1 \leq i \leq N$ is an L-dimensional pixel vector, and N is the total number of pixels in the image. Suppose that $\{\mathbf{t}_1, \mathbf{t}_2, \ldots, \mathbf{t}_p\}$ is a set of p targets of interest present in the image and $\mathbf{m}_1, \mathbf{m}_2, \ldots, \mathbf{m}_p$ are their corresponding spectral signatures. We further assume without loss of generality that among the p targets of interest, there are k desired targets denoted by $\{\mathbf{t}_1, \mathbf{t}_2, \ldots, \mathbf{t}_k\}$ where $k \leq p$ and $\{\mathbf{t}_{k+1}, \mathbf{t}_{k+2}, \ldots, \mathbf{t}_p\}$ are undesired targets. An LCMV-based classifier is to design an FIR linear filter with an L-dimensional weight vector $\mathbf{w} = (w_1, w_2, \ldots, w_L)^T$ that minimizes the filter output energy subject to the following constraint:

$$\mathbf{M}^T\mathbf{w} = \mathbf{c}, \qquad \text{where } \mathbf{m}_j^T\mathbf{w} = \sum_{l=1}^{L} m_{jl}\, w_l = c_j \text{ for } 1 \leq j \leq p \tag{3.9}$$

where $\mathbf{M} = \lfloor \mathbf{m}_1 \mathbf{m}_2 \cdots \mathbf{m}_p \rfloor$ is a signature matrix formed by target signatures of interest and $\mathbf{c} = (c_1, c_2, \ldots, c_k)^T$ is a constraint vector that is imposed on $\mathbf{M}$.

Now, let y_i denote the output of the designed FIR filter resulting from the input $\mathbf{r}_i$. Then y_i can be expressed by

$$y_i = \sum_{l=1}^{L} w_l r_{il} = \mathbf{w}^T\mathbf{r}_i = \mathbf{r}_i^T\mathbf{w} \tag{3.10}$$

and the average energy of the filter output is given by

$$\frac{1}{N}\left[\sum_{i=1}^{N} y_i^2\right] = \frac{1}{N}\left[\sum_{i=1}^{N} (\mathbf{r}_i^T\mathbf{w})^T(\mathbf{r}_i^T\mathbf{w})\right] = \mathbf{w}\left(\frac{1}{N}\left[\sum_{i=1}^{N} \mathbf{r}_i\mathbf{r}_i^T\right]\right)\mathbf{w} = \mathbf{w}^T\mathbf{R}_{L\times L}\mathbf{w} \tag{3.11}$$

where $\mathbf{R}_{L\times L} = \frac{1}{N}\left[\sum_{i=1}^{N} r_i r_i^T\right]$ is the sample autocorrelation matrix of the image.

The goal of the FIR filter is to pass through the desired targets, $\{\mathbf{t}_1, \mathbf{t}_2, \ldots, \mathbf{t}_k\}$ with constraining the undesired targets $\{\mathbf{t}_{k+1}, \mathbf{t}_{k+2}, \ldots, \mathbf{t}_p\}$, while minimizing the average filter output energy. This constrained filter design problem is equivalent to solving the following linearly constrained optimization problem:

$$\min_{\mathbf{w}}\{\mathbf{w}^T\mathbf{R}_{L\times L}\mathbf{w}\} \quad \text{subject to} \quad \mathbf{M}^T\mathbf{w} = \mathbf{c} \tag{3.12}$$

Let $\mathbf{w}^{\text{LCMV}}$ be the solution to Eq. (3.12), which can be obtained in [4] by

$$\mathbf{w}^{\text{LCMV}} = \mathbf{R}_{L\times L}^{-1}\mathbf{M}(\mathbf{M}^T\mathbf{R}_{L\times L}^{-1}\mathbf{M})^{-1}\mathbf{c} \tag{3.13}$$

Using Eqs. (3.10) and (3.13), an LCMV filter, denoted by $\delta^{\text{LCMV}}(\mathbf{r})$, is given by [4]

$$\delta^{\text{LCMV}}(\mathbf{r}) = (\mathbf{w}^{\text{LCMV}})^T\mathbf{r} = \mathbf{r}^T\mathbf{w}^{\text{LCMV}} \tag{3.14}$$

3.3.1. Constrained Energy Minimization (CEM)

If we are only interested in a single target signature $\mathbf{d}$—that is, $\mathbf{M} = \mathbf{d}$—the constraint in Eq. (3.9) is reduced to $\mathbf{d}^T\mathbf{w} = \sum_{l=1}^{L} d_l w_l = 1$ where the constraint vector $\mathbf{c}$ becomes a constraint scalar 1. In this specific case, Eq. (3.12) is simplified to

$$\min_{\mathbf{w}}\{\mathbf{w}^T\mathbf{R}_{L\times L}\mathbf{w}\} \quad \text{subject to} \quad \mathbf{d}^T\mathbf{w} = 1 \tag{3.15}$$

with the optimal solution $\mathbf{w}^{\text{CEM}}$ given by

$$\mathbf{w}^{\text{CEM}} = \frac{\mathbf{R}_{L\times L}^{-1}\mathbf{d}}{\mathbf{d}^T\mathbf{R}_{L\times L}^{-1}\mathbf{d}} \tag{3.16}$$

Substituting the weight vector $\mathbf{w}^{\text{CEM}}$ given by Eq. (3.16) for $\mathbf{w}$ in Eq. (3.14) results in the CEM filter, $\delta^{\text{CEM}}(\mathbf{r})$, given by

$$\delta^{\text{CEM}}(\mathbf{r}) = (\mathbf{w}^{\text{CEM}})^T\mathbf{r} = \left(\frac{\mathbf{R}_{L\times L}^{-1}\mathbf{d}}{\mathbf{d}^T\mathbf{R}_{L\times L}^{-1}\mathbf{d}}\right)^T\mathbf{r} = \frac{\mathbf{d}^T\mathbf{R}_{L\times L}^{-1}\mathbf{r}}{\mathbf{d}^T\mathbf{R}_{L\times L}^{-1}\mathbf{d}} \tag{3.17}$$

The approach to solving Eq. (3.15) for $\mathbf{w}^{\text{CEM}}$ using Eq. (3.16) is referred to as the CEM in Harsanyi's dissertation [5] with its block diagram given in Figure 3.4 by replacing $P_{\mathbf{U}}^{\perp}$ in Figure 3.4 with $\mathbf{R}_{L\times L}^{-1}$ as follows.

3.3.2. Target-Constrained Interference-Minimized Filter (TCIMF)

In many practical applications, the targets of interest can be categorized into two classes: one for desired targets and the other for undesired targets. In this case, we can break up the target signature matrix $\mathbf{M}$ into a desired target signature matrix, denoted by $\mathbf{D} = \lfloor \mathbf{d}_1\mathbf{d}_2 \cdots \mathbf{d}_{n_{\mathbf{D}}} \rfloor$, and an undesired target signature matrix, denoted

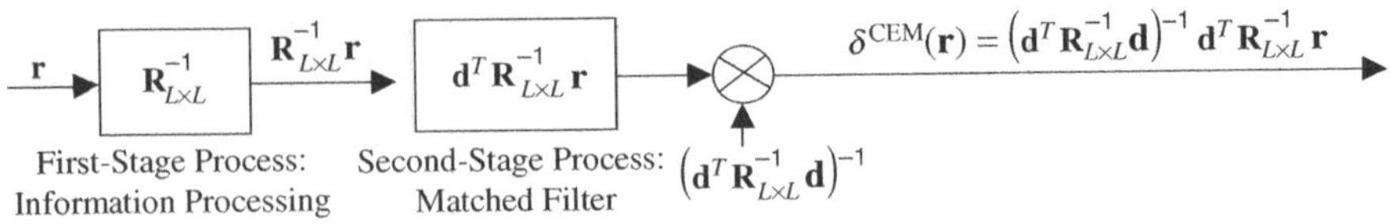

Figure 3.4. A block diagram of $\delta^{\text{CEM}}(\mathbf{r})$.

by $\mathbf{U} = \lfloor \mathbf{d}_1\mathbf{d}_2 \cdots \mathbf{d}_{n_U} \rfloor$, where n_D and n_U are the number of the desired target signatures and the number of the undesired target signatures, respectively. In this case, the signature matrix $\mathbf{M}$ can be further expressed by $\mathbf{M} = [\mathbf{DU}]$. Now, we can design an FIR filter to pass through the desired targets in $\mathbf{D}$ using an $n_D \times 1$ unity constraint vector $\mathbf{1}_{n_D\times 1} = \underbrace{(1, 1, \ldots, 1)}_{n_D}$ while annihilating the undesired targets in $\mathbf{U}$ using an $n_U \times 1$ zero constraint vector $\mathbf{0}_{n_D\times 1} = \underbrace{(0, 0, \ldots, 0)}_{n_U}$. In order to do so, the constraint in Eq. (3.9) is replaced by

$$[\mathbf{DU}]^T\mathbf{w} = \begin{bmatrix} \mathbf{1}_{n_D\times 1} \\ \mathbf{0}_{n_U\times 1} \end{bmatrix} \tag{3.18}$$

which results in the following linearly constrained optimization problem:

$$\min_{\mathbf{w}}\{\mathbf{w}^T\mathbf{R}_{L\times L}\mathbf{w}\} \quad \text{subject to} \quad [\mathbf{DU}]^T\mathbf{w} = \begin{bmatrix} \mathbf{1}_{n_D\times 1} \\ \mathbf{0}_{n_U\times 1} \end{bmatrix} \tag{3.19}$$

The filter solving Eq. (3.19) is called the target-constrained interference-minimized filter (TCIMF) in Ren and Chang [6] with the weight vector, $\mathbf{w}^{\text{TCIMF}}$, given by

$$\mathbf{w}^{\text{TCIMF}} = \mathbf{R}_{L\times L}^{-1}[\mathbf{DU}]\left([\mathbf{DU}]^T\mathbf{R}_{L\times L}^{-1}[\mathbf{DU}]\right)^{-1}\begin{bmatrix} \mathbf{1}_{n_D\times 1} \\ \mathbf{0}_{n_U\times 1} \end{bmatrix} \tag{3.20}$$

Substituting the weight vector $\mathbf{w}^{\text{TCIMF}}$ given by Eq. (3.20) for $\mathbf{w}$ in Eq. (3.14) yields the TCIMF, $\delta^{\text{TCIMF}}(\mathbf{r})$, given by

$$\delta^{\text{TCIMF}}(\mathbf{r}) = (\mathbf{w}^{\text{TCIMF}})^T\mathbf{r} \tag{3.21}$$

3.4. ANOMALY DETECTION

In contrast to the OSP and LCMV filters, anomaly detection does not need target information at all. It utilizes the spectral correlation among data samples to find targets whose signatures spectrally distinct from their surroundings. Two types of anomaly detection are of particular interest. One was developed by Reed and Yu [7], referred to as RX filter, and the other is called low probability detection (LPD), developed in Harsanyi [5].

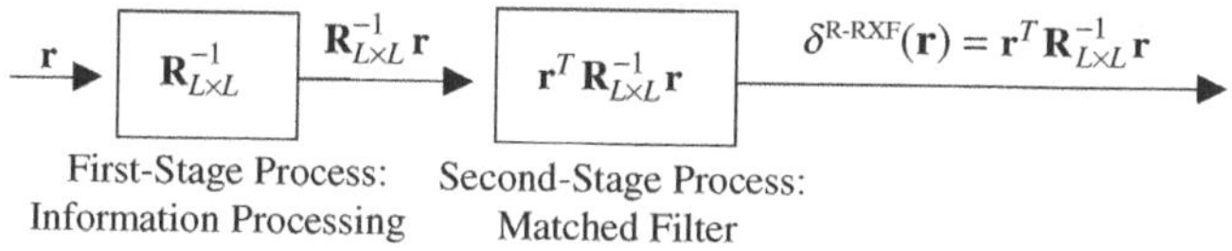

Figure 3.5. A block diagram of $\delta^{\text{R-RXF}}(\mathbf{r})$.

3.4.1. RX Filter-Based Techniques

The Reed–Yu RX filter implements a filter, referred to in this chapter as a K-RX filter (K-RXF), which is specified by

$$\delta^{\text{K-RXF}}(\mathbf{r}) = (\mathbf{r} - \boldsymbol{\mu})^T \mathbf{K}_{L\times L}^{-1}(\mathbf{r} - \boldsymbol{\mu}) \tag{3.22}$$

where $\boldsymbol{\mu}$ is the sample mean and the "K" in the superscript of $\delta^{\text{K-RXF}}(\mathbf{r})$ indicates that the sample covariance matrix, $\mathbf{K}_{L\times L}$, is used in the filter design. The form of $\delta^{\text{K-RXF}}(\mathbf{r})$ in (3.22) is actually the well-known Mahalanobis distance. Replacing $\mathbf{K}_{L\times L}$ with the sample correlation matrix $\mathbf{R}_{L\times L}$ and replacing $\mathbf{r} - \boldsymbol{\mu}$ with the pixel vector $\mathbf{r}$ in Eq. (3.22) results in a sample correlation matrix-based RX filter (R-RXF) given by

$$\delta^{\text{R-RXF}}(\mathbf{r}) = \mathbf{r}^T \mathbf{R}_{L\times L}^{-1}\mathbf{r} \tag{3.23}$$

where the "R" in the subscript of $\delta^{\text{R-RXF}}(\mathbf{r})$ indicates that the sample correlation matrix, $\mathbf{R}_{L\times L}$ is used in the filter design. Its block diagram is given in Figure 3.5.

3.4.2. Low Probability Detection (LPD)

Another type of anomaly detection is called the low probability detection (LPD) and was previously derived in Harsanyi [5]. It uses the sample correlation matrix $\mathbf{R}_{L\times L}$ to implement a filter, $\delta_{\text{LPD}}(\mathbf{r})$, given by

$$\delta^{\text{LPD}}(\mathbf{r}) = \mathbf{1}^T \mathbf{R}_{L\times L}^{-1}\mathbf{r} \tag{3.24}$$

where $\mathbf{1} = \underbrace{(1, 1, \ldots, 1)}_{L}$ is an $(L \times 1)$-dimensional unity vector with ones in all the components. A block diagram of the LPD is delineated in Figure 3.6.

Since there is no prior information available, the best scenario is not to introduce any information into the filter structure in Eq. (3.24). In this case, the anomalous

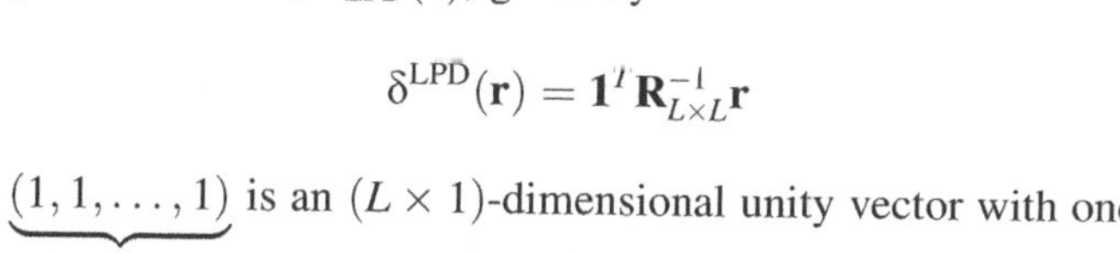

Figure 3.6. A block diagram of LPD, $\delta^{\text{LPD}}(\mathbf{r})$.

targets are assumed to have radiance values uniformly distributed over all the spectral bands. Such targets may occur with low probabilities in an image scene. It has been demonstrated in Chang [9] that the most likely targets that will be extracted by the LPD are image background signatures. This is because background signatures are generally wide spread with a large range of spectral variation. As a result, background signatures can considered as targets with occurrence of low probabilities, and their histograms may be distributed uniformly with low magnitudes. Because of that, the LPD was referred to as the uniform target detector (UTD) in Chang [9].

3.5. INFORMATION-PROCESSED MATCHED-FILTERS

The three types of techniques described in Sections 3.2–3.4 were developed from the utilization of different degrees of target knowledge. It seems that there is no strong tie among them. In this section, we will show that they are indeed closely related through the interpretation of the IPMF described in Figure 3.1. In other words, they all operate some form of an information-processed filter and a matched filter with different matched signals, which are determined by the target information used in their filters.

3.5.1. Relationship Between OSP and CEM

If we compare $\delta^{\text{OSP}}(\mathbf{r}) = \mathbf{d}^T P_{\mathbf{U}}^{\perp}\mathbf{r}$ specified by Eq. (3.6) against $\delta^{\text{CEM}}(\mathbf{r}) = \left(\frac{1}{\mathbf{d}^T\mathbf{R}_{L\times L}^{-1}\mathbf{d}}\right)\mathbf{d}^T\mathbf{R}_{L\times L}^{-1}\mathbf{r}$ specified by Eq. (3.17), we will discover that there is an interesting relationship between $P_{\mathbf{U}}^{\perp}$ used in the $\delta^{\text{OSP}}(\mathbf{r})$ and $\mathbf{R}_{L*L}^{-1}$ used in $\delta^{\text{CEM}}(\mathbf{r})$. They both are matched filters operated by the same matched signal **d**. However, there is also a significant difference in the constant κ, which has been overlooked in the past. In the matched filter derivation of Eq. (3.5), the constant κ resulting from Schwarz's inequality generally plays no role in signal detection and has been assumed to be 1 for signal detectability—for example, *a priori* OSP, $\delta^{\text{OSP}}(\mathbf{r})$—but it does matter for abundance estimation such as *a posteriori* OSP, $\delta^{\text{OBC}}(\mathbf{r})$. As noted in references 8 and 10–12, the constant κ determines the estimation error of the abundance vector α. Thus, the relationship between $\delta^{\text{OSP}}(\mathbf{r})$ and $\delta^{\text{CEM}}(\mathbf{r})$ can be described from two viewpoints, namely, signal detection and signal estimation.

In signal detection, the knowledge of $P_{\mathbf{U}}^{\perp}$, which is assumed to be known in $\delta^{\text{OSP}}(\mathbf{r})$, is not available in $\delta^{\text{CEM}}(\mathbf{r})$. The $\delta^{\text{CEM}}(\mathbf{r})$ must estimate $P_{\mathbf{U}}^{\perp}$ directly from the image data. One way of doing so is to approximate the $P_{\mathbf{U}}^{\perp}$ in the sense of minimum least-squares error by the spectral information that can be obtained by the inverse of the sample correlation matrix, $\mathbf{R}_{L\times L}^{-1}$, from the image data. More specifically, $\delta^{\text{CEM}}(\mathbf{r})$ makes use of the *a posteriori* information, $\mathbf{R}_{L\times L}^{-1}$, obtained from the image data to approximate the *a priori* target information, $P_{\mathbf{U}}^{\perp}$, to accomplish what $\delta^{\text{OSP}}(\mathbf{r})$ does. Since $\delta^{\text{OSP}}(\mathbf{r})$ assumes the prior knowledge of the abundance vector $\boldsymbol{\alpha}$, there is no need of abundance estimation, in which case $\kappa = 1$.

But when it comes to signal estimation, the *a posteriori* OSP-based classifiers $\delta^{GMLC}(\mathbf{r})$, $\delta^{OBC}(\mathbf{r})$, and $\delta^{CEM}(\mathbf{r})$ all include $\kappa = (\mathbf{d}^T\mathbf{R}_{L\times L}^{-1}\mathbf{d})^{-1}$ in their filter designs given by Eqs. (3.7), (3.8), and (3.17) to account for abundance estimation error. Because these three filters are designed based on the same least-squares error criterion, it is not surprising to see that they all generate the same $\kappa = (\mathbf{d}^T\mathbf{R}_{L\times L}^{-1}\mathbf{d})^{-1}$ as shown in Chang [8].

As shown in Figures 3.2 and 3.4, the information-processed matched filters for $\delta^{OSP}(\mathbf{r})$ and $\delta^{CEM}(\mathbf{r})$ are accomplished by $P_{\mathbf{U}}^{\perp}$ in $\delta^{OSP}(\mathbf{r})$ and $\mathbf{R}_{L\times L}^{-1}$ in $\delta^{CEM}(\mathbf{r})$, respectively. The matched signal used for $\delta^{OSP}(\mathbf{r})$ and $\delta^{CEM}(\mathbf{r})$ is the desired target signature $\mathbf{d}$. However, in the case of $\delta^{CEM}(\mathbf{r})$, the output of the matched filter is further scaled by the constant $\kappa = (\mathbf{d}^T\mathbf{R}_{L\times L}^{-1}\mathbf{d})^{-1}$ to produce more accurate abundance estimation. Similarly, Figure 3.3 shows the block diagram of *a posteriori* OSP detectors, $\delta^{OBC}(\mathbf{r})$ or $\delta^{GMLC}(\mathbf{r})$, which also include the scale constant $\kappa = (\mathbf{d}^T P_{\mathbf{U}}^{\perp}\mathbf{d})^{-1}$ to account for abundance estimation.

As noted above, $\delta^{CEM}(\mathbf{r})$ approximates $P_{\mathbf{U}}^{\perp}$ by $\mathbf{R}_{L\times L}^{-1}$. In terms of information approximation, it may be not a best approximation because the sample correlation matrix $\mathbf{R}_{L\times L}$ includes the desired signature $\mathbf{d}$, whereas the undesired target signature matrix $\mathbf{U}$ in $P_{\mathbf{U}}^{\perp}$ excludes $\mathbf{d}$. Thus, a better approximation can be achieved by replacing $\mathbf{R}_{L\times L}^{-1}$ with $\tilde{\mathbf{R}}_{L\times L}^{-1}$, where $\tilde{\mathbf{R}}_{L\times L}$ is the sample correlation matrix formed by the image data, which exclude all the pixels specified by the desired target signature $\mathbf{d}$. More specifically, let $\Delta(\mathbf{d})$ denote the set of pixels in the image data that are specified by $\mathbf{d}$. Then $\tilde{\mathbf{R}}_{L\times L} = \frac{1}{N-|\Delta(\mathbf{d})|}\left[\sum_{i=1}^{N}\mathbf{r}_i\mathbf{r}_i^T - \sum_{\mathbf{r}_k\in\Delta(\mathbf{d})}\mathbf{r}_k\mathbf{r}_k^T\right]$, where $|\Delta(\mathbf{d})|$ is the number of pixels in $\Delta(\mathbf{d})$. As will be demonstrated in computer simulations, using $\tilde{\mathbf{R}}_{L\times L}$ in place of $\mathbf{R}_{L\times L}$ can improve the performance of $\delta^{CEM}(\mathbf{r})$ [8, 10]. For this case it is also true that removing target pixels from $\mathbf{R}_{L\times L}$ whose signatures are close and similar to the desired target signatures can further enhance the performance as well. This was also demonstrated in Chang [10] where the background subspace was generated by removing those signatures which are close to target signatures so that background subspace can be effectively annihilated by an orthogonal projector.

Similarly, the same conclusion can be also applied to LCMV filter and TCIMF. However, in many real applications, the number of pixels specified by $\mathbf{d}$ is generally small compared to the entire image. By taking this into account, using the entire sample correlation $\mathbf{R}_{L\times L}$ instead of $\tilde{\mathbf{R}}_{L\times L}$ may be simple to do and may not have appreciable impact in performance. The real hyperspectral image experiments conducted in this chapter show that there is little visual difference between using $\mathbf{R}_{L\times L}$ and using $\tilde{\mathbf{R}}_{L\times L}$.

3.5.2. Relationship Between CEM and RX Filter

We have seen in the previous section that $\delta^{OSP}(\mathbf{r})$ and $\delta^{CEM}(\mathbf{r})$ performed some sort of a matched filter with different values of the constant κ appearing in front of their matched filters. Following the same IPMF interpretation in Figure 3.1, the R-RXF defined by Eq. (3.23), $\delta^{R\text{-}RXF}(\mathbf{r})$, can be also expressed in terms of a matched form. Unlike $\delta^{CEM}(\mathbf{r})$, which requires prior knowledge of the desired target $\mathbf{d}$ to be used as the matched signal, anomaly detection must be implemented without prior target knowledge at all. Under such circumstance, it is intuitive to choose the currently being processed image pixel $\mathbf{r}$ as its matched signal. Also discussed above, since anomaly detection is primarily for target detection and not abundance estimation, the constant κ is set to $\kappa = 1$. So, if we replace $\mathbf{d}$ with $\mathbf{r}$ as the matched signal in $\delta^{\text{CEM}}(\mathbf{r})$ while discarding $(\mathbf{d}^T P_{\mathbf{U}}^{\perp}\mathbf{d})^{-1}$ by setting $\kappa = 1$, $\delta^{\text{CEM}}(\mathbf{r})$ becomes the R-RXF, $\delta^{\text{R-RXF}}(\mathbf{r})$, specified by Eq. (3.23).

Using such a matched filter approach, two variants of the RX filter—referred to as normalized RXF (NRXF), denoted by $\delta^{\text{R-NRXF}}(\mathbf{r})$, and modified RXF (MRXF), denoted by $\delta^{\text{R-MRXF}}(\mathbf{r})$—can also be derived from Eq. (3.23) in references 10 and 13 for anomaly detection as follows.

$$\begin{aligned}\delta^{\text{R-NRXF}}(\mathbf{r}) &= (\mathbf{r}/\parallel \mathbf{r} \parallel)^T \mathbf{R}_{L\times L}^{-1}(\mathbf{r}/\parallel \mathbf{r} \parallel)\\ &= (1/\parallel \mathbf{r} \parallel^2)\mathbf{r}^T\mathbf{R}_{L\times L}^{-1}\mathbf{r} = (\mathbf{r}/\parallel \mathbf{r} \parallel^2)^T\mathbf{R}_{L\times L}^{-1}\mathbf{r} \qquad (3.25)\end{aligned}$$

$$\delta^{\text{R-MRXF}}(\mathbf{r}) = \parallel \mathbf{r} \parallel^{-1} \mathbf{r}^T\mathbf{R}_{L\times L}^{-1}\mathbf{r} = (\mathbf{r}/\parallel \mathbf{r} \parallel)^T\mathbf{R}_{L\times L}^{-1}\mathbf{r} \qquad (3.26)$$

where $\parallel \mathbf{r} \parallel = \sqrt{\mathbf{r}^T\mathbf{r}}$ is the norm (vector length) of $\mathbf{r}$.

The $\delta^{\text{R-NRXF}}(\mathbf{r})$ specified by Eq. (3.25) can be interpreted in three different ways. One is viewed as the normalized version of the R-RXF. Another interpretation for $\delta^{\text{R-NRXF}}(\mathbf{r})$ is that it can be regarded as a matched filter with the matched signal $\mathbf{d} = \mathbf{r}$ as used in the R-RXF but using a different scale constant $\kappa = \parallel \mathbf{r} \parallel^{-2}$. Or equivalently, $\delta^{\text{R-NRXF}}(\mathbf{r})$ is a matched filter with the matched signal given by $\mathbf{d} = \mathbf{r}/\parallel \mathbf{r} \parallel^2$ with $\kappa = 1$. Similarly, the $\delta^{\text{R-MRXF}}(\mathbf{r})$ specified by Eq. (3.26) can be interpreted as an R-RXF with the constant $\kappa = \parallel \mathbf{r} \parallel^{-1}$ or a matched filter with the matched signal $\mathbf{d} = \mathbf{r}/\parallel \mathbf{r} \parallel$ and $\kappa = 1$.

Similarly, the LPD, $\delta^{\text{LPD}}(\mathbf{r})$, given by Eq. (3.24) can be also interpreted by a matched filter with the matched signal specified by an $(L \times 1)$-dimensional unity vector $\mathbf{1}$ with $\kappa = 1$. In analogy with $\delta^{\text{R-RXF}}(\mathbf{r})$, the constant κ is set to 1 since $\delta^{\text{LPD}}(\mathbf{r})$ is only used for target detection.

In the light of interpretation of Figure 3.1, Figure 3.5 shows the block diagram of the R-RXF where the information-processed filter and the matched filter are specified by $\mathbf{R}_{L\times L}^{-1}$ and $\mathbf{M}_{\mathbf{r}}$, respectively, with the scale constant $\kappa = 1$, and Figure 3.6 shows the block diagram of LPD detector where the information-processed filter and the matched filter are specified by $\mathbf{R}_{L\times L}^{-1}$ and $\mathrm{M}_{\mathbf{1}}$, respectively, with the scale constant $\kappa = 1$. However, it should be noted that the matched signals used in the block of the matched filter of R-RXF and LPD are different where $\delta^{\text{R-RXF}}(\mathbf{r})$ uses $\mathbf{r}$ as its matched signal while $\delta^{\text{LPD}}(\mathbf{r})$ using $\mathbf{1}$ as its matched signal. Because they are anomaly detectors and not abundance estimators, there is no need to use the constant κ to scale the output of the matched filter.

3.5.3. Relationship Between OSP and RX Filter

The *a priori* OSP, $\delta^{\text{OSP}}(\mathbf{r})$, and the Reed–Yu K-RXF, $\delta^{\text{K-RXF}}(\mathbf{r})$, can be related through the GMLC, $\delta^{\text{GMLC}}(\mathbf{r})$, specified by Eq. (3.8). If the noise in Eq. (3.1) is assumed to be a Gaussian process with mean $\boldsymbol{\mu}$ and the covariance matrix $\mathbf{K}_{L\times L}$, the maximum likelihood classifier for the abundance vector $\boldsymbol{\alpha}$ becomes the Gaussian classifier, which has the same form of $\delta^{\text{K-RXF}}(\mathbf{r})$ given in Eq. (3.22), that is, Mahalanobis distance $(\mathbf{r}-\boldsymbol{\mu})^T\mathbf{K}_{L\times L}^{-1}(\mathbf{r}-\boldsymbol{\mu})$. If the noise is further assumed to be a zero-mean Gaussian process, then $\delta^{\text{K-RXF}}(\mathbf{r})$ is reduced to $\delta^{\text{R-RXF}}(\mathbf{r})$. Moreover, if we make an additional assumption that the noise is white Gaussian, then $\mathbf{K}_{L\times L} = \sigma^2\mathbf{I}_{L\times L}$ where the $\mathbf{I}_{L\times L}$ is the $L \times L$ identity matrix and σ^2 is the noise variance. It was shown in references 8 and 10–12 that $\hat{\boldsymbol{\alpha}}$, the estimation of the abundance vector $\boldsymbol{\alpha}$, can be obtained by $\hat{\boldsymbol{\alpha}} = (\mathbf{M}^T\mathbf{M})^{-1}\mathbf{M}^T\mathbf{r}$. In particular, $\hat{\alpha}_p$, the estimate of the pth abundance fraction of $\hat{\boldsymbol{\alpha}}$ contained in the $\mathbf{r}$, is given by $\hat{\alpha}_p = (\mathbf{d}^T\mathbf{R}_{L\times L}^{-1}\mathbf{d})^{-1}\mathbf{d}^T P_{\mathbf{U}}^{\perp}$, which is identical to $\delta^{\text{OBC}}(\mathbf{r})$ given in Eq. (3.8). As a result, $\delta^{\text{K-RXF}}(\mathbf{r})$ is essentially equivalent to $\delta^{\text{OSP}}(\mathbf{r})$. By contrast, $\delta^{\text{OSP}}(\mathbf{r})$ utilizes the spectral information $P_{\mathbf{U}}^{\perp}$ provided by the image pixel $\mathbf{r}$ for target detection. Thus, in order for $\delta^{\text{K-RXF}}(\mathbf{r})$ to be able to compete against the $\delta^{\text{OSP}}(\mathbf{r})$, $\delta^{\text{K-RXF}}(\mathbf{r})$ must use *a posteriori* information provided by $\mathbf{K}_{L\times L}^{-1}$ to approximate the *a priori* information used in $\delta^{\text{OSP}}(\mathbf{r})$ that is provided by $P_{\mathbf{U}}^{\perp}$. This leads to the $\delta^{\text{K-RXF}}(\mathbf{r})$ given by Eq. (3.22). In addition, since no target information is available, the $\delta^{\text{K-RXF}}(\mathbf{r})$ uses the currently processed image pixel $\mathbf{r}$ as the matched signal to extract the target pixels that produce high peak values of $(\mathbf{r}-\boldsymbol{\mu})^T\mathbf{K}_{L\times L}^{-1}(\mathbf{r}-\boldsymbol{\mu})$. As noticed, $\delta^{\text{K-RXF}}(\mathbf{r})$ only accounts for second-order statistics. In many applications, the noise may not be stationary. In this case, $\delta^{\text{R-RXF}}(\mathbf{r})$ can be used to take care of the first-order statistics. As a result, $(\mathbf{r}-\boldsymbol{\mu})^T\mathbf{K}_{L\times L}^{-1}(\mathbf{r}-\boldsymbol{\mu})$ is replaced by $\mathbf{r}^T\mathbf{R}_{L\times L}^{-1}\mathbf{r}$. A detailed discussion between $\delta^{\text{K-RXF}}(\mathbf{r})$ and $\delta^{\text{R-RXF}}(\mathbf{r})$ can be found in references 8, 10 and 13. As an alternative application of $\delta^{\text{R-RXF}}(\mathbf{r})$, the low probability anomaly detector $\delta^{\text{LPD}}(\mathbf{r})$ replaces $\mathbf{r}$ in $\delta^{\text{R-RXF}}(\mathbf{r})$ with the unity vector $\mathbf{1}$ to extract target pixels that produce high peak values of $\mathbf{1}^T\mathbf{R}_{L\times L}^{-1}\mathbf{r}$. As demonstrated in the experiments, the target pixels extracted by $\delta^{\text{LPD}}(\mathbf{r})$ are most likely background pixels.

As a final concluding remark, many linear spectral unmixing methods developed in the literature can be interpreted through the IPMF described in Figure 3.1, particularly the results derived in Chang [8], where the $\delta^{\text{CEM}}(\mathbf{r})$ and the $\delta^{\text{R-RXF}}(\mathbf{r})$ have been shown to be variants of the *a priori* OSP, $\delta^{\text{OSP}}(\mathbf{r})$, using different matched signals.

3.6. EXPERIMENTS

This section presents an experiment-based analysis of effects of information used in target classification for hyperspectral imagery.

3.6.1. Computer Simulations

The data set to be used in the following simulations is Airborne Visible/Infrared Imaging Spectrometer (AVIRIS) reflectance data with spectral coverage from 0.4 μm to 2.5 μm which were considered in Harsanyi and Chang [1]. There are five spectra shown in Figure 3.7 among which only three spectra, creosote leaves, dry grass and red soil are used for experiments. There are 158 bands after water bands are removed.

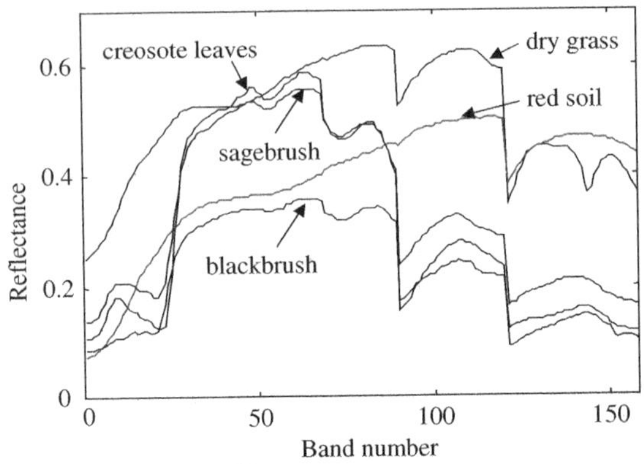

Figure 3.7. Five AVIRIS spectra.

We simulate 400 mixed pixel vectors, $\{\mathbf{r}_i\}_{i=1}^{400}$, as follows. We started the first pixel vector with 100% red soil and 0% dry grass, and then we began to increase 0.25% dry grass and decrease 0.25% red soil every pixel vector until the 400th pixel vector, which contained 100% dry grass. We then added creosote leaves to pixel vector numbers 198–202 at abundance fractions 10% while reducing the abundance of red soil and dry grass evenly. For example, after addition of creosote leaves, the resulting pixel vector 200 contained 10% creosote leaves, 45% red soil, and 45% dry grass. White Gaussian noise is also added to each pixel vector to achieve a 30:1 signal-to-noise ratio as was defined as 50% reflectance divided by the standard deviation of the noise in Harsanyi and Chang [1]. Figures 3.8a–c show the detection results of $\delta^{OSP}(\mathbf{r})$, $\delta^{SSC}(\mathbf{r})$, and $\delta^{OBC}(\mathbf{r})$, respectively, with complete target knowledge of creosote leaves, dry grass, and red soil.

As we can see, the three filters performed very well in terms of detection. However, if we examine their detected abundance fractions, the δ^{SSC} $(\mathbf{r})$ and δ^{OBC} $(\mathbf{r})$ produced abundance fractions close to the true amount, 0.1, while the $\delta^{OSP}(\mathbf{r})$ performed poorly in detection of about 0.25 fraction, which was 2.5 times the true

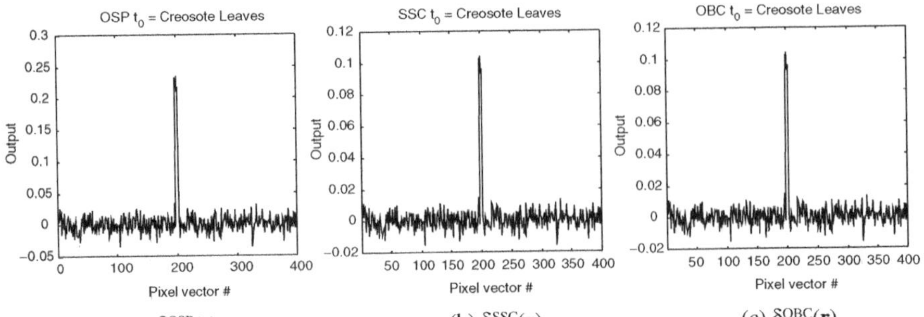

Figure 3.8. Detection results of $\delta^{OSP}(\mathbf{r})$, $\delta^{SSC}(\mathbf{r})$, and $\delta^{OBC}(\mathbf{r})$ with complete target knowledge of creosote leaves, dry grass, and red soil.

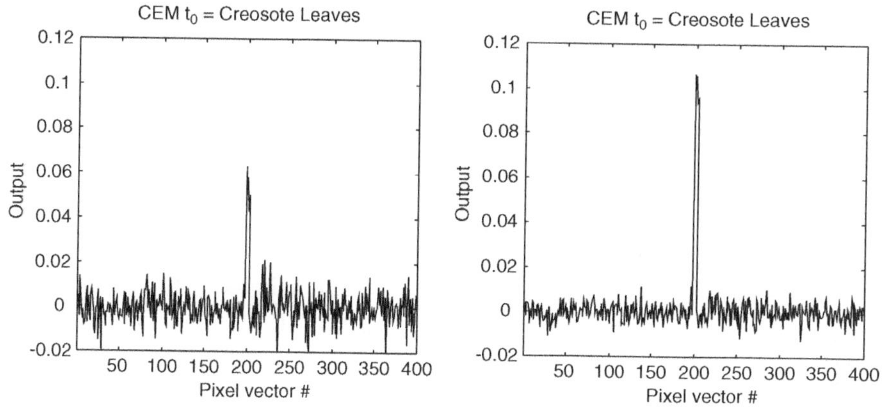

$\mathbf{R}_{L\times L}$ $\tilde{\mathbf{R}}_{L\times L}$

Figure 3.9. Results of $\delta^{\text{CEM}}(\mathbf{r})$ using $\mathbf{R}_{L\times L}$ and $\tilde{\mathbf{R}}_{L\times L}$, respectively, in detection of creosote leaves.

amount. This is because $\delta^{\text{OSP}}(\mathbf{r})$ did not use any *a posteriori* information $(\mathbf{d}^T P_{\mathbf{U}}^{\perp}\mathbf{d})^{-1}$ that was used in the $\delta^{\text{SSC}}(\mathbf{r})$ and $\delta^{\text{OBC}}(\mathbf{r})$. Now suppose that the creosote leaves comprise the only target knowledge available; that is, $\mathbf{d}$ = creosote leaves. Figures 3.9a and 3.9b show the results of $\delta^{\text{CEM}}(\mathbf{r})$ using $\mathbf{R}_{L\times L}$ and $\tilde{\mathbf{R}}_{L\times L}$, respectively, in detection of creosote leaves where the former detected 0.07 of abundance fraction compared to 0.1 of abundance fraction detected by the latter. Thus, the $\delta^{\text{CEM}}(\mathbf{r})$ using $\tilde{\mathbf{R}}_{L\times L}$ performed better than $\delta^{\text{CEM}}(\mathbf{r})$ using $\mathbf{R}_{L\times L}$ in terms of abundance fraction detection where $\tilde{\mathbf{R}}_{L\times L}$ did not include the desired target pixels from 198 to 202, $\{\mathbf{r}_i\}_{i=198}^{202}$.

Similar results shown in Figure 3.10 were also obtained by the $\delta^{\text{TCIMF}}(\mathbf{r})$ using $\mathbf{R}_{L\times L}$ and $\tilde{\mathbf{R}}_{L\times L}$ where $\delta^{\text{TCIMF}}(\mathbf{r})$ was implemented by letting $\mathbf{d}$ = creosote leaves

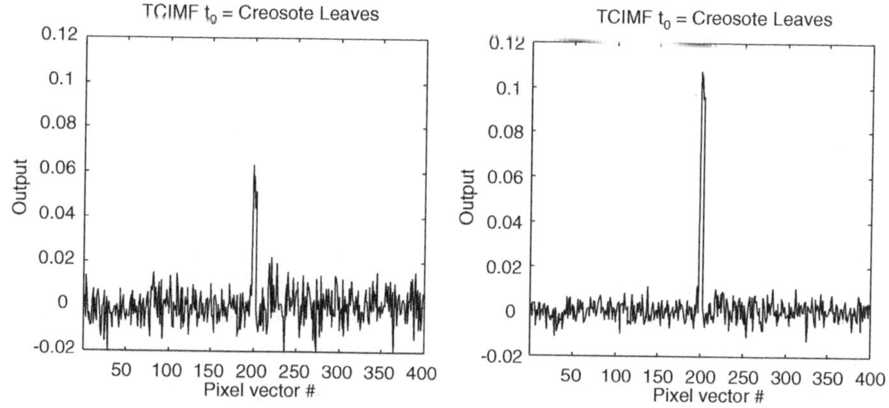

$\mathbf{R}_{L\times L}$ $\tilde{\mathbf{R}}_{L\times L}$

Figure 3.10. Results of $\delta^{\text{TCIMF}}(\mathbf{r})$ using $\mathbf{R}_{L\times L}$ and $\tilde{\mathbf{R}}_{L\times L}$, respectively, in detection of creosote leaves.

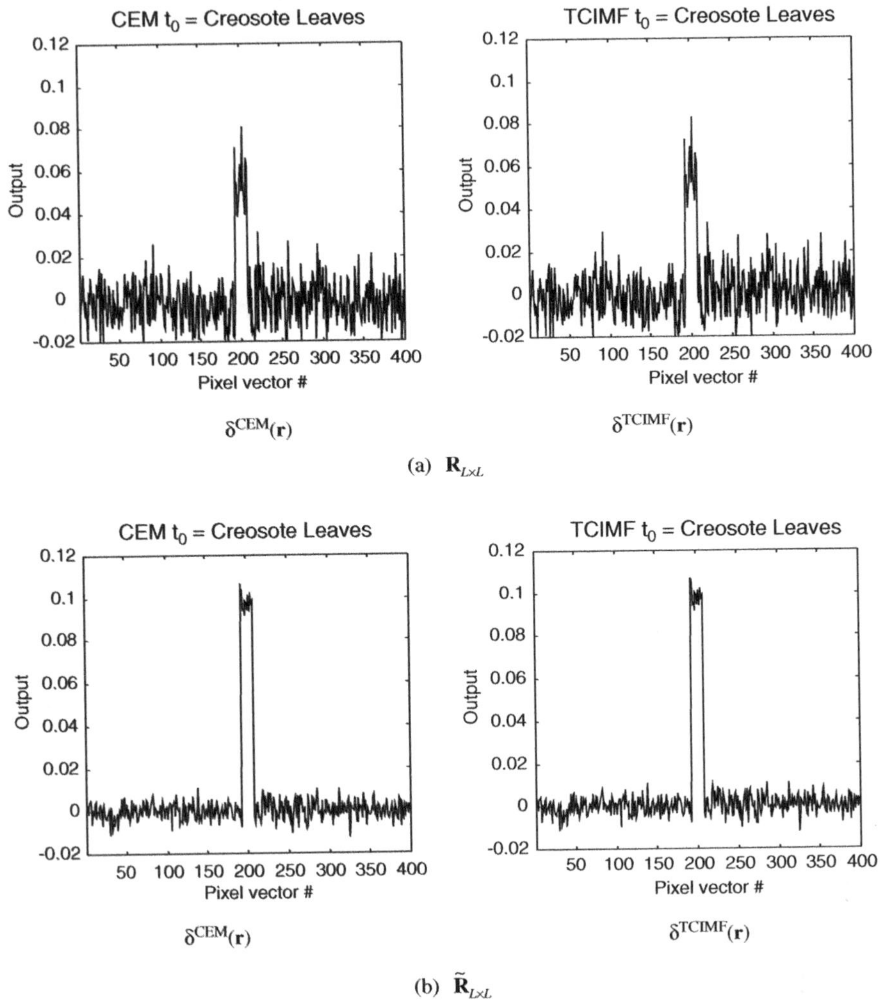

Figure 3.11. Detection results of 15 creosote leaves pixels, with $\{\mathbf{r}_i\}_{i=193}^{207}$ starting from pixel number 193 to pixel number 207. (a) $\delta^{\mathrm{CEM}}(\mathbf{r})$ and $\delta^{\mathrm{TCIMF}}(\mathbf{r})$ using $\mathbf{R}_{L\times L}$; (b) $\delta^{\mathrm{CEM}}(\mathbf{r})$ and δ^{TCIMF} $(\mathbf{r})$ using $\tilde{\mathbf{R}}_{L\times L}$.

and $\mathbf{U} =$ [grass, soil]. As we can see, $\delta^{\mathrm{CEM}}(\mathbf{r})$ and $\delta^{\mathrm{TCIMF}}(\mathbf{r})$ using $\tilde{\mathbf{R}}_{L\times L}$ performed better than their counterparts using $\mathbf{R}_{L\times L}$. The advantage of using $\tilde{\mathbf{R}}_{L\times L}$ becomes more evident as shown in Figures 3.11 and 3.12 where the number of target pixels was increased and expanded from 5 pixels to 15 pixels, with $\{\mathbf{r}_i\}_{i=193}^{207}$ starting from pixel number 193 to pixel number 207 in Figure 3.11, and increased and expanded to 25 pixels, with $\{\mathbf{r}_i\}_{i=188}^{212}$ starting from pixel number 188 to pixel number 212 in Figure 3.12.

A detailed study on the target knowledge sensitivity of CEM can be found in Chang and Heinz [14]. Finally, Figure 3.13 shows anomaly detection of 5 creosote

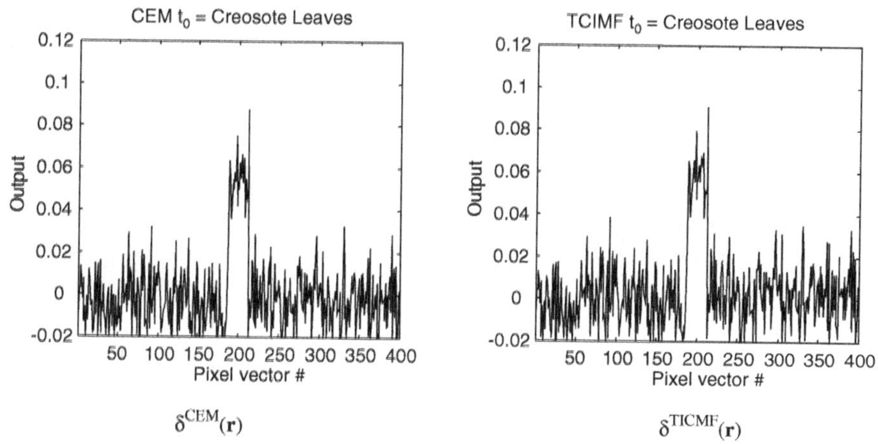

(a) $\mathbf{R}_{L\times L}$

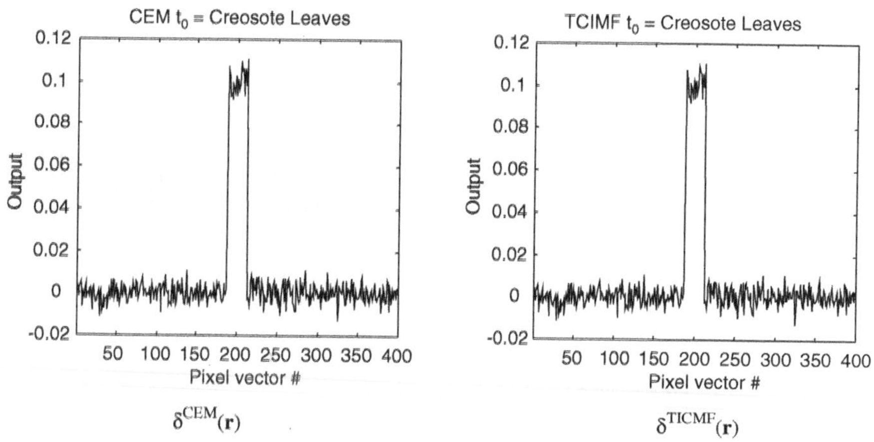

(b) $\tilde{\mathbf{R}}_{L\times L}$

Figure 3.12. Detection results of 25 creosote leaves pixels, with $\{\mathbf{r}_i\}_{i=188}^{212}$ starting from pixel number 188 to pixel number 212. (a) $\delta^{\text{CEM}}(\mathbf{r})$ and $\delta^{\text{TICMF}}(\mathbf{r})$ using $\mathbf{R}_{L\times L}$; (b) $\delta^{\text{CEM}}(\mathbf{r})$ and $\delta^{\text{TCIMF}}(\mathbf{r})$ using $\tilde{\mathbf{R}}_{L\times L}$.

leaves pixels by the $\delta^{\text{R-RXF}}(\mathbf{r})$, $\delta^{\text{R-NRXF}}(\mathbf{r})$, $\delta^{\text{R-MRXF}}(\mathbf{r})$, and $\delta^{\text{LPD}}(\mathbf{r})$, where $\delta^{\text{R-RXF}}(\mathbf{r})$ performed better than $\delta^{\text{R-NRXF}}(\mathbf{r})$ and $\delta^{\text{R-MRXF}}(\mathbf{r})$ in detection of creosote leaves, while $\delta^{\text{LPD}}(\mathbf{r})$ basically extracted most background pixels.

3.6.2. AVIRIS Data

An AVIRIS image that was studied in Harsanyi and Chang [1] and used for the following experiments is shown in Figure 3.9. It is a scene of size 200 $\times$ 200 and is a part of the Lunar Crater Volcanic Field (LCVF) in Northern Nye County, Nevada, where five signatures of interest in these images were demonstrated in Harsanyi and

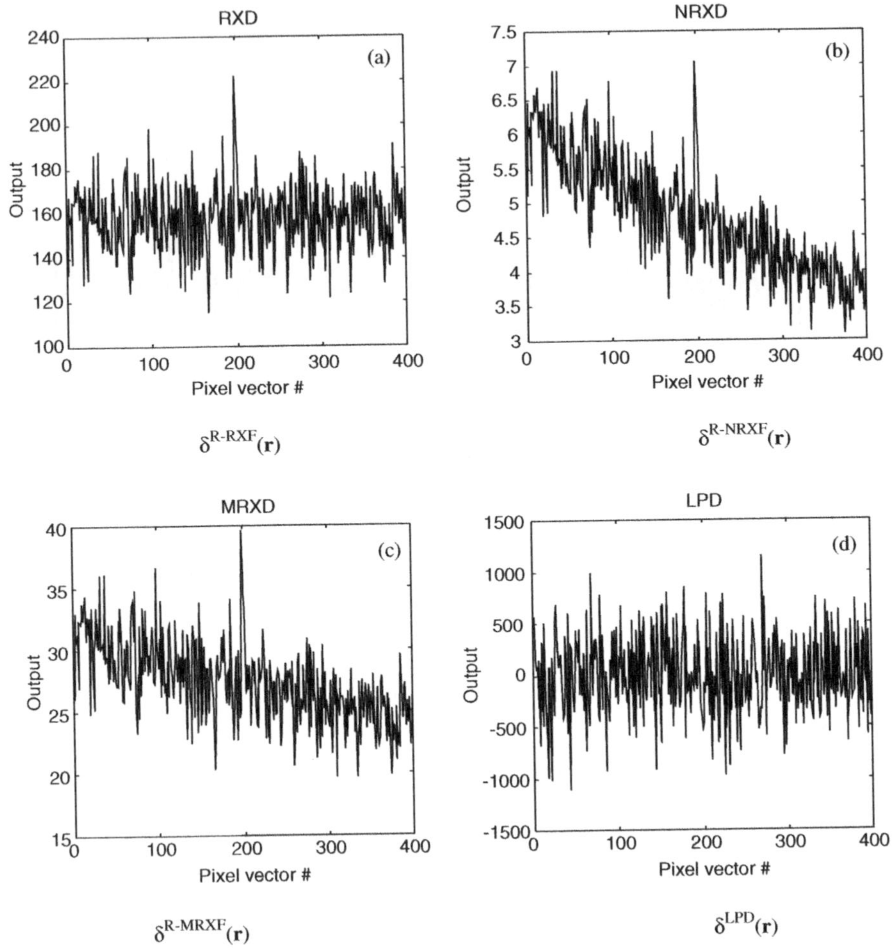

Figure 3.13. Detection results of (a) $\delta^{\text{R-RXF}}(\mathbf{r})$; (b) $\delta^{\text{R-NRXF}}(\mathbf{r})$; (c) $\delta^{\text{R-MRXF}}(\mathbf{r})$; (d) $\delta^{\text{LPD}}(\mathbf{r})$.

Chang [1]: "cinders," "rhyolite," "playa (dry lakebed)," "shade," and "vegetation." Additionally, it was also shown in Chang et al. [13] that there was a single two-pixel anomaly located at the top edge of the lake marked by a dark circle in Figure 3.14.

Since the gray scale images produced by the $\delta^{\text{OSP}}(\mathbf{r})$, $\delta^{\text{SSC}}(\mathbf{r})$, and $\delta^{\text{OBC}}(\mathbf{r})$ have been scaled to 256 gray level values for monitor display, the effect of the constant κ has been offset. As a result, there is no visible difference among their detection results as shown in Chang and Ren [15]. However, $\delta^{\text{CEM}}(\mathbf{r})$ is very sensitive to the target information $\mathbf{d}$ used in the filter. Figures 3.10a and 3.10b show the detection results of $\delta^{\text{CEM}}(\mathbf{r})$ with the desired target signature $\mathbf{d}$ specified by a single pixel and by averaging all the pixels in their specific areas according to the ground truth, respectively.

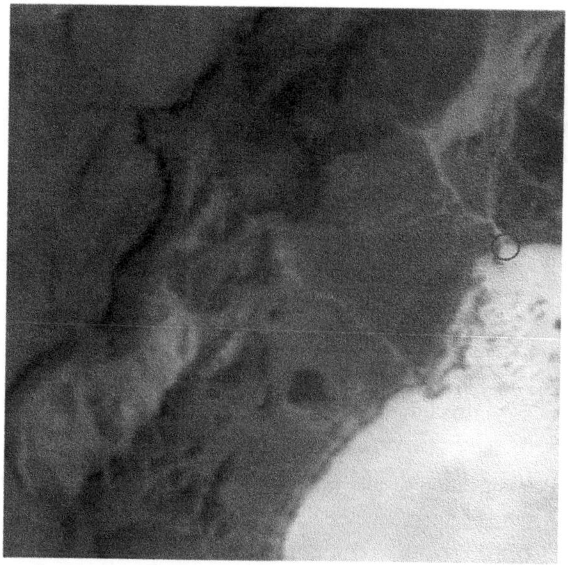

Figure 3.14. An AVIRIS LCVF image scene.

If a single pixel is used for **d**, $\delta^{CEM}(\mathbf{r})$ would only detect that particular pixel as shown in Figure 3.15a, where the brightest pixels in the images were those used as **d**. However, if **d** is obtained from averaging all the pixels in their areas, the detection results are shown in Figure 3.15b. In order to make comparison, Figures 3.16a and 3.16b show the detection results of $\delta^{OSP}(\mathbf{r})$ where the target signatures were

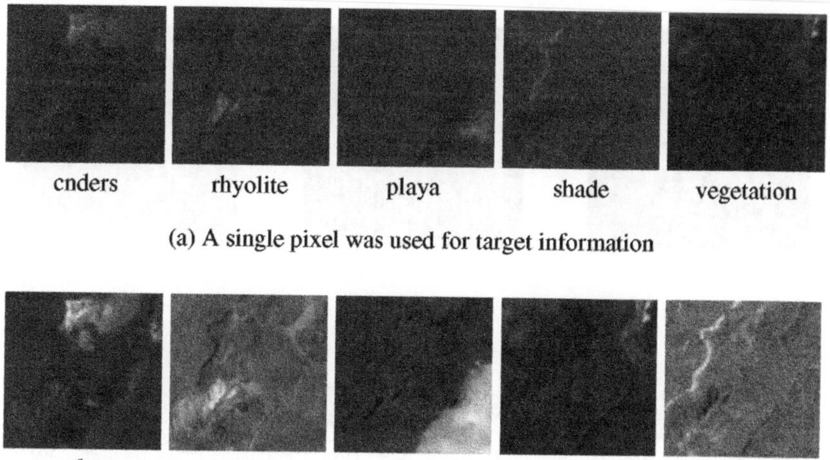

(b) Averaged area pixels used for target information

Figure 3.15. CEM Detection of (a) a single pixel; (b) averaged area pixels.

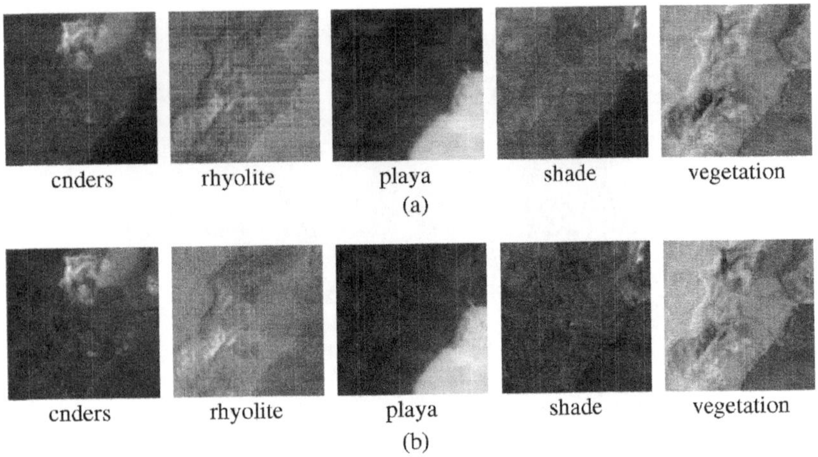

Figure 3.16. OSPclassification results (a) using a single pixel; (b) using the averaged pixel.

obtained by a single pixel and by averaging all the pixels in their specific areas according to the ground truth, respectively.

Apparently there was no visible difference between Figures 3.16a and 3.16b. If we compare Figure 3.15b to Figure 3.16b, $\delta^{\mathrm{CEM}}(\mathbf{r})$ performed as well as did $\delta^{\mathrm{OSP}}(\mathbf{r})$. The significant difference between Figures 3.15 and 3.16 demonstrates how much impact the target information **d** could have on the detection performance of $\delta^{\mathrm{CEM}}(\mathbf{r})$ shown in Figures 3.15a and 3.15b and very little impact on $\delta^{\mathrm{OSP}}(\mathbf{r})$ shown in Figures 3.16a and 3.16b, which implies that $\delta^{\mathrm{OSP}}(\mathbf{r})$ is less sensitive to

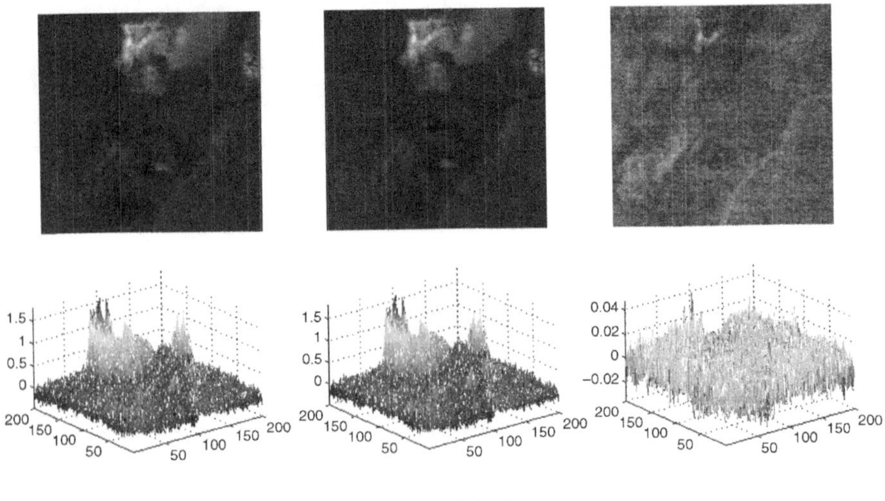

Figure 3.17. Detection results of $\delta^{\mathrm{CEM}}(\mathbf{r})$using the average of a set of pixels as **d** and $\tilde{\mathbf{R}}_{L\times L}$.

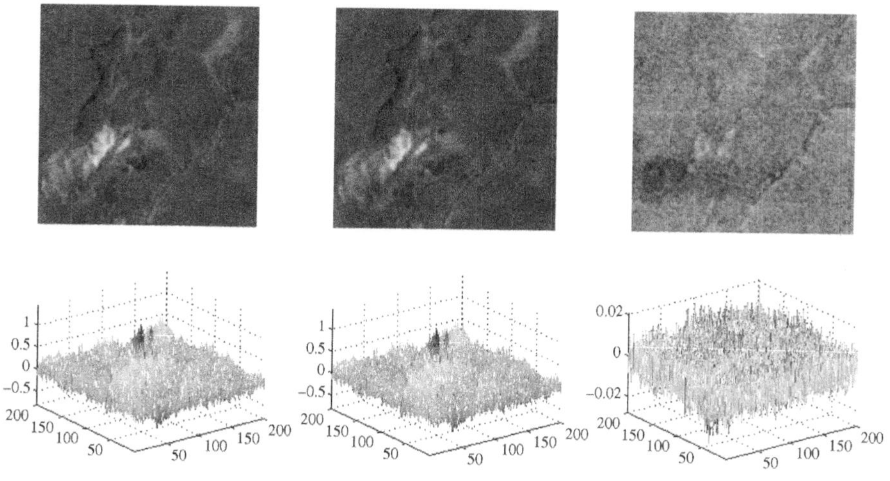

detection of rhyolite

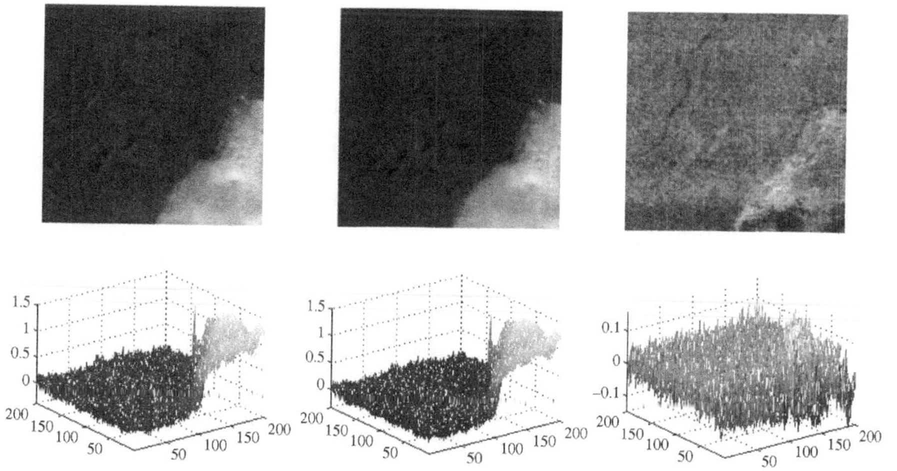

detection of playa

Figure 3.17. (*Continued*).

target knowledge and robust to different levels of target information used in the target signature matrix $\mathbf{M}$.

In order to see how $\tilde{\mathbf{R}}_{L\times L}$ affects the performance of $\delta^{\mathrm{CEM}}(\mathbf{r})$, we repeated the same experiments done for Figure 3.15b with $\mathbf{R}_{L\times L}$ replaced with $\tilde{\mathbf{R}}_{L\times L}$, where all the target pixels used to form $\mathbf{d}$ were removed from the $\mathbf{R}_{L\times L}$. The detection results are shown in Figure 3.17, where the results in the first column labeled by (a) were obtained by $\mathbf{R}_{L\times L}$, and the results in the second column labeled by (b) were obtained

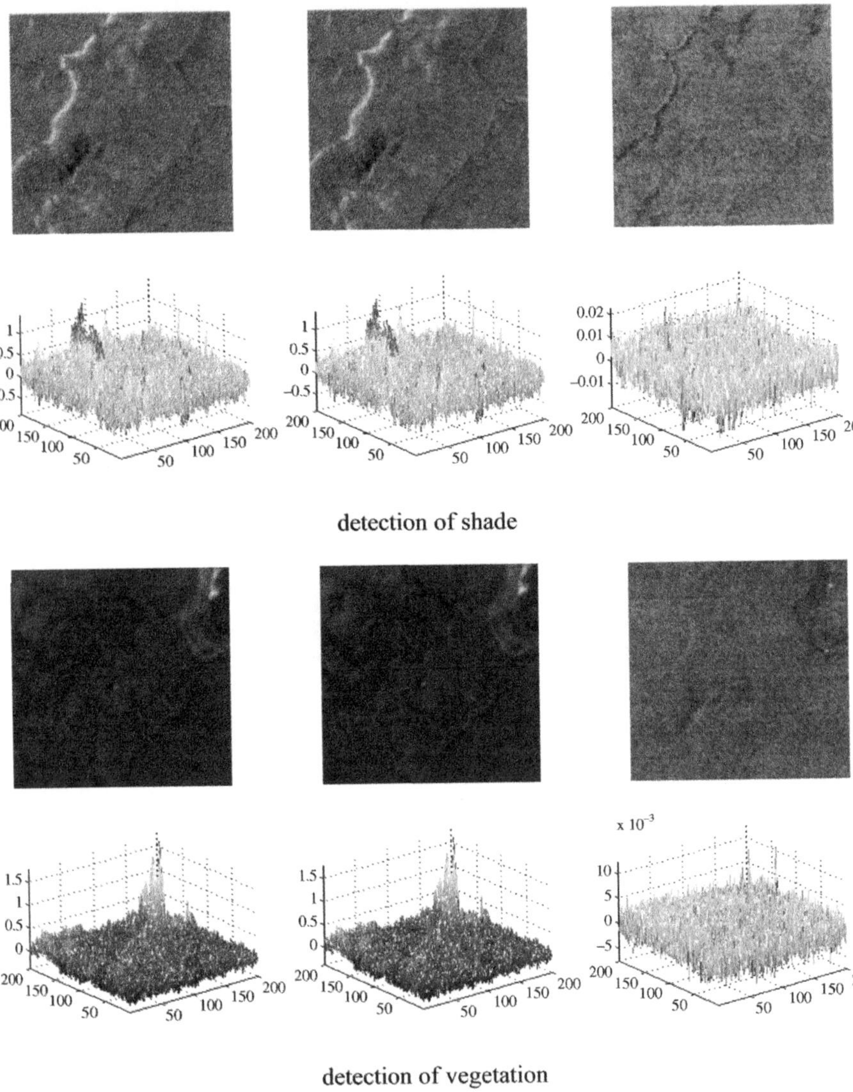

(a) $\mathbf{R}_{L\times L}$ (b) $\tilde{\mathbf{R}}_{L\times L}$ (c) images (b)-(a)

Figure 3.17. (*Continued*).

by $\tilde{\mathbf{R}}_{L\times L}$, while the results in the third column labeled by (c) were obtained by subtracting the results in column (a) from the results in column (b).

Despite the fact that there was subtle difference between them as shown in column (c), there was no visible difference between the results obtained by $\mathbf{R}_{L\times L}$ and by $\tilde{\mathbf{R}}_{L\times L}$ due to limitation of monitor display capability. It should be noted that we did not repeat the experiments for Figure 3.15a using $\tilde{\mathbf{R}}_{L\times L}$ because the removal of

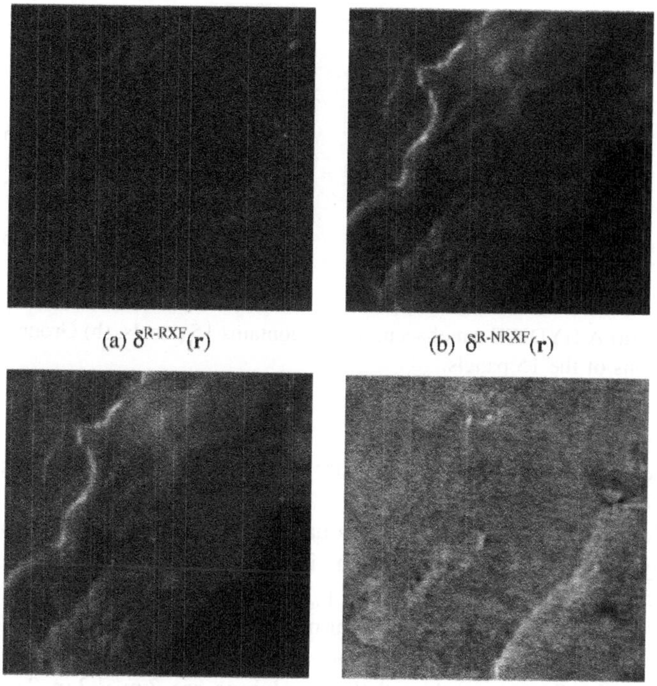

(a) $\delta^{\text{R-RXF}}(\mathbf{r})$ (b) $\delta^{\text{R-NRXF}}(\mathbf{r})$

(c) $\delta^{\text{R-MRXF}}(\mathbf{r})$ (d) $\delta^{\text{LPD}}(\mathbf{r})$

Figure 3.18. Anomaly detection of Figure 3.15 produced by (a) $\delta^{\text{R-RXF}}(\mathbf{r})$, (b) $\delta^{\text{R-NRXF}}(\mathbf{r})$, (c) $\delta^{\text{R-MRXF}}(\mathbf{r})$, and (d) $\delta^{\text{LPD}}(\mathbf{r})$.

a single target pixel from $\mathbf{R}_{L\times L}$ did not change detection results compared to the large number of image pixels. However, if the number of desired target pixels is comparable to that of image size, it will make a significant difference between using $\mathbf{R}_{L\times L}$ and using $\tilde{\mathbf{R}}_{L\times L}$. Similar phenomena were also applied to $\delta^{\text{TCIMF}}(\mathbf{r})$, and their results are not included here. Finally, Figures 3.18a–d show the detection results of $\delta^{\text{R-RXF}}(\mathbf{r})$, $\delta^{\text{R-NRXF}}(\mathbf{r})$, $\delta^{\text{R-MRXF}}(\mathbf{r})$, and $\delta^{\text{LPD}}(\mathbf{r})$. As we can see, the detection results were very different.

Unlike Figure 3.18a, where $\delta^{\text{R-RXF}}(\mathbf{r})$ detected the anomaly, both NRXF $\delta^{\text{R-NRXF}}(\mathbf{r})$ and MRXF, $\delta^{\text{R-MRXF}}(\mathbf{r})$ detected the shade in Figures 3.18b and 3.18c, and $\delta^{\text{LPD}}(\mathbf{r})$ detected mainly image background. In addition to the shade, the MRXF also detected the anomaly.

3.6.3. HYDICE Data

The data to be used in the following experiments were HYDICE data after low-signal/high-noise bands (bands 1–3 and bands 202–210); and water vapor absorption bands (bands 101–112 and bands 137–153 have been removed). The HYDICE image scene to be studied is shown in Figure 3.19a with size of 64×64. There

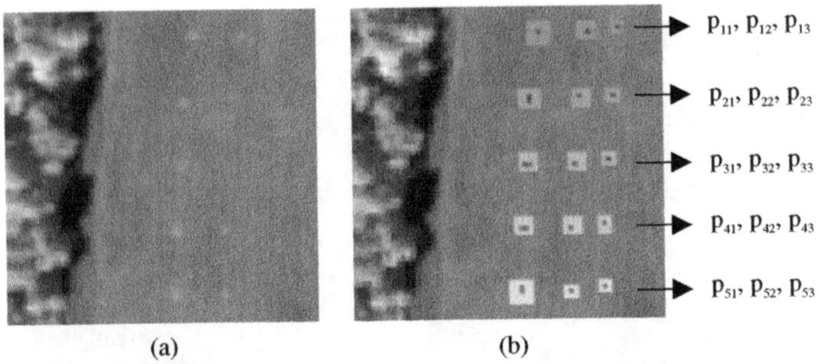

Figure 3.19. (a) A HYDICE panel scene which contains 15 panels. (b) Ground truth map of spatial locations of the 15 panels.

are 15 panels located on the field and are arranged in a 5×3 matrix as shown in Figure 3.19b.

Each element in this matrix is a square panel and denoted by p_{ij} with row indexed by $i = 1, \ldots, 5$ and column indexed by $j = a, b, c$. For each row $i = 1, \ldots, 5$, the three panels p_{ia}, p_{ib}, p_{ic} were made by the same material but have three different sizes. For each column $j = a, b, c$, the five panels p_{1j}, p_{2j}, p_{3j}, p_{4j}, p_{5j} have the same size but were made by five different materials. The sizes of the panels in the first, second, and third columns are $3\,\text{m} \times 3\,\text{m}, 2\,\text{m} \times 2\,\text{m}$, and $1\,\text{m} \times 1\,\text{m}$, respectively. Thus, the 15 panels have five different materials and three different sizes. The ground truth of the image scene provides the precise spatial locations of these 15 panels. Red (R) pixels in Figure 3.19b are the center pixels of all the 15 panels. The 1.56-m spatial resolution of the image scene suggests that except for p_{2a}, p_{3a}, p_{4a}, p_{5a}, which are two-pixel panels, all the remaining panels are only one pixel wide. From Figure 3.19a, the panels in the third column, p_{1c}, p_{2c}, p_{3c}, p_{4c}, p_{5c}, are almost invisible and the first three panels in the second column, p_{1b}, p_{2b}, p_{3b}, are barely visible. Apparently, without ground truth there is no way to locate these panels in the scene. As mentioned in the AVIRIS data experiments, the effect of the constant κ has been offset by 256 gray level values for visual display. There is no visible difference among the detection results produced by $\delta^{\text{OSP}}(\mathbf{r})$, $\delta^{\text{SSC}}(\mathbf{r})$, and $\delta^{\text{OBC}}(\mathbf{r})$,. Experiments similar to those done for AVIRIS data were conducted for the 15-panel image scene. Figures 3.20a and 3.20b show the detection results of $\delta^{\text{CEM}}(\mathbf{r})$ using the average of B pixels as $\mathbf{d}$ and the average of black plus white pixels as $\mathbf{d}$, respectively, where the detection results of $\delta^{\text{OSP}}(\mathbf{r})$ are shown in Figure 3.20c for comparison with the target signatures obtained by averaging all B pixels.

Since the difference between $\delta^{\text{CEM}}(\mathbf{r})$ and $\delta^{\text{TCIMF}}(\mathbf{r})$ was already demonstrated in Ren and Chang [6], no experiments are included here. Figures 3.21a–d show the detection results of $\delta^{\text{R-RXF}}(\mathbf{r})$, $\delta^{\text{R-NRXF}}(\mathbf{r})$, $\delta^{\text{R-MRXF}}(\mathbf{r})$, and $\delta^{\text{LPD}}(\mathbf{r})$.

Comparing Figure 3.21a to Figure 3.21c, both $\delta^{\text{R-RXF}}(\mathbf{r})$ and $\delta^{\text{R-MRXF}}(\mathbf{r})$ performed very closely in terms of panel detection except that $\delta^{\text{R-MRXF}}(\mathbf{r})$ extracted more background signatures than did $\delta^{\text{R-RXF}}(\mathbf{r})$. Figure 3.21b shows that the

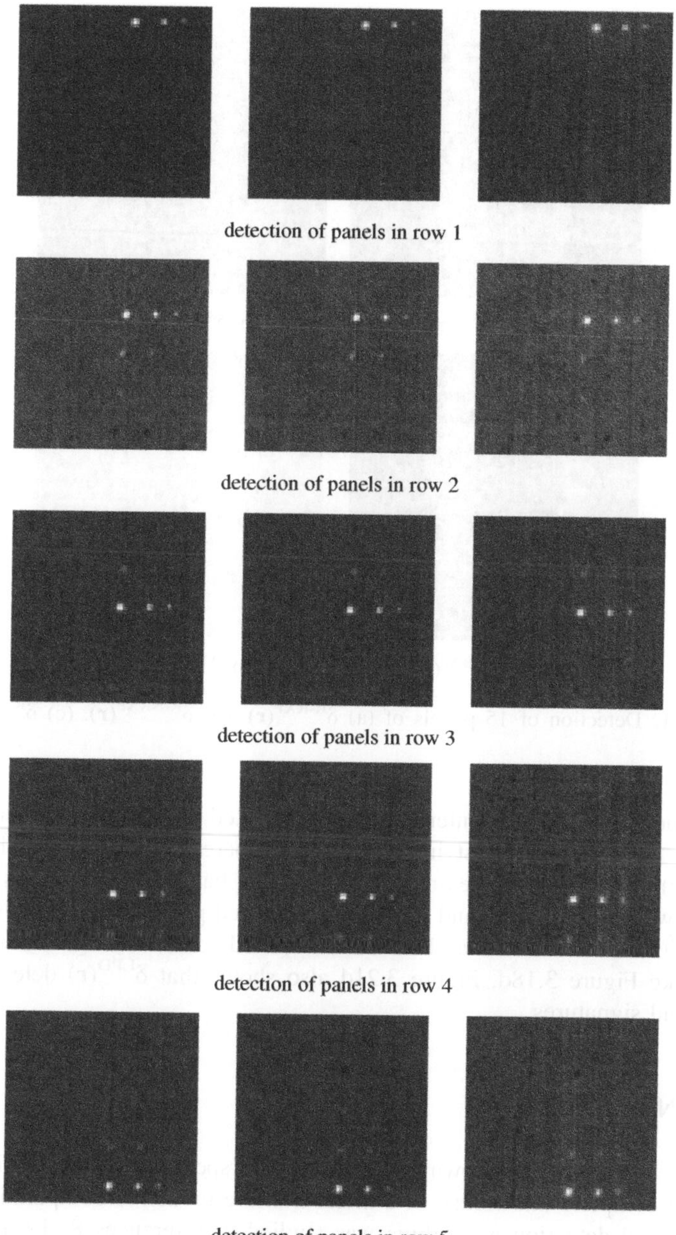

Figure 3.20. Detection results of $\delta^{CEM}(\mathbf{r})$ (a) using the average of black pixels in all 15 panels as **d**, (b) using the average of black pixels in only 3×3 and 2×2 panels as **d**, and (c) using the average of all black plus white pixels as **d**.

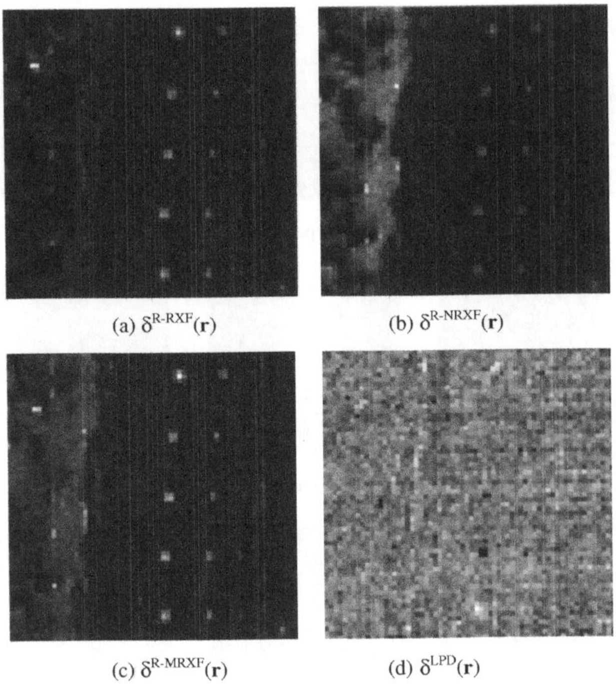

Figure 3.21. Detection of 15 panels of (a) $\delta^{R\text{-}RXF}(\mathbf{r})$, (b) $\delta^{R\text{-}NRXF}(\mathbf{r})$, (c) $\delta^{R\text{-}MRXF}(\mathbf{r})$ and (d) $\delta^{LPD}(\mathbf{r})$.

background signatures and interferers that extracted $\delta^{R\text{-}NRXF}(\mathbf{r})$ were more dominant than panel pixels. Most interestingly, compared to Figures 3.18b–c, where $\delta^{R\text{-}NRXF}(\mathbf{r})$ and $\delta^{R\text{-}MRXF}(\mathbf{r})$ extracted only image background, Figures 3.21b and 3.21c show that $\delta^{R\text{-}NRXF}(\mathbf{r})$ and $\delta^{R\text{-}MRXF}(\mathbf{r})$ detected panels that were also detected by $\delta^{R\text{-}RXF}(\mathbf{r})$. In addition, they both also extracted some tree signatures and interferers. Like Figure 3.18d, Figure 3.21d also shows that $\delta^{LPD}(\mathbf{r})$ detected mainly background signatures.

3.7. CONCLUSION

The OSP approach has shown success in hyperspectral image classification. It requires a complete knowledge of a linear mixture model. A comparative analysis of OSP-based detection algorithms was studied in references 8, 10 and 15. The LCMV is a target-constrained technique that only requires partial knowledge of targets of interest. As a special case of LCMV, CEM has been also shown to be very effective in subpixel detection. The issue of sensitivity to level of target information for CEM was also investigated in Chang [10] and Chang and Heinz [14]. Anomaly detection requires no prior knowledge and detects targets whose signatures are spectrally distinct from their neighborhood. A detailed analysis on anomaly detection was reported in Chang et al. [13]. It seems that there is no strong tie among these three techniques. This chapter presents an information-processed matched filter interpretation to explore the relationship among OSP, LCMV, and anomaly detection techniques. An alternative interpretation using the concept of the OSP was also presented in Chang [8]. It demonstrates that they all perform some sort of a matched filter with the matched signal determined by the level of target information used in the filter. Additionally, it also shows that when the prior target information is not available, this *a priori* information can be approximated by *a posteriori* information obtained directly from the image data. A series of experiments are conducted to illustrate the effects of different levels of information used in target detection.

REFERENCES

1. J. C. Harsanyi and C.-I Chang, Hyperspectral image classification and dimensionality reduction: an orthogonal subspace projection approach, *IEEE Transactions on Geoscience and Remote Sensing*, vol. 32, no. 4, pp. 779–785, 1994.
2. T. M. Tu, C.-H. Chen, and C.-I Chang, *A posteriori* least squares orthogonal subspace projection approach to weak signature extraction and detection, *IEEE Transactions on Geoscience and Remote Sensing*, vol. 35, no. 1, pp. 127–139, 1997.
3. C.-I Chang, X. Zhao, M. L. G. Althouse, and J.-J. Pan, Least squares subspace projection approach to mixed pixel classification in hyperspectral images, *IEEE Transactions on Geoscience and Remote Sensing*, vol. 36, no. 3, pp. 898–912, 1998.
4. C.-I Chang, Target signature-constrained mixed pixel classification for hyperspectral imagery, *IEEE Trans. on Geoscience and Remote Sensing*, vol. 40, no. 5, pp. 1065–1081, 2002.
5. J. C. Harsanyi, *Detection and Classification of Subpixel Spectral Signatures in Hyperspectral Image Sequences*, Department of Electrical Engineering, University of Maryland Baltimore County, Baltimore, MD, August 1993.
6. H. Ren and C.-I Chang, Target-constrained interference-minimized approach to subpixel target detection for hyperspectral imagery, *Optical Engineering*, vol. 39, no. 12, pp. 3138–3145, 2000.
7. I.S. Reed and X. Yu, Adaptive multiple-band CFAR detection of an optical pattern with unknown spectral distribution, *IEEE Transactions on Acoustic, Speech and Signal Process.*, vol. 38, no. 10, pp. 1760–1770, 1990.
8. C.-I Chang, Orthogonal subspace projection revisited: A comprehensive study and analysis, *IEEE Transactions on Geoscience and Remote Sensing*, vol. 43, no. 3, pp. 502–518, 2005.
9. H. V. Poor, *An Introduction to Detection and Estimation Theory*, 2nd edition, Springer-Verlag, New York, 1994.
10. C.-I Chang, *Hyperspectral Imaging: Techniques for Spectral Detection and Classification*, Kluwer/Plenum Academic Publishers, New York, 2003.
11. J.J. Settle, On the relationship between spectral unmixing and subspace projection, *IEEE Transactions on Geoscience and Remote Sensing*, vol. 34, no. 4, pp. 1045–1046, 1996.

12. C.-I Chang, Further results on relationship between spectral unmixing and subspace projection, *IEEE Transactions on Geoscience and Remote Sensing*, vol. 36, no. 3, pp. 1030–1032, 1998.
13. C.-I Chang, S.-S. Chiang, and I. W. Ginsberg, Anomaly detection in hyperspectral imagery, *SPIE Conference on Geo-Spatial Image and Data Exploitation II*, Orlando, Florida, 20–24 April, 2001.
14. C.-I Chang and D. Heinz, Subpixel spectral detection for remotely sensed images, *IEEE Transactions on Geoscience and Remote Sensing*, vol. 38, no. 3, pp. 1144–1159, 2000.
15. C.-I Chang and H. Ren, An experiment-based quantitative and comparative analysis of hyperspectral target detection and image classification algorithms, *IEEE Transactions on Geoscience and Remote Sensing*, vol. 38, no. 2, pp. 1044–1063, 2000.

PART II

THEORY

CHAPTER 4

AN OPTICAL REAL-TIME ADAPTIVE SPECTRAL IDENTIFICATION SYSTEM (ORASIS)

JEFFREY H. BOWLES AND DAVID B. GILLIS
Remote Sensing Division, Naval Research Laboratory, Washington, DC 20375

4.1. INTRODUCTION

A look at the number of publications for the last 7 years involving hyperspectral imagery (HSI) collection and exploitation shows that the field is growing rapidly. (A rough count of the number of published articles with the word hyperspectral in the title or abstract gives the following: in 1997, 24 articles; in 1998, 49 articles; in 1999, 56 articles; in 2000, 62 articles; in 2001, 110 articles; in 2002, 113 articles; in 2003, 181 articles; in 2004, 215 articles; and in 2005 264 articles.) Today, HSI is used for mineral exploration [1], crop assessment [2], determination of vegetative conditions[3], ice and snow measurements [4, 5], land management, characterization of coastal ocean areas, [6, 7], and other environmental efforts[8, 9]. Other uses include finding small targets in cluttered backgrounds [10, 11], medical uses [12], industrial inspection [13], and many others. In short, in situations where spectral information (in the broadest sense) provides information on the physical state of a scene, many hyperspectral imagery exploitation methods portend the ability to automatically make actionable determinations on the state of the scene. Truly, the automation made possible by the additional information content of HSI is a major advantage over more conventional optical imaging techniques.

The use of hyperspectral data is a logical and obvious extension to multispectral data, which has been used extensively since the launch of the first LandSat system in 1972. In simple scenes, such as the open ocean, algorithms used for analysis of multispectral data may be used to find endmembers, or explicit scene constituents, because the number of constituents is still less than the number of spectral bands measured. That is, it is possible to find endmembers, or use models, that can be used to determine the amount of a particular material in an unambiguous manner. This

Hyperspectral Data Exploitation: Theory and Applications, Edited by Chein-I Chang

makes multispectral systems sufficient to attack problems where only limited information is present. However, as the complexity of the scene increases (such as the coastal ocean and urban land scenes), the number of scene constituents increases beyond the number of bands being measured in a multispectral system. In these scenes, it is no longer possible to find endmembers that explicitly determine the presence of a particular constituent, but instead it is necessary to use additional techniques, such as clustering, to determine the different class populations, which portends ambiguity in the results.

Over the last decade, the data produced by modern sensors have been increasing in quality, and therefore, information content, as ground sample distance (GSD) has been shrinking and signal-to-noise ratios (SNR) have been rising. Currently, it is possible for hyperspectral data to have SNR of well over 100:1 in the visible/near-infrared (VNIR) portion of the spectrum (outside the strong absorption features) at GSD sizes of 1–30 m and with spectral bandwidths of 5–10 nm. The exact SNR varies as a function of albedo, solar angles, atmospheric conditions, and measurement parameters (here, noise is considered to be the total of shot noise and "dark" sensor noise). As the quality of the data increases, the information content increases because of the ability of the systems to distinguish increasingly similar materials. (For example, a multispectral system can clearly determine the presence of vegetation, while an HSI can identify a particular species of plants.) The increase in information content portends the ability to extract additional useful information from the data. However, enhanced information content also makes it more difficult to sift through the data to find desired information and present it in a concise and meaningful way.

Many features in hyperspectral data are larger than the 5- or 10-nm spacing used by most systems today. Thus, for many applications, it is not clear that additional spectral resolution will help with the retrieval of information from the data. However, this point does not argue for multispectral systems. A hyperspectral data set has so many bands that the data are, in most cases, spectrally over-sampled. In order to have a versatile multipurpose instrument, the hyperspectral approach is desirable. Said another way, given a random scene with unknown lighting and atmospheric conditions, one should fly a system that over-samples the scene so that the spectral shapes that are important are measured. The exact shapes/bands can be determined later by analysis. In complicated scenes, variation in lighting conditions, scene content, and atmospheric conditions are usually too variable for it to be possible to know ahead of time what bands would be necessary to produce a effective multispectral system for a particular task.

4.2. LINEAR MIXING MODEL

For the analysis of multispectral data, there are analysis methods such as false color RGB, band ratios, and indexes such as the normalized difference vegetation index (NDVI). These approaches could be simply extended to hyperspectral data analysis. However, the incredible information content of hyperspectral imagery argues for different approaches and the development of new analysis techniques.

Analysis of hyperspectral data broadly falls into two classes: statistical and model-based. The statistical methods used to analyze hyperspectral data are large in number. They include Principal Components Analysis (PCA) and its variations, classification methods such as ISODATA, minimum distances, Reed–Xiaoli (RX), and Stochastic Expectation Maximization (SEM), which are used to find locally or globally anomalous spectra in the scene.

PCA is strictly statistical in nature—finding the directions of largest variance—and the outputs, including the eigenvectors and data projected into the reduced space defined by the eigenvectors, depend not only on what is in the scene but also on how much of it is there. The eigenvectors are not physically meaningful and cannot be directly compared to library spectra. It is possible to project a library into the PCA space in order to do comparisons; however, when doing so, one is dependent on the eigenvectors to be able to provide the dimensions needed to differentiate the materials. The number (and shape) of significant eigenvectors (i.e., the dimensions of the subspace) are not significantly influenced by the presence of a small number of outlying spectra and, thus, the eigenvectors may not describe a subspace that encompasses the rare spectra. This can be an important issue in target identification as targets tend to be relatively rare. Because there is no physical model behind these methods, it is hard to make significant statements about the nature of the results without significant additional information.

Model-based approaches, on the other hand, while not a panacea, offer the prospect of making more definitive physical statements about the analyzed scene without additional information. As a whole, these methods would include, but not be limited to, atmospheric radiative transfer models, in-water radiative transfer models, the Linear Mixing Model (LMM), and nonlinear models. The ability to derive physical parameters from the data using these models relies on the accuracy to which the model represents reality and often relies on broad assumptions about the scene. For example, many atmospheric correction schemes make assumptions about the ground illumination that may be inaccurate because of the presence of clouds.

The LMM is based on the assumption that measured spectra are linear mixtures of the scene constituents. In addition to the linearity constraint, a convexity constraint is often also imposed. The convexity constraint implies that only fractional abundances between 0 and 1.0 are found in the data. Thus, a measured spectrum, $\mathbf{S}_j$, would be written as

$$\mathbf{S}_j = \sum_{i=i}^{n} \alpha_{ij}\mathbf{E}_i + \mathbf{N} \tag{4.1}$$

where the $\mathbf{E}_i$ are the scene constituents (often called endmembers), the α_{ij} are the fractional abundance of endmember i in spectrum j, and $\mathbf{N}$ represents the noise in the measurement. The assumption here is that $\sum_i a_{ij} = 1.0$ for each j and again the convexity constraint imposes the condition that the α_{ij} are between 0 and 1. It is a very simple model that still retains the ability to make statements about the physical makeup of the measured spectra. It is, very strictly, on a pixel-by-pixel basis always a true model. That is, the light emanating from a pixel is a sum of the backscatter from each constituent in the pixel. Where the model is limited is in the assumption that a single (e.g., "grass") endmember can accurately represent grass in every pixel in the scene. Even if that were possible, there is no accounting in the model for variations in the spectral content of the lighting throughout the scene, or differing view and illumination angles, or multiple reflections that, for example, may occur in trees and even in grass. In addition to all of that, the sensors are not perfect, and optical distortions, stray light, and other effects led to spectral/spatial mixing that is hard to untangle.

The Optical Real-Time Adaptive Spectral Identification System (ORASIS) [14, 15] was developed at the Naval Research Laboratory. It is based on the LMM and was developed from the beginning as a system capable of working in real time on average computer systems of the time. Other examples of such methods that are either implicitly or explicitly based on the LMM are Pixel Purity (PP) [16, 17], NFINDR [18], and others methods that determine endmembers through either supervised or unsupervised classification.

As opposed to PCA, PP and NFINDR will find an individual outlying spectrum and their output is not significantly affected by the amount of a particular material in the scene. In the PP method, the entire data set is repeatedly projected onto random vectors. The spectra that produce the most positive and most negative values in each of the random directions (mathematically, these spectra are referred to as extreme points) are most likely to be the purest examples of a particular material in the data set. Repeating the projection with different random vectors and counting the number of times a spectrum is extreme gives a measure of the likelihood that the spectrum is the purest example of a particular material in the set. NFINDR also attempts to find the most pure pixels in the data set. The method behind NFINDR is that of finding the largest N-dimensional simplex possible using the spectra in the scene where the dimensionality N is determined previously. Note that in both of these methods, the determined endmembers come from within the data set. ORASIS determines endmembers (that may not be represented in the data) by exploiting the scene data using an approach significantly different from either PP or NFINDR.

4.3. ORASIS

The Optical Real-Time Adaptive Spectral Imaging System (ORASIS) presented in this section is a collection of a number of algorithms that work together to produce a set of endmembers that are not necessarily from within the data set. The algorithm that determines the endmembers, called the shrinkwrap, intelligently extrapolates outside the data set to find endmembers that may be closer to pure substances than any of the spectra that exist in the data. Large hyperspectral data sets provide the algorithm with many different mixtures of the materials present, and each mixture gives a clue as to the makeup of the endmembers. As discussed below in more detail, this allows the extrapolation to a "pure" endmember easier. The family of algorithms that make up ORASIS is described in the following pages. Applications of these algorithms, such as automatic target recognition and data compression, are discussed in Section 4.4.

4.3.1. Prescreener

The prescreener module in ORASIS has two main functions: to replace the (large) set of original image spectra with a (smaller) representative set, known as the exemplars, and to associate each image spectrum with exactly one member of the exemplar set. The reasons for doing this are twofold. First, by choosing a small set of exemplars that faithfully represents the image data, further processing can be greatly sped up, often by orders of magnitude, with little loss of precision of the output. Second, by replacing each original spectrum with an exemplar, the amount of data that must be stored to represent the image can be greatly reduced. Such a technique is known in the compression community as vector quantization (VQ).

The basic methodology of the prescreener is intuitively quite simple. We begin by assigning the first spectrum of a given scene to the exemplar set. Each subsequent spectrum in the image is then compared to the current exemplar set. If the image spectrum is "sufficiently similar" (meaning within a certain spectral "error" angle), then the spectrum is considered redundant and is replaced, by reference, by a member of the exemplar set. If not, the image spectrum is assumed to contain new information and is added to the exemplar set. In this way, the entire original image is modeled using only a small subset of the original data. Remarkably, we have found that natural scenes (similar to those of Cuprite, NV) we have examined can be well represented (with ~1-degree spectral difference) by using less than 10% of the original data. However, there are complicated scenes (e.g., urban areas) that would require significantly more than 10% of the pixels to be represented at the 1-degree level. A 1-degree spectral difference is pretty small and therefore it appears reasonable to argue that the exemplar selection process simply removes the large spatial redundancy that appears in most hyperspectral images with little to no loss of information. Figure 4.1 provides examples of how the number of exemplars scales with the error angle.

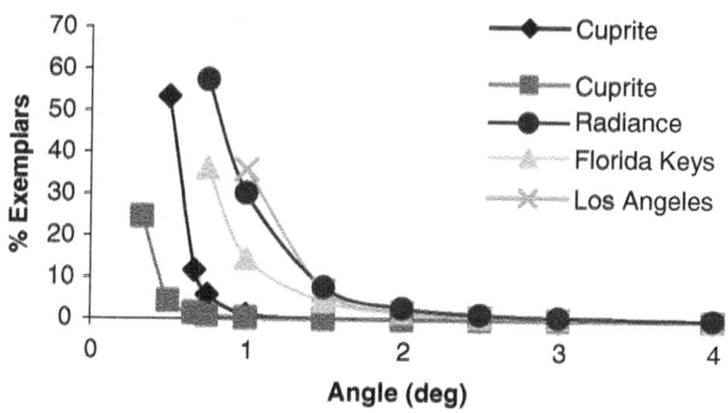

Figure 4.1. A plot of the percentage of pixels that become exemplars versus error angle for a number of different data cubes.

Although the basic idea of the prescreener is simple, ORASIS was designed, as its name implies, to be fast enough to work in a real-time environment. Given that modern hyperspectral imagers are easily able to generate 30,000+ spectra over several hundred wavelengths every second, it is clear that a simple, "brute force" searching routine would be quickly overwhelmed. For this reason, it has been necessary to create algorithms that can quickly perform "near-neighbor"-type searches in high-dimensional spaces. The remainder of this subsection describes the various algorithms that are used in the prescreener.

The prescreener module can be thought of as a two-step problem; first, deciding whether or not a given image spectrum is "unique" (i.e., an exemplar), and then, if not, finding the "best" exemplar to represent the spectrum. We refer to the first step as the exemplar selection process and refer to the second step as the replacement process. In ORASIS, the two steps are intimately related; however, for ease of exposition, we begin by examining the exemplar selection step separately, followed by a discussion of the replacement process.

4.3.1.1. Exemplar Selection. At each step in the process, an image spectrum X_i is read in, and a quick "sanity check" is performed. If the spectrum is deemed too "noisy" (i.e., having excessive dropouts, multiple spikes, etc.), then it is simply rejected and the reason for its rejection is recorded. Otherwise, the spectrum X_i is compared to the current set of exemplars. If the set is empty, X_i automatically becomes the first exemplar, and we move on to the next image spectra. If not, X_i is compared to the set of exemplars $E_1, \ldots, E_m$ to see if it is "sufficiently similar" to any one of them. If not, we add the image spectra to the exemplar set: $E_{m+1} = X_i$. Otherwise, the spectrum is considered "redundant" and is replaced by a reference to one of the exemplars. (The process of replacing the image spectrum with an exemplar is discussed in the next subsection.) This process continues until every spectrum in the image has been assigned either to the exemplar set or to an index into this set.

By "sufficiently similar," we simply mean that the angle $\theta(X_i, E_j)$ between the image spectrum X_i and the exemplar E_j must be smaller than some predetermined error angle θ_T. Recall that the angle between any two vectors is defined as

$$\theta(X_i, E_j) = \cos^{-1}\left(\frac{|\langle X_i, E_j\rangle|}{\| X_i \| \cdot \| E_j \|}\right)$$

where <#,#> is the usual (Euclidean) dot product of vectors, and ||#|| represents the (Euclidean) norm. If we assume that the vectors have been normalized to unit norm, then the condition for "rejecting" (i.e., not adding to the exemplar set) an incoming spectrum becomes

$$|\langle X_i, E_j\rangle| \geq \cos\theta_T \tag{4.2}$$

where we define $\varepsilon_T = 1 - \cos\theta_T$. Note that the inequality sign is reversed since the cosine is decreasing on the interval $(0,\pi)$. We also use the term "matching" to describe any two spectra that satisfy Eq. (4.2).

If we let $X_1, \ldots X_n$ denote the set of spectra in a given image and let $\{E_i\}$ denote the set of exemplars, then the above discussion may be formalized as follows:

Given a set of vectors $X_1, \ldots X_n$ and a threshold θ, set

$$E_1 = X_1, I(1) = 1, \qquad i = 1$$

For $j = 2, \ldots, n$:

Let $\theta_k = \min_i\{\angle(X_j, E_i)\}$ be the minimum of the angles between X_j and each exemplar E_i, and let E_k be the corresponding exemplar.

If $\theta_k \leq \theta_T$:

Assign the index k to the vector j, $I(j) = k$

else

Add the vector j to the set of exemplars

$$i = i + 1, \qquad E_i = X_j, \qquad I(j) = i$$

end

end

We note for completeness that the set $X_1, \ldots, X_n$ actually represents only those image spectra that have passed the "sanity check." Any spectra that have been rejected as being too noisy/damaged are simply ignored from here on out. Also, note that the exemplar set is not unique: There exist many exemplar sets that satisfy the stated conditions. From a practical point of view, it is dependent on the order in which the spectra are processed.

The only part of the above algorithm that takes any computational effort at all is finding the smallest angle between the candidate image spectrum X_j and the exemplars. The simplest approach would be to simply calculate all of the relevant angles and then find the minimum. Unfortunately, as discussed earlier, this would simply take too long, and faster methods are needed. The basic approach that ORASIS uses to speed things up is to try to reduce the actual number of exemplars that must be checked in order to decide if a match is possible. In order to do this, we use a set of "reference vectors" that allow us to quickly (i.e., in fewer processing steps) decide whether a given exemplar can possibly match a given image spectrum. As we will show, the actual number of exemplars that must be checked can often be significantly reduced by imposing bounds on the values of the reference vector projections.

We begin with the following simple lemma.

Lemma. Let X, E, and R be vectors with $\| X \| = \| E \| = \| R \| = 1$, and let $0 \leq \tau \leq 1$. Then

$$|\langle X, E \rangle| \geq \tau \Rightarrow |\langle X, R \rangle - \langle E, R \rangle| \leq \sqrt{2(1 - \tau)}.$$

Proof. Let $\sigma = |\langle X, E \rangle|$. By assumption, we have $0 \leq \tau \leq \sigma \leq 1$ and, therefore,

$$\sqrt{2(1 - \sigma)} \leq \sqrt{2(1 - \tau)}$$

By definition, and using the fact that each vector has unit norm,

$$\| X - E \| = \sqrt{\langle X - E, X - E \rangle} = \sqrt{\langle X, X \rangle + \langle E, E \rangle - 2\langle X, E \rangle} = \sqrt{2(1 - \sigma)}$$

Finally, by the Cauchy–Schwarz inequality we obtain

$$|\langle X, R \rangle - \langle E, R \rangle| = |\langle X - E, R \rangle| \leq \| X - E \| \cdot \| R \| = \| X - E \|$$

which proves the lemma.■

Suppose now that we wish to check if the angle between two vectors, X and E, is below some threshold θ_T. Symbolically, we wish to check if

$$\theta = \cos^{-1}\left(\frac{\langle X, E \rangle}{\| X \| \cdot \| E \|}\right) \leq \theta_T$$

If we assume that X and E are nonnegative (which will always be true for wavelength space hyperspectral data, i.e., not Fourier transform data that can be analyzed directly using the other parts of ORASIS) and that $\| X \| = \| E \| = 1$, then this is equivalent to checking if

$$|\langle X, E \rangle| \geq \cos(\theta_T) \equiv \tau$$

as long as for the angle we have $\theta_T \leq \pi$. Combining this with the lemma, we have the following rejection condition:

$$\theta(X, E) \leq \theta_T \quad \text{only if} \quad \sigma_{\min} \leq \langle E, R \rangle \leq \sigma_{\max}$$

where

$$\sigma_{\min} = \langle X, R \rangle - \sqrt{2(1 - \cos(\theta_T))}$$
$$\sigma_{\max} = \langle X, R \rangle + \sqrt{2(1 - \cos(\theta_T))}$$

To put it another way, if we want to test whether the angle between two vectors is sufficiently small, we could first choose a reference vector R, next calculate $\sigma_{\min}, \sigma_{\max}$, and $\langle E, R \rangle$, and test whether or not the rejection condition holds. If not, then we know that the vectors X and E cannot be within angle θ_T. We note that the converse is not true; even if the rejection condition holds, the vectors are not necessarily within angle θ_T. We would still need to actually calculate the angle to be sure.

It should be clear that the preceding discussion is not particularly helpful if one only needs to check the angle between two given vectors one time. However, we claim that the above method can be very useful when checking a large number of vectors against a (relatively) small test set. In particular, suppose we are given a set of (test) vectors $X_1, \ldots, X_n$ and a second set of (exemplar) vectors $E_1, \ldots, E_m$. For each X_i we would like to check if there exists an exemplar E_j such that the angle between them is smaller than some threshold θ_T. To do this, pick a reference vector R such that $\| R \| = 1$ and define

$$\sigma_i = \left\langle \frac{E_i}{\| E_i \|}, R \right\rangle$$

By renumbering the exemplars, if necessary, we may assume that

$$\sigma_1 \leq \sigma_2 \leq \cdots \leq \sigma_m$$

For a given test vector X_i we then calculate

$$\sigma^i_{\text{min}} = \langle X_i, R \rangle - \sqrt{2(1 - \cos(\theta_T))}$$
$$\sigma^i_{\text{max}} = \langle X_i, R \rangle + \sqrt{2(1 - \cos(\theta_T))}$$

By the rejection condition, it follows that the only exemplars that we need to check are those whose sigma value lies in the interval $[\sigma^i_{\text{min}}, \sigma^i_{\text{max}}]$; we call this interval the possibility zone for the test vector X_i. Assuming that the possibility zone is not too wide and that the sigma values are sufficiently "spread," it is often possible to significantly reduce the number of exemplars that need to be checked. There is, of course, the overhead of calculating the sigma values. However, we note that they only need to be calculated once for the entire set of test vectors; as the number of test vectors grows, this overhead quickly becomes insignificant.

As an example, we show in Figure 4.2 a histogram of a typical hyperspectral scene. The x axis is the value of the exemplars projected onto the reference vector,

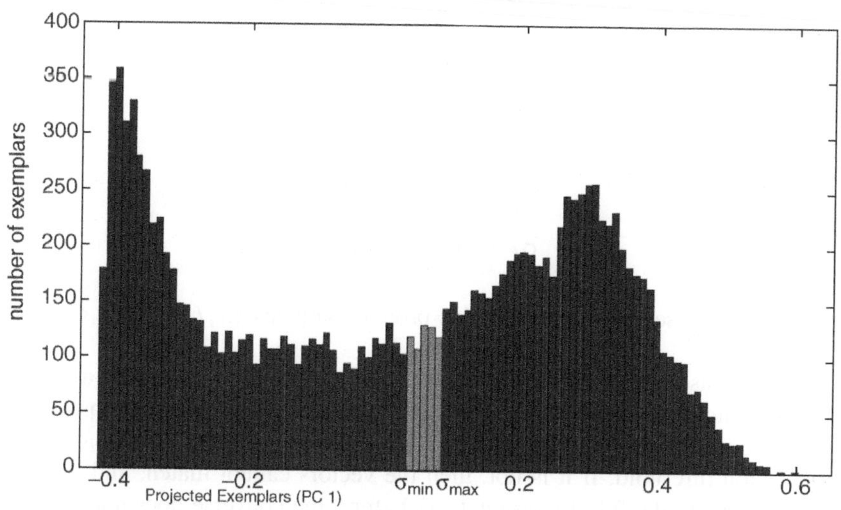

Figure 4.2. Histogram showing the small number of exemplars that are in the possibility zone when using a single reference vector.

and the y axis shows the number of exemplars. Using a brute force search, each of the (thousands of) exemplars would have to be searched. Using the preceding ideas, however, only those exemplars in the possibility zone (shown as light colored in the figure) would have to be searched. Clearly, the majority of exemplars are excluded by the possibility zone test.

We can extend the preceding idea to multiple reference vectors as follows. Suppose $R_1, \ldots, R_k$ is an orthonormal set of vectors, and let $\| X \| = \| E \| = 1$. Then we can decompose X and E as

$$X = \sum_{i=1}^{k} \alpha_i R_i + \alpha^{\perp} R^{\perp}$$

$$E = \sum_{i=1}^{k} \sigma_i R_i + \sigma^{\perp} S^{\perp}$$

where $\alpha_i = \langle X, R_i \rangle, \sigma_i = \langle E, R_i \rangle$, and $R^{\perp}, S^{\perp}$ are the residuals of X and E, respectively. In particular, $R^{\perp}, S^{\perp}$ have unit norm and are orthogonal to the subspace defined by the R_i's. It is easy to check that

$$\langle X, E \rangle = \sum \alpha_i \sigma_i + \alpha^{\perp} \sigma^{\perp} \langle R^{\perp}, S^{\perp} \rangle$$

By Cauchy–Schwartz, we have $\langle R^{\perp}, S^{\perp} \rangle \leq \| R^{\perp} \| \cdot \| S^{\perp} \| = 1$, and by the assumption that X and E have unit norm we obtain

$$\alpha^{\perp} = \sqrt{1 - \sum \alpha_i^2}$$

$$\sigma^{\perp} = \sqrt{1 - \sum \sigma_i^2}$$

If we define the projected vectors

$$\alpha_P = (\alpha_1, \ldots, \alpha_k, \alpha^{\perp})$$

$$\sigma_P = (\sigma_1, \ldots, \sigma_k, \sigma^{\perp})$$

then it is easily seen that the full dot product satisfies $\langle X, E \rangle \leq \sum \alpha_i \sigma_i + \alpha^{\perp} \sigma^{\perp} \equiv \langle \alpha_P, \sigma_P \rangle$.

As in the case with a single reference vector, this allows us to define a multizone rejection condition before doing a full dot product comparison. The process then becomes one of first checking that the projected dot product $\langle \alpha_P, \sigma_P \rangle$ is below the rejection threshold. If it is not, then the vectors cannot match, and there is no need to calculate the full dot product. As before, the converse does not hold; if the projected dot product is less than the threshold, the full dot product must still be checked. The advantage of this method is that, by using a small number of reference vectors, the total number of full band dot products, which may take hundreds of multiplications and additions, can be reduced by checking the projected dot products (which only take on the order of 10 operations). The trade-off is that this method first requires that the reference vector products must first be taken. In our experience, the number of these products (we generally use only three or four reference vectors) is usually much smaller than the number of full-band exemplar/image spectra dot products that are saved, thus justifying the use of the multizone rejection criterion. The overhead limits the number of reference vectors that should be used. In Figure 4.3, we show the same exemplars from Figure 4.2 now projected onto two reference vectors. The two-dimensional possibility zone is shown in light gray. It is clear from the figure that a large majority of the exemplars do not need to be searched in this example. It is also clear that the total number of comparisons has been significantly decreased from the single reference vector case shown in Figure 4.2.

The final part of the exemplar selection process is the "popup stack." Hyperspectral images generally contain a large amount of spatial homogeneity. As a result, neighboring pixels tend to be very similar. In terms of the prescreener, this implies that if two consecutive pixels are rejected, then there is a reasonable chance that they were both matched to the same exemplar. For this reason, we keep a dynamic list of the most recently matched exemplars. Before checking

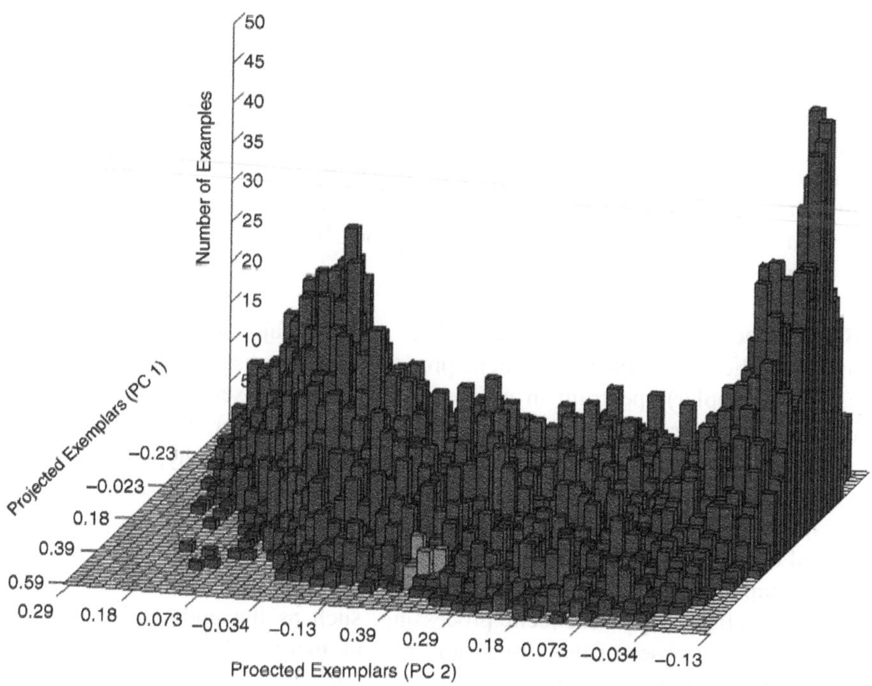

Figure 4.3. Histogram showing the number of exemplars that would need to go through the full test (in light gray).

the full set of exemplars, a candidate image spectrum is first compared to this list to see if it matches any of the recent exemplars. This list is continuously updated and should be small enough to be quickly searched, but large enough to capture the natural scene variation. In our experience, a size of four to six works well; the current version of ORASIS uses a five-element stack.

To put it all together, the prescreener works as follows: First, an incoming image spectrum is checked to make sure it is not too noisy, and then it is compared to the most recent matching exemplars (the popup stack). If no match is found, the reference vector dot products and the limits of the possibility zone are calculated. Starting from the center of the zone, the candidate is compared to the exemplars, first by comparing the projected dot products and then, if necessary, by comparing the full dot products. This search continues until a matching exemplar is found (and the image spectrum rejected) or all elements within the possibility zone have been checked. If no match is found, the candidate is considered to be new and unique and is added to the exemplar list.

We conclude this subsection by noting that the shape of the reference vectors is important in determining the size of the possibility zone, and therefore in the overall speed of the prescreener. Initially, random vectors were used, but it was soon discovered that using the PCA eigenvectors produced the best results. Perhaps this is not a surprise because the PCA eigenvectors provide the directions within the data that provide for the most variance, or separation. The PCA directions are calculated on the fly using a weighted exemplar substitution method to calculate the covariance matrix and from there the noncentered PCA directions are found. Experience has shown that sufficient directions can be determined after just a couple of hundred exemplars are found. On occasion, for very long or complicated scenes, it is relatively simple to update the covariance matrix and recalculate the eigenvectors and the corresponding value of $\langle E, R \rangle$ for each exemplar. Conceptually, the use of PCA eigenvectors for the reference vectors ensures that a grass spectrum is compared only to exemplars that look like grass and not to exemplars that are mostly water, for example.

4.3.1.2. Codebook Replacement. The second major function of the prescreener is the codebook replacement process, which substitutes each redundant (i.e., non-exemplar) spectrum in a given scene with an exemplar spectrum. The primary reason for doing this is that, as mentioned earlier, ORASIS was also designed to be able to do fast (near real-time) compression of hyperspectral imagery. The full details of the compression algorithm are discussed in a later section. This subsection discusses only the various methods in which non-exemplars are replaced during the prescreener processing. This process only affects the matches to the exemplars and does not change the spectral content of the exemplar set. Thus, it does not affect any subsequent processing, such as the endmember selection stage. For the purposes of this discussion, assume that the data scene is fully processed and, therefore, the exemplar set is complete.

As discussed in Section 4.3.1.1, each new candidate image spectrum is compared to the list of “possible” matching exemplars. Each candidate spectrum that becomes an exemplar must, by necessity, be checked against every exemplar in the possibility zone. However, in the vast majority of cases, the candidate will "match" one of the exemplars and be rejected as redundant. In this case, we would like to replace the candidate with the "best" exemplar, for some definition of best.

In ORASIS, there are three different ways of doing this replacement. The first case, which we refer to as "first match," simply replaces the candidate with the first exemplar that it matches. This is by far the easiest and fastest method.

The trade-off for the speed of the first match method is that the first matching exemplar may not be the best, in the sense that there may be another exemplar that is closer (in terms of difference angles) to the candidate spectrum. As it turns out, each spectrum may have multiple exemplars that match it within the error angle. However, one of them will be the best match but may not be found in first match mode because a match with another exemplar has stopped the search. In compression terms, this implies that the distortion between the spectrum and the exemplar replacement could be lowered if we continued to search through the possibility zone, looking for a better match. A second issue, which is not immediately obvious, is that the first match algorithm tends to choose the same exemplars over and over because of the popup stack. Since ORASIS works on a line-by-line basis, this means that homogeneous areas within a given line tend to exhibit "streaking" as shown in Figure 4.4.

One way to overcome the shortcomings of the first match method is to simply check every exemplar in the possibility zone and choose the exemplar that is closest to the candidate. This method, which we denote "true best fit," guarantees that the distortion between the original and compressed image will be minimized, and it also reduces the streaking effect of first match, as shown in Figure 4.4. The true best fit method is slower than the first match method; however, for compression

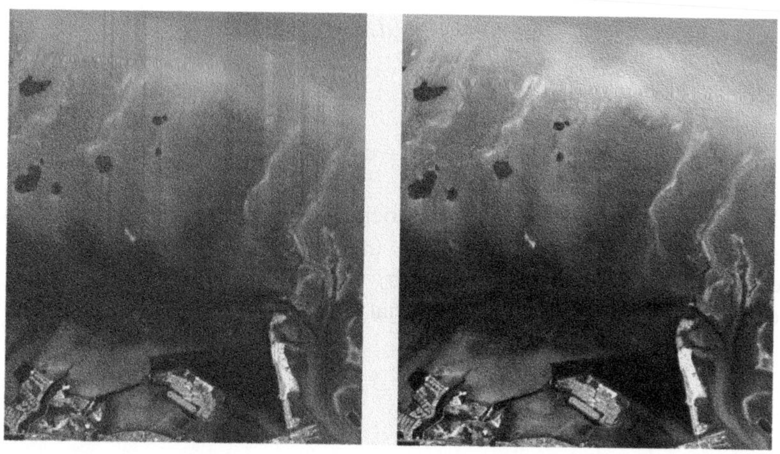

Figure 4.4. (Left) Streaking that was a side effect of the compression approach. (Right) The method that fixed the problem is demonstrated, which was calculated in the same manner as in the figure on the left except for the fix described in the text.

methods that depend on the codebook, it can be worthwhile. This is especially true in scenes with a large number of exemplars and in those that exhibit a large amount of homogeneity, since the possibility zones in these cases can be relatively large.

A third method, which attempts to balance the quality of the true best fit method with the speed of the first match method, is a "modified best fit method." This method is called the Efficient Near-Neighbor Search (ENNS) algorithm[19]. In modified best fit, a candidate spectrum is compared to the exemplars until a first matching exemplar is found. At this point, the (approximate) probability of finding a better matching exemplar is found, and a decision is made on whether to continue the search. If the calculated probability is lower than some (user-defined) threshold, then the search is stopped and the candidate is replaced by the current matching exemplar. Conversely, if the probability is above the threshold, the error parameter is set to the angle between the candidate and the current best matching exemplar. A new, smaller possibility zone is constructed, and the search continues. In this way, a candidate continues to search the exemplar set until either all possible exemplars have been examined, or the probability of finding a better match than the current one is too low to continue the search. In many cases, this process has the effect of finding a better matching exemplar than just the first match, thereby decreasing the distortion as well as eliminating the streaking effect, while still keeping the total number of comparisons relatively small, thus keeping the computation time reasonably fast.

The remainder of this subsection describes the method in which ORASIS estimates the probability of finding a better match in the possibility zone for a given candidate spectrum. We begin with some notation. Assume that $E_1, \ldots, E_n$ are the current exemplars and that d is the candidate image spectrum to be tested. By assumption, d must already match one of the exemplars E_j. It follows that there exists ε_j such that

$$\varepsilon_j = 1 - \langle d, E_j \rangle \leq \varepsilon_d$$

where

$$\varepsilon_d = 1 - \cos(\theta_T)$$

and θ_T is the threshold (error) angle associated with the candidate spectrum d.

As in the previous discussion, let $\alpha = \langle d, R \rangle$, $\sigma_j = \langle E_j, R \rangle$, where R is the (first) reference vector, and define $\delta_j = \alpha - \sigma_j$ and $\Delta_j = d - E_j$. Let ψ be the angle between Δ_j and R. It is easy to check that ψ satisfies

$$\cos\psi = \frac{\delta_j}{\sqrt{2\varepsilon_j}}$$

Intuitively, ψ measures how close the projections α and σ_j are, assuming that the vectors d and E_j match. We argue that $\cos\psi$ can be treated as a symmetric random variable that is independent of the difference $\| \Delta_j \|$, at least when $\| \Delta_j \|$ is small (of the order of the noise). Note that since by assumption we are only dealing with matching candidate/exemplar vectors, we may always assume that $\| \Delta_j \|$ is small. Let $\mu(\psi)$ be the mean of the random variable $\cos\psi$. It follows that $\mu(\psi)$ represents the "average" value for those exemplars and candidate vectors that are matches and that, as the value of $\cos\psi$ moves further away from $\mu(\psi)$, the likelihood of there being a match decreases. To put it another way, once we find a matching exemplar for a given candidate, then the probability of finding a better match decreases as we move further away from $\mu(\psi)$. If we let $\sigma(\psi)$ be the standard deviation of $\cos\psi$, then we can define a neighborhood $\mu(\psi) \pm n(\psi)\sigma(\psi)$ outside of which the probability of finding a better match is lower than some defined threshold. Here $n(\psi)$ is simply a measure of how many "standard deviations" we would like to use, and it is set by the user. It follows that, for a desired level of probability, we need only check those exemplars that satisfy

$$\mu(\psi) - n(\psi)\sigma(\psi) \le \cos\psi = \frac{\delta_j}{\sqrt{2\varepsilon_j}} \le \mu(\psi) + n(\psi)\sigma(\psi)$$

Recalling that $\delta_j = \alpha - \sigma_j$, we can rewrite this as

$$\alpha - \sigma_j \in \sqrt{2\varepsilon_j}\mu(\psi) \pm \sqrt{2\varepsilon_j}n(\psi)\sigma(\psi)$$

We refer to the interval $\sqrt{2\varepsilon_j}\mu(\psi) \pm \sqrt{2\varepsilon_j}n(\psi)\sigma(\psi)$ as the "probability zone" for the given candidate vector. Note that, as this zone is searched, a better match than the current matching exemplar may be found. As a result, the ε_j will decrease, and the probability zone will become smaller.

The last step is to estimate the mean and standard deviation of the random variable $\cos\psi$. We do this experimentally. For the first 100 exemplars, we search the entire possibility zone to find all possible matches, and calculate $\delta_j/\sqrt{2\varepsilon_j}$ for each exemplar/candidate match. This set of numbers samples the $\cos\psi$ distribution and can be used to calculate a sample mean and standard deviation, which are then used as estimates for $\mu(\psi)$ and $\sigma(\psi)$.

4.3.2. Basis Selection

Once the prescreener has been run and the exemplars calculated, the next step in the ORASIS algorithm is to project the set of exemplars into an appropriate, lower-dimensional subspace. The reason for doing this is that the linear mixing model implies that, if we ignore the noise, the data must lie in a subspace that is spanned by the endmembers themselves. Reasoning backwards, it follows that if we can find a low-dimensional subspace that contains the data, then we simply need to find the "right" basis for that subspace to find the endmembers. Moreover, by projecting the data into this subspace, we both reduce the computational complexity (by working in much lower dimensions), as well as reduce the noise. The trick, of course, is finding the right subspace to work with. There have been a number of different methods suggested in the literature [20] on how to choose this subspace, though these methods are not always discussed in these terms. In this section, we discuss the two methods that are available in ORASIS for determining the optimal subspace.

The original ORASIS basis selection algorithm uses a Gramm–Schmidt-like procedure to sequentially build a set of orthonormal basis vectors. Individual basis vectors are added until the largest residual of the projected exemplars is smaller than some user-defined threshold. (We note that previous ORASIS publications have referred to this algorithm as a "modified Gramm–Schmidt procedure." This term has a standard meaning in mathematics that is unrelated to our procedure, and we have stopped using the "modified" term.)

The algorithm begins by finding the two exemplars, $S_{i(1)}$ and $S_{i(2)}$, that have the largest angle between them. These exemplars are known as "salients," and the indices $i(1)$ and $i(2)$ are stored for use in the endmember selection stage. The two salients are then orthonormalized to form the first two basis vectors B_1 and B_2. Next, the set of exemplars are projected down into the two-dimensional subspace defined by B_1 and B_2, and the residuals are calculated. If the value of the largest residual is smaller than some predefined threshold, then the process terminates. Otherwise, the exemplar with the largest residual (S_j, say) is added to the salient set, and the index is saved. This exemplar is then orthonormalized to the current basis set (using Gramm–Schmidt) to form the third basis vector B_3. The exemplars are then projected into the three-dimensional space defined by B_1, B_2, B_3, and the process is repeated. This continues until either the threshold is reached or a predetermined maximum number of basis vectors are chosen.

At the end of the basis module, the exemplars have been projected into some k-dimensional subspace spanned by the basis vectors $B_1, \ldots, B_k$. By the assumptions of the linear mixing model, the endmembers must also span this same space, so we are free to use the projected exemplars in order to find the endmembers. The salient exemplars $S_{i(1)}, \ldots, S_{i(k)}$ form the "first guess" of the endmember selection algorithm. In recent versions, we have also added an option for improving the salients by using a maximum-simplex method similar to the NFINDR [21] algorithm.

We note that the basis algorithm described above guarantees that the largest residual (or error) is smaller than some predefined threshold. Thus, the algorithm may be thought of as a "mini-max" type of algorithm. The reason for doing so is mainly to make sure that small, relatively rare objects in the scene are included in the projected space. To put it another way, the ORASIS basis algorithm is designed to include outliers, which are oftentimes (e.g., target and/or anomaly detection) the objects of most interest. By comparison, most statistically based methods (such as PCA) are designed to exclude outliers. One problem with our approach is that it can be sensitive to noise effects and sensor artifacts. This problem can be mitigated by using the sanity check of the prescreener module described above.

There are, of course, certain times when one would like to exclude outliers and simply find the subspace that minimizes the total residual (or error). This is mainly of use in compression, but may also be used in other cases, such as rough terrain categorization. For this reason, ORASIS also includes the option of using standard principal components as a basis selection algorithm. In this case, the principal components form the basis vectors, and the exemplars are then projected into this space before being passed to the endmember selection stage. Since there is no way to define the salient vectors via PCA, the endmember stage can be seeded either with a random set or via the simplex (NFINDR-like) method mentioned above. At the current time, the number of PCA basis vectors to use must be decided in advance by the user.

4.3.3. Endmember Selection

The next step in the ORASIS package is the endmember selection module. Over the years, there have a number of different algorithms used to find the endmembers. Though the basic ideas have stayed the same, the actual implementations have changed. For this reason, we begin with a discussion of the general plan in ORASIS for finding endmembers and then discuss the actual implementation of the current algorithm.

In very broad terms, ORASIS defines the endmembers to be the vertices of some "optimal" simplex that encapsulates the data. This is similar to a number of other "geometric" endmember algorithms, such as pixel purity index (PP) and NFINDR, and is a direct consequence of the linear mixing model. Unlike PP and NFINDR, however, ORASIS does not assume that the endmembers must be actual data points. The reason for this is that there is no *a priori* reason for believing that "pure" endmembers exist in the data. To be more specific, by assuming that endmembers must be one of the image spectra, it is implicitly assumed that there exists at least one pixel in the scene that contains each given endmember material, and nothing else. We believe that this assumption fails in a large number of scenes. For example, in images with large GSDs, the individual objects may be too small to fill an entire pixel. Similarly, in mineral-type scenes, the individual mineral types will often be "mixed in" with other types of materials. The reader can no doubt think of many other examples, but the key point is that, if the given "pure" constituents exist only in mixed pixels, then, by definition, there is no point in the data set that corresponds to that pure material, and, hence, the true endmember itself cannot be one of the data points.

As an example, consider Figure 4.5. The figure is a cartoon representation of a scene with two background materials (grass and dirt, say) and a third "target"-type material (e.g., tanks, mines, or Improvised Explosive Devices (IED)). Assume that the majority of the pixels in the scene are some combination (including pure examples) of the two background materials, and there are relatively few pixels, each of which contains the target. If we assume that the target itself is smaller than the GSD of the sensor, then it is clear there will not be any spectra in the scene that are equal to the target signature. However, by the linear mixing model, each spectrum will be a convex combination of the three materials (grass, dirt, and target) and will, therefore, lie in the simplex generated by these signatures. In this case, the simplex of interest has three vertices: Two of these appear in the data, whereas the third one (corresponding to the target) does not. We call the "missing" third endmember a "virtual" endmember. The simplex generated using both mixtures of the target/

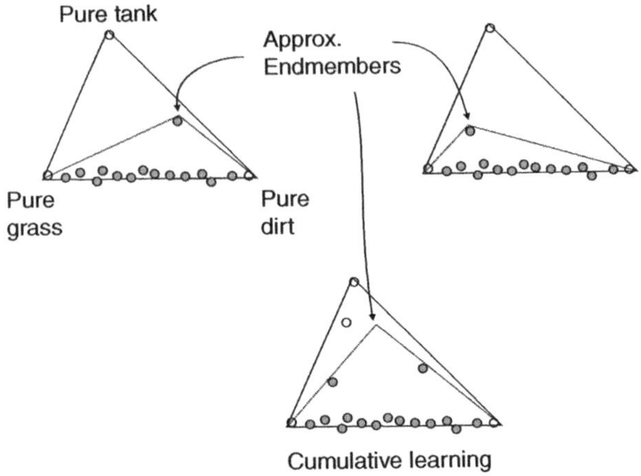

Figure 4.5. Diagram showing the effect of "seeing" multiple instances of mixtures of a target material against a simple background.

background (on dirt and on grass) results in an endmember that is closer to the real endmember than is achieved by using only one of the mixtures.

One of the major goals in the ORASIS endmember algorithm has always been to be able to detect and construct these "virtual" endmembers. The advantage to this approach is the ability to be able to detect endmembers that may not be present in pure form in the data itself. Moreover, if there are no virtual endmembers—that is, every endmember does appear in the data set—then the simplex method will continue to work as well. The main disadvantage to this approach is that it is very difficult to decide when a constructed "virtual" endmember is viable, in the sense that it is entirely possible that a given endmember may not be physically realistic. In certain situations, the endmembers may even have negative components.

We conclude this section with an overview of the ORASIS endmember algorithm as it is currently implemented. As discussed in the previous section, the inputs to the endmember module are the exemplars from the prescreener, projected down into some k-dimensional subspace, as well as an initial set of vectors known as the salients. The salients form a k-simplex within the subspace. The basic idea is to "push" the vertices of this simplex outwards until all the exemplars lie inside the new simplex. If all the exemplars are already inside the simplex defined by the salients, then we assume that the salients are in fact the endmembers, and we are finished. In practice, however, this never happens, and the original simplex must be expanded in order to encapsulate the exemplars. To do this, we begin by finding the exemplar $S_{\max}$ that lies the furthest out of the current simplex—this is found by demixing with the current endmembers and looking for the most negative abundance coefficient. The vertex $v_{\max}$ that is the furthest from the most outlaying exemplar is held stationary, and the remaining vertices are moved outward until the $S_{\max}$ exemplar is inside the new simplex. The process is then simply repeated until all exemplars are within the simplex. The vertices of this final encompassing simplex are then defined to be the current endmembers.

4.3.4. Demixing

Once the endmembers have been determined, the last step in the ORASIS algorithm is to estimate the abundance of each endmember in each scene spectra. This process is generally known as demixing in the hyperspectral community. ORASIS allows for two separate methods for demixing, depending on which (if any) of the constraints are imposed. We note that ORASIS may be used to demix the exemplar set, the entire image cube, or both. In this section, we discuss the two demixing algorithms (constrained and unconstrained) that are available in ORASIS.

Demixing the data produces "maps" of the α_{ij}'s. Remembering that physically the α_{ij}'s represent an abundance of material j in spectrum i, it is clear that one can refer to these maps as abundance maps. Exactly what the abundance maps measure physically depends on what calibrations/normalizations are performed during the processing. If the data were calibrated and the endmembers are normalized, then the abundance maps represent the radiance associated with each endmember. Other interpretations are possible such as relating the abundance maps to the fraction of radiance from each endmember. In this case, the abundance maps are sometimes called the fraction planes.

4.3.4.1. Unconstrained Demix. The simplest (and fastest) method for demixing the data occurs when no constraints are placed on the abundance coefficients. Note that even in this simplest case, the measured image spectra will rarely lie exactly in the subspace defined by the endmembers; this is due to both modeling error and various types of noise in the sensor. It follows that the demixing process will not be exactly solvable, and the abundance coefficients must be estimated. We define X as a matrix where the columns are given by the endmembers. If we let $\mathbf{P}$ be the $k \times n$ matrix (k = number of endmember and n = number of bands) defined by

$$\mathbf{P} = (X^t X)^{-1} X^t \tag{4.3}$$

then it is straightforward to show that the least-squares estimate (LSE) $\hat{\alpha}$ to the true mixing coefficients α for a given spectrum Y is given by

$$\hat{\alpha} = \mathbf{P}Y \tag{4.4}$$

In geometrical terms, the LSE defines a vector $\hat{Y} = X\hat{\alpha}$ in the endmember subspace that is closest (in the Euclidean sense) to the measured spectrum Y. It can also be shown that the LSE is "best" in a statistical sense. In particular, under the correlated and equal noise assumption, then the maximum likelihood estimate (MLE) of the abundance coefficient vector α is exactly the same as the LSE $\hat{\alpha}$.

The matrix $\mathbf{P}$ defined in Eq. (4.3) is known by a number of names, including the Moore–Penrose inverse and the pseudo-inverse. It follows that once the matrix

P has been calculated (which needs to be done only once), the unconstrained demixing process reduces to a simple matrix–vector product, which can be done very quickly.

It is worth noting that the columns $P_1, \ldots, P_l$ of the pseudo-inverse P form a set of "matched filters" for the endmembers. In particular, it is easy to check that

$$\langle P_i, E_j \rangle = \delta_{i,j} = \begin{cases} 1, & i = j \\ 0, & i \neq j \end{cases}$$

Therefore,

$$\langle P_i, Y \rangle = \langle P_i, \sum \alpha_j E_j + N \rangle = \sum \alpha_j \langle P_i, E_j \rangle + \langle P_i, N \rangle \cong \alpha_i$$

where the last approximation follows by assuming that the noise is uncorrelated with the columns, and so $\langle P_i, N \rangle \cong 0$. It follows that the individual components of the abundance coefficients may be calculated by a simple dot product. In previous papers describing ORASIS, the columns P_i were called "filter vectors," and the entire demixing process was described in terms of the filter vectors.

4.3.4.2. Constrained Demix. The constrained demixing algorithm applies the non-negativity constraints to the abundance coefficients. In this case, there is no known analytical solution, and numerical methods must be used. Our approach is based on the well-known Nonnegative Least-Squares (NNLS) method of Lawson and Hanson [22]. The NNLS algorithm is guaranteed to converge to the unique solution that is closest (in the least-squares sense) to the original spectrum. The FORTRAN code for the NNLS algorithm is freely available from NETLIB (www.netlib.org). We note that, compared to the unconstrained demixing algorithm, the NNLS can be significantly (orders of magnitude) slower. At the current time, ORASIS does not implement the sum-to-one constraint, either with or without the nonnegativity constraint.

4.4. APPLICATIONS

The combination of algorithms discussed above can be arranged to perform various tasks such as automatic target recognition (ATR), terrain categorization (TERCAT), and compression. The next couple of subsections discuss these applications.

4.4.1. Automatic Target Recognition

One of the more popular and useful applications of hyperspectral imagery is automatic target recognition (ATR). In very broad terms, ATR algorithms attempt to find unusual or interesting spectra in a given scene, using spectral and/or spatial properties of the scene. More precisely, anomaly detection algorithms attempt to find spectra whose signatures are significantly different from the majority of the main background components. Generally, the anomaly detection algorithms do not look for specific types of spectra. However, once an anomaly is found, further postprocessing (such as library lookup or cuing high-resolution imagers) may be performed to try to identify the material present. By contrast, target detection algorithms attempt to find pixels that contain specific materials. This is usually done by matching a given library spectrum to image spectra and labeling each spectra as a "hit" (if the target material is present) or "miss" (if not).

Over the last several years, there have been a number of different algorithms published in the literature [23–25]. ATR algorithms have been used in both military (e.g., to detect mines, tanks, etc.) and nonmilitary (e.g., tumor detection) applications. In this section, we discuss two different algorithms that have been developed using ORASIS; the first is an anomaly detector, and the second is a target detection system.

4.4.1.1. ORASIS Anomaly Detection. The ORASIS Anomaly Detection (OAD) algorithm [26] was originally developed as part of the Adaptive Spectral Reconnaissance Program (ASRP). The OAD algorithm uses a three-step approach to identify anomalous pixels within a given scene. The first step is to run ORASIS as usual, to create a set of exemplars and identify endmembers. Next, each exemplar is assigned a measure of "anomalousness," and a target map is created, with each pixel being assigned a score equal to that of its corresponding exemplar. Finally, the target map is thresholded and segmented, to create a list of distinct objects. The various spatial properties (i.e., width, height, aspect ratio) of the objects are calculated and stored. Spatial filters may then be applied to reduce false alarms by removing those objects that are not relevant.

In order to define an anomaly measure, OAD first divides the endmembers into "target" and "background" classes. In broad terms, background endmembers are those endmembers in which a relatively large number of exemplars have a relatively high abundance coefficient. This condition (for example, 80% of the spectra in the scene are composed of at least 50% of one endmember) implies that many spectra in the scene have at least part of this endmember present, and this endmember is unlikely to be a target endmember. Conversely, target endmembers are those where most exemplars have very small abundances, with a relatively small number of exemplars having high abundances. In terms of histograms, the abundance coefficients of background endmembers will have relatively wide histograms, with a relatively large mean value (see Figure 4.6a), while target endmembers will have relatively thin histograms, with small means and a few pixels with more extreme abundance values (see Figure 4.6b). Once the endmembers have been classified, the background dimensions are discarded. A final measure is calculated, based on a combination of how "target-like" (i.e., how much target abundance is present) a given exemplar is along with how "isolated" (i.e., how many other exemplars are nearby, in target space) it is. Once the final measure has been calculated, an anomaly map is created by assigning to each pixel the anomaly measure of its corresponding exemplar. The anomaly map is then thresholded, and the spectra that survive are considered anomalies. The anomalies are then clumped into "objects"

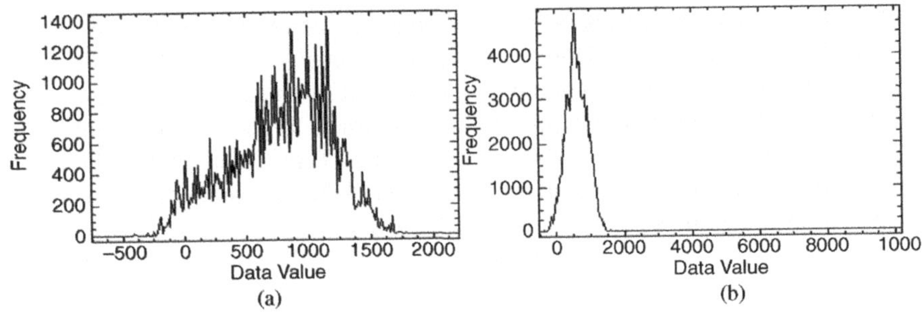

Figure 4.6. (a) The histogram of a background endmember. (b) The histogram of a target endmember. This graph has only a small number of pixels with values much higher than the average.

by identifying spectra that are spatially neighboring, and spectrally similar, to the anomalies. After segmenting the image, various spatial properties of the individual objects are calculated. The object list is then filtered by removing those objects that do not fit given, user-supplied spatial characteristics. Figure 4.7 shows the results of applying the ORASIS anomaly detection algorithm to the HYDICE Forest Radiance I data set.

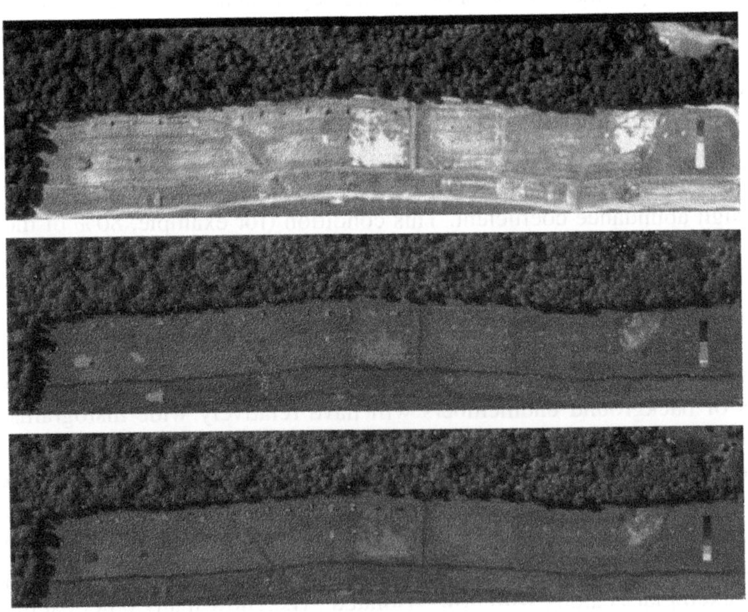

Figure 4.7. Top image is a grayscale image of a HyDICE forest radiance cube. The middle image shows where the pixels are thought to be targets and not background. The lower image shows the results of spatial filtering when looking only for small targets. Note that the larger tarps prominently shown in the middle image are not highlighted in the lower image.

4.4.1.2. Demixed Spectral Angle Mapper (D-SAM). In addition to the anomaly detection discussed above, ORASIS has also been used as part of a target detection system. In target detection, the user generally has a predetermined signature of some desired material, and the algorithm attempts to identify any pixels in the scene that contains the given material. It is worth noting that the library spectrum is properly kept as a reflectance spectrum; therefore, in order to compare the image and library spectrum, either the library spectrum must be converted to radiance or the image spectra must be converted to reflectance. In this subsection, we assume that one of these methods has already been used to transform the spectra to the same space.

One of the earliest, and still very popular, target detection algorithms for hyperspectral imagery is the spectral angle mapper (SAM). SAM attempts to find target pixels by simply calculating the angle $\theta(X_i, T)$ between each image pixel X_i and the given target signature T, where $\theta(X_i, T)$ is defined as

$$\theta(X_i, T) = \cos^{-1}\left(\frac{|\langle X_i, T\rangle|}{\| X_i \| \cdot \| T \|}\right) \tag{4.5}$$

We note that many users (such as in the ENVI implementation) skip taking the inverse cosine and simply define SAM as

$$\hat{\theta}(X_i, T) = \frac{|\langle X_i, T\rangle|}{\| X_i \| \cdot \| T \|} \tag{4.6}$$

Since the inverse cosine is (decreasing) monotonous on the interval $(0,\pi)$, this merely results in a (inverse) rescaling of Eq. (4.5). In the resulting discussion, either Eq. (4.5) or Eq. (4.6) may be used, however, to be precise, we always mean Eq. 4.5 when referring to SAM.

The SAM algorithm tends to work well in relatively easy scenes, especially those in which the target material is fully illuminated and large enough to fill entire pixels. In these cases, the measured pixel will contain only the desired target, and thus the image spectrum should be relatively close (small angle) to the target spectrum. Of course, the match will rarely be exact, because of modeling error and noise.

It is well known, however, that SAM does not do well in scenes that contain mixed pixel targets. Such situations can occur when the target is too small (relative to the sensor GSD) to occupy an entire pixel, or when the target itself is not clearly visible (e.g., due to shadowing or overhang). In such cases, the measured spectrum will be an additive mixture of both the target and some nontarget signature, such as the background or a "shadow" spectrum. Intuitively, the addition of the nontarget spectrum will cause the measured image spectrum to be pushed away from the library spectrum, and the angle between the two will increase. It is easily seen in real-world imagery that the addition of even a small amount of nontarget material can lead to a large angular separation. As a consequence, to identify the target pixels, either the angular threshold must be increased (leading to an increase in the number of false alarms) or the mixed pixel targets will not be identified (leading to an increase in missed detections).

In order to address the problems that SAM has with mixed pixels, we have developed the Demixed Spectral Angle Mapper (D-SAM) [27]. The basic idea behind D-SAM is quite simple. First, ORASIS is run on a scene and the endmembers are extracted. We note that this step is independent of the given target spectrum (or spectra) and needs to be run only once. Next, the image spectra and the target spectrum are demixed, and then the angle between the *demixed* image and target spectra is calculated. In symbols, let $E_1, \ldots E_n$ be the endmembers, and suppose the target T and image spectra X are demixed (ignoring noise) as

$$T = \sum \beta_j E_j$$
$$X = \sum \alpha_j E_j$$

Then the D-SAM angle $\Theta(X, T)$ between them is defined as

$$\Theta(X, T) = \cos^{-1}\left(\frac{|\langle X_{\text{proj}}, T_{\text{proj}}\rangle|}{\| X_{\text{proj}} \| \cdot \| T_{\text{proj}} \|}\right)$$
$$= \cos^{-1}\left(\frac{|\sum \alpha_j \beta_j|}{\sqrt{\sum \alpha_j^2 \sum \beta_j^2}}\right)$$

where $X_{\text{proj}} = (\alpha_1, \ldots, \alpha_n)$ and $T_{\text{proj}} = (\beta_1, \ldots, \beta_n)$ are the demixed image and target spectra, respectively.

Is it important to note that the D-SAM angle $\Theta(X, T)$ is not just a simple rescaling of the SAM angle $\theta(X, T)$. This is because the demixing process, which is mathematically nothing more than a projection operator, does not preserve angles, nor is it monotonic. In fact, it can be shown that in many cases involving mixed pixels, the D-SAM angle will cause target-containing pixels to move closer to the target, while forcing pixels that do not contain the target to move further from the target. As a result, the threshold needed to identify targets can be lowered, resulting in a reduction of false alarms while keeping the detection level constant.

An example of D-SAM in practice is shown in the next couple of figures. Figure 4.8 shows a grayscale image of part of a HyDICE forest radiance scene. In this scene are five rows of targets, and within each row are three targets. The three targets are different sizes with the smallest targets smaller than a pixel and the largest being big enough that the measurement can be considered close to a pure example of the target. Using the spectrum from the pixel of the largest target, both SAM and D-SAM were run. Histograms of the spectral angle with the target spectrum for both SAM and the D-SAM approaches are shown in Figure 4.9. The small target pixel location within these histograms is marked with a vertical line. Note that the target line is inside the bulk distribution of the background for the SAM approach, implying that a large number of false alarms will occur before the small target is detected. However, using the D-SAM approach, the small target

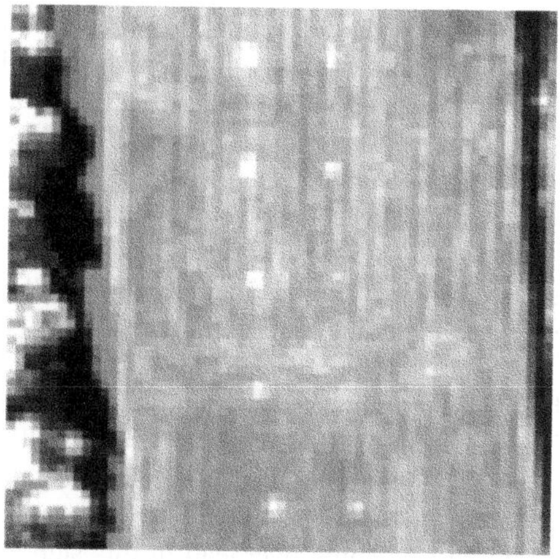

Figure 4.8. HYDICE data from forest radiance.

location is outside the bulk background distribution. In fact, in this example the D-SAM algorithm will detect the smallest target before any false alarms are generated.

4.4.2. Compression

As mentioned earlier, one of the design goals of ORASIS from the beginning was to be able to compress hyperspectral imagery, in order to reduce the large amount of space needed to store a given cube and also to reduce transmission times. In this section, we discuss one of the compression algorithms that has been developed.

The first algorithm uses the prescreener as a vector quantization scheme. In particular, each image spectrum in the scene is replaced by exactly one member

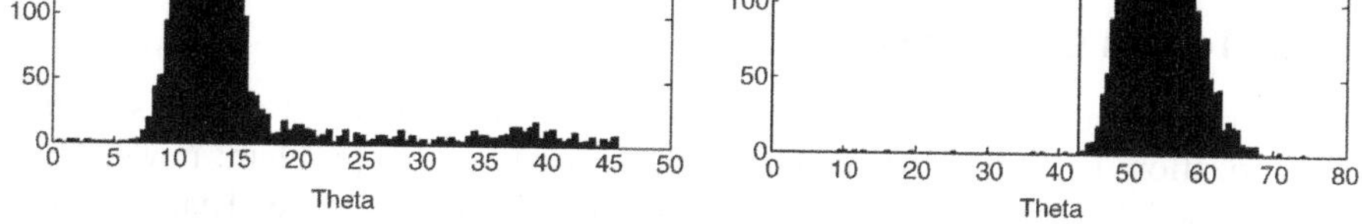

Figure 4.9. Histograms of the spectral angles for all pixels in the scene. The small target pixel is marked with the line. The left-hand figure is from SAM, and the right-hand figure is from D-SAM. Note that the marked target is outside of bulk pixels for the D-SAM but not for SAM.

of the exemplar set. In this way, the total number of spectra that must be stored is reduced from (number of image samples × number of image lines) to (number of exemplars). In addition, a "codebook" or index map, which contains the exemplar index for each pixel in the scene, must be stored. For example, if 5% of the image spectra are chosen as exemplars, then the total compression ratio is a little less than 20:1, after including the codebook overhead. In practice, the ratio will be slightly lower, since ORASIS saves only the unit normalized exemplars and thus must also store the original magnitudes of the exemplar vectors. This has the consequence of adding one extra "band" to each of the exemplar vectors.

The image can be further compressed by demixing the exemplars. In this way, the total number of bands that must be stored can be reduced; as a result, the compression ratio is increased. For example, if the original image data consisted of 200 bands, but only 20 endmembers were used, then the data can be further compressed by a factor of 10. There will be some overhead in this step as well, since the original, full-band space endmembers must be stored. In practice, ORASIS stores the endmembers as a sum of basis vectors, as well as the basis vectors themselves. This involves a very slight increase in the overhead, but makes certain post-processing routines slightly easier.

Putting it all together, the final ORASIS output consists of the set of projected exemplars, the original exemplar magnitudes, the projected endmembers, the full-band space basis vectors, and the codebook.

As an example of the preceding, consider a typical AVIRIS image consisting of 614 samples, 1024 lines, and 224 bands. At 16 bits of precision, the total storage space for this cube is 2,253,389,824 bits, or approximately 268 megabytes. Using a constant noise angle of 0.65 degrees, ORASIS produces 4829 exemplars and 20 endmembers. The final size of the output file, including all overhead, is 44,542,784 bits, or approximately 5.3 megabytes. The final compression ratio for this example is, thus, approximately 50:1 (Table 4.1).

Of course, no compression scheme is of much use if the data itself are not accurately preserved. This is especially true in scientific data, where small changes to the data could, in principle, lead to large changes in processing. Unfortunately, it is quite difficult to define meaningful distortion metrics for scientific data. This is especially true in hyperspectral imagery, where both the spectral and spatial integrity needs to be preserved.

4.4.3. Terrain Categorization

Another application of ORASIS (and similar algorithms) is that of coarse terrain categorization. Classification methods can be used for this problem; however, the mixture model approaches have at least one significant advantage. That is, using a mixture model approach allows the handling of mixed pixels. For example, ORASIS can tell how much of a pixel is covered with vegetation and how much is sand. ORASIS is not required to place the spectrum in either the sand or vegetation class. A coarse terrain categorization is done by running ORASIS (prescreener,

TABLE 4.1. Various Compression Parameters and the Resulting Compression Ratios Obtained with ORASIS Compression

File	Exemplar Angle	Basis Angle	Original File Size(bytes)	Compressed File Size (bytes)	Compression Ratio
Cuprite reflectance	2	2	116,316,160	2,554,552	45.53
	2	1	116,316,160	2,574,232	45.18
	0.75	2	116,316,160	3,856,384	30.16
	0.75	1	116,316,160	4,599,304	25.29
Cuprite radiance	2	2	116,316,160	2,542,300	45.75
	2	0.5	116,316,160	2,552,464	45.57
	0.5	2	116,316,160	3,343,480	34.79
	0.5	0.5	116,316,160	3,837,976	30.31
Florida Keys	2	2	140,836,864	3,010,272	46.79
	2	1	140,836,864	3,189,664	44.15
	1	2	140,836,864	6,600,848	21.34
	1	1	140,836,864	8,457,584	16.65
Los Angeles	3	3	140,836,864	2,617,604	53.80
	3	1	140,836,864	2,707,840	52.01
	1.5	3	140,836,864	4,151,728	33.92
	1.5	1	140,836,864	5,877,680	23.96
Forest radiance	2	2	31,987,200	874,668	36.57
	2	1	31,987,200	982,952	32.54
	1	2	31,987,200	3,125,928	10.23
	1	1	31,987,200	4,524,768	7.07

dimensionality and endmember determination, demixing) using a large angle of, say, 10 degrees.

An example of this is shown in Figure 4.10. The data are from the PHILLS sensor, which is a VNIR pushbroom spectrometer, and the data are approximately 1.5 m GSD. The three endmembers, shown in the plot, that were found are visually identifiable as vegetation, water, and sand. A grayscale image (see the upper right-hand side) was made of the abundance maps with dark gray set equal to the "water" map, white set equal to the vegetation maps, and gray set equal to the sand map. The classification is subpixel capable in that mixtures will be shown to be members of multiple classes. This effect cannot be viewed easily without color images.

ACKNOWLEDGMENTS

This work was funded by the Office of Naval Research. The work on ORASIS presented in this chapter represents the efforts of many people over the last 12 years

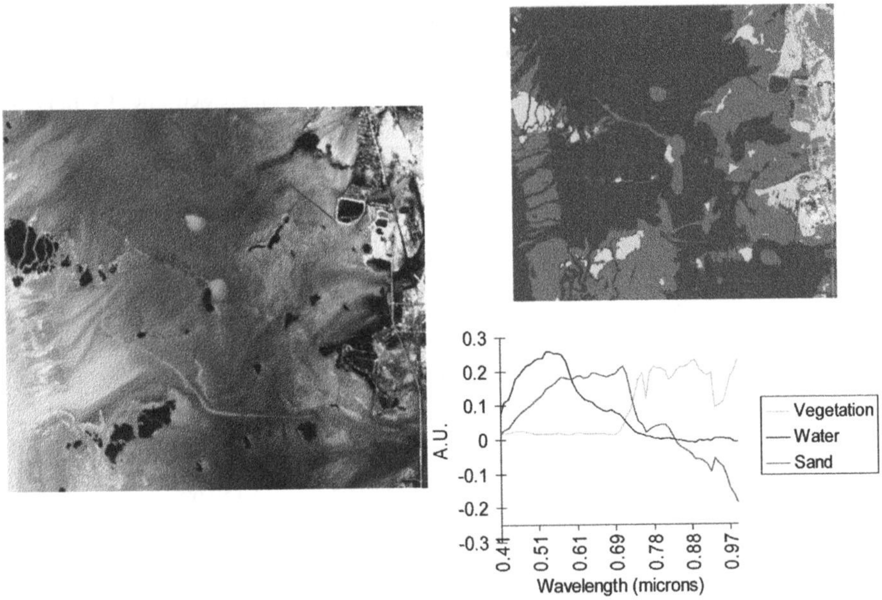

Figure 4.10. Example of doing a coarse terrain categorization. A gray scale image of an area in the Bahamas is shown on the left. An image of an abundance maps is shown above on the right.

as evidenced by the references. The authors wish to acknowledge the substantial contributions of the following people: John Antoniades, Mark Baumback, Mark Daniel, John Grossmann, Daniel Haas, Peter Palmadesso, and Jeffrey Skibo.

REFERENCES

1. H. M. Rajesh, Application of remote sensing and GIS in mineral resource mapping—An overview, *Journal of Mineralogical and Petrological Sciences*, vol. 99, pp. 83–103, 2004.
2. Y. L. Tang, R. C. Wang, and J. F. Huang, Relations between red edge characteristics and agronomic parameters of crops, *Pedosphere*, vol. 14, pp. 467–474, 2004.
3. L. Estep, G. Terrie, and B. Davis, Crop stress detection using AVIRIS hyperspectral imagery and artificial neural networks, *International Journal of Remote Sensing*, vol. 25, pp. 4999–5004, 2004.
4. J. Dozier and T. H. Painter, Multispectral and hyperspectral remote sensing of alpine snow properties, *Annual Review of Earth and Planetary Sciences*, vol. 32, pp. 465–494, 2004.
5. A. W. Nolin and J. Dozier, A hyperspectral method for remotely sensing the grain size of snow, *Remote Sensing of Environment*, vol. 74, pp. 207–216, 2000.
6. V. E. Brando and A. G. Dekker, Satellite hyperspectral remote sensing for estimating estuarine and coastal water quality, *IEEE Transactions on Geoscience and Remote Sensing*, vol. 41, pp. 1378–1387, 2003.

7. P. R. Schwartz, Future directions in ocean remote sensing, *Marine Technology Society Journal*, vol. 38, pp. 109–120, 2004.
8. R. L. Phillips, O. Beeri, and E. S. DeKeyser, Remote wetland assessment for Missouri Coteau prairie glacial basins, *Wetlands*, vol. 25, pp. 335–349, 2005.
9. F. Salem, M. Kafatos, T. El-Ghazawi, R. Gomez, and R. X. Yang, Hyperspectral image assessment of oil-contaminated wetland, *International Journal of Remote Sensing*, vol. 26, pp. 811–821, 2005.
10. H. Kwon and N. M. Nasrabadi, Kernel RX-algorithm: A nonlinear anomaly detector for hyperspectral imagery, *IEEE Transactions on Geoscience and Remote Sensing*, vol. 43, pp. 388–397, 2005.
11. R. N. Feudale and S. D. Brown, An inverse model for target detection, *Chemometrics and Intelligent Laboratory Systems*, vol. 77, pp. 75–84, 2005.
12. A. Chung, S. Karlan, E. Lindsley, S. Wachsmann-Hogiu, and D. L. Farkas, *In vivo* cytometry: A spectrum of possibilites, *Cytometry Part A*, vol. 69A, pp. 142–146, 2006.
13. P. Tatzer, M. Wolf, and T. Panner, Industrial application for inline material sorting using hyperspectral imaging in the NIR range, *Real-Time Imaging*, vol. 11, pp. 99–107, 2005.
14. P. J. Palmadesso and J. A. Antoniades, Intelligent hypersensor processing system (IHPS). US Patent No. 6038344: The United States of America as represented by the Secretary of the Navy, Washington, DC, 2000.
15. J. Bowles, J. Antoniades, M. Baumback, J. Grossmann, D. Haas, P. Palmadesso, and J. Stracka, Real time analysis of hyperspectral data sets using NRL's ORASIS algorithm, *Proceedings of the SPIE*, vol. 3118, p. 38, 1997.
16. J. Bowles, M. Daniel, J. Grossmann, J. Antoniades, M. Baumback, and P. Palmadesso, Comparison of output from ORASIS and pixel purity calculations, *Proceedings of the SPIE*, vol. 3438, p. 148,1998.
17. J. Boardman, Automating spectral unmixing of AVIRIS data using convex geometry concepts, *Summaries of the Fourth Annual JPL Airborne Geoscience Workshop*, vol. 1, pp. 11–14, 1998.
18. M. Winter, Fast autonomous spectral end-member determination in hyperspectral data, *Proceedings of the Thirteenth International Conference on Applied Geologic Remote Sensing*, vol. II, pp. 337–344, 1999.
19. P. J. Palmadesso, J. H. Bowles, and D. B. Gillis, Efficient near neighbor search (ENNsearch) method for high dimensional data sets with noise. US Patent No. 6947869: The United States of America as represented by the Secretary of the Navy, Washington, DC, 2005.
20. C.-I. Chang, *Hyperspectral Imaging: Techniques for spectral detection and classification*, Kluwer Academic/Plenum Publishers, New York, 2003.
21. E. Winter and M. M. Winter, Autonomous hyperspectral end-member determination methods, *Proceedings of the SPIE*, vol. 3870, p. 150, 1999.
22. C. Lawson and R. Hanson, Solving least squares problems, *Classics in Applied Mathematics*, vol. 15, SIAM, Philadelphia, 1995.
23. D. Manolakis and G. Shaw, Detection algorithms for hyperspectral Imaging applications, *IEEE Signal Processing Magazine*, vol. 19, pp. 29–43, 2002.

24. D. Manolakis, C. Siracusa, and G. Shaw, Hyperspectral subpixel target detection using the linear mixing model, *IEEE Transactions on Geoscience and Remote Sensing*, vol. 39, pp. 1392–1409, 2001.
25. D. W. J. Stein, S. G. Beaven, L. E. Hoff, E. M. Winter, A. P. Schaum, and A. D. Stocker, Anomaly detection from hyperspectral imagery, *IEEE Signal Processing Magazine*, vol. 19, pp. 58–69, 2002.
26. J. M. Grossmann, J. H. Bowles, D. Haas, J. A. Antoniades, M. R. Grunes, P. J. Palmadesso, D. Gillis, K. Y. Tsang, M. M. Baumback, M. Daniel, J. Fisher, and I. A. Triandaf, Hyperspectral analysis and target detection system for the Adaptive Spectral Reconnaissance Program (ASRP), *Proceedings of the SPIE*, vol. 3372, p. 2, 1998.
27. D. Gillis and J. Bowles, Target detection in hyperspectral Imagery using demixed spectral angles, *Proceedings of the SPIE*, vol. 5238, p. 244, 2004.

CHAPTER 5

STOCHASTIC MIXTURE MODELING

MICHAEL T. EISMANN
AFRL's Sensors Directorate, Electro Optical Technology Division, Electro Optical Targeting Branch, Wright-Patterson AFB, OH 45433

DAVID W. J. STEIN
MIT Lincoln Laboratory, Lexington, MA 02421

5.1. INTRODUCTION

As described elsewhere in this book, hyperspectral imaging is emerging as a powerful remote sensing tool for a variety of applications, ranging from the detection of low-contrast objects in complex background clutter to the classification of subtle man-made and natural terrain features. In order to effectively extract such information, hyperspectral signal processing algorithms usually depend on mathematical models of the spectral behavior of the scene. For terrain classification algorithms, for example, such mathematical models form the basis on which the terrain types are categorized. Similarly, target detection algorithms use such models to characterize the background clutter against which particular target spectra of interest are to be separated. In both applications, the algorithm performance is dependent on how well the underlying background models fit the actual variations in the data.

In this chapter, the issue of hyperspectral data modeling will be addressed primarily from the perspective of characterizing "background" scene content as opposed to "target" signatures. That is, it will focus on the characterization of the variance of abundant materials in a hyperspectral image (for example, natural terrain types, roads, bodies of water, etc.) as opposed to describing the nature of rare materials (vehicles, etc., that are often the focus of detection algorithms). Complementary work to that discussed in this chapter has been performed from the perspective of describing expected target spectra behavior, and examples can be found in references 1 and 2.

A variety of background modeling approaches have been applied to hyperspectral data, and this chapter will not attempt to provide a comprehensive treatment of

Hyperspectral Data Exploitation: Theory and Applications, Edited by Chein-I Chang

this extensive subject matter. However, it is probably fair to claim that there are two primary approaches to this problem, one based on a linear mixing model (LMM) and the other based on a statistical representation of the data. In simple terms, linear mixing models describe scene variation as random mixtures of a discrete number of pure deterministic material spectra, while statistical models describe the scene variation by distributions of random vectors. Such models are based on simplifying assumptions of the nature of hyperspectral data, which limits not only the efficacy of detection and classification algorithms based on them, but also parametric models for estimating the performance of such algorithms. The basic formulation and implementation of these models will be reviewed in Sections 5.2 and 5.3 to form a foundation for the remainder of the chapter, and some of these simplifying assumptions will be discussed.

The focus of this chapter is to describe the mathematical basis, implementation, and applications of a stochastic mixing model (SMM) that attempts to combine the fundamental mixing character of the LMM, which has a strong physical basis, with a statistical representation that is needed to capture variations in the data that are not well-described by linear mixing. The underlying formulation of this model is described in Section 5.4. Since the SMM is a more complex representation of hyperspectral data compared to more conventional LMM or statistical approaches, it poses a more challenging problem with regard to parameter estimation. Research in this area has been directed in two fundamentally different directions. These two approaches, called the discrete-class SMM and normal compositional model (NCM), are described and example applications are given that illustrate the effectiveness of the modeling approach.

5.2. LINEAR MIXING MODEL

The linear mixing model (LMM) is widely used to analyze hyperspectral data and has become an integral part of a variety of hyperspectral classification and detection techniques [3]. The physical basis of the linear mixing model is that hyperspectral image measurements often capture multiple material types in an individual pixel and that measured spectra can be described as a linear superposition of the spectra of the pure materials from which the pixel is composed. The weights of the superposition correspond to the relative abundances of the various pure materials. This assumption of linear superposition is physically well-founded in situations, for example, where the sensor response is linear, the illumination across the scene is uniform, and there is no scattering. Nonlinear mixing can occur, for example, when there is multiple scattering of light between elements in the scene.

Despite the potential for nonlinear mixing in real imagery, the linear mixing model has been found to be a fair representation of hyperspectral data in many situations. An example two-dimensional scatterplot of (simulated) data conforming to an LMM is depicted in Figure 5.1 in order to describe the basic method. These data simulate mixtures of three pure materials where the relative mixing is uniformly distributed. They form a triangular region where the vertices of the

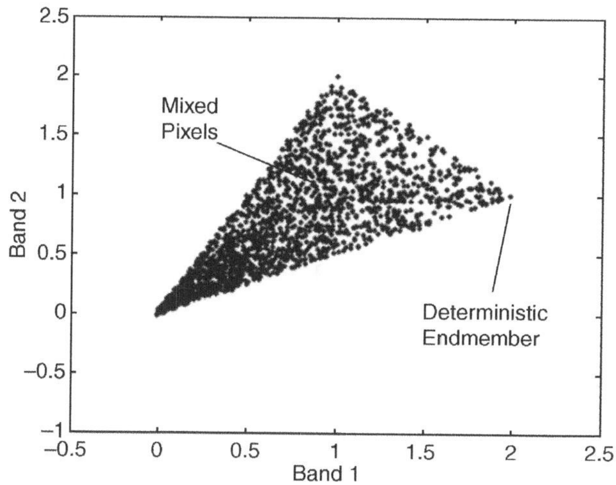

Figure 5.1. Scatterplot of dual-band data that conforms well to a linear mixing model.

triangle represent the spectra (two bands, in this case) of the pure materials (called endmembers), the edges represent mixtures of two endmembers, and interior points represent mixtures of more than two endmembers. The abundances of any particular sample are the relative distances of the sample data point to the respective endmembers. Theoretically, no data point can reside outside the triangular region, because this would imply an abundance of some material that exceeds unity, which violates the physical model.

The example scatterplot shown in Figure 5.1 can be conceptually extended to the multiple dimensions of hyperspectral imagery by recognizing that data composed of M endmembers will reside within a subspace of dimension $M - 1$, at most, and will be contained within a simplex with M vertices. The relative abundances of a sample point within this simplex are again given by the relative distances to these M endmembers.

5.2.1. Mathematical Formulation

To mathematically describe the linear mixing model, consider a hyperspectral image as a set of K-element vector measurements $\{\mathbf{x}_i, i = 1, 2, \ldots, N\}$, where N is the number of image spectra (or image pixels) and K is the number of spectral bands. According to the linear mixing model, each spectrum $\mathbf{x}_i$ can be described as a linear mixture of a set of endmember spectra $\{\varepsilon_m, m = 1, 2, \ldots, M\}$ with abundances $a_{i,m}$ and the addition of sensor noise described by the random vector process $\mathbf{n}$, or

$$\mathbf{x}_i = \sum_{m=1}^{M} a_{i,m}\, \varepsilon_m + \mathbf{n} \tag{5.1}$$

The noise process $\mathbf{n}$ is a stationary process according to the fundamental model description and is generally assumed to be zero mean. Because of the random nature of $\mathbf{n}$, each measured spectrum $\mathbf{x}_i$ should also be considered a realization of a random vector process. It may also be appropriate to consider the abundances as a random component; however, an inherent assumption of the LMM is that the endmembers spectra $\boldsymbol{\varepsilon}_m$ are deterministic.

Equation (5.1) can be written in matrix form,

$$\mathbf{X} = \mathbf{E}\mathbf{A} + \mathbf{N} \tag{5.2}$$

by arranging the image spectra as columns in the $K \times N$ data matrix $\mathbf{X}$, arranging the endmember spectra as columns in the $K \times M$ endmember matrix $\mathbf{E}$, defining the $M \times N$ abundance matrix $\mathbf{A}$ as $(a_{i,m})^{\mathrm{T}}$, and representing the noise process by $\mathbf{N}$. This notation is used to simplify the mathematical description of the estimators for the unknown endmember spectra and abundances.

5.2.2. Endmember Determination

The fundamental problem of establishing a linear mixing model, called spectral unmixing, is to determine the endmember and abundance matrices in Eq. (5.2) given a hyperspectral image. While there are methods that attempt to estimate these unknown model components simultaneously, for example, by positive matrix factorization [4], this is usually performed by first determining the endmember spectra and then estimating the abundances on a pixel-by-pixel basis. Often, endmember determination is performed in a manual fashion either by selecting areas in the image that are known or suspected to contain pure materials [5] or by selecting spectra that appear to form vertices of a bounding simplex of the data cloud using pixel purity filtering and multidimensional data visualization methods [6].

Automated methods typically either select spectra from the data as simplex vertices that maximize its volume [7, 8] or determine a minimum-volume simplex with arbitrary vertices that circumscribes the data [9, 10]. The former approach is detailed here as it forms the basis for initialization of the stochastic mixture modeling methods to be discussed later.

With reference again to Figure 5.1, consider the approach of selecting the M spectra from the columns of $\mathbf{X}$ that most closely represent vertices of the scatter plot. Under the assumptions that (1) pure spectra for each endmember exist in the data and (2) there is no noise, such spectra should be able to be exactly determined. The presence of sensor noise will, of course, cause an error in the determined endmembers on the order of the noise variance. If there are no pure spectra in the image for a particular endmember, then this approach will, at best, select vertices that do not quite capture the full extent of the true LMM simplex. This is referred to as an inscribed simplex.

From a mathematical perspective, endmember determination in this manner is a matter of maximizing the volume of a simplex formed by any subset of M spectra selected from the full data set [7]. The volume of a simplex formed by a set of vectors $\{\mathbf{y}_i, \mathrm{j} = 1, 2, \ldots, M\}$ in a spectral space of dimension $M - 1$ is proportional to

$$V = \begin{vmatrix} 1 & \cdots & 1 \\ \mathbf{y}_1 & \cdots & \mathbf{y}_M \end{vmatrix} \tag{5.3}$$

According to the procedure described in Winter [8], also known as N-FINDR, all the possible image vector combinations are recursively searched to find the subset of spectra from the image that maximize the simplex volume given in Eq. (5.3). To use this approach, however, the dimensionality of the data must be reduced from the K-dimensional hyperspectral data space to an $(M - 1)$-dimensional subspace within which the simplex volume can be maximized. Furthermore, this also implies some reasonable way to estimate the number of endmembers needed to describe the data. This dimensionality reduction issue is discussed later in Section 5.2.4.

While the procedure described above is sufficient for ultimately arriving at the M spectra in the image that best represent the endmembers according to the simplex volume metric, it can be significantly improved in terms of computational complexity by using a recursive formula for updating the determinant in (5.3) upon substitution of a column.

5.2.3. Abundance Estimation

Given the determination of the endmember matrix $\mathbf{E}$ by one of the methods described, the abundance matrix $\mathbf{A}$ in (5.2) is generally solved using least squares methods. This can be performed on a pixel-by-pixel basis. Defining $\mathbf{a}_i$ as the M-element abundance vector associated with image spectrum $\mathbf{x}_i$, the least-squares estimate is given by

$$\hat{\mathbf{a}}_i = \min_{\mathbf{a}_i} \| \mathbf{x}_i - \mathbf{E}\mathbf{a}_i \|_2^2 \tag{5.4}$$

The estimated abundance vectors are usually formed into images corresponding to the original hyperspectral image and are termed the abundance images for each respective endmember. According to the physical model underlying the LMM, the abundances are constrained by positivity,

$$a_{i,m} \leq 0 \qquad \forall i, m \tag{5.5}$$

and full additivity,

$$\mathbf{u}^T \mathbf{a}_i = 1 \qquad \forall i \tag{5.6}$$

where $\mathbf{u}$ is an M-dimensional vector whose entries are all unity.

Different least-squares solution methods are used based on if and how these constraints are applied. The unconstrained least-squares solution ignores both

Eqs. (5.5) and (5.6), and can be expressed in closed form [11]. A closed-form solution can also be derived in the case using only (5.6) using the method of Lagrange variables [12]. An alternative approach uses nonlinear least squares [13] to solve for the case using only Eq. (5.5). Strictly, however, a quadratic programming approach must be used to find the optimum solution with respect to both constraints. This can be done using the active set method described in Fletcher [14] or some other constrained optimization method.

5.2.4. Dimensionality Reduction

As previously discussed, spectral unmixing is generally performed within a subspace of dimension $M - 1$, where M is the number of endmembers used to represent the data. This raises the issue of dimensionality reduction—namely, what the correct number of endmembers is for a given hyperspectral image, and how the data should be transformed to this low dimensionality subspace.

A standard method of dimensionality reduction is based on the principal components transformation [15]. This method linearly transforms the data into a subspace for which the component images are uncorrelated and ranked in decreasing order of variance. The sample covariance matrix $\mathbf{C}$ is diagonalized into $\mathbf{V}^{\mathrm{T}}\mathbf{D}\mathbf{V}$, where $\mathbf{V}$ is the eigenvector matrix and $\mathbf{D}$ is a diagonal matrix containing the eigenvalues in decreasing order along its diagonal. The original spectra are transformed into the principal component domain by the linear transformation

$$\mathbf{z}_i = \mathbf{V}^T \mathbf{x}_i \tag{5.7}$$

This form of the principal component transformation is based on an assumption that the system noise is uncorrelated. An alternative form, called the minimum noise transformation [16], performs an initial linear transformation to whiten the noise based on an estimate of the noise covariance matrix prior to performing the transformation described above.

Since the eigenvalues correspond to the variance of their respective principal component images, the ordering of the principal components implies that most the hyperspectral image variance will be captured in the leading principal components. The trailing components will be dominated by system noise, as well as by rare spectra that do not contribute much to the global image statistics. For an image that adhered to the linear mixing model of Eq. (5.1) with an additive noise variance that was small relative to the endmember separation, all of the principal components from the Mth to the Kth would only include noise. Therefore, it would be not only adequate but also advantageous to perform spectral unmixing in the leading subspace of $M - 1$ principal components.

If there is no *a priori* information upon which to make a decision concerning the number of endmembers M in an image, then a fair estimate can be made by comparing the distribution of eigenvalues to that expected based on some knowledge of the system noise level. Given the true covariance matrix and white noise, the eigenvalue distribution becomes constant at and beyond the Mth eigenvalue. However, because the eigenvalues are only estimated based on a sample covariance matrix, the resulting distribution due to white noise will be of the form of Silverstein's asymptotic distribution [17] and making a good estimate of M is actually a bit more difficult. A methodology is given in Stocker et al. [18] for estimating a Silverstein model fit of the low-order modes from the sample covariance matrix of vector data. The data dimensionality can be estimated from the principal component at which the Silverstein model fit begins to match the actual data.

5.2.5. Limitations of the Linear Mixing Model

The linear mixing model is widely used because of the strong tie between the mathematical foundations of the model and the physical processes of mixing that result in much of the variance seen in hyperspectral imagery. However, situations exist where the basic assumptions of the model are violated and it fails to accurately represent the nature of hyperspectral imagery. The potential for nonlinear mixing is one situation that has already been mentioned. Another, which is the focus of this section, is the assumption that mixtures of a small number of deterministic spectra can be used to represent all of the non-noise variance in hyperspectral imagery.

Common observations would support the case that there is a natural variation to almost all materials; therefore, one would expect a certain degree of variability to exist for any materials that would be selected as endmembers for a particular hyperspectral image. Few materials in a scene will actually be pure, and endmembers usually only represent classes of materials that are spectrally very similar. Given the great diversity of materials of which a typical hyperspectral image is likely to be composed, the variation in endmember characteristics is further increased because of the practical limit to how many endmembers can be used to represent a scene. When the intra-class variance (within endmember classes) is very small relative to the inter-class variance (between endmembers), the deterministic endmember assumption upon which the linear mixing model is based may remain valid. However, when the intra-class variance becomes appreciable relative to the inter-class variance, the deterministic assumption becomes invalid and the endmembers themselves should be treated as random vectors.

Another source of variance in the endmember spectra is illumination. In reflective hyperspectral imagery, this is generally treated by capturing a dark endmember and assuming that illumination variation is equivalent to mixing between the dark endmember and each respective fully illuminated endmember. In thermal imagery, the situation is more complicated because the up-welling radiance is driven by both material temperature and down-welling radiance. Temperature variations will cause a nonlinear change in the up-welling radiance spectra, and the concept of finding a dark endmember (analogous to the reflective case) does not carry over into the thermal domain because of the large radiance offset that exists. Also, the down-welling illumination has a much stronger local component (i.e., emission from nearby objects) and is not adequately modeled as a simple scaling of the endmember spectra. Stochastic modeling of the endmember spectra is one way to deal with variations of this nature that are not well represented in a linear mixing model.

5.3. NORMAL MIXTURE MODEL

Given the limitations of the deterministic component of the linear mixing model, the motivation for adding a stochastic component to the model is hopefully clear. Prior to specifically describing how this can be achieved within the construct of the linear mixing model, however, it is first necessary to provide some background on purely statistical representations of hyperspectral imagery. Again, this section will only focus on those elements in the broad literature on this subject that are directly relevant to understanding the formulation of the stochastic mixing model discussed in the next section.

5.3.1. Statistical Representation

Setting the linear mixing model aside for a moment, an alternative approach to modeling hyperspectral imagery is to assume that the image, denoted again by the set of K-element vector measurements $\{\mathbf{x}_i, i = 1, 2, \ldots, N\}$, is a realization of a random vector process that will be denoted by $\mathbf{x}$. In general, the probability density function that defines $\mathbf{x}$ will be both spatially and spectrally dependent. However, the assumption of spatial independence that is often made in hyperspectral image modeling will be employed throughout this section. In that case, the random process is completely described by a probability density function $p(\mathbf{x})$ in the multidimensional spectral space.

Although other distributions have been employed [19], use of the multidimensional normal distribution is the most common in representing $p(\mathbf{x})$. To deal with spatial variation in scene statistics, however, the use of a simple normal distribution with global statistics is insufficient. Instead, such variation can be captured by either employing spatially variant parameters [20] or through the use of a normal mixture model. The latter is discussed in this section, because one methodology of stochastic mixture modeling is founded on this approach.

The normal mixture model is based on an assumption that each spectrum $\mathbf{x}_i$ in an image is a member of one of Q classes defined by a class index q where $q = 1, 2, \ldots, Q$. Each spectrum has a prior probability $P(q)$ of belonging to each respective class, and the class-conditional probability density function $p(\mathbf{x}|q)$ is normally distributed and completely described by a mean vector $\mathbf{m}_{\mathbf{x}|q}$ and covariance matrix $\mathbf{C}_{\mathbf{x}|q}$. The probability density function $p(\mathbf{x})$ is then given by

$$p(\mathbf{x}) = \sum_{q=1}^{Q} P(q)\, p(\mathbf{x}|q) \tag{5.8}$$

where

$$p(\mathbf{x}|q) = \frac{1}{(2\pi)^{K/2}} \frac{1}{|\mathbf{C}_{\mathbf{x}|q}|^{1/2}} \exp\left\{ -\frac{1}{2} [\mathbf{x} - \mathbf{m}_{\mathbf{x}|q}]^T \mathbf{C}_{\mathbf{x}|q}^{-1} [\mathbf{x} - \mathbf{m}_{\mathbf{x}|q}] \right\} \tag{5.9}$$

The statistical representation in (5.8) and (5.9) is referred to as a normal mixture distribution. The term *mixture* in this context, however, is used in a very different manner than its use with respect to the linear mixing model, and it is very important to understand the distinction. In the linear mixing model, the mixing represents the modeling of spectra as a combination of the various endmembers. In the normal mixture model, however, each spectrum is ultimately assigned to a single class, and mixing of this sort (i.e., weighted combinations of classes) is not modeled. The term *mixture*, in this case, merely refers to the representation of the total probability density function $p(\mathbf{x})$ as a linear combination of the class conditional distributions.

The prior probability values $P(q)$ are subject to the constraints of positivity,

$$P(q) > 0 \qquad \forall q \tag{5.10}$$

and full additivity,

$$\sum_{q=1}^{Q} P(q) = 1 \tag{5.11}$$

in a similar manner to the abundances in the linear mixing model. This common characteristic is somewhat circumstantial, however, since these parameters represent very different quantities in the respective models. Figure 5.2 illustrates a dual-band scatterplot of data that conforms well to a normal mixture model. The ovals indicate contours of the class-conditional distributions for each class.

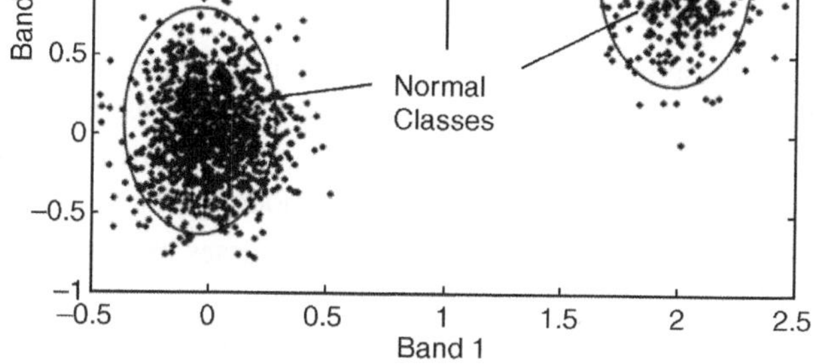

Figure 5.2. Scatterplot of dual-band data that conforms well to a normal mixture model.

5.3.2. Spectral Clustering

A normal mixture model is fully represented by the set of parameters $\{P(q), \mathbf{m}_{\mathbf{x}|q}, \mathbf{C}_{\mathbf{x}|q}, q = 1, 2, \ldots, Q\}$. When these parameters are known, or estimates of these parameters exist, each image spectrum $\mathbf{x}_i$ can be assigned into a class using a distance metric from the respective classes. Some alternatives include the Euclidean distance,

$$d_q = \| \mathbf{x}_i - \mathbf{m}_{\mathbf{x}|q} \|_2 \tag{5.12}$$

and the Mahalanobis distance,

$$d_q = (\mathbf{x}_i - \mathbf{m}_{\mathbf{x}|q})^T \mathbf{C}_{\mathbf{x}|q}^{-1} (\mathbf{x}_i - \mathbf{m}_{\mathbf{x}|q}) \tag{5.13}$$

The assignment process is *classification*, with these two alternatives referred to as linear and quadratic classification, respectively.

Normal mixture modeling includes not only the classification of the image spectra but also the concurrent estimation of the model parameters. This process is typically called *clustering*, and a variety of methods are described in the literature based on linear and quadratic classifiers as well as maximum likelihood methods [21]. One problem that often arises with these clustering approaches is that they tend to bias against overlapping class-conditional distributions. This occurs because of the hard decision rules in classification. That is, even though a spectrum may have comparable probability of being a part of more than one class, it is always assigned to the class that minimizes the respective distance measure.

5.3.3. Stochastic Expectation Maximization

Stochastic expectation maximization (SEM) is a quadratic clustering algorithm that addresses the bias against overlapping class-conditional probability density functions by employing a Monte Carlo class assignment [22]. The SEM algorithm is detailed here and is employed in the discrete stochastic mixing model algorithm discussed later. Functionally, it is composed of the following steps:

1. Initialize the parameters of the Q classes using uniform priors and global sample statistics (i.e., the same parameters for all classes).
2. Estimate the posterior probability for the combinations of the N image spectrum and Q classes based on the current model parameters.
3. Assign the N image spectra amongst the Q classes by a Monte Carlo method based on the posterior probability estimates.
4. Estimate the parameters of the Q classes based on the sample statistics of the spectra assigned to each respective class.
5. Repeat steps 2 to 4 until the algorithm converges.

6. Repeat step 2 and classify the image spectra to the classes exhibiting the maximum posterior probability.

Steps 2 to 4 are detailed below.

5.3.3.1. Posterior Class Probability Estimation. The posterior class probability values for an iteration n are estimated for all combinations of image spectra $\{\mathbf{z}_i, i = 1, 2, \ldots, N\}$ and mixture classes $q = 1, 2, \ldots, Q$ by

$$\hat{P}^{(n)}(q|\mathbf{z}_i) = \frac{\hat{P}^{(n)}(q)\hat{p}^{(n)}(\mathbf{z}_i|q)}{\sum_{q=1}^{Q} \hat{P}^{(n)}(q)\hat{p}^{(n)}(\mathbf{z}_i|q)} \tag{5.14}$$

The principal component representation of the spectra (i.e., $\mathbf{z}$ as opposed to $\mathbf{x}$) is used in Eq. (5.14) because SEM is typically performed in the leading principal component subspace to reduce computational complexity and estimation errors due to limited sample populations. Equation (5.14) represents the probability that an image spectrum is contained in any of the mixture classes based on the current model parameters.

5.3.3.2. Monte Carlo Class Assignment. The stochastic characteristic of SEM arises due to the Monte Carlo assignment of spectra into classes based on the posterior probability values. For each spectrum $\mathbf{z}_i$, a random number R is generated with a probability distribution of

$$P_R(R) = \begin{cases} \hat{P}^{(n-1)}(q|\mathbf{z}_i), & q = 1,\ 2, \ldots, Q \\ 0 & \text{otherwise} \end{cases} \tag{5.15}$$

The class index estimate is then given by

$$\hat{q}^{(n)}(\mathbf{z}_i) = R \tag{5.16}$$

This is independently repeated for each image spectrum. Because of the Monte Carlo nature of this class assignment, the same spectrum may be assigned to different classes on successive iterations even when the model parameters do not change. This is what allows the resulting classes to overlap. This property, however, means that the change in class assignments is no longer an appropriate metric for algorithm convergence, and metrics based on convergence of the model parameters must be used.

5.3.3.3. Parameter Estimation. The normal mixture model parameters are estimated according to

$$\hat{P}^{(n)}(q) = \frac{N_q^{(n)}}{N} \tag{5.17}$$

$$\hat{\mathbf{m}}_q^{(n)} = \frac{1}{N_q^{(n)}} \sum_{\substack{i=1 \\ i \in \Omega_q^{(n)}}}^{N_q^{(n)}} \mathbf{z}_i \tag{5.18}$$

and

$$\hat{\mathbf{C}}_q^{(n)} = \frac{1}{N_q^{(n)} - 1} \sum_{\substack{i=1 \\ i \in \Omega_q^{(n)}}}^{N_q^{(n)}} [\mathbf{z}_i - \hat{\mathbf{m}}_q^{(n)}] \, [\mathbf{z}_i - \hat{\mathbf{m}}_q^{(n)}]^T \tag{5.19}$$

where $N_q^{(n)}$ is the number of spectra assigned to the qth class and $\Omega_q^{(n)}$ refers to the set of indices i for all samples $\mathbf{z}_i$ assigned to the qth class.

5.4 STOCHASTIC MIXING MODEL

In the simplest terms, the stochastic mixing model is a linear mixing model that treats the endmembers as random vectors as opposed to deterministic spectra. While the potential exists for statistically representing the endmembers through a variety of probability density functions, only the use of normal distributions is detailed in this section. As in the normal mixture model, the challenge in stochastic mixture modeling lies more in the estimation of the model parameters than in the representation of the data. The subsequent two sections detail two fundamentally different approaches to this estimation challenge, the first of which is strongly tied to the SEM algorithm discussed in Section 5.2. Before embarking on these two paths, however, this section addresses the underlying model formulation in more general terms.

5.4.1. Model Formulation

The stochastic mixing model is similar to the well-developed linear mixing model in that it attempts to decompose spectral data in terms of a linear combination of endmember spectra. However, in the case of an SMM, the data are represented by an underlying random vector $\mathbf{x}$ with endmembers $\boldsymbol{\varepsilon}_m$ that are $K \times 1$ normally distributed random vectors, parameterized by their mean vector $\mathbf{m}_m$ and covariance matrix $\mathbf{C}_m$. The variance in the hyperspectral image is then interpreted by the model as a combination of both endmember variance and subpixel mixing, which contrasts with the LMM for which all the variance is modeled as subpixel mixing. Figure 5.3 illustrates dual-band data that conforms well to a stochastic mixing model. The ovals indicate contours of the endmember class distributions that linearly mix to produce the remaining data scatter.

The hyperspectral data $\{\mathbf{x}_i, i = 1, 2, \ldots, N\}$ are treated as a spatially independent set of realizations of the underlying random vector $\mathbf{x}$, which is related to the

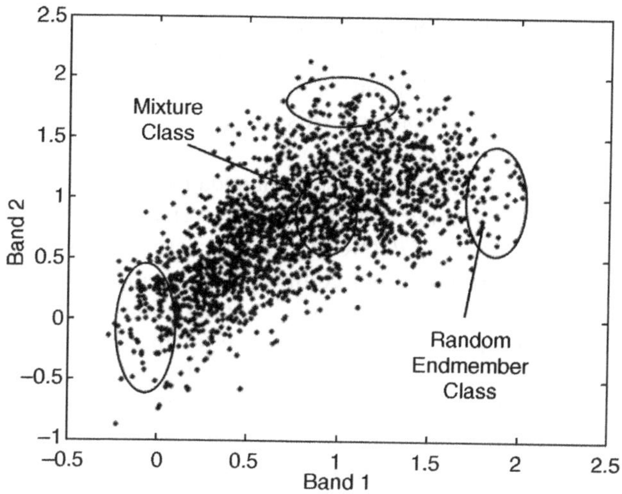

Figure 5.3. Scatterplot of dual-band data that conforms well to a stochastic mixing model.

random endmembers according to the linear mixing relationship

$$\mathbf{x} = \sum_{m=1}^{M} a_m \, \boldsymbol{\varepsilon}_m + \mathbf{n} \tag{5.20}$$

where a_m are random mixing coefficients that are constrained by positivity and full additivity as in Eqs. (5.5) and (5.6), and $\mathbf{n}$ represents the sensor noise as in (5.1). Because of the random nature of the endmembers, the sensor noise can easily be incorporated as a common source of variance to all endmembers. This allows the removal of noise process $\mathbf{n}$ in Eq. (5.20) without losing any generality in the model.

5.4.2. Parameter Estimation Challenges

While the SMM provides the benefit relative to the LMM of being able to model inherent endmember variance, this benefit comes at the cost of turning a fairly standard least-squares estimation problem into much more complicated estimation problem. This arises due to the doubly stochastic nature of (5.20)—that is, the fact that $\mathbf{x}$ is the sum of products of the two random variables a_m and $\boldsymbol{\varepsilon}_m$.

Two fundamentally different approaches are presented in this chapter for solving the problem of estimating the random parameters underlying the SMM in (5.20) from a hyperspectral image. In the first, the problem is constrained by requiring the mixing coefficients to be quantized to a discrete set of mixing levels. By performing this quantization, the estimation problem can be turned into a quadratic clustering problem similar to that described in Sections 5.3.2 and 5.3.3. The difference in the SMM case is that the classes are interdependent due to the linear mixing relationships. A variant of SEM is employed to self-consistently estimate the endmember statistics and the abundance images. As will be seen, the abundance quantization that is fundamental to this method provides an inherent limitation in model fidelity, and the achievement of finer quantization results in significant computational complexity problems. This solution methodology is termed the *discrete stochastic mixing model.*

The second approach employs no such abundance quantization or restriction (sum to one) on the constraints. The continuous version of the SMM is termed the *normal compositional model* (NCM). In this terminology, *normal* refers to the underlying class distribution and compositional to the property of the model that represents each datum as a (convex) combination of underlying classes. The term *compositional* is in contrast to mixture, as in normal mixture model, in which each datum emanates from one class that is characterized by a normal probability distribution. Rather than imposing the additivity constraint (5.6), the NCM may be employed with more general constraints as described in Section 5.6. Furthermore, the NCM makes no assumptions about the existence of "pure pixels." The NCM is identified as a hierarchical model, and the Monte Carlo expectation maximization algorithm is used to estimate the parameters. The Monte Carlo step is performed by using Monte Carlo Markov chains to sample from the posterior distribution (i.e., the distribution of the abundance values given the observation vector and the current estimate of the class parameters).

5.5. DISCRETE STOCHASTIC MIXTURE MODEL

Use of the discrete SMM for hyperspectral image modeling was first reported by Stocker and Schaum [23] and was further refined by Eismann and Hardie [24, 25]. This section details a fairly successful embodiment of the discrete SMM parameter estimation strategy based on these references. For a more thorough treatment of the variations that have been investigated, these references should be consulted.

5.5.1. Discrete Mixture Class Formulation

In the discrete SMM, a finite number of mixture classes Q are defined as linear combinations of the endmember random vectors,

$$\mathbf{x}|q = \sum_{m=1}^{M} a_m(q)\boldsymbol{\varepsilon}_m \tag{5.21}$$

where $a_m(q)$ are the fractional abundances associated with the mixture class, and $q = 1, 2, \ldots, Q$ is the mixture class index. The fractional abundances corresponding to each mixture class are assumed to conform to the physical constraints of positivity,

$$a_m(q) \geq 0 \qquad \forall m, q \tag{5.22}$$

and full additivity,

$$\sum_{m=1}^{M} a_m(q) = 1 \qquad \forall q \tag{5.23}$$

Also, the abundances are quantized into L levels (actually $L+1$ levels when zero abundance is included) such that

$$a_m(q) \in \left\{0, \frac{1}{L}, \frac{2}{L}, \ldots, 1\right\} \qquad \forall q \tag{5.24}$$

This is referred to as the quantization constraint. For a given number of endmembers M and quantization levels L, there is a finite number of combinations of abundances $\{a_1(q), a_2(q), \ldots, a_M(q)\}$ that simultaneously satisfy the constraints (5.22), (5.23), and (5.24). This defines the Q discrete mixture classes and is discussed further in Section 5.5.2.2.

Due to the linear relationship in (5.22), the mean vector and covariance matrix for the qth mixture class are functionally dependent on the corresponding endmember statistics by

$$\mathbf{m}_{\mathbf{x}|q} = \sum_{m=1}^{N_e} a_m(q)\, \mathbf{m}_m \tag{5.25}$$

and

$$\mathbf{C}_{\mathbf{x}|q} = \sum_{m=1}^{N_e} a_m^2(q)\, \mathbf{C}_m \tag{5.26}$$

The probability density function is described by the Gaussian mixture distribution given in Eqs. (5.8) and (5.9). As in the case of the normal mixture model described in Section 5.3, the mixing model parameters that must be estimated from the measured data are the prior probability values for all the mixture classes and the mean vectors and covariance matrices of the endmember classes. In this case, however, the relationship between the mixture class statistics is constrained through (5.25) and (5.26). Therefore, the methodology used for parameter estimation must be modified to account for this fact.

It is important to recognize the difference between the prior probability values P(q) and the mixture-class fractional abundances $a_m(q)$ used in this model. The prior probability characterizes the expectation that a randomly selected image pixel will be classified as part of the qth mixture class, while the fractional abundances define the estimated proportions of the endmembers of which the pixel is composed if it is contained in that mixture class. All of the mixture classes and not just the endmember classes, are represented with prior probability values.

5.5.2. Stochastic Unmixing Algorithm

A modified Stochastic Expectation Maximization (SEM) approach is used to self-consistently estimate the model parameters, $\mathbf{m}_{\mathbf{x}|q}$, $\mathbf{C}_{\mathbf{x}|q}$, and $P(q)$, and the abundance vectors $a_m(q)$ for the measured hyperspectral scene data. Estimation in a lower-dimensional subspace of the hyperspectral data, using one of the dimensionality reduction methods discussed in Section 5.2.4, is preferred (if not necessary) for reasons of reducing computational complexity and improving robustness against system noise. The primary steps of this stochastic unmixing algorithm are detailed below.

5.5.2.1. Endmember Class Initialization. Through Eqs. (5.25) and (5.26), the mean vectors and covariance matrices of all the mixture classes are completely dependent on the corresponding statistical parameters of the endmember classes, as well as on the set of feasible combinations of mixture class abundances. To initialize the SMM, therefore, it is only necessary to initialize these parameters. Several methods for initialization of the endmember class parameters were investigated in Eismann and Hardie [24]. For endmember mean vector initialization, a methodology based on the simplex maximization algorithm presented in Section 5.2.2.1 provided the best results. That is, M image spectra are selected from the image data that maximize the simplex volume metric given in Eq. (5.3), and these are used to define the initial endmember class mean vectors.

In practice, the data dimensionality K may not be precisely equal to $M - 1$, which is needed to use (5.3). To accommodate this, the initialization strategy is slightly modified. If K is greater than $M - 1$, then the first $M - 1$ components of the initial endmember mean spectra are determined using Eq. (5.3) in the leading subspace (principal components 1 through $M - 1$), and the remaining $K - M + 1$ components are then determined as the global mean vector of the trailing susbspace (principal components M through K). The case where K is less than $M - 1$ has not been considered.

Use of the global covariance matrix of the data for all the initial endmember classes has been found to be an adequate approach for covariance initialization. Empirical observations reported in Eismann and Hardie [24] indicate some advantage to using the global covariance matrix scaled by some factor between one and two, but the impacts on algorithm convergence were not significant.

5.5.2.2. Feasible Abundance Vectors. Initialization also involves specifying the set of feasible abundance vectors that are to represent mixed classes, where an abundance vector is defined as

$$\mathbf{a}(q) = [a_1(q), a_2(q), \ldots, a_M(q)]^T \tag{5.27}$$

Note that a unique abundance vector is associated with each mixture class index q, and the set of all abundance vectors over the Q mixture classes encompass all the feasible abundance vectors subject to the constraints given by Eqs. (5.22) to (5.24).

Therefore, these abundance vectors only need to be defined and associated with mixture class indices once. As the iterative SEM algorithm progresses, these fixed abundance vectors will maintain the relationship between endmember and mixture class statistics given by Eqs. (5.25) and (5.26). The final estimated abundances for a given image spectrum will be determined directly by the associated abundance vector for the mixture class into which that spectrum is ultimately classified.

A structured, recursive algorithm has been used to determine the complete, feasible set of abundance vectors. It begins by determining the complete, unique set M-dimensional vectors containing ordered (nonincreasing) integers between zero and the number of levels L whose elements add up to L. With $M = L = 4$, for example, these would be

$$[4,0,0,0], \quad [3,1,0,0], \quad [2,2,0,0], \quad [2,1,1,0], \quad \text{and} \quad [1,1,1,1]$$

A matrix is then constructed where the rows consist of all possible permutations of each of the ordered vectors. The rows are sorted and redundant rows are removed by comparing adjacent rows. After dividing the entire matrix by L, the remaining rows contain the complete, feasible set of abundance vectors for the given quantization level and number of endmembers.

The resulting number of total mixture classes Q (or unique abundance vectors) is equivalent to the unique combinations of placing L items in M bins [26], or

$$Q = \binom{L+M-1}{L} = \frac{(L+M-1)!}{(L)!(M-1)!} \tag{5.28}$$

An extremely large number of mixture classes can occur when the number of quantization levels and/or number of endmembers increases, and this can make the stochastic mixing model impractical in these instances. Even when the number of endmembers in a scene is large, however, it is unexpected for real hyperspectral imagery that any particular pixel will contain a large subset of the endmembers. Rather, it is more reasonable to expect that they will be composed of only a few constituents, although the specific constituents will vary from pixel to pixel. This physical expectation motivates the idea of constraining the fraction vectors to allow only a subset (say, $M_{\max}$) of the total set of endmembers to be used to represent any pixel spectrum. This can be performed by adding this additional constraint in the first step of determining the feasible abundance vector set. The equivalent combinatorics problem then becomes placing L items in M endmember classes where only $M_{\max}$ classes can contain any items. As shown in the Appendix, the resulting number of mixture classes in this case is

$$Q = \sum_{j=1}^{\min(M_{\max},L)} \binom{M}{j}\binom{L-1}{j-1} \tag{5.29}$$

Figure 5.4 illustrates how limiting the mixture classes in this way reduces the total number of mixture classes in the model for an example six-endmember case. As the

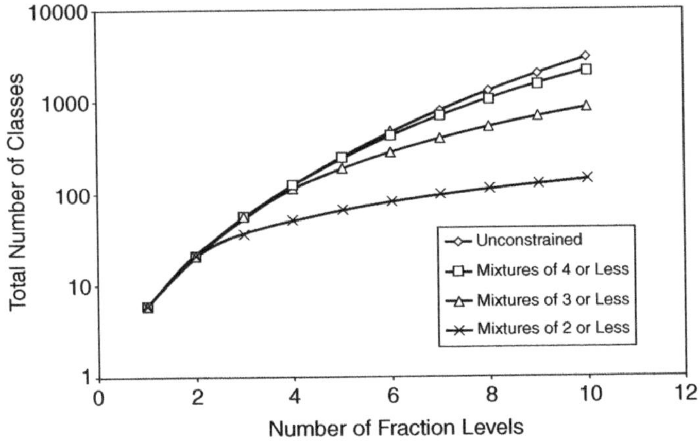

Figure 5.4. Impact of a mixture constraint on the number of SMM mixture classes.

number of endmembers increases, the reduction is even more substantial. Results presented in Eismann and Hardie [24, 25] indicate that employing such a constraint does not tend to provide a significant degradation in modeling performance.

5.5.2.3. Parameter Estimation. After initialization, the model is updated by repeating the same three SEM steps described in Section 5.3.3 with some modification to account for the linear mixing structure of the model. Posterior class probability estimation is performed identically, with estimates computed for combinations of all N image spectra and Q mixture classes. Monte Carlo class assignment is also performed identically, with the resulting class index estimates associated with one of all Q mixture classes.

The modifications of the standard SEM algorithm concern parameter estimation. The first involves estimation of the class mean vectors and covariance matrices. Because of the fixed relationship between the endmember class statistics and all the other mixture class statistics, it is sufficient to only estimate the endmember parameters using (5.18) and (5.19) from spectra assigned to the M endmember classes (i.e., those for which one abundance vector element is unity), and then to estimate the parameters of all the mixture classes based on these estimates through (5.25) and (5.26). This is the approach that has been reported in the literature [23–25]. An alternative method would be to (a) estimate the parameters individually for each mixture class based on the full data set and (b) derive estimates for the endmember classes by solving the linear system of equations that results from (5.25) and (5.26). This latter approach has not been pursued to date.

The second modification involves estimation of the prior probability values $P(q)$. In the traditional SEM clustering algorithm, these parameters are iteratively updated and have the effect of increasing the number of pixels assigned to the highly populated classes and decreasing the number assigned to the sparsely populated classes. This is a positive feature for a clustering algorithm because sparsely populated classes do not represent good clusters unless they are distantly located from other, more highly populated clusters. In a spectral mixing model, however, there is no reason to bias against sparsely populated mixture classes because they represent legitimate mixtures of endmembers that just happen to be somewhat uncommon. In fact, empirical results show that biasing against such mixture classes has the negative effect of forcing the model toward a situation where it interprets all the image variance as endmember variance and not variance due to mixing.

To remedy this undesired feature, the algorithm was modified such that the random class assignments used for estimating the class statistics are based on uniformly distributed priors as opposed to those estimated using Eq. (5.17). The rationale for this modification is to preserve sparsely populated mixture classes and to prevent the model from migrating toward an end state where almost all of the spectra are classified into a small number of pure classes with large variance. This modification was found to have a significant positive effect on algorithm convergence [24, 25].

5.5.2.4. Abundance Map Estimation. After the SEM iterations are terminated, final pixel class assignments are made by making a final posterior probability estimate according to Eq. (5.14) and then assigning each pixel to the class $\hat{q}_i$ that maximizes the posterior probability. The estimated prior probability values for the classes are used in this step, unlike the uniformly distributed values used during parameter estimation. The estimated abundance vector for each sample $\mathbf{z}_i$ is then given by the abundances $a_m(\hat{q}_i)$ corresponding to this mixture class. These abundance vectors can be formatted into abundance images corresponding to each endmember, and they are quantized at the initially selected level.

5.5.3. Data Representation Metrics

Statistical metrics are used to understand (a) the behavior of the iterative algorithm as it progresses and (b) the characteristics of the resulting mixing model. These metrics are intended to measure two key aspects of the SMM: (1) how well the data are represented by the mixture distribution described by the estimated SMM parameters and (2) how separated the endmember classes are from each other. Achieving both characteristics provides a statistical model that interprets as much of the image variance as possible by endmember mixing (i.e., increasing relative endmember separability) while interpreting only the remaining variance as endmember variability in order to maintain a good statistical fit. Two statistical metrics for these purposes are described here.

5.5.3.1. Log-Likelihood Metric. The fit of an SMM on the nth iteration can be characterized by the log-likelihood [23, 27]

$$L^{(n)} = \frac{1}{N}\sum_{i=1}^{N} \ln\left[\sum_{q=1}^{Q} \hat{P}^{(n)}(q)\hat{p}^{(n)}(\mathbf{z}_i|q)\right] \tag{5.30}$$

where the estimated class-conditional probability density function is of the form given in (5.9) and the mixture class parameters are derived from the endmember class parameters through the relationships given in Eqs. (5.25) and (5.26). A higher value of the log-likelihood represents a better fit of the data to the model because the posterior probability is maximized in a cumulative sense for the given set of spectra relative to the SMM parameter estimates.

5.5.3.2. Endmember Class Separability Metric. A standard separability metric [28] is the ratio of the between-class to within-class scatter for the pure endmember classes:

$$J^{(n)} = \text{trace}[(\mathbf{S}_w^{(n)})^{-1}\mathbf{S}_b^{(n)}] \tag{5.31}$$

where $\mathbf{S}_\text{w}$ is the within-class scatter matrix,

$$\mathbf{S}_w^{(n)} = \sum_{m=1}^{N_e} \hat{P}^{(n)}(q_m)\, \hat{\mathbf{C}}_m^{(n)} \tag{5.32}$$

$\mathbf{S}_\text{b}$ is the between-class scatter matrix,

$$\mathbf{S}_b^{(n)} = \sum_{m=1}^{N_e} \hat{P}^{(n)}(q_m)\, [\, \hat{\mathbf{m}}_m^{(n)} - \hat{\mathbf{m}}_0^{(n)}]\, [\, \hat{\mathbf{m}}_m^{(n)} - \hat{\mathbf{m}}_0^{(n)}]^T \tag{5.33}$$

$\hat{\mathbf{m}}_0$ is the overall mean vector for the pure endmember classes,

$$\hat{\mathbf{m}}_0^{(n)} = \sum_{m=1}^{N_e} \hat{P}^{(n)}(q_m)\, \hat{\mathbf{m}}_m^{(n)} \tag{5.34}$$

and $\{q_m\}$ is the set of pure endmember class indices,

$$\{q_m : a_m(q_m) = 1,\; a_p(q_m) \neq 1,\, p \neq m,\, m = 1, 2, \ldots, M\} \tag{5.35}$$

For the same log-likelihood, higher pure class separability represents a better SMM because the image variance is more the result of endmember mixing and less the result of endmember variance. On the other hand, maximizing the endmember class separability at the expense of the log-likelihood means that the image variance is overly attributed to endmember mixing in the model and represents a poor statistical fit. Thus, a result is sought that simultaneously maximizes (or balances) both metrics.

5.5.4. Example Results

To illustrate the basic characteristics of an SMM, example results for simple test image data are shown in this section and briefly compared to those derived using

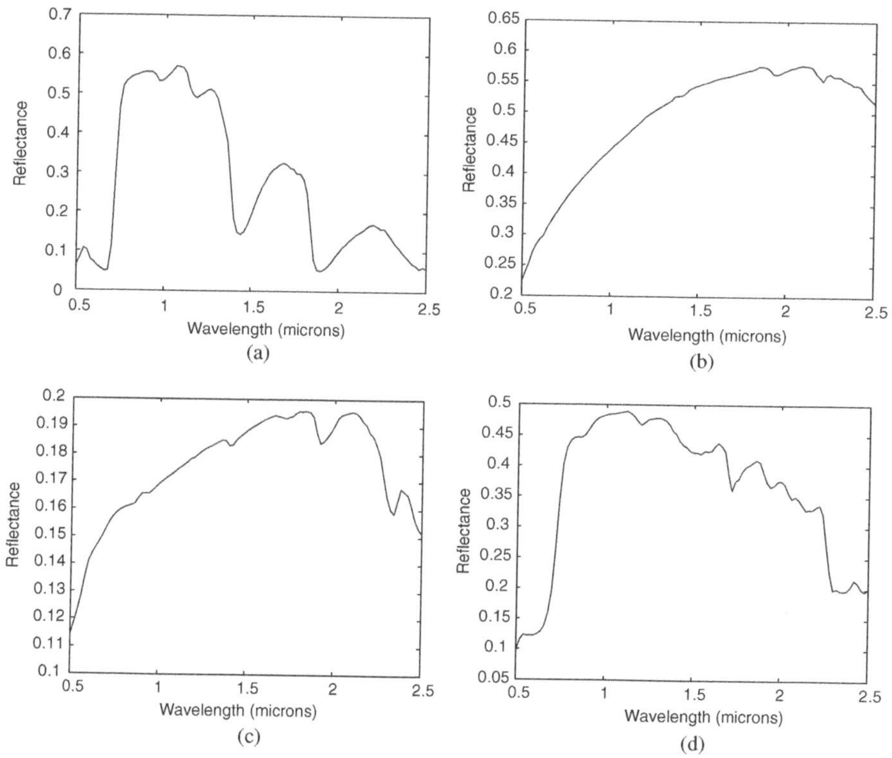

Figure 5.5. Endmember spectra for reflective test data.

LMM. The objective of this section is merely illustrative. The efficacy of the SMM for various realistic applications is left for discussion in Section 5.7.

5.5.4.1. Reflective Test Data. As a first example of how the SMM works, consider a test hyperspectral image consisting of the four endmembers shown in Figure 5.5 that are linearly mixed based on the true abundance images shown in Figure 5.6 with the endmembers located in the image corners. The abundance images are displayed such that white corresponds to an abundance of one and black

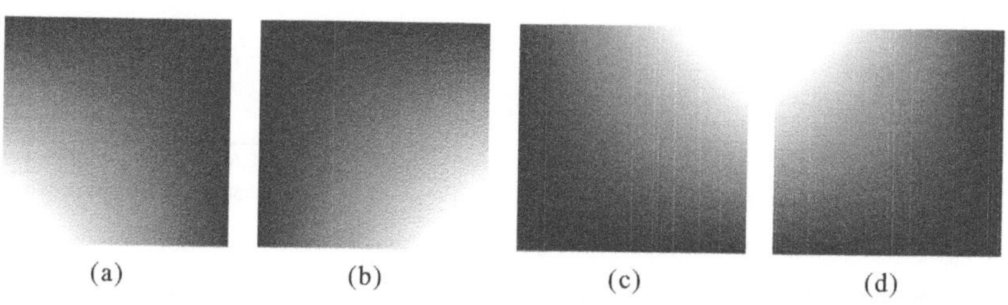

Figure 5.6. Abundance images for reflective test data.

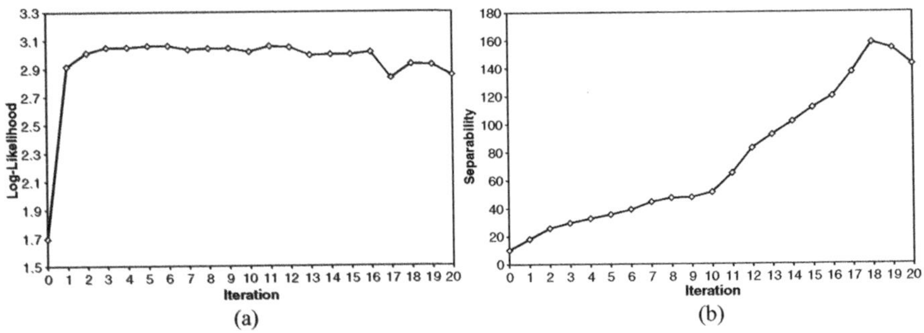

Figure 5.7. Convergence of SMM (a) log-likelihood and (b) endmember separability.

corresponds to an abundance of zero. This particular example consists of 101 spectral bands over the 0.4- to 2.5-μm spectral range (in reflectance units) and 51 × 51 spatial pixels. Spatially independent, normally distributed, zero-mean noise was added to the data, and the images were transformed into a three-dimensional subspace for LMM and five-dimensional subspace for SMM. Both LMM and SMM results were produced for four endmembers, with the SMM cases consisting of eight abundance quantization levels and 20 SEM iterations.

Figure 5.7 illustrates the convergence of the log-likelihood and separability metrics over the 20 SEM iterations for an image noise standard deviation of 0.05 reflectance units. Although not implemented as part of this example, these metrics could be used in an automated stopping criterion for the algorithm. The pixels assigned to the SMM endmember classes are indicated in the scatterplots for the leading three principal components displayed in Fig. 5.8. Such classes are positioned in the general vicinity of the true endmembers and exhibit a variance that is indicative of the noise level of the data.

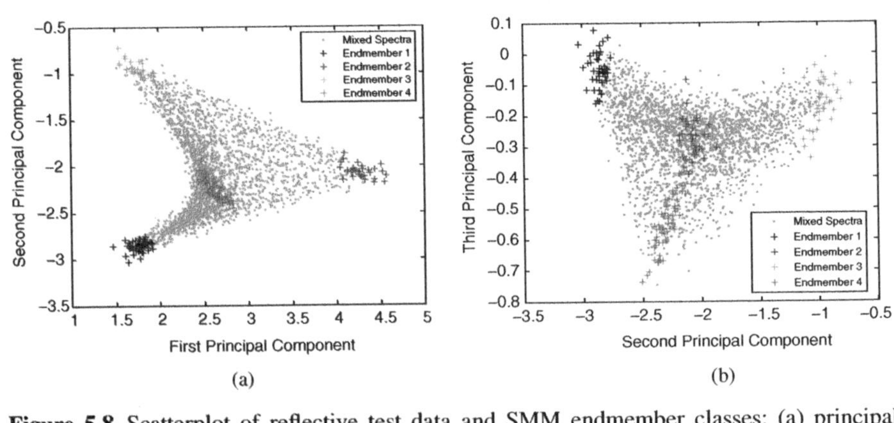

Figure 5.8 Scatterplot of reflective test data and SMM endmember classes: (a) principal components 1 and 2 and (b) principal components 2 and 3.

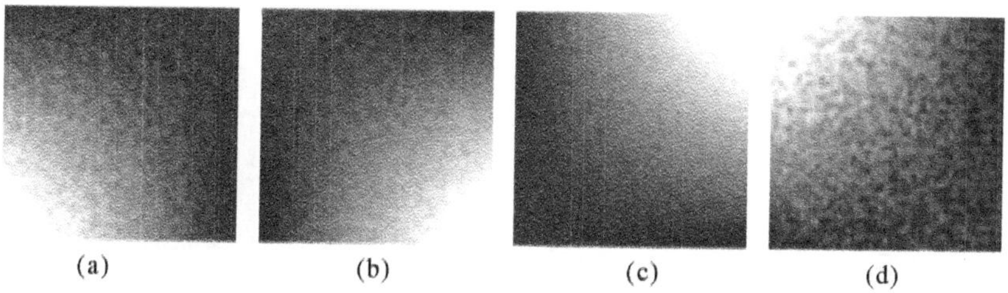

Figure 5.9. LMM abundance images for reflective test data.

For low noise levels, this example conforms very well to the assumptions of the LMM. As the noise level increases, it also conforms to the basic model assumptions, but the errors in endmember estimation result in inaccuracies in abundance estimates. This was investigated by comparing the error in the LMM and SMM abundance images relative to the truth as a function of the noise level. Figures 5.9 and 5.10 illustrate the estimated LMM and SMM abundance images at the noise level of 0.05 reflectance units. Note the more random nature of the LMM result as compared to the tiered abundance profile for the SMM. This is due to the quantized nature of the discrete SMM. In Figure 5.11, the root-mean-square error between the estimated abundance images and the truth is plotted as a function of the noise level for both approaches. The LMM error is roughly equal to the noise level. At low noise levels, the SMM error is indicative of the abundance quantization. At higher noise levels, it is on the order of the noise level, but slightly lower than that of the LMM.

5.5.4.2. Thermal Test Data. Even at the higher noise levels, the reflective example conforms to the LMM assumptions because the noise is additive and independent of the endmembers from which the image spectra are composed. In this section, a thermal hyperspectral example is provided for which there is a large inherent variance to the endmembers, and the characteristics of the variance are endmember-dependent. This case illustrates the type of situation for which stochastic mixture modeling is required.

Consider an image for which all of the spectra are composed of mixtures of two endmembers with linear abundance images depicted in Figure 5.12. A dual-band

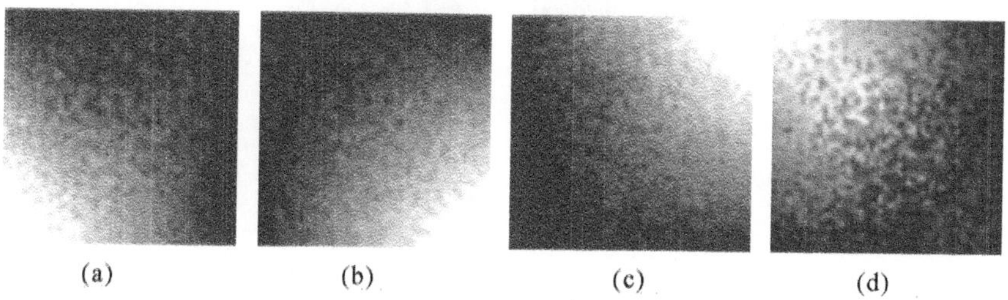

Figure 5.10. SMM abundance images for reflective test data.

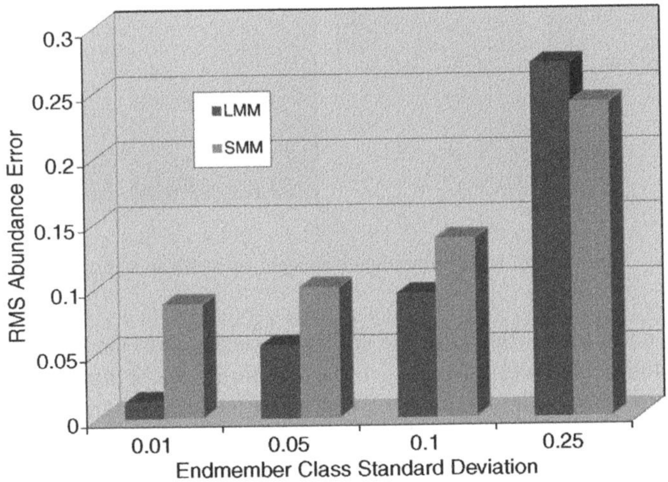

Figure 5.11. SMM and LMM root-mean-square abundance error as a function of noise level for reflective test data.

example is given here in order to simplify the illustration. Each image spectrum can be modeled as a linear superposition of two emission profiles [29] given by

$$\mathbf{x}_i = a_{i,1}\begin{bmatrix} \dfrac{2hc^2}{\lambda_1^5}\dfrac{\varepsilon_1(\lambda_1)}{e^{hc/\lambda_1 kT_{i,1}}-1} \\ \dfrac{2hc^2}{\lambda_2^5}\dfrac{\varepsilon_1(\lambda_2)}{e^{hc/\lambda_2 kT_{i,1}}-1} \end{bmatrix} + a_{i,2}\begin{bmatrix} \dfrac{2hc^2}{\lambda_1^5}\dfrac{\varepsilon_2(\lambda_1)}{e^{hc/\lambda_1 kT_{i,2}}-1} \\ \dfrac{2hc^2}{\lambda_2^5}\dfrac{\varepsilon_2(\lambda_2)}{e^{hc/\lambda_2 kT_{i,2}}-1} \end{bmatrix} \tag{5.36}$$

where h is Planck's constant, c is the speed of light in a vacuum, k is Boltzmann's constant, ε_1 and ε_2 are the emissivities of the two endmembers, λ_1 and λ_2 are the

(a) (b)

Figure 5.12. Abundance images for thermal test data.

(a) (b)

Figure 5.13. SMM abundance images for thermal test data.

center wavelengths for the two bands, $a_{i,1}$ and $a_{i,2}$ are the abundances for the ith pixel, and $T_{i,1}$ and $T_{i,2}$ are the temperatures of the two endmember components of the ith pixel. The two temperatures are modeled as normal random variables of a specified means, μ_1 and μ_2, and standard deviations, σ_1 and σ_2. This endmember-dependent temperature variance is the component that necessitates the use of SMM over LMM for effective modeling, and the approximate linearity of the blackbody function over a typical variance in temperature supports the inherent assumptions of the SMM.

A 50×50 pixel example image was formed based on the abundances from Figure 5.13 and the following characteristics: $\lambda_1 = 9\,\mu\text{m}$, $\lambda_2 = 10\,\mu\text{m}$,

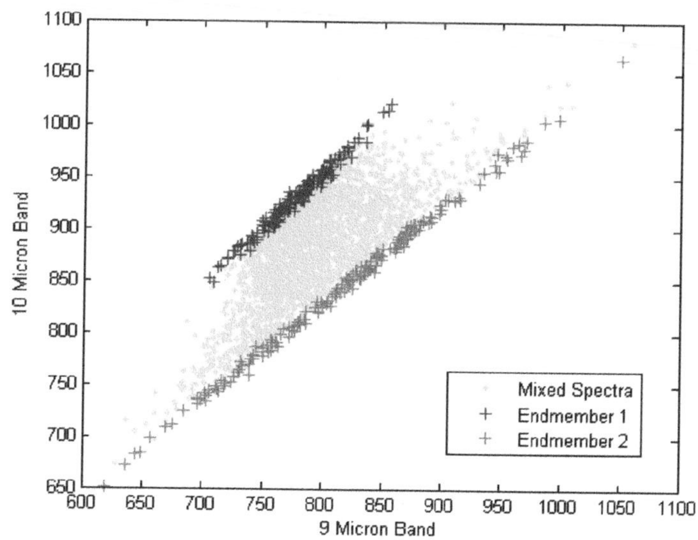

Figure 5.14. Scatterplot of thermal test data and SMM endmember classes.

$\varepsilon_1(\lambda_1) = 0.8$, $\varepsilon_1(\lambda_2) = 0.95$, $\varepsilon_2(\lambda_1) = 1.0$, $\varepsilon_2(\lambda_2) = 1.0$, $\mu_1 = 300\,\text{K}$, $\mu_2 = 290\,\text{K}$, $\sigma_1 = 2\,\text{K}$, $\sigma_2 = 5\,\text{K}$, and a noise standard deviation of 2 $\mu\text{W/cm}^2\mu\text{m}\cdot\text{sr}$. An SMM was developed based on 2 endmembers, 16 abundance quantization levels, and 30 SEM iterations. The resulting abundance images shown in Figure 13 indicate good correspondence to the truth. A scatterplot of the data indicating the pixels assigned to the two endmember classes is shown in Figure 5.14. This illustrates that the SMM is able to capture the random temperature variance properly as endmember variance while also effectively modeling the linear mixing in the abundance images. In this case, the RMS abundance error is about 0.032.

5.6. NORMAL COMPOSITIONAL MODEL

The normal compositional model (NCM) provides an alternative approach to the SMM for stochastic mixture modeling. The fundamental difference between the NCM and SMM is that the NCM does not constrain the abundance levels to be quantized as in the SMM. In this sense, NCM is a more accurate model; however, it also poses a more challenging parameter estimation problem. Like the SMM, the NCM can be applied either to data in the original spectral space or to a lower-dimensional subspace. The notation in this section will be based on the former.

5.6.1. Model Formulation

The normal compositional model (NCM) describes the random observation vector by

$$\mathbf{x} = c\boldsymbol{\varepsilon}_0 + \sum_{m=1}^{M} a_m \boldsymbol{\varepsilon}_m \tag{5.37}$$

subject to constraints

$$a_m \geq 0 \tag{5.38}$$

and either

$$\sum_{m=1}^{M} a_m = C \tag{5.39}$$

or

$$\sum_{m=1}^{M} a_m \leq C \tag{5.40}$$

where $c = 0$ or 1. This formulation fundamentally differs from the SMM in that the a_m are arbitrary real numbers subject to the constraints, and not quantized (except for machine precision). Also, the NCM introduces an explicit additive term that can be incorporated with the other class parameters, and it exhibits more general constraints. Using Eq. (6.4) rather than Eq. (5.39) allows for scalar variation in received radiance as might occur if surfaces are not Lambertian.

The NCM is a two-stage hierarchical model [30] of the random vector **x**, which is defined assuming the existence of unknown random effects vector **u**, sets of parameters θ_1 and θ_2, a conditional probability density function $f(\mathbf{x}|\mathbf{u}, \theta_1)$, and a prior probability density function $g(\mathbf{u}, \theta_2)$. Then,

$$f(\mathbf{x}|\theta_1, \theta_2) = \int f(\mathbf{x}|\mathbf{u}, \theta_1)\, g(\mathbf{u}, \theta_2)\, d\mathbf{u} \tag{5.41}$$

is a two-stage hierarchical model for **x**. For the NCM, the abundance vector **a** represents the random effects (i.e., $\mathbf{u} = \mathbf{a}$), the parameter θ_1 represents the endmember class statistics,

$$\theta_1 = \{\mathbf{m}_1, \mathbf{C}_1, \mathbf{m}_2, \mathbf{C}_2, \ldots, \mathbf{m}_2, \mathbf{C}_2\} \tag{5.42}$$

and the parameter θ_2 represents the parameters of a prior distribution of **a**. With these considerations, the conditional probability density function is normal,

$$f(\mathbf{x}|\mathbf{a}, \theta_1) = \frac{1}{(2\pi)^{K/2}} \frac{1}{|\,\mathbf{C}(\mathbf{a})\,|^{1/2}} \exp\left\{-\frac{1}{2}[\mathbf{x} - \mathbf{m}(\mathbf{a})\,]^T \mathbf{C}^{-1}(\mathbf{a})\,[\mathbf{x} - \mathbf{m}(\mathbf{a})\,]\right\} \tag{5.43}$$

with a mean vector

$$\mathbf{m}(\mathbf{a}) = \sum_{m=1}^{M} a_m \mathbf{m}_m \tag{5.44}$$

and covariance matrix

$$\mathbf{C}(\mathbf{a}) = \sum_{m=1}^{M} a_m^2 \mathbf{C}_m \tag{5.45}$$

Equations (5.43) to (5.45) directly parallel the SMM; however, the abundance vector **a** is treated as a continuous random variable.

Assume that independent observations $\mathbf{x}_i$ are available and that their probability density function is given by Eq. (5.41). Then the density function for the set of observation vectors corresponding to the entire image,

$$\mathbf{X} = (\mathbf{x}_1, \ldots, \mathbf{x}_N) \tag{5.46}$$

is given by

$$f(\mathbf{X}|\theta_1,\theta_2) = \prod_{i=1}^{N} \int f(\mathbf{x}_i|\mathbf{a}_i,\theta_1)\, g(\mathbf{a}_i,\theta_2)\, d\mathbf{a}_i \tag{5.47}$$

5.6.2. Parameter Estimation

A number of techniques are available to estimate the parameters of a statistical model including maximum likelihood, simulated maximum likelihood, expectation maximization, and Monte Carlo expectation maximization [31–35]. Maximum likelihood parameter estimation seeks to maximize the log-likelihood function

$$L(\mathbf{X},\theta_1,\theta_2) = \log f(\mathbf{X}|\theta_1,\theta_2) \tag{5.48}$$

or after inserting (5.46),

$$L(\mathbf{X},\theta_1,\theta_2) = \sum_{i=1}^{N} \log \int f(\mathbf{x}_i|\mathbf{a}_i,\theta_1)\, g(\mathbf{a}_i,\theta_2)\, d\mathbf{a}_i \tag{5.49}$$

Assuming a uniform prior probability density $g\ (\mathbf{a},\theta_2)$, this reduces to

$$L(\mathbf{X},\theta_1,\theta_2) = \sum_{i=1}^{N} \log \int \frac{1}{(2\pi)^{K/2}} \frac{1}{|\mathbf{C}(\mathbf{a}_i)|^{1/2}} \times \exp\left\{-\frac{1}{2}[\mathbf{x}_i - \mathbf{m}(\mathbf{a}_i)]^T \mathbf{C}^{-1}(\mathbf{a}_i)\, [\mathbf{x}_i - \mathbf{m}(\mathbf{a}_i)]\right\} d\mathbf{a}_i \tag{5.50}$$

plus a constant.

Due to the complexity of Eq. (5.50), NCM parameter estimation does not lend itself well to a direct application of standard numerical optimization techniques, such as quasi-Newton methods or expectation maximization (EM). Consider the EM algorithm, which produces a sequence of parameter values that converge to a stationary point (i.e., a local maximum or saddle point) of the likelihood function [31, 32]. Given a current set of parameter estimates $\theta^{(n)}$ where $\theta = \{\theta_1,\theta_2\}$, the update equation is given for the likelihood function in (5.49) as

$$\hat{\theta}^{(n+1)} = \arg \max_{\theta=\{\theta_1,\theta_2\}} \sum_{i=1}^{N} \log \int f(\mathbf{x}_i|\mathbf{a}_i,\theta_1)\, g(\mathbf{a}_i,\theta_2) \times \frac{f(\mathbf{x}_i|\mathbf{a}_i,\hat{\theta}_1^{(n)})g(\mathbf{a}_i,\hat{\theta}_2^{(n)})}{\int f(\mathbf{x}_i|\mathbf{a}_i,\hat{\theta}_1^{(n)})\, g(\mathbf{a}_i,\hat{\theta}_2^{(n)})d\mathbf{a}_i}\, d\mathbf{a}_i \tag{5.51}$$

The intractability of the integrals in (5.51) suggests that Monte Carlo methods be used to estimate the parameters. The simulated maximum likelihood (SML) and the

Monte Carlo Expectation Maximization (MCEM) algorithms as described in [33–35] are both considered here. In the SML approach, the integral in Eq. (5.49) is replaced with a Monte Carlo approximation [34]. For each i, a set of n_i random effects samples,

$$\{\mathbf{u}_{i,1}, \mathbf{u}_{i,2}, \ldots, \mathbf{u}_{i,n_i}\} \tag{5.52}$$

are obtained using an importance-sampling probability density function $h(\mathbf{u}_{i,j})$, and the integral in Eq. (5.49) for the ith sample is approximated as

$$\int f(\mathbf{x}_i|\mathbf{a}_i, \theta_1)\, g(\mathbf{a}_i, \theta_2)\, d\mathbf{a}_i \approx \frac{1}{n_i}\sum_{j=1}^{n_i} f(\mathbf{x}|\mathbf{u}_{i,j}, \theta_1)\frac{g(\mathbf{u}_{i,j}, \theta_2)}{h(\mathbf{u}_{i,j})} \tag{5.53}$$

Substituting Eq. (5.52) into Eq. (5.50) and identifying the random effects samples $\mathbf{u}_{i,j}$ as NCM abundance samples $\mathbf{a}_{i,j}$, the SML update equation is given by

$$\hat{\theta}_{\text{SML}}^{(n+1)} = \arg\max_{\theta=\{\theta_1,\theta_2\}} \sum_{i=1}^{N} \log\left[\frac{1}{n_i}\sum_{j=1}^{n_i} f(\mathbf{x}_i|\mathbf{a}_{i,j}, \theta_1)\frac{g_{\mathbf{u},\theta_2}(\mathbf{a}_{i,j}, \theta_2)}{h(\mathbf{a}_{i,j})}\right] \tag{5.54}$$

For each i corresponding to the SML method discussed above, MCEM [33, 34] proceeds by selecting a set of n_i samples $\{\tilde{\mathbf{a}}_{i,j}\}$ having probability density function

$$p(\tilde{\mathbf{a}}_{i,j}|\mathbf{x}_i, \hat{\theta}^{(n)}) = \frac{f(\mathbf{x}_i|\tilde{\mathbf{a}}_{i,j}, \hat{\theta}_1^{(n)})g(\tilde{\mathbf{a}}_{i,j}, \hat{\theta}_2^{(n)})}{\int f(\mathbf{x}_i|\tilde{\mathbf{a}}_{i,j}, \hat{\theta}_1^{(n)})\, g(\tilde{\mathbf{a}}_{i,j}, \hat{\theta}_2^{(n)})\, d\mathbf{a}_i} \tag{5.55}$$

The update equation then becomes

$$\hat{\theta}_{\text{MCEM}}^{(n+1)} = \arg\max_{\theta=\{\theta_1,\theta_2\}} \sum_{i=1}^{N}\sum_{j=1}^{n_i} \log f(\mathbf{x}_i|\tilde{\mathbf{a}}_{i,j}, \theta_1) + \log g(\tilde{\mathbf{a}}_{i,j}, \theta_2) \tag{5.56}$$

Jank and Booth [35] conclude that the MCEM approach is often more efficient than SML, particularly if the importance distribution $h(\mathbf{u})$ used to obtain the SML samples is not sufficiently close to the probability density function

$$p(\mathbf{u}|\mathbf{x}, \theta) = \frac{f(\mathbf{u}|\mathbf{x}, \theta_1)g(\mathbf{u}|\mathbf{x}, \theta_2)}{p(\mathbf{x})} \tag{5.57}$$

As this density depends on the unknown parameter set θ, it is difficult in practice to obtain a suitable importance-sampling function. Thus the MCEM algorithm is applied to the problem of estimating the NCM parameters. The sampling is achieved using Monte Carlo Markov chains as described in Stein [36].

5.6.3. Maximization Step

MCEM produces a sequence of parameter estimates that approximate (up to sampling error) a saddle point or local maximum of the likelihood function. Assume that parameter estimates θ_1 have been obtained. The application of MCEM to estimate the parameters of the NCM requires, after obtaining the samples $\{\tilde{\mathbf{a}}_{i,j}\}$ based on Eq. (5.54) and assuming a uniform prior probability density function $g\ (\mathbf{a},\theta_2)$, the solution of

$$\{\mathbf{m}_m^{(n+1)}, \mathbf{C}_m^{(n+1)}\} = \arg \max_{\{\mathbf{m}_m^{(n+1)},\mathbf{C}_m^{(n+1)}\}} \sum_{i=1}^{N} \sum_{j=1}^{n_{i,j}} -\frac{1}{2}\log \| \mathbf{C}\left(\tilde{\mathbf{a}}_{i,j}^{(n)}\right) \|$$

$$-\frac{1}{2}\left[\mathbf{x}_i - \mathbf{m}\left(\tilde{\mathbf{a}}_{i,j}^{(n)}\right)\right]^T \mathbf{C}\left(\tilde{\mathbf{a}}_{i,j}^{(n)}\right)^{-1}\left[\mathbf{x}_i - \mathbf{m}\left(\tilde{\mathbf{a}}_{i,j}^{(n)}\right)\right] \qquad (5.58)$$

The EM algorithm is applied to solve Eq. (5.56) with the hidden variables being the values of the random vectors $\boldsymbol{\varepsilon}_m$ in Eq. (5.37). The update equations are given in the following theorem which is proven in [36].

Theorem. Given a current parameter estimate

$$\theta_1^{(n)} = \left\{\mathbf{m}_1^{(n)}, \mathbf{C}_1^{(n)}, \mathbf{m}_2^{(n)}, \mathbf{C}_2^{(n)}, \cdots, \mathbf{m}_M^{(n)}, \mathbf{C}_M^{(n)}\right\} \qquad (5.59)$$

and abundance values

$$\{\, a_{i,m,j} : 1 \le i \le N,\ 0 \le m \le M,\ 1 \le j \le n_i\} \qquad (5.60)$$

define

$$\begin{aligned}
\delta_{i,j}^{(n)} &= [\mathbf{C}^{(n)}(\mathbf{a}_{i,j})]^{-1}[\mathbf{x}_i - \mathbf{m}^{(n)}(\mathbf{a}_{i,j})] \\
\psi_{i,m,j}^{(n)} &= a_{i,m,j}\delta_{i,j}^{(n)} \\
\bar{\delta}^{(n)} &= \frac{1}{N}\sum_{i=1}^{N}\frac{1}{n_i}\sum_{j=1}^{n_i}\delta_{i,j}^{(n)} \\
\bar{\psi}_m^{(n)} &= \frac{1}{N}\sum_{i=1}^{N}\frac{1}{n_i}\sum_{j=1}^{n_i}\psi_{i,m,j}^{(n)}
\end{aligned} \qquad (5.61)$$

The EM update equations for the model are

$$\mathbf{m}_m^{(n+1)} = \mathbf{m}_m^{(n)} + \mathbf{C}_m^{(n)}\bar{\psi}_m^{(n)} \qquad (5.62)$$

and

$$\mathbf{C}_m^{(n+1)} = \mathbf{C}_m^{(n)} - \mathbf{C}_m^{(n)}\left[\frac{1}{N}\sum_{i=1}^{N}\frac{1}{n_i}\sum_{j=1}^{n_i}(a_{i,m,j})^2\left[\mathbf{C}_m^{(n+1)}(\mathbf{a}_{i,j})\right]^{-1}\right]\mathbf{C}_m^{(n)}$$
$$+\,\mathbf{C}_m^{(n)}\left[\frac{1}{N}\sum_{i=1}^{N}\frac{1}{n_i}\sum_{j=1}^{n_i}\psi_{i,m,j}^{(n)}\left(\psi_{i,m,j}^{(n)}\right)^T - \bar{\psi}_m^{(n)}\left(\bar{\psi}_m^{(n)}\right)^T\right]\mathbf{C}_m^{(n)} \tag{5.63}$$

A variation on the update equations with faster convergence is provided in Stein [36].

5.6.4. Unmixing the NCM

Two techniques, maximum likelihood and mean estimation, are available for solving for the abundance values given the class parameters. Both are described in Stein [36]. The maximum likelihood approach amounts to maximizing Eq. (5.50) subject to the selected constraints, for example, by sequential quadratic programming. Alternatively, Monte Carlo Markov chains (MCMC) can be used to provide a set of samples of the generating the posterior distribution. As is shown in Stein [36], the mean of the sample converges to the expected value of the posterior distribution.

5.6.5. Example Results

The NCM and the LMM are applied to two-dimensional, synthetic test data shown in Figure 5.15 that was generated as a convex combination of two Gaussian classes such that the abundance values sum to one and have a uniform distribution on [0,1].

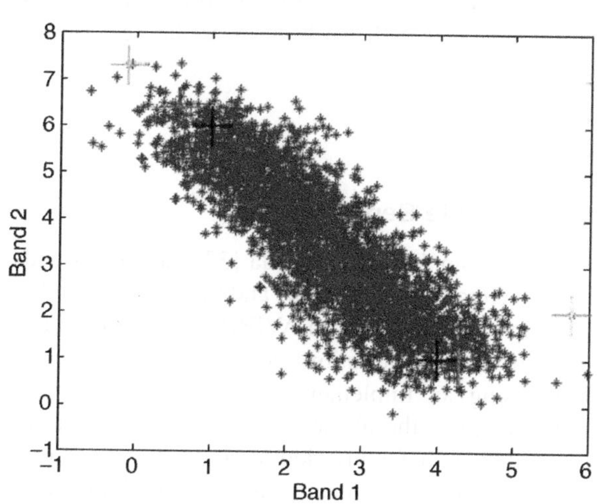

Figure 5.15. Scatterplot of two-class test data with LMM and NCM endmember classes.

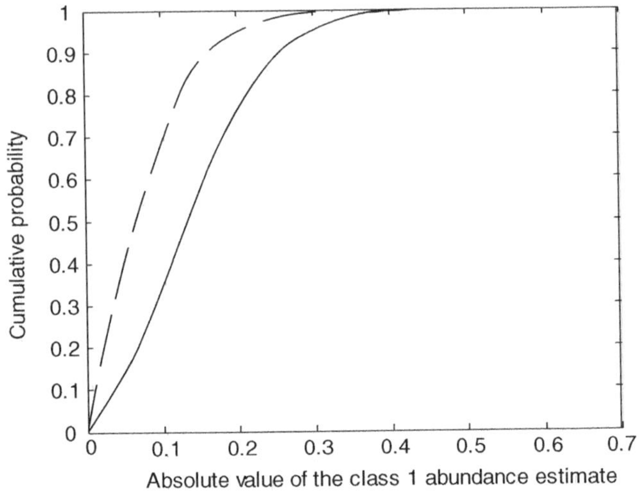

Figure 5.16. Cumulative distribution functions of LMM and NCM abundance error.

The true mean values of the classes are indicated by black crosses. Endmembers are estimated for the data using the maximal volume inscribed simplex method and are indicated by gray crosses. Estimates of the NCM class means obtained using MCMC sampling techniques are shown by gray crosses. The NCM class means are considerably closer to the true values of the class means than the endmembers.

The cumulative probability distribution functions of the absolute value of the abundance estimation error for the first class associated with the LMM (solid) and NCM (dashed) are shown in Figure 5.16. The NCM abundance estimates are mean values of sets of MCMC samples. Errors produced using various implementations of the NCM are presented in [36], where the accuracy of the covariance estimates is also studied.

5.7. APPLICATIONS

5.7.1. Geological Remote Sensing

Spectral data collected from Cuprite, Nevada [37] have been widely used to evaluate remote sensing technology. The LMM and the NCM, initialized with eighteen classes, were applied to a 2.0- to 2.5-μm AVIRIS hyperspectral cube over a portion containing two acid-sulfate alteration zones. The endmembers and classes were identified by using an implementation of the Tetracorder algorithm [38, 39]. The algorithm identifies the absorption features of a test spectrum, removes the continuum, and defines a feature depth. A depth ratio and correlation score are then computed for similarly located absorption features of elements of a spectral library, where the match to a library spectrum is a weighted sum of the correlation coefficients.

TABLE 5.1. Identification of LMM Endmembers and NCM Mean Vectors for Cuprite AVIRIS Data

Model Class Index	Linear Mixture Model Depth Ratio	Linear Mixture Model Correlation Coefficient	Linear Mixture Model Species	Normal Compositional Model Depth Ratio	Normal Compositional Model Correlation Coefficient	Normal Compositional Model Species
1	0.81	0.98	Alunite-2	0.77	0.96	Alunite-2
2	0.77	0.93	Halloysite	0.77	0.99	Muscovite-1
3	0.67	0.99	Calcite	0.56	0.98	Calcite
4	0.53	0.98	Muscovite-2	0.47	0.98	Muscovite-2
7	0.61	0.99	Dickite	0.59	0.99	Dickite
9	0.60	0.92	Chalcedony	0.47	0.87	Chalcedony
10	0.46	0.99	Alunite-1	0.47	0.99	Alunite-1
11	0.47	0.95	Chalcedony	0.53	0.94	Chalcedony
12	0.47	0.97	Chalcedony	0.27	0.93	Chalcedony
15	0.5	0.97	Buddingtonite	0.30	0.94	Buddingtonite
17	0.9	0.99	Kaolinite	0.87	0.99	Kaolinite

As detailed in Stein [40, 41] and summarized in Table 5.1, the NCM and LMM identify similar geological species. These are also generally consistent with previously-published results from Swayze [38], Clark et al. [39], and Kruse et al. [42]. One difference is the class 2 species, identified as halloysite by LMM and high aluminum muscovite by NCM. These species have a similar feature at 2.2 μm, but are distinguished by the 2.35 μm feature. As shown in Figure 5.17, this feature is better developed in the NCM class mean vector than it is in the LMM endmember, leading to a more distinct identification.

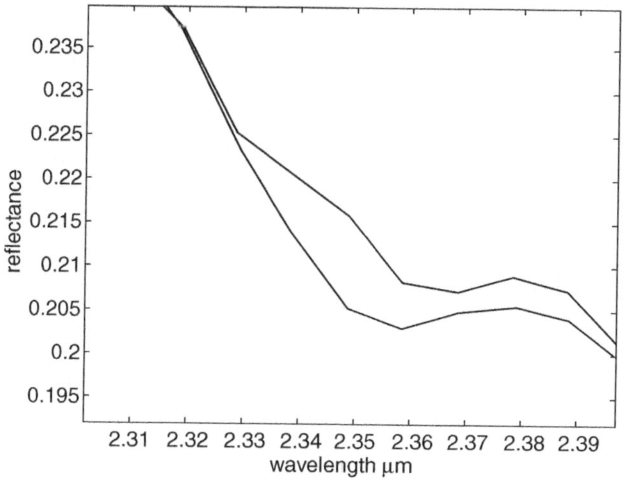

Figure 5.17. Comparison of 2.35-μm feature in LMM endmember and NCM mean vector.

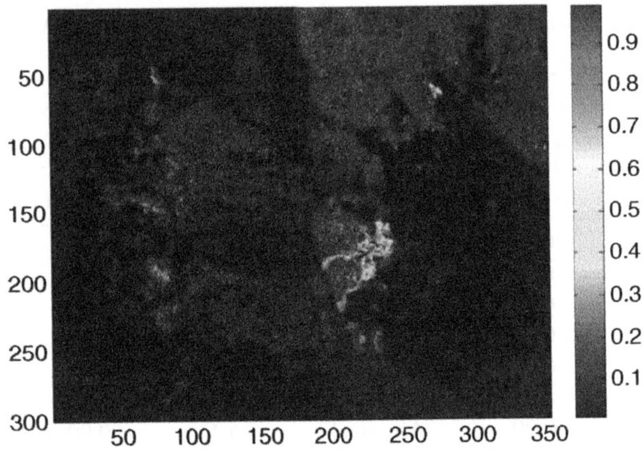

Figure 5.18. NCM abundance map for Buddingtonite class.

NCM-based abundance maps also show many of the same features as previously published mineral maps. For example, Figure 5.18 is the average value of the MCMC sample coefficients of the buddingtonite class and is consistent with the findings in Kruse et al. [42].

5.7.2. Coastal Remote Sensing

Estimating water constituents is an important remote sensing application. Shallow water remote sensing reflectance is often expressed as a sum of two terms: one due to scattering from the water column and one that is a reflection off of the bottom. Lee et al. [43, 44] use this approach to develop a semianalytical model that expresses shallow water remote sensing reflectance, assuming that the bottom reflectivity spectrum is known up to a scalar multiple, as a function of five scalar parameters: bottom reflectance at 550 nm (B), absorption at 440 nm due to gelbstoff (G), absorption at 440 nm due to phytoplankton (P), backscatter at 400 nm (X), and water depth (H).

The LMM and the NCM (initialized with 20 classes) are applied to remote sensing reflectance data derived from 4-m AVIRIS data collected over Tampa Bay, FL. The resulting classes are evaluated using the semianalytical model to determine the contribution from the water column, the contribution from the bottom (assuming either a grass or sand bottom), and the water column parameters (G, P, and X). If the contribution from the bottom is insignificant for both bottom models, then the class is declared to be a water class, and the parameters are obtained without presupposing a bottom.

The relative contribution of the water column to the total remote sensing reflectance was determined for each of the LMM and NCM classes. Setting a threshold for the water column contribution under both bottom models at 95%, the LMM produced no unambiguous water classes, whereas the NCM produced four water

TABLE 5.2. Estimate of Water Quality Parameters Based on LMM and NCM with Two Bottom Assumptions

	Class Index/20			
Parameters	5	12	16	20
$a_p(440)\ \mathrm{m}^{-1}$	0.58–0.60	0.70	0.78–1.07	0.64
	(0.47–0.55)	(0.51–0.53)	(0.47–0.97)	(0.55–0.56)
$a_g(440)\ \mathrm{m}^{-1}$	2.9–3.1	0.07	2.45–2.78	0.00
	(1.8–2.8)	(0.32–0.33)	(1.31–2.28)	(0.00–0.01)
$b_{bp}(440)\ \mathrm{m}^{-1}$	0.87–0.93	0.08	0.44–0.53	0.07
	(0.47–0.70)	(0.08)	(0.18–0.27)	(0.05–0.07)

classes: classes 5, 12, 16, and 20. Estimates of P, G, and X were obtained by least-squares fitting to the semianalytical model from these classes, and they are summarized Table 5.2 based on the LMM and NCM class spectra with both bottom assumptions. Where two numbers are listed, black refers to the sand bottom and green to the grass bottom. The parameter estimates based on NCM class means are less affected by the bottom assumption.

Figure 5.19 shows the dominant water class for pixels such that the scaled abundance of the class is greater than or equal to 0.1. Pixels in white in the figure did not have a dominant water class due perhaps to the stronger influence of the bottom. This analysis is able to identify regions of clearer (dark gray) and murkier (light gray) water without reference to a specific model or bottom assumption. The identification of the murkiest water in the upper-left-hand corner is consistent with the findings in Lee et al. [45] based on the semianalytical model. Techniques that can extract water parameters without assuming a bottom may be valuable, and

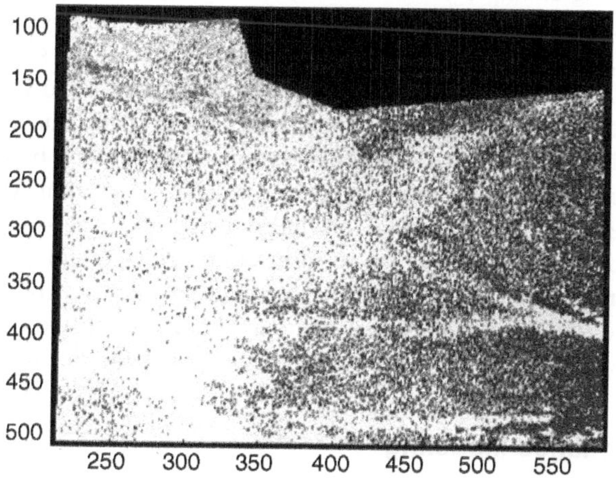

Figure 5.19. Dominant water class map based on NCM results.

mixed pixel classification techniques that do not depend upon a physical model may also be useful in those situations where the inputs to the parametric model are not known or where the model may not be applicable because of peculiarities of the data. Furthermore, features of the class means may be useful for the identification of water constituents.

Further analysis is carried out in Stein [41] to extract in-scene estimates of the bottom reflectivity and to use these estimates with a multicomponent bottom model to estimate, using the SA model, values of B, P, G, X, and H at each pixel. Fitting errors using in-scene estimates of the bottom and library grass and sand bottom spectra are compared.

5.7.3. Resolution Enhancement

Hyperspectral imaging sensors are often employed in conjunction with higher-resolution panchromatic (or multispectral) sensors, and several researchers investigated using the higher spatial resolution content of such an auxiliary sensor to enhance the spatial resolution of hyperspectral imagery. Such methods, which include principal component substitution [46] and least-squares estimation [47], generally have the effect of only enhancing the first principal component of the hyperspectral image. This occurs because it is most highly correlated with the panchromatic image. Achieving resolution enhancement in lower-order principal components requires some underlying model for the spectral mixing that occurs between the higher and lower resolutions. A maximum *a posteriori* (MAP) estimation approach based on the SMM has been developed for this problem [48], where an SMM derived from the hyperspectral observation is used as the underlying statistical mixture model for the unknown high-resolution hyperspectral image. This was found to provide improved resolution enhancement in the lower-order principal components of the test imagery shown relative to prior methods.

One interesting example of this resolution enhancement method is the application to Hyperion imagery [49]. Hyperion is a hyperspectral imaging camera on the NASA Earth Observer 1 satellite that covers the 0.4- to 2.5-μm spectral range with nominally a 30-m ground sample distance. The Advanced Land Imager (ALI) onboard the same satellite can provide concurrent panchromatic imagery at nominally a 10-m ground sample distance. The MAP estimation approach based on the SMM used to enhance example Hyperion imagery using the concurrent ALI imagery, and the results were compared to that produced using the least-squares estimation approach. Figure 5.20 compares the resulting fifth principal components, a typical example illustrating that the MAP/SMM approach is able to reconstruct finer resolution structure than least-squares estimation. This is due to the underlying SMM that forces the solution into a direction consistent with the linear mixing process.

5.7.4. Performance Modeling

Accurate modeling of hyperspectral data and sensing systems has been a goal of the remote sensing community. The work in references 50 and 51 takes a

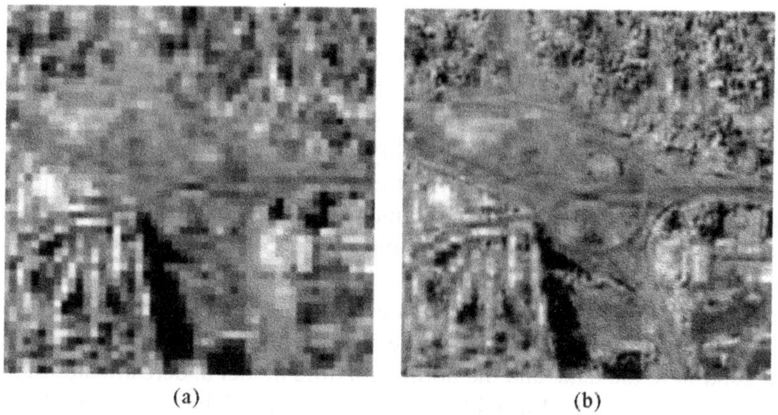

Figure 5.20. Resolution enhancement of the fifth principal component of a Hyperion image: (a) least-squares estimation, and (b) MAP/SMM estimation.

physics-based approach to creating synthetic data to which processing methods can be applied to predict system performance. Alternatively, Kerekes and Baum [52] take a statistical approach to modeling hyperspectral data. The reflectivity of each material within the scene is modeled with one or more multivariate normal probability distributions, and the probability distribution of the reflectivity of the scene is then modeled with a Gaussian mixture probability distribution

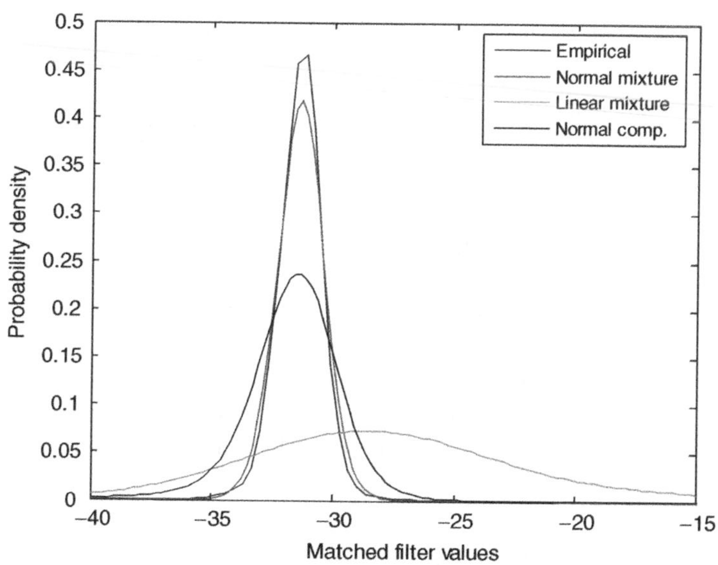

Figure 5.21. Comparison of empirical and theoretical density functions for the HYDICE forest radiance example.

function. Transformations that model illumination, atmospheric transmission, and the sensor spectral response function are then applied to these parameters to obtain a Gaussian mixture model of the data at the output of the sensor. Finally, the sensor sampling function and noise characteristics are then used to obtain a Gaussian mixture probability density function of the output of the sensor. The linear mixture and normal compositional models, combined with a probability density function of the abundance values, can be used to extend the statistical approach to hyperspectral image modeling.

This modeling approach was tested using LMM, NCM, and a normal mixture model for HYDICE forest radiance data using 10 classes. The resulting probability density functions are compared in Figure 5.21. In this case, the LMM does not provide a good fit to the data, and the normal mixture model underestimates the tails of the distribution. The NCM provides a better estimate of the tails of the distribution [41], and it could be generalized to provide a better overall fit by using non-Gaussian endmember class distributions.

APPENDIX: Proof of Equation (5.9)

Theorem 1. Let $S(L,M,k) = \{(a_1,\ldots,a_m) | 1 \leq a_j \leq L, \sum_{j=1}^{M} a_j = k \leq L\}$. Let $N(L,M,k) = |S(L,M,k)|$, the number of elements in $S(L,M,k)$. Then

$$N(L,M,k) = \binom{k-1}{M-1} \tag{A.1}$$

Lemma 1. $N(L,M,k) = N(L,M,k-1) + N(L,M-1,k-1)$.

Proof. Define the mapping $t : S(L,M,k) \to \cup_{\ell=1}^{k-1} S(L,M-1,\ell)$ by $t(a_1,\ldots,a_m) = (a_1,\ldots,a_{m-1}) \in S(L,M-1,k-a_m)$. Define $h : \cup_{\ell=1}^{k-1} S(L,M,\ell) \to S(L,M,k)$ by $h(a_1,\ldots,a_{m-1}) = (a_1,\ldots,a_{m-1},k-\ell)$. Then $t \circ h = 1$ and $h \circ t = 1$. Therefore

$$\begin{aligned} N(L,M,k) &= \sum_{\ell=M-1}^{k-1} N(L,M-1,\ell) \\ &= N(L,M,k-1) + N(L,M-1,k-1). \end{aligned}$$

Proof of Theorem 1. Binomial coefficients satisfy the recursion $\binom{k-1}{M-1} + \binom{k-1}{M} = \binom{k}{M}$. For $M = 1$, $N(L,1,k) = 1 = \binom{k-1}{0}$. Also, $N(L,M,j) = 0$ if $j < M$, and $N(L,M,M) = 1 = \binom{M-1}{M-1}$. Therefore, by induction on M and k, $N(L,M,k) = \binom{k-1}{M-1}$.

Theorem 2. Assume that M classes are available. Define a restricted sequence as one satisfying the following: At most $M_{\max}$ abundance values are greater than zero; the sum of the abundance values at each pixel is 1; and each abundance value is of the form *k/L*. Then there are

$$Q = \sum_{j=1}^{M_{\max}} \binom{M}{j} N(L,j,L) = \sum_{j=1}^{M_{\max}} \binom{M}{j} \binom{L-1}{j-1} \tag{A.2}$$

restricted sequences.

Proof. The number of sequences with exactly j nonzero terms is given by $\binom{M}{j}N(L,j,L)$. Therefore the number of sequences with not more than $M_{\max}$ terms is as stated. Note that $M_{\max} \leq L$. Let $(a_1, \ldots, a_M\max)$ be a sequence of abundance values such that for all $1 \leq i \leq M_{\max}$, $a_i > 0$. Since $a_i \geq \frac{1}{L}$, $1 = \sum_{i=1}^{M_{\max}} a_i \geq M_{\max}\left(\frac{1}{L}\right)$. Therefore, $L \geq M_{\max}$.

REFERENCES

1. G. Healey and D. Slater, Models and methods for automated material identification in hyperspectral imagery acquired under unknown illumination and atmospheric conditions, *IEEE Transactions on Geoscience and Remote Sensing*, vol. 37, pp. 2706–2717, 1999.
2. A. P. Schaum, Matched affine joint subspace detection in remote hyperspectral reconnaissance, *Proceedings of the 31st Applied Imagery Pattern Recognition Workshop*, pp. 1–6, 2002.
3. N. Keshava and J. F. Mustard, Spectral unmixing, *IEEE Signal Processing Magazine*, pp. 44–57, 2002.
4. Y. Masalmah, S. Rosario-Torres, and M. Velez-Reyes, An algorithm for unsupervised unmixing of hyperspectral imagery using positive matrix factorization, *Proceedings of the SPIE*, vol. 5806, 2005.
5. S. Tompkins, J. F. Mustard, C. Pieters, and D. W. Forsyth, Optimization of endmembers for spectral mixture analysis, *Remote Sensing of the Environment*, vol. 59, pp. 472–489, 1997.
6. J. W. Boardman, F. A. Kruse, and R. O. Green, Mapping target signatures via partial unmixing of AVIRIS data, *Summaries, Fifth JPL Airborne Earth Science Workshop*, vol. 1, pp. 23–26, 1995.
7. S. R. Lay, *Convex Sets and Their Applications*, John Wiley & Sons, New York, 1982.
8. M. E. Winter, Fast autonomous endmember determination in hyperspectral data, *Proceedings of the 13th International Conference on Applied Geological Remote Sensing*, Vol. II, pp. 337–344, 1999.
9. M. D. Craig, Minimum-volume transforms for remotely sensed data, *IEEE Transactions on Geoscience and Remote Sensing*, vol. 32, pp. 542–552, 1994.
10. D. Gillis, P. Palmedesso, and J. Bowles, An automatic target recognition system for hyperspectral imagery using ORASIS, *Proceedings of the SPIE*, vol. 4381, pp. 34–43, 2001.
11. G. Strang, *Linear Algebra and Its Applications*, 2nd edition, Chapter 3, Academic Press, Orlando, FL, 1980.

12. S. M. Kay, *Fundamentals of Statistical Signal Processing*, Prentice Hill, Englewood Cliffs, NJ, 1993.
13. C. L. Lawson and R. J. Hanson, *Solving Least Squares Problems*, Prentice Hill, Englewood Cliffs, NJ, 1974.
14. R. Fletcher, *Practical Methods of Optimization*, 2nd edition, Chapter 10, John Wiley & Sons, New York, 1987.
15. R. A. Schowengerdt, *Remote Sensing: Models and Methods for Image Processing*, Academic Press, Orlando, FL, 1997.
16. A. A. Green, M. Berman, P. Schwitzer, and M. D. Craig, A transformation for ordering multispectral data in terms of image quality with implications for noise removal, *IEEE Transactions on Geoscience and Remote Sensing*, vol. 26, pp. 65–74, 1988.
17. J. W. Silverstein and Z. D. Bai, On the empirical distribution of eigenvalues for a class of large dimensional random matrices, *Journal of Multivariate Analysis*, vol. 54, pp. 175–192, 1995.
18. A. D. Stocker, E. Ensafi, and C. Oliphint, Applications of eigenvalue distribution theory to hyperspectral processing, *Proceedings of the SPIE*, vol. 5093, pp. 651–665, 2003.
19. S. Catterall, Anomaly detection based on the statistics of hyperspectral imagery, *Proceedings of the SPIE*, vol. 5546, pp. 171–178, 2004.
20. A. D. Stocker, I. S. Reed, and X. Yu, Multi-dimensional signal processing for electro-optical target detection, *Proceedings of the SPIE*, vol. 1305, pp. 218–231, 1990.
21. Y. Linde et al., An algorithm for vector quantization, *IEEE Transactions on Communication Theory*, vol. 28, no. 1, pp. 84–95, 1980.
22. P. Masson and W. Pieczynski, SEM algorithm and unsupervised statistical segmentation of satellite images, *IEEE Transactions on Geoscience and Remote Sensing*, vol. 31, no. 3, pp. 618–633, 1993.
23. A. D. Stocker and A. P. Schaum, Application of stochastic mixing models to hyperspectral detection problems, *Proceedings of the SPIE*, vol. 3071, pp. 47–60, 1997.
24. M. T. Eismann and R. C. Hardie, Initialization and convergence of the stochastic mixing model, *Proceedings of the SPIE*, vol. 5159, pp. 307–318, 2003.
25. M. T. Eismann and R. C. Hardie, Improved initialization and convergence of a stochastic spectral unmixing algorithm, *Applied Optics*, vol. 43, pp. 6596–6608, 2004.
26. G. R. Fowles, *Introduction to Modern Optics*, 2nd edition, p. 213, Dover, 1975.
27. R. A. Redner and H. F. Walker, Mixture densities, maximum likelihood, and the EM algorithm, *SIAM Review*, vol. 26, pp. 195–239, 1984.
28. K. Fukunaga, *Introduction to Statistical Pattern Recognition*, Academic Press, New York, 1990.
29. G. R. Fowles, *Introduction to Modern Optics*, 2nd edition, Chapter 7, New York, Dover, 1975.
30. J. P. Hobert, Hierarchical models: A current computational perspective, *Journal of the American Statistical Association*, vol. 95, no. 452, pp. 1312–1316, 2000.
31. A. P. Dempster, N. M. Laird, and D. B. Rubin, Maximum likelihood from incomplete data via the EM algorithm, *Journal of the Royal Statistical Society*, vol. 39, pp. 1–38, 1977.
32. C. F. J. Wu, On the convergence properties of the EM algorithm, *The Annals of Statistics*, vol. 11, pp. 95–103, 1983.

33. J. Diebolt and E. H. S. Ip, Stochastic EM: method and application, in *Markov Chain Monte Carlo in Practice*, edited by W. R. Gilks, S. Richardson, and D. J. Spiegelhalter, pp. 259–274, Chapman and Hall/CRC, New York, 1996.

34. J. G. Booth, J. P. Hobert, and W. Jank, A survey of Monte Carlo algorithms for maximizing the likelihood of a two-stage hierarchical model, *Statistical Modeling*, vol. 4, pp. 333–339, 2001.

35. W. Jank and J. G. Booth, Efficiency of Monte Carlo EM and simulated maximum likelihood in two-stage hierarchical models, *Journal of Computational and Graphical Statistics*, vol. 12, pp. 214–229, 2003.

36. D. W. J. Stein, The normal compositional model, to be submitted to *Journal of the Royal Statistical Society.*

37. F. A. Kruse, K. S. Kierein-Young, and J. W. Boardman, Mineral mapping at Cuprite, Nevada with a 63-channel imaging spectrometer, *Photogrammetric Engineering and Remote Sensing*, vol. 56, pp. 83–92, 1990.

38. G. A. Swayze, *The Hydrothermal and Structural History of the Cuprite Mining District, Southwestern Nevada: An Integrated Geological and Geophysical Approach,* Ph.D. thesis, University of Colorado, 1997.

39. R. N. Clark et al., Imaging spectroscopy: Earth and planetary remote sensing with the USGS Tetracorder and expert systems, *Journal of Geophysical Research*, vol. 108, p. 5-1-44, December 2003.

40. D. W. J. Stein, Material identification and classification from hyperspectral imagery using the normal compositional model, *Proceedings of the SPIE*, vol. 5093, pp. 559–568, 2003.

41. D. W. J. Stein, Application of the NCM to classification and modeling of hyperspectral imagery, *IEEE Transactions on Geoscience and Remote Sensing*, submitted.

42. F. A. Kruse, J. W. Boardman, and J. F. Huntington, Comparison of airborne and satellite hyperspectral data for geologic mapping, *Proceedings of the SPIE*, vol. 4725, pp. 128–139, 2002.

43. Z. Lee et al., Hyperspectral remote sensing for shallow waters. I. A semi-analytical model, *Applied Optics*, vol. 37, pp. 6329–6338, 1998.

44. Z. Lee et al., Hyperspectral remote sensing for shallow waters. 2 Deriving bottom depths and water properties by optimization, *Applied Optics*, vol. 38, pp. 3831–3843, 1999.

45. Z. Lee, C. L. Carder, R. F. Chen, and T. G. Peacock, Properties of the water column and bottom derived from airborne visible infrared imaging spectrometer (AVIRIS) data, *Journal of Geophysical Research*, vol. 106, pp. 11639–11651, 2001.

46. V. K. Shettigara, A generalized component substitution technique for spatial enhancement of multispectral images using a higher resolution data set, *Photogrammetric Engineering and Remote Sensing*, vol. 58, pp. 561–567, 1992.

47. J. C. Price, Combining panchromatic and multispectral imagery from dual resolution satellite instruments, *Remote Sensing of the Environment*, vol. 21, pp. 119–128, 1987.

48. M. T. Eismann and R.C. Hardie, Application of the stochastic mixing model to hyperspectral resolution enhancement, *IEEE Transactions on Geoscience and Remote Sensing*, vol. 42 pp. 1924–1933, 2004.

49. J. Pearlman et al., Overview of the Hyperion imaging spectrometer for the NASA EO-1 mission, *Proceedings of the 2002 International Geoscience and Remote Sensing Symposium*, pp. 3036–3038, 2002.

50. B. Shetler et al., A comprehensive hyperspectral system simulation I: Integrated sensor scene modeling and the simulation architecture, *Proceedings of the SPIE*, vol. 4049, pp. 94–104, 2000.
51. R. B. Bartell et al., A comprehensive hyperspectral system simulation II: Hyperspectral sensor simulation and preliminary VNIR testing results, *Proceedings of the SPIE*, vol. 4049, pp. 105–119, 2000.
52. J. P. Kerekes and J. E. Baum, Spectral imaging system analytical model for subpixel object detection, *IEEE Transactions on Geoscience and Remote Sensing*, vol. 40, pp. 1088–1101, 2002.

CHAPTER 6

UNMIXING HYPERSPECTRAL DATA: INDEPENDENT AND DEPENDENT COMPONENT ANALYSIS*

JOSÉ M. P. NASCIMENTO
Instituto Superior de Engenharia de Lisboa, Lisbon 1049-001, Portugal

JOSÉ M. B. DIAS
Instituto de Telecomunicações, Lisbon 1049-001, Portugal

6.1. INTRODUCTION

The development of high spatial resolution airborne and spaceborne sensors has improved the capability of ground-based data collection in the fields of agriculture, geography, geology, mineral identification, detection [2, 3], and classification [4–8].

The signal read by the sensor from a given spatial element of resolution and at a given spectral band is a mixing of components originated by the constituent substances, termed *endmembers*, located at that element of resolution. This chapter addresses hyperspectral unmixing, which is the decomposition of the pixel spectra into a collection of constituent spectra, or spectral signatures, and their corresponding fractional abundances indicating the proportion of each endmember present in the pixel [9, 10].

Depending on the mixing scales at each pixel, the observed mixture is either linear or nonlinear [11, 12]. The linear mixing model holds when the mixing scale is macroscopic [13]. The nonlinear model holds when the mixing scale is microscopic (i.e., intimate mixtures) [14, 15]. The linear model assumes negligible interaction among distinct endmembers [16, 17]. The nonlinear model assumes that incident solar radiation is scattered by the scene through multiple bounces involving several endmembers [18].

Under the linear mixing model and assuming that the number of endmembers and their spectral signatures are known, hyperspectral unmixing is a linear problem, which can be addressed, for example, under the maximum likelihood setup [19], the

*Work partially based on reference 1, copyright © 2005 IEEE.

Hyperspectral Data Exploitation: Theory and Applications, Edited by Chein-I Chang

constrained least-squares approach [20], the spectral signature matching [21], the spectral angle mapper [22], and the subspace projection methods [20, 23, 24].

Orthogonal subspace projection [23] reduces the data dimensionality, suppresses undesired spectral signatures, and detects the presence of a spectral signature of interest. The basic concept is to project each pixel onto a subspace that is orthogonal to the undesired signatures. As shown in Settle [19], the orthogonal subspace projection technique is equivalent to the maximum likelihood estimator. This projection technique was extended by three unconstrained least-squares approaches [24] (signature space orthogonal projection, oblique subspace projection, target signature space orthogonal projection). Other works using maximum *a posteriori* probability (MAP) framework [25] and projection pursuit [26, 27] have also been applied to hyperspectral data.

In most cases the number of endmembers and their signatures are not known. *Independent component analysis* (ICA) is an unsupervised source separation process that has been applied with success to blind source separation, to feature extraction, and to unsupervised recognition [28, 29]. ICA consists in finding a linear decomposition of observed data yielding statistically independent components.

Given that hyperspectral data are, in given circumstances, linear mixtures, ICA comes to mind as a possible tool to unmix this class of data. In fact, the application of ICA to hyperspectral data has been proposed in reference 30, where endmember signatures are treated as sources and the mixing matrix is composed by the abundance fractions, and in references 9, 25, and 31–38, where sources are the abundance fractions of each endmember. In the first approach, we face two problems: (1) The number of samples are limited to the number of channels and (2) the process of pixel selection, playing the role of mixed sources, is not straightforward. In the second approach, ICA is based on the assumption of mutually independent sources, which is not the case of hyperspectral data, since the sum of the abundance fractions is constant, implying dependence among abundances. This dependence compromises ICA applicability to hyperspectral images. In addition, hyperspectral data are immersed in noise, which degrades the ICA performance.

IFA [39] was introduced as a method for recovering independent hidden sources from their observed noisy mixtures. IFA implements two steps. First, source densities and noise covariance are estimated from the observed data by maximum likelihood. Second, sources are reconstructed by an optimal nonlinear estimator. Although IFA is a well-suited technique to unmix independent sources under noisy observations, the dependence among abundance fractions in hyperspectral imagery compromises, as in the ICA case, the IFA performance.

Considering the linear mixing model, hyperspectral observations are in a simplex whose vertices correspond to the endmembers. Several approaches [40–43] have exploited this geometric feature of hyperspectral mixtures [42]. Minimum volume transform (MVT) algorithm [43] determines the simplex of minimum volume containing the data.

The MVT-type approaches are complex from the computational point of view. Usually, these algorithms first find the convex hull defined by the observed data and then fit a minimum volume simplex to it. Aiming at a lower computational complexity, some algorithms such as the *vertex component analysis* (VCA) [44], the *pixel purity index* (PPI) [42], and the N-FINDR [45] still find the minimum volume simplex containing the data cloud, but they assume the presence in the data of at least one pure pixel of each endmember. This is a strong requisite that may not hold in some data sets. In any case, these algorithms find the set of *most pure pixels* in the data.

Hyperspectral sensors collects spatial images over many narrow contiguous bands, yielding large amounts of data. For this reason, very often, the processing of hyperspectral data, included unmixing, is preceded by a dimensionality reduction step to reduce computational complexity and to improve the signal-to-noise ratio (SNR). *Principal component analysis* (PCA) [46], *maximum noise fraction* (MNF) [47], and *singular value decomposition* (SVD) [48] are three well-known projection techniques widely used in remote sensing in general and in unmixing in particular. The newly introduced method [49] exploits the structure of hyperspectral mixtures, namely the fact that spectral vectors are nonnegative. The computational complexity associated with these techniques is an obstacle to real-time implementations. To overcome this problem, band selection [50] and nonstatistical [51] algorithms have been introduced.

This chapter addresses hyperspectral data source dependence and its impact on ICA and IFA performances. The study consider simulated and real data and is based on mutual information minimization. Hyperspectral observations are described by a generative model. This model takes into account the degradation mechanisms normally found in hyperspectral applications—namely, signature variability [52–54], abundance constraints, topography modulation, and system noise. The computation of mutual information is based on fitting mixtures of Gaussians (MOG) to data. The MOG parameters (number of components, means, covariances, and weights) are inferred using the minimum description length (MDL) based algorithm [55]. We study the behavior of the mutual information as a function of the unmixing matrix. The conclusion is that the unmixing matrix minimizing the mutual information might be very far from the true one. Nevertheless, some abundance fractions might be well separated, mainly in the presence of strong signature variability, a large number of endmembers, and high SNR.

We end this chapter by sketching a new methodology to blindly unmix hyperspectral data, where abundance fractions are modeled as a mixture of Dirichlet sources. This model enforces positivity and constant sum sources (full additivity) constraints. The mixing matrix is inferred by an expectation-maximization (EM)-type algorithm. This approach is in the vein of references 39 and 56, replacing independent sources represented by MOG with mixture of Dirichlet sources. Compared with the geometric-based approaches, the advantage of this model is that there is no need to have pure pixels in the observations.

The chapter is organized as follows. Section 6.2 presents a spectral radiance model and formulates the spectral unmixing as a linear problem accounting for abundance constraints, signature variability, topography modulation, and system noise. Section 6.3 presents a brief resume of ICA and IFA algorithms. Section 6.4 illustrates the performance of IFA and of some well-known ICA algorithms with experimental data. Section 6.5 studies the ICA and IFA limitations in unmixing hyperspectral data. Section 6.6 presents results of ICA based on real data. Section 6.7 describes the new blind unmixing scheme and some illustrative examples. Section 6.8 concludes with some remarks.

6.2. SPECTRAL RADIANCE MODEL

Figure 6.1 schematizes a typical passive remote sensing scenario. The sun illuminates a random media formed by the earth surface and by the atmosphere; a sensor (airborne or spaceborne) reads, within its *instantaneous field of view* (IFOV), the scattered radiance in the solar-reflectance region extending from 0.3 to 2.5 μm, encompassing the visible, near-infrared, and shortwave infrared bands. Angles θ and ϕ, with respect to the normal **n** on the ground, are the colatitude and the longitude, respectively. The solar and sensor directions are (θ_0, ϕ_0) and (θ_s, ϕ_s), respectively.

The total radiance at the surface level is the sum of three components, as schematized in Figure 6.1: the sunlight (ray 1), the skylight (ray 2), and the light due to the *adjacency effect* (ray 3)—that is, due to the successive reflections and scattering between the surface and the atmosphere. Following references 57 and 58, the spectral radiances of these components are, at a given wavelength λ, respectively, given by

1. $L_1 = \mu_0 E_0 T_\downarrow$, where E_0 is the solar flux at the top of the atmosphere, $\mu_0 = \cos(\theta_0)$, and $T_\downarrow = T_\downarrow(\theta_0)$ is the downward transmittance.
2. $L_2 = \mu_0 E_0 t_\downarrow$, where $t_\downarrow = t_\downarrow(\theta_0)$ is the downward diffuse transmittance factor.
3. $L_3 = \mu_0 E_0 T'_\downarrow([\rho_t S + (\rho_t S)^2 + (\rho_t S)^3 \ldots]$, where $T'_\downarrow = [T_\downarrow + t_\downarrow]$, ρ_t is the mean reflectance of the surroundings with respect to the atmospheric point spread function, and S is the spherical albedo of the atmosphere.

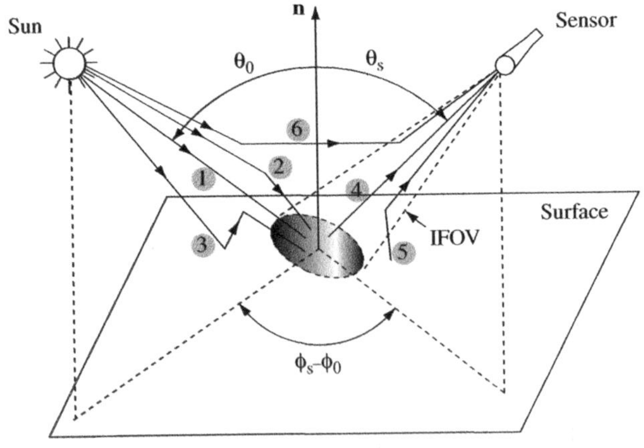

Figure 6.1. Schematic diagram of the main contributions to the radiance read by the sensor in the solar spectrum. Copyright © 2005 IEEE.

The total radiance incident upon the sensor location is the sum of three components: the light scattered by the surface (ray 4), the light scattered by the surface and by the atmosphere (ray 5), and light scattered by the atmosphere (ray 6), the so-called *path radiance*. Assuming a Lambertian surface, and again following references 57 and 58, these radiances at the top of the atmosphere are, at wavelength λ, respectively, given by

1 $L_4 = \frac{\mu_0 E_0}{\pi} \frac{T'_\downarrow T_\uparrow}{1-\rho_t S} \rho$, where ρ is the surface reflectance and $T_\uparrow = T_\uparrow(\theta_0)$ is the upward transmittance.
2. $L_5 = \frac{\mu_0 E_0}{\pi} \frac{T'_\downarrow t_\uparrow}{1-\rho_t S} \rho_t$, where $t_\uparrow = t_\uparrow(\theta_s)$ is the upward diffuse transmittance factor.
3. $L_6 = \frac{\mu_0 E_0}{\pi} \rho_a$, where $\rho_a(\theta_0, \theta_s, \phi_s - \phi_0)$ is the atmosphere reflectance.

Thus, the total radiance, L, incident upon the sensor location is

$$L = a\rho + b$$

where

$$a = \frac{\mu_0 E_0}{\pi} \frac{T'_\downarrow T_\uparrow}{1 - \rho_t S} \tag{6.1}$$

$$b = \frac{\mu_0 E_0}{\pi} \left(\frac{T'_\downarrow t_\uparrow}{1 - \rho_t S} \rho_t + \rho_a \right) \tag{6.2}$$

Let us assume that the sensor has B channels (wavebands). Assuming linear receivers and narrow wavebands, the signal at the output the ith channel (waveband centered at wavenumber λ_i) is given by

$$r_i = c_i \rho + d_i + n_i$$

where c_i and d_i are proportional to $a(\lambda_i)$ and $b(\lambda_i)$, respectively, and n_i denotes the receiver electronic noise at channel i plus the Poisson (photonic) signal noise (see, e.g., Jain [59]).

Terms a and b in Eqs.(6.1) and (6.2) depend, in a complex way, on the sun and sensor directions, on the atmosphere composition, on the topography, and on the scene materials and configurations [57, 58, 60]. The compensation for this term, the so-called atmospheric correction, is a necessary step in many quantitative algorithms aimed at extracting information from multispectral or hyperspectral imagery [58, 61, 62].

Herein, linear unmixing of fractional abundances at the pixel level is addressed. The term *linear* means that the observed entities are linear combinations of the end-members spectral signatures weighted by the correspondent fractional abundances. Therefore, we assume that atmospheric correction has been applied to a degree ensuring a linear relation between the radiance L and the reflectance ρ that is, for each channel, the relation between the radiance and the reflectivity is linear with coefficients not depending on the pixel.

The details of the atmospheric correction necessary to achieve such a linear relation are beyond the scope of this work. Note, however, that no correction may be necessary. That is the case when the scene is a surface of approximately constant altitude, the atmosphere is horizontally homogeneous, and ρ_t, the mean reflectance of the surroundings, exhibits negligible variation.

6.2.1. Linear Spectral Mixture Model

In spectral mixture modelling, the basic assumption is that the surface is made of a few number of endmembers of relatively constant spectral signature, or, at least, constant spectral shape. If the multiple scattering among distinct endmembers is negligible and the surface is partitioned according to the fractional abundances, then the spectral radiance upon the sensor location is well approximated by a linear mixture of endmember radiances weighted by the correspondent fractional abundances [9, 11, 12, 63, 64].

Under the linear mixing model and assuming that the sensor radiation pattern is ideal (i.e., constant in the IFOV and zero outside), the output of channel i from a given pixel is

$$r_i = c_i \sum_{j=1}^{p} \rho_{ij}\alpha_j + d_i + n_i \tag{6.3}$$

where ρ_{ij} denotes the reflectance of endmember j at wavenumber λ_i, α_j denotes the fractional abundance of endmember j at the considered pixel, and p is the number of endmembers.

Fractional abundances are subject to

$$\sum_{j=1}^{p} \alpha_j = 1, \quad \alpha_j \geq 0, \ j = 1, \ldots, p \tag{6.4}$$

termed full additivity and positivity constraints, respectively. For a real sensor, the output of channel i is still formally given by Eq. (6.3), but α_j depends on sensor point spread function (PSF) $h_{x,y}(u, v)$ according to

$$\alpha_j = \frac{\int_{A_i} h_{x,y}(-u, -v)\, dudv}{\int_{\Re^2} h_{x,y}(-u, -v)\, dudv}$$

where A_i denotes the set of points on the surface belonging to the ith endmember. The PSF $h_{x,y}(u, v)$ may be spatially variant (i.e., it depends on surface coordinates x and y) and it includes the effect of finite aperture and receiver impulse response. Note that the meaning of α_j depends not only on the true fractional abundance, but also on the endmember distribution inside the IFOV and on the surface point

(x, y). Wu and Schowengerdt [65] propose an image-restoration-based approach to mitigate the errors introduced by the nonideal nature of the PSF.

Herein, we assume that the signal at the output of the channel i is given by Eq. (6.3). A usual procedure is to remove low-SNR channels from the spectral vectors. Thus, since only high-SNR channels are considered, Poisson noise is neglected. Concerning atmospheric correction, we assume that the images have been processed to remove path radiance (ray 5 in Figure 6.1) and the light scattered by interaction between surface and the atmosphere (ray 6 in Figure 6.1). Note, however, that the corrected image is still in radiance units.

Let $\mathbf{r}$ be a $B \times 1$ vector, where B is the total number of bands and $\mathbf{m}_j^0 \equiv [c_1\rho_{1j}, c_2\rho_{2j}, \ldots, c_B\rho_{Bj}]^T$ is the so-called signature of the jth endmember. Equation (6.3) can be written as

$$\mathbf{r} = \mathbf{M}\boldsymbol{\alpha} + \mathbf{w}' \tag{6.5}$$

where $\mathbf{M} \equiv [\mathbf{m}_1^0, \mathbf{m}_2^0, \ldots, \mathbf{m}_p^0]$ is a matrix with the signatures of the endmembers present in the covered area, $\boldsymbol{\alpha} \equiv [\alpha_1, \alpha_2, \ldots, \alpha_p]^T$, and $\mathbf{w}'$ models additive receiver electronic noise. The notation $(\cdot)^T$ indicates vector transposed.

Model (6.5) is an oversimplification of reality, because it does not take into account signature variability (from pixel to pixel) due to changes in the configuration and in the composition of substances, surface contaminants, variation in the substances such as age-induced color fading due to oxidation or bleaching, uncompensated atmospheric and environmental effects, and uncompensated errors in the sensor. Signature variability has been studied and accounted for in a few unmixing algorithms (see, e.g., references 66–68).

Signature variability is primarily characterized by spectral shape invariance [52]; that is, while the spectral shapes of the endmembers are fairly consistent, their amplitude varies considerably over the scene. Based on this rationale, we model spectral variability of the ith endmember at a given pixel as

$$\mathbf{m}_i^0 = \psi_i \mathbf{m}_i + \mathbf{w}_i \tag{6.6}$$

where $\psi_i \geq 0$ is a scale factor and $\mathbf{w}_i$ is a zero-mean random vector. Noise $\mathbf{w}_i$ accounts for signature variability not modelled by ψ_i. Introducing Eq. (6.6) into Eq. (6.5), we obtain, for a given pixel,

$$\mathbf{r} = \mathbf{M}\boldsymbol{\psi}\boldsymbol{\alpha} + \sum_{i=1}^{p} \alpha_i \mathbf{w}_i + \mathbf{w}' \tag{6.7}$$

where $\boldsymbol{\psi} \equiv \text{diag}\,(\psi_1, \ldots, \psi_p)$ is a $p \times p$ diagonal matrix.

Having in mind that illumination variability due to surface topography affects equally all bands and that we are assuming additive system noise, then we can write

$$\mathbf{r} = \mathbf{M}\underbrace{\gamma\boldsymbol{\psi}\boldsymbol{\alpha}}_{\mathbf{s}} + \underbrace{\gamma\sum_{i=1}^{p} \alpha_i \mathbf{w}_i + \mathbf{w}'}_{\mathbf{w}} = \mathbf{M}\mathbf{s} + \mathbf{w} \tag{6.8}$$

where γ models surface topography. Model (6.8) is still linear. If $\mathbf{M}$ is known, estimating $\mathbf{s}$ is an ordinary linear problem. If $\mathbf{M}$ is unknown, the problem is more difficult. Considering independent sources $\mathbf{s}$ and no noise, unmixing (6.8) can be solved under the ICA framework. This has been addressed in references 9, 25, and 30–38. Linear unmixing of independent sources in the presence of noise has been addressed in Moulines et al. [56] and in Attias [39], where the IFA algorithm was introduced. Herein, and in the presence of noise we adopt this algorithm.

Signature variability introduces randomness among sources and thereby attenuates their statistical dependencies. This can be understood by computing the correlation factor between sources $s_i = \psi_i \alpha_i$ and $s_j = \psi_j \alpha_j$. Assuming that ψ_i is independent of ψ_j and that $\boldsymbol{\psi}$ is independent of $\boldsymbol{\alpha}$, we have

$$
\begin{aligned}
|\varrho_{s_i s_j}| &= \frac{|E[\psi_i]E[\psi_j]||C_{\alpha_i \alpha_j}|}{\sqrt{E[\psi_i^2]E[\alpha_i^2] - E^2[\psi_i]E^2[\alpha_i]}\sqrt{E[\psi_j^2]E[\alpha_j^2] - E^2[\psi_j]E^2[\alpha_j]}} \\
&\leq \frac{|E[\psi_i]E[\psi_j]|}{\sqrt{E[\psi_i^2]E[\psi_j^2]}} \frac{|C_{\alpha_i \alpha_j}|}{\sigma_{\alpha_i}\sigma_{\alpha_j}} \qquad (6.9) \\
&\leq |\varrho_{\alpha_i \alpha_j}| \qquad (6.10)
\end{aligned}
$$

$E[\cdot]$ denotes the expectation operator where $E[\psi_k^2] \geq E^2[\psi_k]$ was invoked to obtain the right-hand side of Eqs. (6.9) and (6.10). We conclude then that signature variability does not increase source correlation. Of course, decorrelation does not imply independency. It is, however, plausible that increasing decorrelation means increasing independency. In fact, in the next sections we give experimental evidence that increasing signature variability improves ICA and IFA results.

6.3. INDEPENDENT COMPONENT ANALYSIS AND INDEPENDENT FACTOR ANALYSIS

ICA [28, 29, 69] is an unsupervised source separation process, which has been applied to linear blind separation problems [70–72]. The goal of ICA is to recover independent sources, given only sensor observations that are unknown linear mixtures of the unobserved independent sources.

Let $\mathbf{r}$ be a $B \times 1$ observation column vector, such as

$$\mathbf{r} = \mathbf{M}\mathbf{s} \qquad (6.11)$$

where $\mathbf{M}$ is an unknown $B \times p$ ($B \geq p$) mixing matrix and $\mathbf{s} \equiv [s_1\ s_2\ \ldots\ s_p]^T$ is an unknown random data vector of mutually independent sources having unknown distributions, although, at most, one might be Gaussian distributed. ICA finds a $p \times B$ separating matrix $\mathbf{W}$, such that

$$\mathbf{y} = \mathbf{W}\mathbf{r} = \mathbf{P}\mathbf{C}\mathbf{s} \qquad (6.12)$$

where $\mathbf{y}$ is a vector of independent components, and $\mathbf{P}$ and $\mathbf{C}$ are permutation and scale matrices, respectively.

ICA looks for a linear representation that maximizes a nongaussianity measure [73]. A commonly objective function used in ICA algorithms is the mutual information [28] of vector $\mathbf{y} \equiv [y_1 \; y_2 \; \ldots \; y_p]^T$ given by

$$\begin{aligned} I(y_1, y_2, \ldots, y_p, \mathbf{W}) &\equiv \sum_i H(y_i) - H(\mathbf{y}) \\ &= \sum_i H(y_i) - H(\mathbf{r}) - \log |\mathbf{W}| \end{aligned} \tag{6.13}$$

where $H(y_i)$, $H(\mathbf{y})$, and $H(\mathbf{r})$ are the entropy of random variable y_i, of random vector $\mathbf{y}$, and of random vector $\mathbf{r}$, respectively (see, e.g., references 71 and 72). The mutual information is a measure of dependence between random variables. It is nonnegative and is equal to zero if and only if variables are statistically independent.

Most ICA algorithms find the separating matrix by minimizing Eq. (6.13), or an equivalent objective function, with respect to $\mathbf{W}$. The *Negentropy* (see, e.g., references 70 and 73), an entity closely related with the mutual information, has also been used as an objective function to obtain $\mathbf{W}$. It is defined as

$$J(\mathbf{y}) \equiv H(\mathbf{y}_{gauss}) - H(\mathbf{y}) \tag{6.14}$$

where $\mathbf{y}_{gauss}$ is a Gaussian random vector with the same mean and covariance as $\mathbf{y}$ [74]. *Negentropy* is nonnegative and is equal to zero if and only if $\mathbf{y}$ has Gaussian distribution. Assuming that components y_i, $i = 1, \ldots, p$, are uncorrelated, it follows that

$$I(y_1, y_2, \ldots, y_p) = J(\mathbf{y}) - \sum_i J(y_i) \tag{6.15}$$

which means that finding maximum of *Negentropy* directions (i.e., maximizing $\sum_i J(y_i)$ with respect to $\mathbf{W}$) is equivalent to minimize the mutual information.

Well-known ICA methods are *fastica* [70], *jade* [71], and the Bell and Sejnowski algorithm [72]. *Fastica* is based on a fixed-point procedure and uses the absolute value of kurtosis as a measure of nongaussianity. *Jade* uses the fourth-order cross-cumulants of the data to separate sources. The Bell and Sejnowski algorithm uses the stochastic gradient ascent learning rule to minimize the mutual information.

IFA [39] was proposed as a method for recovering independent hidden sources from their observed mixtures immersed in additive noise. IFA implements two steps. First, source densities and noise covariance are estimated from the observed data by maximum likelihood. Second, sources are reconstructed by an optimal nonlinear estimator. IFA assumes the observation model

$$\mathbf{r} = \mathbf{M}\mathbf{s} + \mathbf{w} \tag{6.16}$$

where sources are independent with unknown distributions (at most one is Gaussian) and $\mathbf{w}$ is Gaussian noise with covariance C_w, a matrix not necessarily diagonal. To make the model analytically tractable, each source density is modeled by a MOG [56]. An expectation-maximization (EM) algorithm [75, 76] is applied to compute the maximum-likelihood estimate of the noise covariance and the Gaussian mixture parameters.

The classical Principal Component Analysis (PCA) [46] seeks for a linear decomposition that best represents data in a least-squares sense. PCA finds a linear transformation $\mathbf{y} = \mathbf{Wr}$, where each row vector of $\mathbf{W}$ corresponds to the normalized orthogonal eigenvector of the data covariance matrix. While PCA uses only second-order statistics, ICA looks for components that are statistically independent rather than uncorrelated; thus, it requires statistics of orders higher than the second [73].

In the next section, we give evidence that the most well-known ICA and IFA algorithms do not correctly unmix hyperspectral data. This study is based on several experiments, where the degradation mechanisms normally found in hyperspectral applications—namely, signature variability, abundance constraints, topography modulation, and system noise—are simulated in our model.

6.4. ICA AND IFA EVALUATION WITH SIMULATED DATA

In this section we apply ICA (*fastica* [70], *jade* [71], and Bell and Sejnowski [72] and the IFA [39] algorithms to simulated data. Three experiments are conducted: In the first, modeling a canonical ICA scenario, the abundance fractions are independent; in the second, modeling an ideal hyperspectral scenario, we enforce only constraints (6.4), meaning that abundance fractions are dependent; in the third, modeling a real hyperspectral scenario, we generate abundance fractions according to Eq. (6.8), thus modeling abundance fraction dependence, signature variability, topography modulation, and system noise. We have adopted the cross-correlation factor as a performance measure, because ICA and IFA unmixes abundance fractions up to a constant factor.

In all experiments, the scene dimension is of 30×30 pixels and endmember signatures were extracted from a hyperspectral subimage of Indian Pine Test Site in Northwestern Indiana acquired by an AVIRIS instrument in June 1992.* Noisy channels and water absorption channels were removed (channels 1–4, 107–113, and 150–166). Concerning atmospheric correction, this image has been processed to remove path radiance and the light scattered by the interaction between surface and the atmosphere.

Figure (6.2) shows endmember signatures used in these three experiments.

In the first experiment, the abundance fractions are mutually independent each one following a Beta distribution† given by

$$p(\alpha_i) = \frac{\Gamma(\beta_1 + \beta_2)}{\Gamma(\beta_1)\Gamma(\beta_2)} \alpha_i^{\beta_1 - 1}(\alpha_i - 1)^{\beta_2 - 1} \tag{6.17}$$

The expected value of the ith endmember fraction is $\mu_i = E[\alpha_i] = \beta_1/(\beta_1 + \beta_2)$, $E[\cdot]$ denotes the expectation operator, and $\Gamma(\cdot)$ denotes the Gamma function. Mean

*Available at http://dynamo.ecn.purdue.edu/~biehl/MultiSpec/

†Beta is a particular case of Dirichlet distribution.

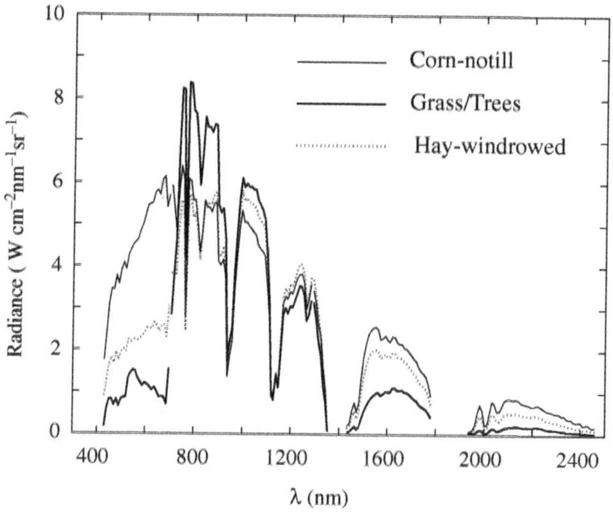

Figure 6.2. Radiance spectra of the three substances used in experiments I, II, and III. Copyright © 2005 IEEE.

values are set to $\mu_1 = 0.4, \mu_2 = 0.3$, and $\mu_3 = 0.3$. In the remaining experiments, abundance fractions follow a joint Dirichlet distribution given by

$$D(\alpha_1, \alpha_2, \ldots, \alpha_p|\theta_1, \theta_2, \ldots, \theta_p) = \frac{\Gamma(\sum_{i=1}^{p} \theta_i)}{\Gamma(\theta_1)\Gamma(\theta_2) \ldots \Gamma(\theta_p)} \prod_{i=1}^{p} \alpha_i^{\theta_i - 1} \tag{6.18}$$

Such distribution constrains abundance fractions to $0 \leq \alpha_i \leq 1$ and $\sum_i \alpha_i = 1$. The expected value of the ith endmember fraction is $\mu_i = E[\alpha_i] = \theta_i / \sum_{i=1}^{p} \theta_i$.

Table 6.1 presents the sample mean of the cross-correlation coefficients between the abundance fractions and their estimates. These coefficients are computed from

TABLE 6.1. Sample Cross-Correlation Between Abundance Fractions and the Correspondent Estimates. Results based on 256 Monte Carlo runs. © 2005 IEEE

Algorithm	Source	IFA $\overline{\varrho_\alpha}$	*fastica* $\overline{\varrho_\alpha}$	*jade* $\overline{\varrho_\alpha}$	Bell and Sejnowski $\overline{\varrho_\alpha}$
Experiment I	1	0.9767	0.9978	0.9994	0.7817
	2	0.9926	0.9970	0.9985	0.8553
	3	0.9797	0.9895	0.9911	0.5738
Experiment II	1	—	—	0.1978	0.3332
	2	0.8310	0.7902	0.4778	0.8352
	3	0.4913	0.3281	0.2482	0.4490
Experiment III	1	0.9755	0.8780	0.9304	0.9727
	2	0.9670	0.9641	0.9752	0.9692
	3	0.9065	0.8454	0.8981	0.8778

256 Monte Carlo runs. We can see that under this condition (independent abundance fractions) the first three algorithms work very well, whereas Bell and Sejnowski algorithm fails to correctly unmix the endmembers.

In the second experiment, abundance fractions are dependent, following a Dirichlet distribution with parameters $\mu_1 = 0.4, \mu_2 = 0.3$, and $\mu_3 = 0.3$. It is clear that none of the algorithms correctly unmix the original dependent data. IFA and *fastica* algorithms only estimate two abundance fractions because they implement a pre-processing step to whiten the observed data and to reduce the dimension (note that the constraint $\sum_i \alpha_i = 1$ decreases by one the dimension of the observed data).

In the third experiment, abundance fractions are dependent following a Dirichlet distribution with parameters $\mu_1 = 0.4, \mu_2 = 0.3$, and $\mu_3 = 0.3$ for each endmember. Scale ψ_i, controlling signature variability, is uniformly distributed within the interval [0.9 , 1.1]; parameter γ is Beta distributed with parameters $\beta_1 = 1$, $\beta_2 = 0.8$. Noise in Eq. (6.8) is zero-mean white Gaussian with variance σ^2 in each band such that the SNR defined as the ratio between the signature power and the noise power, that is,

$$\text{SNR} = 10 \log_{10} \frac{E[(\mathbf{Ms})^T(\mathbf{Ms})]}{E[\mathbf{w}^T\mathbf{w}]} \tag{6.19}$$

is set to 30 dB. As in the previous experiments, ICA and IFA algorithms do not correctly unmix the three abundance fractions. IFA, however, yields the best results, being able to approximately unmix two abundance fractions.

The pattern of behavior exhibited in experiments I to III was systematically replicated regardless of the source statistics [constraint (6.4) is understood]. Basically, we conclude that the accuracy of ICA applied to hyperspectral data tends to increase with the signature variability, the number of sources, and the SNR. There are, however, always endmembers incorrectly unmixed, regardless the unmixing scenario. A more exhaustive study of ICA algorithms is addressed in Nascimento and Dias [1]. The study is based on model (6.8) and it evaluates the mean magnitude of the cross-correlation factor between sources and their estimates by *fastica* algorithm, as a function of the SNR, of signature variability, and of the number of sources.

6.5. LIMITATIONS OF ICA AND IFA IN UNMIXING HYPERSPECTRAL DATA

In this section, we study the ICA and IFA limitations in unmixing hyperspectral data. To this end, we characterize the behavior of the mutual information in the neighborhood of the true unmixing matrix. The aim is to show that when sources are dependent, minimizing mutual information does not yield the true unmixing matrix.

A brief introduction to ICA and IFA was presented in Section 6.3. Given an unknown linear mixture of the unobserved independent sources $\mathbf{r} = \mathbf{Ms}$, ICA and IFA look for a matrix $\mathbf{W}$ that maximizes a nongaussianity measure of the vector $\mathbf{y} = \mathbf{Wr}$. An objective function commonly used in ICA algorithms is the mutual information [28].

Assuming $|\mathbf{W}|$ constant, the minimization of the mutual information reduces to finding [see expression (6.13)]

$$\widehat{\mathbf{W}} = \arg\min_{\mathbf{W}} I(y_1, y_2, \ldots, y_p, \mathbf{W}) = \arg\min_{\mathbf{W}} \sum_i H(y_i) \tag{6.20}$$

where

$$H(y_i) \equiv -\int_{-\infty}^{+\infty} f_{Y_i}(u) \log f_{Y_i}(u)\, du \tag{6.21}$$

with f_{Y_i} being the probability density function of y_i. To compute Eq. (6.21), we need to estimate f_{Y_i}, for $i = 1, \ldots, p$. Herein this estimate is based on fitting a mixture of Gaussians to samples of y_i [77]:

$$f_{Y_i}(y_i) = \sum_{q_i=1}^{n_i} \sigma_{q_i} (2\pi|\Sigma_{q_i}|)^{-1/2} \exp\left[-\frac{1}{2}(y_i - \mu_{q_i})^T \Sigma_{q_i}^{-1} (y_i - \mu_{q_i})\right] \tag{6.22}$$

where n_i is the number of Gaussians modes to fit the ith source and σ_{q_i}, μ_{q_i}, and Σ_{q_i} are the weight, the mean, and the covariance of the q_ith Gaussian mode, respectively. The number of Gaussians modes and respective parameters (means, covariances, and weights) are obtained via the minimum description-length-based expectation-maximization algorithm (MDL-EM) [55]. The entropy (6.21) is computed via numerical integration.

Figure 6.3 shows an example of a Rayleigh probability density function fitted with a Gaussian mixture using algorithm [55].

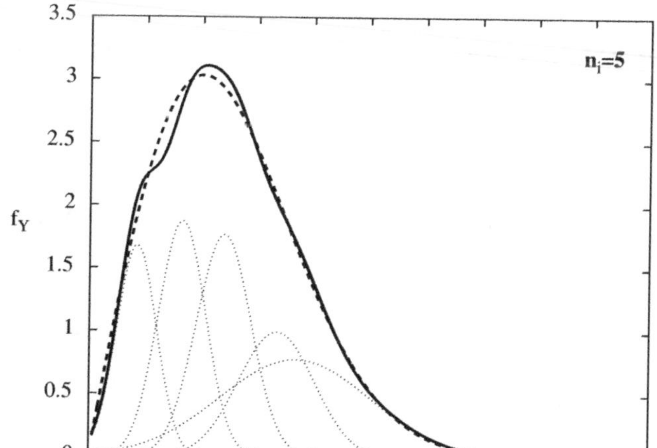

Figure 6.3. Rayleigh probability density function (dashed line) fitted with a Gaussian mixture (solid line) using algorithm [55]. Copyright © 2005 IEEE.

The maxima of $I(y_1, y_2, \ldots, y_p, \mathbf{W})$ with respect to $\mathbf{W}$ is not very sensitive to the shape of f_{Y_i} (see Chapter 9, reference 29). For example, Attias [39] uses only three Gaussian modes to fit whatever density shape. Herein, however, we use all modes given by the MDL-EM algorithm [55], because as we are interested not only in the unmixing matrix $\mathbf{W}$, but also in computing the mutual information $I(y_1, y_2, \ldots, y_p, \mathbf{W})$ as a function of $\mathbf{W}$.

6.5.1. Experimental Results

In the next experiments, we study the behavior of the mutual information $I(y_1, y_2, \ldots, y_p, \mathbf{W})$, for $\mathbf{W}$ in the neighborhood of the true unmixing matrix $\mathbf{W} = \mathbf{M}^{-1}$. In all experiments we assume constraint $|\mathbf{W}| = 1$. This setting does not constraint the unmixed results, as they are defined up to a constant.

Experiment I. The first experiment considers independent abundance fractions with uniform distribution to test our setup under canonical mixing conditions. This experiment assumes $p = 3$ (number of endmembers), $B = 3$ (number of bands), $\mathbf{M} = \mathbf{I}$, $\psi = \mathbf{I}$, $\gamma = 1$, and $\mathbf{w} = \mathbf{0}$.

Figure 6.4a shows the mutual information as function of ϕ_1 and ϕ_2 in a gray scale (ϕ_1 and ϕ_2 define a rotation in $\Re^3$. We term this angles azimuth and elevation, respectively). The minimum is global and occurs for $\phi_1 = 0$ and $\phi_2 = 0$, that is, $\mathbf{W} = \mathbf{I}$.

As mentioned above, abundance fractions in hyperspectral data are not independent. In order to test ICA with this constraint, on next experiment, we generate abundance fractions according to the Dirichlet distribution [see expression (6.18)] parameterized with $\mu_1 = 1/3$, $\mu_2 = 1/3$, and $\mu_3 = 1/3$ (recall that μ_i is the expected value of the ith abundance fraction).

Experiment II. In this experiment we set $p = 3$, $B = 3$, $\mathbf{M} = \mathbf{I}$, $\psi = \mathbf{I}$, $\gamma = 1$, $\mathbf{w} = \mathbf{0}$.

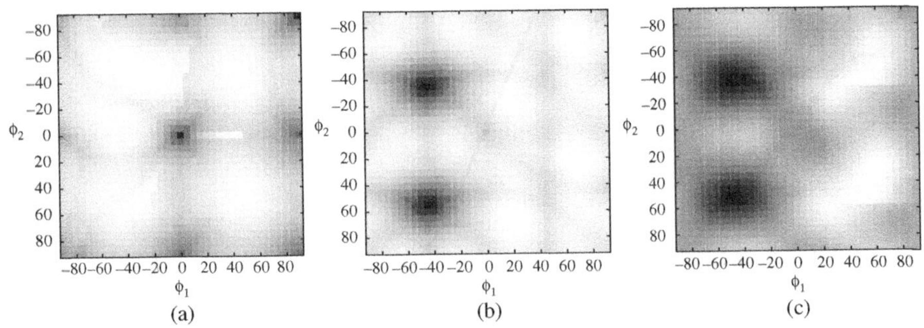

Figure 6.4. Mutual information as function of parameters ϕ_1 (azimuth angle) and ϕ_2 (elevation angle). Black color represents the lower values. (a) Three independent sources; (b) dependent sources; (c) dependent sources with illumination perturbations, signature variability, and SNR = 20 dB. Copyright © 2005 IEEE.

In Figure 6.4b we present the mutual information as function of angles ϕ_1 and ϕ_2. No ICA algorithm could ever correctly unmix the original dependent data, since ($\hat{\phi}_1 \simeq -\pi/4$, $\hat{\phi}_2 \simeq -\pi/5$), far from the true unmixing matrix, that is, ($\phi_1 = 0$, $\phi_2 = 0$).

Experiment III. In this experiment the abundance fractions are dependent and Dirichlet distributed ($\mu_1 = 1/3$, $\mu_2 = 1/3$, $\mu_3 = 1/3$). The remaining parameters are $p = 3$, $B = 3$, $\mathbf{M} = \mathbf{I}$, ψ_i uniformly distributed in the interval $[0.9, 1.1]$, and γ Beta distributed with parameters $\beta_1 = 1$ and $\beta_2 = 0.8$, and the SNR is set to 20 dB.

Figure 6.4c presents the mutual information as function of angles ϕ_1 and ϕ_2. The global minimum occurs at $\hat{\phi}_1 \simeq 10\pi/36$, $\hat{\phi}_2 \simeq -10\pi/36$. Note, however, the presence of a local minimum at $\hat{\phi}_2 \simeq 0$ and $\hat{\phi}_2 \simeq 0$.

The pattern of behavior described in these experiments was systematically observed in a series of experiments with different number of sources, different abundance fraction distributions, different parameters, and different SNR.

> Basically, we concluded that in linear hyperspectral data unmixing, the unmixing matrix **W** minimizing the mutual information might be very far from the true one, at least for a few number of endmembers.

This is basically in agreement with conclusions drawn in Section 6.4.

6.6. FASTICA ALGORITHM APPLIED TO REAL DATA

In this section, we consider a subimage of the hyperspectral data set from Indian Pine Test Site in Northwestern Indiana acquired by an AVIRIS in June 1992. The data set is composed of 220 spectral channels with 10 nm bandwidth acquired in the 0.4 to 2.5 μm region. It contains 145 × 145 pixels (21,025 pixels) with a ground pixel resolution of 17 m [78]. This region contains a mixture of agriculture and forestry. However, due to the early season date of data collection, the cultivated land appears to have very little canopy cover. There is a major dual lane highway (U.S. 52 and U.S. 231), a rail line crossing near the top, a major secondary road (Jackson Highway) near the middle, several other county roads, and houses (Figure 6.5a shows band 29 of the data set). The ground truth of the region [79] classifies the ground covered area into sixteen classes and ignores many small variations within fields that can be seen in the image data (see Figure 6.5a). With respect to atmospheric correction, this data set has been processed to remove path radiance and the light scattered by interaction between surface and the atmosphere. The corrected image is still, however, in radiance units.

A PCA pre-processing step was implemented to whiten the observed data and to reduce dimension: Every pixel vector $\mathbf{r}$ is transformed into a vector of dimension 16 with zero-mean and identity covariance matrix: $\mathbf{x} = \mathbf{U}(\mathbf{r} - E[\mathbf{r}])$, such that $E[\mathbf{x}] = 0$ and $E[\mathbf{x}\mathbf{x}^T] = \mathbf{I}$, where $\mathbf{I}$ the identity matrix.

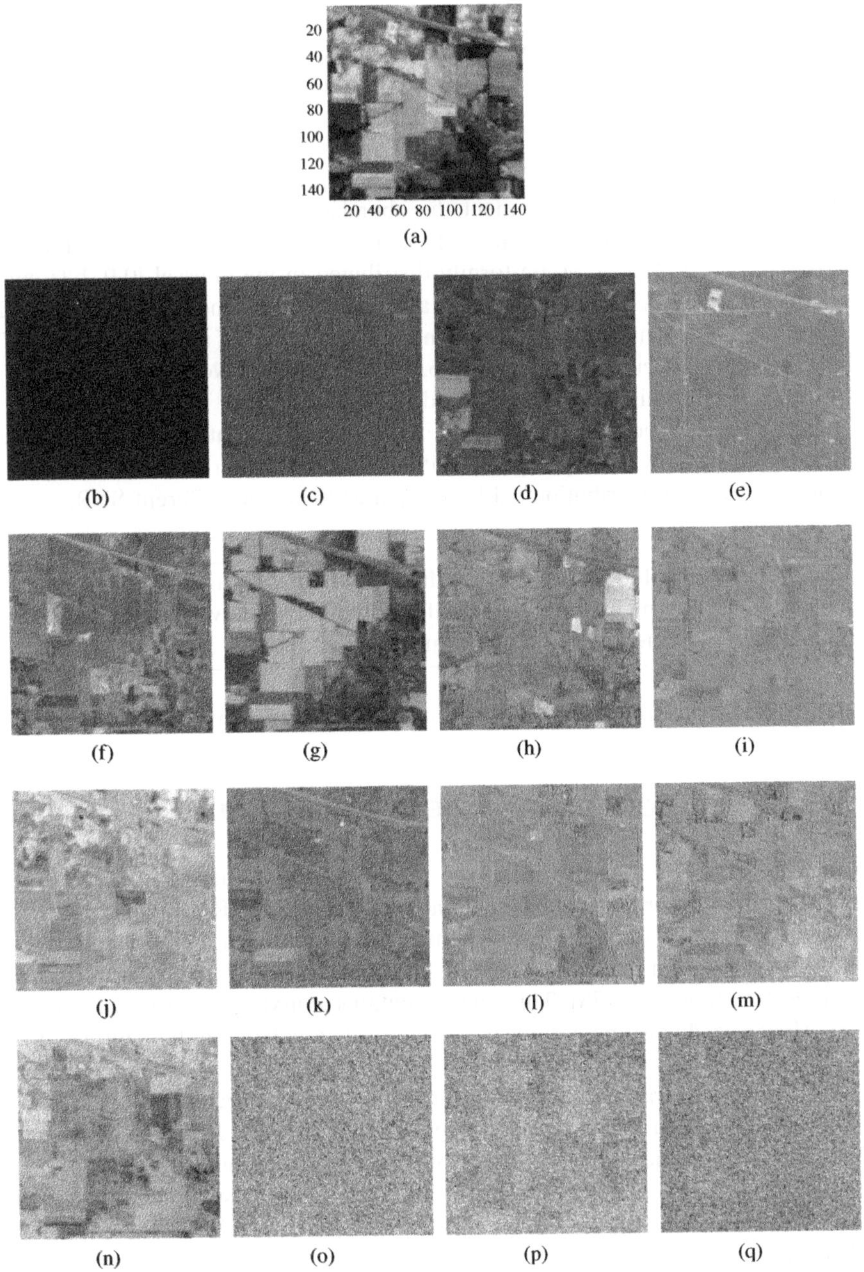

Figure 6.5. (a) Subimage of Indiana Pines test site, band 29($\lambda = 667.3$nm) and independent components extracted from data set, with *fastica* algorithm (b) Not classified; (c) houses; (d) grass/pasture; (e) stone-steel towers, rail line, and highways; (f) grass/trees; (g) grass, soybeans, and corn; (h) hay-windrowed; (i) not classified; (j) vegetation mowed; (k) houses; (l) houses; (m) not classified; (n) not classified; (o) noise; (p) noise; (q) noise. Copyright © 2005 IEEE.

Fastica algorithm was applied to the processed data set. Figure 6.5 presents the first 16 components. The first, second, fourth, tenth, and eleventh components correspond to houses or man made materials that exist in a few locations in the scene (see Figures 6.5c, 6.5e, 6.5k, and 6.5l). We can see that maximum contrast of sources occurs in house locations, stone-steel towers, Jackson highway, and rail line. The third and fifth components are a mixture of grass with pasture and trees, respectively (see Figures. 6.5d and 6.5f). The sixth component (see Figure 6.5g) represents a mixture of grass, soybeans, and corn. Apparently, the seventh component is hay-windrowed (see Figure 6.5h). Ninth component represents vegetation mowed (see Figure 6.5j). Components presented in Figures 6.5b, 6.5i, 6.5m, and 6.5n do not represent any class of the available ground truth [79]. Figures 6.5o, 6.5p, and 6.5q are mainly noise. We conclude that six sources are unmixed, and ten are a mixture of several materials present in the scene or are manly noise. This is basically in accordance with our findings based on simulated data.

A pertinent question is, What sources are correctly unmixed? To address this question, we computed the entropy of each component, normalized to unit variance, following the procedure described in Section 6.5. The underlying idea is that a mixed source tends to exhibit higher entropy, the maximum value being $H_{\max} = 1.419$ nats achieved by a Gaussian source.

Table 6.2 shows the entropy for each component shown in Figure 6.5b–q. By inspection of Table 6.2, we can identify two subsets: The first, with smaller values of entropy, corresponds to components correctly unmixed (see Figures 6.5c–h); the second, with larger values of entropy, corresponds to components with mixed sources (see Figures 6.5b and 6.5i–q). The larger values, near $H_{\max}$, correspond to the last three components, which are mostly noise.

6.7. DEPENDENT COMPONENT ANALYSIS

Previous sections addressed the impact of the source dependence on unmixing hyperspectral data with ICA/IFA. It was shown that the most well-known ICA and IFA algorithms do not correctly unmix hyperspectral data. Basically, we concluded that the unmixing matrix minimizing the mutual information might be very far from the true one.

TABLE 6.2. Entropy of Each Component

Component:	(b)	(c)	(d)	(e)	(f)	(g)	(h)	(i)
Entropy:	1.4020	1.3290	1.1834	1.2002	1.3399	1.1300	1.2811	1.4031
Unmixed:		*	*	*	*	*	*	
Component:	(j)	(k)	(l)	(m)	(n)	(o)	(p)	(q)
Entropy:	1.3781	1.3945	1.3842	1.4069	1.4172	1.4188	1.4187	1.4189
Unmixed:								

Source: Copyright © 2005 IEEE.

This section sketched a new direction to blindly unmix hyperpsectral data, where abundance fractions are modeled by a mixture of Dirichlet densities. This family of densities ensures source nonnegativity and additivity constraints while, at the same time, covering a wide range of shape densities.

The new method is of EM type and is in the vein of references 39 and 56, replacing independent sources represented by MOG with mixtures of Dirichlet sources. Compared with the geometric-based approaches, its advantage is that there is no need to have pure pixels in the observations.

Assuming a noiseless linear observation model (6.8), spectral vectors are in a $(p-1)$ -dimensional simplex in $\Re^L$ [constraints (6.4) are understood]. If $p \ll L$, it is worthy to project the observed spectral vectors onto the subspace signal. Let $\mathbf{P}_p \equiv \mathbf{U}_p\mathbf{U}_p^T$ be the projection matrix, where $\mathbf{U}_p$ is a unitary matrix spanning the signal subspace. The coordinates of $\mathbf{x}$ with respect to $\mathbf{U}_p$ are

$$\begin{aligned} \mathbf{x} &\equiv \mathbf{U}_p^T\mathbf{r} \\ &= \mathbf{U}_p^T\mathbf{M}\mathbf{s} \\ &= \mathbf{A}\mathbf{s} \end{aligned} \tag{6.23}$$

where $\mathbf{A}$ is a $(p \times p)$ square mixing matrix and $\mathbf{x} = [x_1, x_2, \ldots, x_p]^T$ is a p-vector.

The Full Additivity constraint [see Eq. (6.4)] implies that $s_p = 1 - \sum_{i=1}^{p-1} s_i$; that is, only $p-1$ components of $\mathbf{s}$ are free. Recall that $\mathbf{s} = [s_1, \ldots, s_p]^T$.

Assuming that $\mathbf{W} \equiv \mathbf{A}^{-1}$ exists,* we may write $\widehat{\mathbf{s}} = \mathbf{W}\mathbf{x}$. Consider that each pixel $\mathbf{x}$ represents one particular outcome of a p-dimensional random variable $\mathbf{X} = [X_1, \ldots, X_p]^T$. Given a set of N independent and identically distributed samples $\mathcal{X} = \{\mathbf{x}^{(1)}, \ldots, \mathbf{x}^{(N)}\}$, the log-likelihood of $\mathbf{W}, \theta$ is

$$\begin{aligned} L(\mathbf{W}, \boldsymbol{\theta}) &\equiv \log p_X(\mathcal{X}|\mathbf{W}, \boldsymbol{\theta}) \\ &= \log p_X(\mathbf{x}^{(1)}, \mathbf{x}^{(2)}, \ldots, \mathbf{x}^{(N)}|\mathbf{W}, \boldsymbol{\theta}) \\ &= \log\left(\prod_{j=1}^{N} p_X(\mathbf{x}^{(j)}|\mathbf{W}, \boldsymbol{\theta})\right) \\ &= \sum_{j=1}^{N} \log p_X(\mathbf{x}^{(j)}|\mathbf{W}, \boldsymbol{\theta}) \\ &= \sum_{j=1}^{N} \log p_S(\mathbf{s}^{(j)}|\boldsymbol{\theta}) + N \log|\det \mathbf{W}| \end{aligned} \tag{6.24}$$

where $\boldsymbol{\theta}$ is a vector with the parameters of density $p_S(\mathbf{s}|\boldsymbol{\theta})$. Denoting $\langle \mathbf{x} \rangle \equiv 1/N \sum_{j=1}^{N} \mathbf{x}^{(j)}$ (i.e., the sample average of $\mathbf{x}$), it follows that

$$L_N(\mathbf{W}, \boldsymbol{\theta}) = \langle \log p_S(\mathbf{s}|\boldsymbol{\theta}) \rangle + \log(|\det \mathbf{W}|) \tag{6.25}$$

Let the abundance fractions follow a K-component Dirichlet finite mixture given by

$$p_S(\mathbf{s}|\boldsymbol{\theta}) = \sum_{q=1}^{K} \epsilon_q D(\mathbf{s}|\boldsymbol{\theta}_q) \tag{6.26}$$

*If $\mathbf{W}$ does not exist, it is always possible to permute the order of the endmembers such that $\mathbf{W}$ exists.

where $\epsilon_1, \ldots, \epsilon_K$ are the mixing probabilities, satisfying $\sum_{q=1}^{K} \epsilon_q = 1$ and $\epsilon_q \geq 0$, for $q = 1 \ldots, K$, $\boldsymbol{\theta}_q$ is the set of parameters defining the q-component, and $D(\mathbf{s}|\boldsymbol{\theta}_q)$ is a Dirichlet distribution given by

$$D(\mathbf{s}|\boldsymbol{\theta}_q) = \frac{\Gamma(\sum_{i=1}^{p} \theta_{qi})}{\prod_{i=1}^{p} \Gamma(\theta_{qi})} \prod_{i=1}^{p} s_i^{\theta_{qi}-1} \tag{6.27}$$

The expected value of the ith endmember fraction is $E[s_i] = \theta_{qi} / \sum_{l=1}^{p} \theta_{ql}$.

The complete set of parameters $\boldsymbol{\theta}$ needed to specify the mixture includes the mixing probabilities $\epsilon_1, \ldots, \epsilon_K$ and the Dirichlet parameters $\boldsymbol{\theta}_q = \{\theta_{q1}, \ldots, \theta_{qp}\}$, for $q = 1, \ldots, K$, that is, $\boldsymbol{\theta} = \{\epsilon_1, \ldots, \epsilon_K, \boldsymbol{\theta}_1, \ldots, \boldsymbol{\theta}_K\}$.

Introducing expression (6.26) into (6.25), we obtain

$$L_N(\mathbf{W}, \boldsymbol{\theta}) = \left\langle \log \sum_{q=1}^{K} \epsilon_q D(\mathbf{s}|\boldsymbol{\theta}_q) \right\rangle + \log(|\det \mathbf{W}|) \tag{6.28}$$

The *maximum likelihood* (ML) estimate of $(\mathbf{W}, \boldsymbol{\theta})$ is given by

$$\left(\widehat{\mathbf{W}}, \widehat{\boldsymbol{\theta}}\right) = \arg \max_{\mathbf{W}, \boldsymbol{\theta}} L_N(\mathbf{W}, \boldsymbol{\theta}) \tag{6.29}$$

which can not be found analytically.

6.7.1. Unmixing and Mixture Estimation with EM Algorithm

Here in we adopt the EM algorithm to infer the parameters of a mixture [77]. The EM algorithm relies on the so-called incomplete data and missing data. In our setup, let us denote $\mathcal{X}$ as incomplete data and denote $\mathcal{Z} = \{\mathbf{z}^{(1)}, \ldots, \mathbf{z}^{(N)}\}$, a set of N labels indicating which component has produced each sample, as the missing data. Each label $\mathbf{z}^{(j)} = [z_1^{(j)}, \ldots, z_K^{(j)}]$ is a binary K-vector, where only one z_q^j is set to one, indicating which component produced the j-sample. The complete log-likelihood is

$$\begin{aligned} L_C(\mathbf{W}, \boldsymbol{\theta}) &= \log(p_{X,Z}(\mathcal{X}, \mathcal{Z}|\boldsymbol{\theta}) \\ &= \left\langle \sum_{q=1}^{K} z_q \log \epsilon_q D(\mathbf{s}|\boldsymbol{\theta}_q) \right\rangle + \log(|\det \mathbf{W}|) \end{aligned} \tag{6.30}$$

The EM algorithm iterates between the E-step and the M-step:

- *E-Step*: Computes the conditional expectation of the complete log-likelihood, given the samples and the current estimate $\widehat{\boldsymbol{\theta}}^{(t)}$. The result is the so-called Q-function

$$\begin{aligned} Q(\boldsymbol{\theta}, \widehat{\boldsymbol{\theta}}^{(t)}) &\equiv E[\log p_{X,Z}(\mathcal{X}, \mathcal{Z}|\boldsymbol{\theta}) |\, \mathcal{X}, \widehat{\boldsymbol{\theta}}^{(t)}] \\ &= \left\langle \sum_{q=1}^{K} \underbrace{E[z_q|\, \mathbf{s}, \widehat{\boldsymbol{\theta}}^{(t)}]}_{\beta_q^{(t)}} \log[\epsilon_q^{(t)} D(\mathbf{s}|\boldsymbol{\theta}_q^{(t)})] \right\rangle \\ &\quad + \log(|\det \mathbf{W}|) \end{aligned} \tag{6.31}$$

where

$$\beta_q^{(t)} \equiv E[z_q|\mathbf{s}, \widehat{\boldsymbol{\theta}}^{(t)}] = \frac{\widehat{\epsilon}_q^{(t)} D(\mathbf{s}|\widehat{\boldsymbol{\theta}}_q^{(t)})}{\sum_{l=1}^{K} \widehat{\epsilon}_l^{(t)} D(\mathbf{s}|\widehat{\boldsymbol{\theta}}_l^{(t)})} \tag{6.32}$$

- *M-Step*: Updates the parameter estimates according to

$$\widehat{\boldsymbol{\theta}}^{(t+1)} = \arg\max_{\boldsymbol{\theta}} \{Q(\boldsymbol{\theta}, \widehat{\boldsymbol{\theta}}^{(t)})\} \tag{6.33}$$

We implement the following learning rules for the mixture of Dirichlet source parameters:

$$\epsilon_q^{(t)} = \langle \beta_q^{(t)} \rangle \tag{6.34}$$

$$\widehat{\theta}_{qi}^{(t+1)} = \text{psi}^{-1}\left[\text{psi}\left(\sum_{l=1}^{p} \widehat{\theta}_{ql}^{(t)}\right) + \frac{\langle \beta_q^{(t)} \log \widehat{s}_i^{(t)} \rangle}{\langle \beta_q^{(t)} \rangle}\right] \tag{6.35}$$

where symbols psi$(\cdot)$ and psi$^{-1}(\cdot)$ denote the psi function* and its inverse, respectively. Equation (6.34) maximizes Q with respect to $\epsilon_q^{(t)}$; Eq.(6.35) ensures that Q do not decrease (see Dias [80] for details).

Since $\partial Q/\partial \mathbf{W} = 0$ is not a linear equation and cannot be solved analytically, herein we derive an iterative learning rule for the unmixing matrix $\mathbf{W}$:

$$\mathbf{W}^{(t+1)} = \mathbf{W}^{(t)} + \eta^{(t)} \left(\frac{\partial Q}{\partial \mathbf{W}}\right)^{(t)} \tag{6.36}$$

where $\eta^{(t)}$ determines the learning rate on iteration t and

$$\left(\frac{\partial Q}{\partial \mathbf{w}_i}\right)^{(t)} = \left\langle \sum_{q=1}^{K} \left[\beta_q^{(t)} \frac{(\widehat{\theta}_{qi}^{(t)} - 1)}{\widehat{s}_i} \mathbf{x}^T - \beta_q^{(t)} \frac{(\widehat{\theta}_{qp}^{(t)} - 1)}{\widehat{s}_p} \mathbf{x}^T\right] \right\rangle + [\mathbf{W}^{-T}]_i - [\mathbf{W}^{-T}]_p \tag{6.37}$$

where $\mathbf{w}_i$, for $i = 1, \ldots, p-1$ denotes the ith row of matrix $\mathbf{W}$ and $[\mathbf{W}^{-T}]_i$ denotes the ith row of the inverse of $\mathbf{W}$ transposed. Rule (6.37) is of the type considered in references 29 and 39. The term $[\mathbf{W}^{-T}]_i - [\mathbf{W}^{-T}]_p$ in Eq. (6.37) results from the constraint $\sum_i s_i = 1$

*psi$(\cdot)$ function is the derivative of the logarithm of the Gamma function.

Algorithm 1

initialize $\eta, \widehat{\mathbf{W}}, \widehat{\boldsymbol{\theta}}$, and L_{best}
while $L_N - L_{best} > threshold$ **do**
  $\widehat{\mathbf{s}} := \widehat{\mathbf{W}}^{(t)}\mathbf{x}$;
  for $q := 1$ to K **do**
    $D\left(\widehat{\mathbf{s}}|\widehat{\boldsymbol{\theta}}_q^{(t)}\right) := \frac{\Gamma(\sum_{i=1}^{p} \widehat{\theta}_{qi}^{(t)})}{\prod_{i=1}^{p} \Gamma(\theta_{qi}^{(t)})} \prod_{i=1}^{p} \widehat{s}_i^{\widehat{\theta}_{qi}^{(t)}-1}$;
    $\beta_q^{(t)} := \frac{\widehat{\epsilon}_q^{(t)} D(\widehat{\mathbf{s}}|\widehat{\boldsymbol{\theta}}_q^{(t)})}{\sum_{l=1}^{K} \widehat{\epsilon}_l^{(t)} D(\widehat{\mathbf{s}}|\widehat{\boldsymbol{\theta}}_l^{(t)})}$;
    $\widehat{\epsilon}_q^{(t)} := \langle \beta_q^{(t)} \rangle$;
  end for
  for $i :=$ to p **do**
    $\theta aux_{qi} := \mathrm{psi}^{-1}\left[\mathrm{psi}(\sum_{l=1}^{p} \widehat{\theta}_{ql}^{(t)}) + \frac{\langle \beta_q^{(t)} \log \widehat{s}_i^{(t)} \rangle}{\langle \beta_q^{(t)} \rangle}\right]$;
    $\frac{\partial Q}{\partial \mathbf{w}_i} := \left\langle \sum_{q=1}^{K} \left[\beta_q^{(t)} \frac{(\widehat{\theta}_{qi}^{(t)}-1)}{\widehat{s}_i} \mathbf{x}^T - \beta_q^{(t)} \frac{(\widehat{\theta}_{qp}^{(t)}-1)}{\widehat{s}_p} \mathbf{x}^T\right] \right\rangle +$
      $+[\widehat{\mathbf{W}}^{-T}]_i - [\widehat{\mathbf{W}}^{-T}]_p$;
  end for
  $\mathbf{W}aux := \widehat{\mathbf{W}}^{(t)} + \eta \frac{\partial Q}{\partial \mathbf{W}}$;
  $L_N := \langle \log \sum_{q=1}^{K} \widehat{\epsilon}_q^{(t)} D(\mathbf{s}|\boldsymbol{\theta} aux_q) \rangle + \log(|\det \mathbf{W}aux|)$;
  if $L_{best} < L_N$ **then**
    $L_{best} = L_N$;
    $\widehat{\mathbf{W}}^{(t+1)} := \mathbf{W}aux$;
    $\widehat{\boldsymbol{\theta}}^{(t+1)} := \boldsymbol{\theta} aux$;
    increment η;
else
    decrement η;
  end if
end while

Algorithm 1 presents pseudo-code aimed at the maximization of Eq. (6.28). It implements a cyclic maximizer algorithm, which splits the estimation of $\mathbf{W}$ and $\boldsymbol{\theta}$ into block maximization operations. The estimation of $\mathbf{W}$ uses a gradient ascendent method with adaptative steps. The estimation of $\boldsymbol{\theta}$ is based on the algorithm [80] derived under the EM framework [75, 81]. This approach is in the vein of references 39 and 56, replacing independent sources represented by MOG with mixture of Dirichlet sources.

Given the set of N observations $\mathbf{x}$, $\mathbf{W}$ is initialized such that $s_i^{(j)} = \mathbf{w}_i\mathbf{x}^{(j)} > 0$, for $i = 1, \ldots, p$ and $j = 1, \ldots, N$.

For illustration purposes, a simulated scene was generated according to Eq. (6.23). Three spectral signatures (Biotite, Carnallite, and Ammonioalunite)

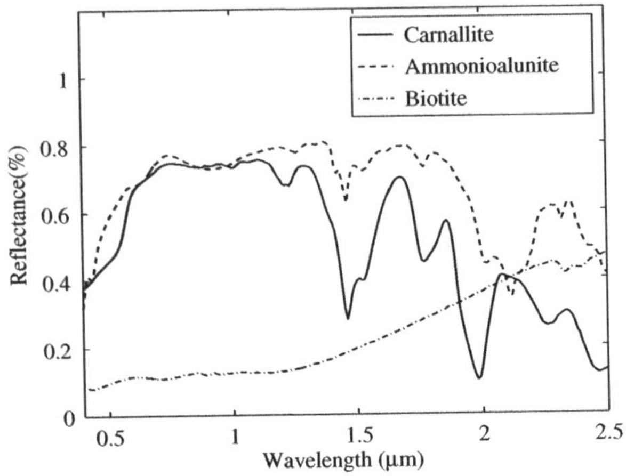

Figure 6.6. Reflectances of carnallite, ammonioalunite, and biotite. Copyright © 2005 IEEE.

were selected from the U.S. Geological Survey (USGS) digital spectral library [82] (see Figure 6.6).

The scene is split into two regions with the same number of pixels. The abundance fractions follows a Dirichlet distribution with $\boldsymbol{\theta}_a = [7,\ 4,\ 4]$ and $\boldsymbol{\theta}_b = [2,\ 5,\ 9]$ for region A and region B of the scene, respectively.

Figure 6.7 presents a scatterplot (bands $\lambda = 827$ nm and $\lambda = 1780$ nm) of a simulated scene, where dots and small circles represent the pixels of the different

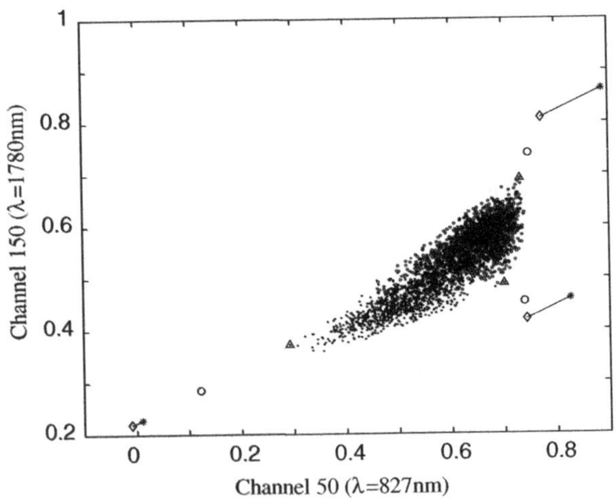

Figure 6.7. Scatterplot (bands $\lambda = 827$ nm and $\lambda = 1780$ nm) of the three endmembers mixture; Large circles: True endmembers; triangles:VCA estimate; Diamonds: Estimation with new method; Lines: Represent the estimation for each step.

regions. For comparison proposes we show the endmembers estimation by this method and by VCA [44]. As expected, the proposed approach yields better results than the VCA algorithm. VCA finds the most pure pixels in data (see triangles in Figure 6.7) which are far from the true endmembers (large circles), since there are no pure pixels in data set. Note that pure pixels-based algorithms find a simplex that may not contain all data, implying negative abundance fractions. Estimates provided by the algorithm are close to the true endmembers. In the same figure we also present the evolution for each step of the algorithm.

Figure 6.8a presents the estimated Carnallite signature, showing a good agreement with the true data. Figure 6.8b presents the estimated Carnallite signature with VCA algorithm.

6.8. CONCLUDING REMARKS

Blind hyperspectral linear unmixing aims at estimating the number of reference substances (also called endmembers), their spectral signatures, and their fractions at each pixel (called abundance fractions), using only the observed data (mixed pixels). Geometric approaches have been used whenever pure pixels are present in data [42–45]. In most cases, however, pure pixels can not be found in data. In such cases, unmixing procedures become a difficult task.

In the recent past, ICA has been proposed as a tool to unmix hyperspectral data [9, 25, 30–38]. ICA consists in finding a linear decomposition of data into statistically independent components. IFA extends ICA concepts when noise is present [39, 56]. Crucial assumptions of ICA and IFA are that each pixel is a linear mixture of endmember signatures weighted by the correspondent abundance fractions, and

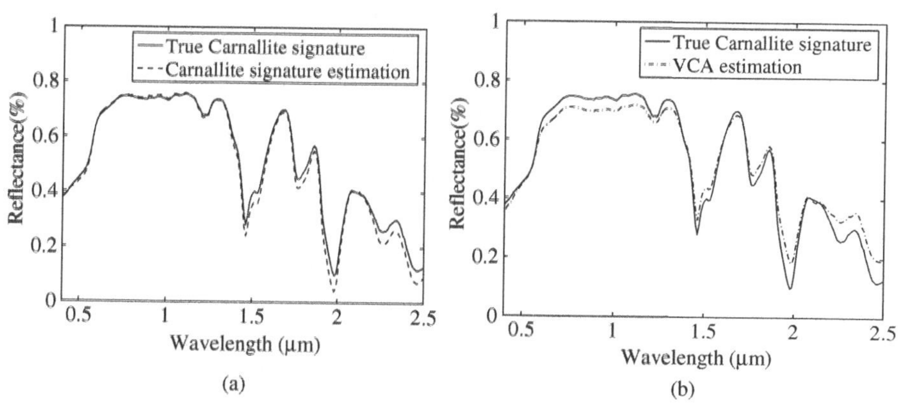

Figure 6.8. (a) Carnallite signature (solid line) and the new method estimation (dashed line). (b) Carnallite signature (solid line) and the VCA method estimation (dashed-dot line).

these abundances are independent. Concerning hyperspectral data, the first assumption is valid whenever the multiple scattering among the distinct endmembers is negligible and the surface is partitioned according to the fractional abundances. The second assumption, however, is not valid due to physical constraints on the acquisition process.

This chapter addresses the impact of the abundance fraction (sources) dependence on unmixing hyperspectral data with ICA/IFA. The study considers simulated and real hyperspectral data. Hyperspectral observations are described by a generative model that includes degradation mechanism such as signature variability, abundance constraints, topography modulation, and system noise.

IFA and three well-known ICA algorithms were tested on simulated data. Our main findings were the following:

1. ICA/IFA performance increases with the SNR.
2. ICA/IFA performance tends to increase with the signature variability and/or with the number of endmembers. The underlying reason is that by increasing the signature variability and/or the number of endmembers the statistical dependence among endmembers is attenuated.
3. There are always endmembers incorrectly unmixed, regardless of the unmixing scenario.

In order to assess the impact of hyperspectral abundance fraction dependence on the ICA/IFA algorithms, we studied the behavior of the mutual information of the unmixed sources in the neighborhood of the true unmixed data. We conclude that in hyperspectral linear unmixing, the unmixing matrix minimizing the mutual information might be very far from the true one, at least for a few number of endmembers.

Herein ICA and IFA algorithms were tested in a subimage of the hyperspectral data set from Indian Pine test site in Northwestern Indiana acquired by an AVIRIS in June 1992. According to the available ground truth of the region, we conclude that 6 sources are correctly unmixed and 10 are incorrectly unmixed. This is in line with the conclusion drawn from simulated data. A method based on the source entropy to sort the output of ICA or IFA algorithms according to the likelihood of being correctly separated was proposed.

Finally, we have proposed a new direction to blindly unmix hyperspectral data, where abundance fractions are modelled as Dirichlet sources. This model forces abundance fractions to be nonnegative and to have constant sum on each pixel. The mixing matrix is inferred by a expectation-maximization (EM)-type algorithm. The main advantage of this model is that there is no need to have pure pixels in the observations.

The performance of the proposed scheme is illustrated with simulated hyperpsectral data. Comparisons with pure pixel estimation methods are conducted. The results achieved indicate that the proposed direction is a promising tool to blindly unmix hyperspectral data.

REFERENCES

1. J. M. P. Nascimento and J. M. B. Dias, Does independent component analysis play a role in unmixing hyperspectral data? *IEEE Transactions on Geoscience and Remote Sensing*, vol. 43, no. 1, pp. 175–187, 2005.
2. D. Manolakis, C. Siracusa, and G. Shaw, Hyperspectral subpixel target detection using linear mixing model, *IEEE Transactions on Geoscience and Remote Sensing*, vol. 39, no. 7, pp. 1392–1409, 2001.
3. C.-I. Chang and D. Heinz, Subpixel spectral detection for remotely sensed images, *IEEE Transactions on Geoscience and Remote Sensing*, vol. 38, no. 3, pp. 1144–1159, 2000.
4. J. Benediktsson, J. Palmason, and J. Sveinsson, Classification of hyperspectral data from urban areas based on extended morphological profiles, *IEEE Transactions on Geoscience and Remote Sensing*, vol. 43, no. 3, pp. 480–491, 2005.
5. F. Melgani and L. Bruzzone, Classification of hyperspectral remote sensing images with support vector machines, *IEEE Transactions on Geoscience and Remote Sensing*, vol. 42, no. 8, pp. 1778–1790, 2004.
6. P. Mantero, G. Moser, and S. B. Serpico, SVM-based density estimation for supervised classification of remotely sensed images with unknown classes, in *Proceedings of the SPIE Conference on Image and Signal Processing for Remote Sensing IX*, Vol. 5238, pp. 386–397, 2004.
7. G. Shaw and D. Manolakis, Signal processing for hyperspectral image exploitation, *IEEE Signal Processing Magazine*, vol. 19, no. 1, pp. 12–16, 2002.
8. D. Landgrebe, Hyperspectral image data analysis, *IEEE Signal Processing Magazine*, vol. 19, no. 1, pp. 17–28, 2002.
9. N. Keshava, J. Kerekes, D. Manolakis, and G. Shaw, An algorithm taxonomy for hyperspectral unmixing, in *Proceedings of the SPIE AeroSense Conference on Algorithms for Multispectral and Hyperspectral Imagery VI*, Vol. 4049, pp. 42–63, 2000.
10. J. A. Richards and X. Jia, *Remote Sensing Digital Image Analysis: An Introduction*, 4th edition, Springer, New York, 2005.
11. S. Liangrocapart and M. Petrou, Mixed pixels classification, in *Proceedings of the SPIE Conference on Image and Signal Processing for Remote Sensing IV*, Vol. 3500, pp. 72–83, 1998.
12. N. Keshava and J. Mustard, Spectral unmixing, *IEEE Signal Processing Magazine*, vol. 19, no. 1, pp. 44–57, 2002.
13. R. B. Singer and T. B. McCord, Mars: Large scale mixing of bright and dark surface materials and implications for analysis of spectral reflectance, in *Proceedings of the 10th Lunar and Planetary Science Conference*, pp. 1835–1848, 1979.
14. R. Singer, Near-infrared spectral reflectance of mineral mixtures: Systematic combinations of pyroxenes, olivine, and iron oxides, *Journal of Geophysical Research*, vol. 86, pp. 7967–7982, 1981.
15. B. Nash and J. Conel, Spectral reflectance systematics for mixtures of powdered hypersthene, labradoride, and ilmenite, *Journal of Geophysical Research*, vol. 79, pp. 1615–1621, 1974.
16. B. Hapke, Bidirection reflectance spectroscopy. I. theory, *Journal of Geophysical Research*, vol. 86, pp. 3039–3054, 1981.

17. R. N. Clark and T. L. Roush, Reflectance spectroscopy: Quantitative analysis techniques for remote sensing applications, *Journal of Geophysical Research*, vol. 89, no. B7, pp. 6329–6340, 1984.

18. C. C. Borel and S. A. Gerstl, Nonlinear spectral mixing models for vegetative and soils surface, *Remote Sensing of the Environment*, vol. 47, no. 2, pp. 403–416, 1994.

19. J. J. Settle, On the relationship between spectral unmixing and subspace projection, *IEEE Transactions of Geoscience and Remote Sensing*, vol. 34, pp. 1045–1046, 1996.

20. C.-I. Chang, *Hyperspectral Imaging: Techniques for Spectral Detection and Classification*, Kluwer Academic, New York, 2003.

21. A. S. Mazer, M. Martin, et al., Image processing software for imaging spectrometry data analysis, *Remote Sensing of the Environment*, vol. 24, no. 1, pp. 201–210, 1988.

22. R. H. Yuhas, A. F. H. Goetz, and J. W. Boardman, Discrimination among semi-arid landscape endmembres using the spectral angle mapper (SAM) algorithm, in *Summaries of the 3rd Annual JPL Airborne Geoscience Workshop*, edited by R. O. Green, Publication, 92-14, Vol. 1, pp. 147–149, 1992.

23. J. C. Harsanyi and C.-I. Chang, Hyperspectral image classification and dimensionality reduction: An orthogonal subspace projection approach, *IEEE Transactions Geoscience and Remote Sensing*, vol. 32, no. 4, pp. 779–785, 1994.

24. C.-I. Chang, X. Zhao, M. L. G. Althouse, and J. J. Pan, Least squares subspace projection approach to mixed pixel classification for hyper-spectral images, *IEEE Transactions on Geoscience and Remote Sensing*, vol. 36, no. 3, pp. 898–912, 1998.

25. L. Parra, K.-R. Mueller, C. Spence, A. Ziehe, and P. Sajda, Unmixing hyperspectral data, *Advances in Neural Information Processing Systems*, vol. 12, pp. 942–948, 2000.

26. A. Ifarraguerri and C.-I. Chang, Unsupervised hyperspectral image analysis with projection pursuit, *IEEE Transactions on Geoscience and Remote Sensing*, vol. 38, no. 6, pp. 127–143, 2000.

27. L. O. Jimenez and D. A. Landgrebe, Hyperspectral data analysis and supervised feature reduction via projection pursuit, *IEEE Transactions on Geoscience and Remote Sensing*, vol. 37, no. 6, pp. 2653–2664, 1999.

28. P. Common, Independent component analysis: A new concept, *Signal Processing*, vol. 36, pp. 287–314, 1994.

29. A. Hyvarinen, J. Karhunen, and E. Oja, *Independent Component Analysis*, John Wiley & Sons, New York, 2001.

30. J. D. Bayliss, J. A. Gualtieri, and R. F. Cromp, Analysing hyperspectral data with independent component analysis, in *Proceedings of the SPIE conference 26th AIPR Workshop: Exploiting New Image Sources and Sensors*, Vol. 3240, pp. 133–143, 1997.

31. C. Chen and X. Zhang, Independent component analysis for remote sensing study, in *Proceedings of the SPIE Symposium on Remote Sensing Conference on Image and Signal Processing for Remote Sensing V*, Vol. 3871, pp. 150–158, 1999.

32. T. M. Tu, Unsupervised signature extraction and separation in hyperspectral images: A noise-adjusted fast independent component analysis approach, *Optical Engineering of SPIE*, vol. 39, no. 4, pp. 897–906, 2000.

33. S.-S. Chiang, C.-I. Chang, and I. W. Ginsberg, Unsupervised hyperspectral image analysis using independent component analysis, in *Proceedings of the IEEE International Geoscience and Remote Sensing Symposium*, 2000.

34. M. Lennon, M. Mouchot, G. Mercier, and L. Hubert-Moy, Spectral unmixing of hyperspectral images with the independent component analysis and wavelet packets, in *Proceedings of the IEEE International Geoscience and Remote Sensing Symposium*, 2001.
35. N. Kosaka and Y. Kosugi, ICA aided linear spectral mixture analysis of agricultural remote sensing images, in *Proceedings of the 4th International Symposium on Independent Component Analysis and Blind Signal Separation*, pp. 221–226, 2003.
36. V. Botchko, E. Berina, Z. Korotkaya, J. Parkkinen, and T. Jaaskelainen, Independent component analysis in spectral images, in *Proceedings of the 4th International Symposium on Independent Component Analysis and Blind Signal Separation*, pp. 203–207, 2003.
37. B. R. Foy and J. Theiler, Scene analysis and detection in thermal infrared remote sensing using independent component analysis, in *Proceedings of SPIE, Conference on Independent Component Analyses, Wavelets, Unsupervised Smart Sensors, and Neural Networks II*, Vol. 5439, pp. 131–139, 2004.
38. N. Kosaka, K. Uto, and Y. Kosugi, ICA-aided mixed-pixel analysis of hyperspectral data in agricultural land, *IEEE Geoscience Remote Sensing Letters*, vol. 2, no. 2, pp. 220–224, 2005.
39. H. Attias, Independent factor analysis, *Neural Computation*, vol. 11, no. 4, pp. 803–851, 1999.
40. A. Ifarraguerri and C.-I. Chang, Multispectral and hyperspectral image analysis with convex cones, *IEEE Transactions on Geoscience and Remote Sensing*, vol. 37, no. 2, pp. 756–770, 1999.
41. A. Plaza, P. Martinez, R. Perez, and J. Plaza, Spatial/spectral endmember extraction by multidimensional morphological operations, *IEEE Transactions on Geoscience and Remote Sensing*, vol. 40, no. 9, pp. 2025–2041, 2002.
42. J. Boardman, Automating spectral unmixing of AVIRIS data using convex geometry concepts, in *Summaries of the Fourth Annual JPL Airborne Geoscience Workshop, JPL Publication 93–26, AVIRIS Workshop*, Vol. 1, pp. 11–14, 1993.
43. M. D. Craig, Minimum-volume transforms for remotely sensed data, *IEEE Transactions on Geoscience and Remote Sensing*, vol. 32, pp. 99–109, 1994.
44. J. M. P. Nascimento and J. M. B. Dias, Vertex component analysis: A fast algorithm to unmix hyperspectral data, *IEEE Transactions on Geoscience and Remote Sensing*, vol. 43, no. 4, pp. 898–910, 2005.
45. M. E. Winter, N-findr: An algorithm for fast autonomous spectral endmember determination in hyperspectral data, in *Proceedings of the SPIE conference on Imaging Spectrometry V*, Vol. 3753, pp. 266–275, 1999.
46. I. T. Jolliffe, *Principal Component Analysis*, Springer-Verlag, New York, 1986.
47. A. Green, M. Berman, P. Switzer, and M. D. Craig, A transformation for ordering multispectral data in terms of image quality with implications for noise removal, *IEEE Transactions on Geoscience and Remote Sensing*, vol. 26, no. 1, pp. 65–74, 1988.
48. L. L. Scharf, *Statistical Signal Processing, Detection Estimation and Time Series Analysis*, Addison-Wesley, Reading, MA, 1991.
49. J. M. B. Dias and J. M. P. Nascimento, Estimation of signal subspace on hyperspectral data, in *Proceedings of SPIE Conference on Image and Signal Processing for Remote Sensing XI*, Vol. 5982, edited by, L. Bruzzone, pp. 191–198, 2005.
50. S. S. Shen and E. M. Bassett, Information-theory-based band selection and utility evaluation for reflective spectral systems, in *Proceedings of the SPIE Conference on Algorithms*

and Technologies for Multispectral, Hyperspectral, and Ultraspectral Imagery VIII, Vol. 4725, pp. 18–29, 2002.

51. J. H. Bowles, J. A. Antoniades, M. M. Baumback, J. M. Grossmann, D. Haas, P. J. Palmadesso, and J. Stracka, Real-time analysis of hyperspectral data sets using NRL's ORASIS algorithm, in *Proceedings of the SPIE Conference on Imaging Spectrometry III*, Vol. 3118, pp. 38–45, 1997.
52. G. Shaw and H. Burke, Spectral imaging for remote sensing, *Lincoln Laboratory Journal*, vol. 14, no. 1, pp. 3–28, 2003.
53. J. S. Tyo, J. Robertson, J. Wollenbecker, and R. C. Olsen, Statistics of target spectra in hsi scenes, in *Proceedings of the SPIE Conference on Imaging Spectrometry VI*, vol. 4132, pp. 306–314, 2000.
54. G. Healey and D. Slater, Models and methods for automated material identification in hyperspectral imagery acquired under unknown illumination and atmospheric conditions, *IEEE Transactions on Geoscience and Remote Sensing*, vol. 37, no. 6, pp. 2706–2717, 1999.
55. M. A. T. Figueiredo and A. K. Jain, Unsupervised learning of finite mixture models, *IEEE Transactions on Pattern Analysis and Machine Intelligence*, vol. 44, no. 3, pp. 381–396, 2002.
56. E. Moulines, J.-F. Cardoso, and E. Gassiat, Maximum likelihood for blind separation and deconvolution of noisy signals using mixture models, in *Proceedings of the IEEE International Conference on Acoustics, Speech, and Signal Processing*, Vol. 5, pp. 3617–3620, 1997.
57. D. Tanre, M. Herman, P. Deschamps, and A. de Leffe, Atmospheric modeling for space measurements of ground reflectances, including bidirectional properties, *Applied Optics*, vol. 18, pp. 3587–3594, 1979.
58. E. Vermote, D. Tanré, J. Deuzé, M. Herman, and J. Morcette, Second simulation of the satellite signal in the solar spectrum 6S: An overview, *IEEE Transactions on Geoscience and Remote Sensing*, vol. 35, no. 3, pp. 675–686, 1997.
59. A. K. Jain, *Fundamentals of Digital Image Processing*, edited by E. Cliffs, Prentice Hall, Englewood Cliffs, NJ, 1989.
60. K. Liou, *An Introduction to Atmospheric Radiation*, 2nd edition, Academic Press, New York, 2002.
61. D. Roberts, Y. Yamaguchi, and R. Lyon, Calibration of airborne imaging spectrometer data to percent reflectance using field spectral measurements, in *Proceeding of the Nineteenth International Symposium on Remote Sensing of Environment* 2, Ann Arbor, Michigan, pp. 679–688, 1985.
62. A. Berk, L. Bernstein, G. Anderson, P. Acharya, D. Robertson, J. Chetwynd, and S. Adler-Golden, MODTRAN cloud and multiple scattering upgrades with application to AVIRIS, *Remote Sensing of the Environment*, vol. 65, pp. 367–375, 1998.
63. J. B. Adams and M. O. Smith, A new analysis of rock and soil types at the viking lander 1 site. *Journal of Geophysical Research*, vol. 91, no. B8, pp. 8098–8112, 1986.
64. B. Hapke, *Theory of Reflectance and Emittance Spectroscopy*, Cambridge University Press, Cambridge, U. K.: 1993.
65. H.-H. Wu and R. A. Schowengerdt, Improved estimation of fraction images using partial image restoration, *IEEE Transactions on Geoscience and Remote Sensing*, vol. 31, no. 4, pp. 771–778, 1993.

66. C. Bateson, G. Asner, and C. Wessman, Endmember bundles: A new approach to incorporating endmember variability into spectral mixture analysis, *IEEE Transactions on Geoscience and Remote Sensing*, vol. 38, pp. 1083–1094, 2000.
67. F. Kruse, Spectral identification of image endmembers determined from AVIRIS data, in *Summaries of the VII JPL Airborne Earth Science Workshop*, 1998.
68. J. Boardman and F. Kruse, Automated spectral analysis: A geological example using AVIRIS data, northern grapevine mountains, Nevada, in *Proceedings of the 10th Thematic Conference, Geologic Remote Sensing*, 1994.
69. A. Hyvärinen, Survey on independent component analysis, *Neural Computing Surveys*, vol. 2, pp. 94–128, 1999.
70. A. Hyvarinen and E. Oja, Independent component analysis: Algorithms and applications, *Neural Networks*, vol. 13, no. 4–5, pp. 411–430, 2000.
71. J. Cardoso, Infomax and maximum likelihood of source separation, *IEEE Signal Processing Letters*, vol. 4, no. 4, pp. 112–114, 1997.
72. A. J. Bell and T. J. Sejnowski, An information-maximization approach to blind separation and blind deconvolution, *Neural Computation*, vol. 10, pp. 215–234, 1995.
73. T.-W. Lee, M. Girolami, A. Bell, and T. Sejnowski, An unifying information-theoretic framework for independent component analysis, *International Journal on Mathematical and Computer Modeling*, vol. 31, pp. 1–21, 2000.
74. T. Cover and J. Thomas, *Elements of Information Theory*, John Wiley & Sons, New York, 1991.
75. A. P. Dempster, N. M. Laird, and D. B. Rubin, Maximum likelihood from incomplete data via the EM algorithm, *Journal of the Royal Statistical Society*, vol. 39, no. B, pp. 1–38, 1977.
76. G. McLachlan and T. Krishnan, *The EM Algorithm and Extensions.*, John Wiley &, Sons, New York, 1997.
77. G. McLachlan and D. Peel, *Finite Mixture Models*, John Wiley & Sons, New York, 2000.
78. P.-F. Hsieh and D. Landgrebe, Classification of high dimensional data, Purdue University, Ph.D. Thesis and School of Electrical & Computer Engineering Technical Report TR-ECE 98-4, 1998.
79. D. Landgrebe, Multispectral data analysis: A signal theory perspective, Purdue University, Technical Report, 1998.
80. J. M. B. Dias, An EM algorithm for the estimation of dirichlet parameters, Instituto de Telecomunicações, http://www.lx.it.pt/~bioucas/, Technical Report, 2005.
81. G. McLachlan and T. Krishnan, *The EM Algorithm and Extensions.*, John Wiley & Sons, New York, 1996.
82. R. N. Clark, G. A. Swayze, A. Gallagher, T. V. King, and W. M. Calvin, The U.S. geological survey digital spectral library: Version 1: 0.2 to 3.0 μm, U.S. Geological Survey, Open File Report 93-592, 1993.

CHAPTER 7

MAXIMUM VOLUME TRANSFORM FOR ENDMEMBER SPECTRA DETERMINATION

MICHAEL E. WINTER
Hawaii Institute of Geophysics and Planetology, University of Hawaii, Honolulu, HI 96822

7.1. INTRODUCTION

Hyperspectral data can convey a great deal of information about the content of a large area by allowing the separation of different materials within a scene based on their spectral properties. This gives an analyst a unique opportunity to perform high-quality remote spectrographic analysis in order to separate plant types, geologic materials, and targets of interest from each other and from background clutter. It is often useful to interpret hyperspectral data in two ways. First, it can be interpreted in a purely physical sense. That is, each channel of each pixel in an image presents a physical variable to the viewer (for example, radiance in the case of a calibrated data directly from a hyperspectral imager). The second interpretation is that the pixel data presented within the image can be interpreted in a purely statistical manner.

With a physical interpretation, an analyst uses scientific methodology to exploit a hyperspectral image. For example, an analyst looking for a certain material within an image will match the spectral properties of the material (e.g., absorption features) with the spectral properties of an individual pixel within the image. This approach offers a great deal of processing potential for hyperspectral imagery; however, it is the most demanding in terms of image quality and analyst expertise. Not all analysts are spectroscopists, that are able to interpret individual spectral features necessary for material identification, and not all images are well-calibrated and atmospherically corrected enough to support accurate spectroscopy (for example, atmospheric effects must be carefully considered). A full spectrographic analysis of a scene requires the examination of each spectrum in an image by a trained

Hyperspectral Data Exploitation: Theory and Applications, Edited by Chein-I Chang

analyst; manual interpretation of this quantity of data is not generally practical. Without data reduction, an image analyst would be required to examine millions of spectra per image in order to produce a typical end product of a map of materials or targets. The USGS Tetracorder system [1] that automatically identifies materials based on a spectrographic rule database is a sophisticated automated approach for physically based hyperspectral image exploitation.

Using a statistical approach offers the potential for completely automated processing of a remotely sensed image. Here the data representing a hyperspectral image is interpreted solely as a set of statistical variables. Examples of this approach are the principal components transform (see, for example, Richards [2], p. 240) commonly used for multispectral and hyperspectral imagery and the RX anomaly detection algorithm [3] used in target detection. While it would seem counterintuitive that a statistical interpretation of imagery is a good means of exploitation, it has proven remarkably successful. Indeed the human visual processing system approximates some of the statistical methods used for spectral image processing. For example, Tyo et al. [4] exploited a metaphor between the statistical "processing" inherent in human vision and hyperspectral data exploitation to produce an automated image-coloring algorithm.

One commonly used statistical approach for reducing the complexity of hyperspectral data is the use of orthogonal subspace projections (OSPs) [2]. Orthogonal subspace projections reduce the statistical dimensionality of the data by finding the combinations of bands which best represent the image in some manner. While OSPs are statistically efficient at reducing the dimensionality of the image, the resulting images have a mathematical rather than physical relationship with the original image and the derived eigenvalue spectra do not correspond to any physical spectra.

The popular spectral image exploitation methodology that can be loosely called "endmember spectra determination and unmixing" relies on a mixture of statistically based and physically based methodologies. In the traditional approach, an analyst will transform an image using a statistical transform (such as principal components) and then manually select a group of spectra that characterize that image. Generally, that group of spectra is representative of the spectra in the scene that are most different from each other and that ideally are pure pixels of constituent materials. Other pixels in the image are assumed to be mixtures of these "pure" pixels so that the image is physically modeled as a combination of these spectra. Material maps are then produced using a mathematical inversion of the selected material spectra (called "unmixing" or "demixing"). The principal advantage of unmixing an image using endmember spectra over performing an OSP-based data reduction is that the reduction inherent in unmixing is based on a physical rather than a statistical model. The resulting images are the fractional abundance of the corresponding endmember spectrum for that pixel. In essence the practice of finding endmember spectra and unmixing amounts to a transformation of the image from a radiance axis to a physical composition axis. The endmember spectra detail the relationship between both axes.

The result of this analysis is a physical interpretation of the hyperspectral image that can approach the ideal of a full map of each material in the scene, much like a mineralogical map maps each mineral across an area. This approach can be completely automated using approaches such as ORASIS ([5], as well as this book) or N-FINDR [6–8]. Automated endmember spectra determination algorithms generally use assumptions as to the distribution of data in the hyperspectral image to remove the need for an expert to collect image spectra based on a visual analysis.

Finding a set of endmember spectra and unmixing an image offers a conceptual benefit to the analyst examining the data. Image analysis is often performed with the objective of identifying and mapping one or more materials within the scene. Endmember spectra represent a set of archetypical spectra that define the image, something like a painters palette of paints defines the colors seen in a painting. For example, for a hyperspectral image containing a mixture of grass, asphalt, and cement, the endmember spectra for that scene are the three things that make it up, and any pixel in the image can be described in terms of these three spectra. The power of endmember analysis is that oftentimes a complex image of a million pixels can be described very compactly as a dozen or so endmember spectra. However, determining these endmember spectra is not necessarily straightforward except in ideal cases where the endmember spectra for an image are known.

While the image may be classified using laboratory spectra, this requires that the constituent materials in the image be known and that the conversion to reflectance be done very accurately. Moreover, the selection of endmembers can be nonunique. A more practical approach is to determine the endmember spectra based solely on the information contained within the image itself. This can be done manually by a trained analyst using a methodology defined by Boardman et al. [9]; however, this process is often arduous.

Despite two decades of research into spectral image reduction, a satisfying "end-to-end" hyperspectral analytic chain whereby an image is interpreted in a fully automated fashion remains unavailable. There are a number of approximate methods for automated image exploitation which are only applicable to specific types of ideal images. One approach for doing this, a maximum volume transform based on the N-FINDR algorithm, will be discussed in this chapter.

7.1.1. Approaches for Automated Unmixing of Hyperspectral Data

Endmember determination and unmixing algorithms broadly fall into two categories: geometric algorithms and statistical models. The geometric methods use the distribution of hyperspectral data to reduce the complexity of the task, while the statistical methods use some sort of error metric. Since hyperspectral data can be modeled using the linear mixture model (cf. Adams et al. [10]), the data tends to be arranged around the vertices and lines of a class of geometric objects [9,11,12]. There are two widely used geometric approaches for determining the endmembers of a hyperspectral image using geometric principles: "inflation" and "shrink-wrap." Both of these methods fit a simplex to a set of hyperspectral data in order to determine image endmember spectra. A simplex is a geometric object similar to a triangle in higher dimensions: a tetrahedron in three dimensions for example. The methods differ in the manner in which the simplex is fit to the data cloud. "Inflation"-based algorithms start from a point and expands a simplex within a data cloud, while "shrink-wrap" algorithms start with a simplex encompassing the data and reduce it to fit. In contrast to these geometric methods for endmember determination, the statistical approaches usually are based on minimizing the error of a statistical model of the hyperspectral data.

The two most widely used commercial algorithms for endmember spectra determination are the Naval Research Lab's ORASIS (discussed in Chapter 4 in this book) and the N-FINDR algorithm (developed by the author of this chapter), both of which represent different geometric approaches to the problem; other approaches, such as IEA [13] from Canadian Centre for Remote Sensing (CCRS), is a purely statistical methodology. While N-FINDR itself is proprietary, the approach it uses is simple enough that it can be examined in detail later in the chapter as an example of how and why these algorithms work.

It should be noted that in general these automated approaches make no requirements as to the spatial presentation of the data. These algorithms operate on per-pixel bases over the entire image and are entirely independent of the spatial distribution of the spectra of interest. All of these algorithms operate without any user knowledge and with limited *a priori* information of what is in the scene. These algorithms use simple approximations to the physical models describing the spectral characteristics of a hyperspectral pixel. While these approximations are demonstrably inaccurate, they generally produce good results.

7.1.2. N-FINDR

The algorithm to be described here is based on the commercial algorithm N-FINDR. N-FINDR [6–8] was designed to work on data after dimensionality reduction via subspace projection. A later implementation has bypassed this step. The input to this process is the full spectral image cube, without either the dimensionality reduction or a data thinning process. The procedure must examine the full data set to find those pure pixels that can be used to describe the various mixed pixels in the scene. This algorithm finds the set of pixels with the largest possible volume by "inflating" a simplex inside the data. The procedure begins with a random set of vectors. In order to refine the estimate of the endmembers, every pixel in the image must be evaluated as to its likelihood of being a pure or nearly pure pixel. To do this, the volume must be calculated with each pixel in place of each endmember. A trial volume is calculated for every pixel in each endmember position by replacing that endmember and finding the volume. If the replacement results in an increase in volume, the pixel replaces the endmember. This procedure is repeated until there are no more replacements of endmembers. Once the endmembers are found, their spectra can be used to unmix the original image using either linear inversion or nonnegatively constrained least squares. This produces a set of images, each of which shows the fractional abundance of an endmember in each pixel. The endmember determination step of N-FINDR can be executed in seconds, and unmixing process also generally takes a matter of seconds [6–8].

7.1.3. Maximum Volume Transform

The purpose of this chapter is to outline a "maximum volume transform" (MVT), a method for hyperspectral and multispectral remotely sensed data. This approach is an approximation to the commercial N-FINDR algorithm (developed by this author) that operates in more or less the same manner with some optimizations for speed and accuracy. The transform described here is a systematic automated approach to deriving endmember estimates from hyperspectral imagery. The MVT produces two outputs: (1) an estimate of the constituent spectra within the scene (the endmember spectra) and (2) an estimate of the abundance of each endmember in each image pixel (the abundance planes). This transform is potentially much more useful than the widely used principal component transform (PCT) [2] and the maximum noise fraction transform (MNFT) [14], because the output is presented in physical units (fraction abundance and constituent material spectra), rather than statistical units (eigenvectors and decorrelated images).

This process represents a data reduction in both practical and analytic term. In practical terms the data set has been reduced by a significant factor: Often about a dozen fraction planes and endmembers can efficiently represent a 224-band AVIRIS image. Analytically, a complex radiance sensor image has been reduced to physical material maps, generally a much easier form of data for human interpretation.

It is impossible to prove that an algorithm of this sort works with every possible image. Indeed, this algorithm can be shown to fail under certain common circumstances. Nonetheless, for a large class of problems this approach will be shown to yield useful results, and it will be shown to behave robustly in the presence of suboptimal data.

The discussion of this algorithm will be presented in four parts. First, a description of the algorithm itself will be presented. Then it will be shown that the algorithm unconditionally converges to the correct solution with perfect data. Subsequent sections describe the behavior of this algorithm with imperfect data and real-world sensor data.

7.2. ALGORITHM DESCRIPTION

7.2.1. The Linear Mixing Model for Hyperspectral Data

Hyperspectral data are often treated as a linear mixture problem [10]. A pixel on the ground, as seen by a sensor, generally contains more than one species of material. For example, a pixel can contain both vegetation and asphalt. In the absence of intimate mixing, the total radiance produced can be modeled as a linear sum of radiance spectra from the constituent materials. In discrete form we have

$$p_{ij} = \sum_{i=1}^{m} e_{ik} c_{kj} + \varepsilon \tag{7.1}$$

where p_{ij} is the radiance or reflectance at sensor at wavelength i for pixel j (the pixel spectrum), m is the number of endmembers in the scene, e_{ik} is the radiance or reflectance spectrum for endmember k at wavelength i with each row corresponding to a sensor spectral channel, c_{ij} is the abundance of endmember i in pixel j (the pixel composition), and ε represents the influence of Gaussian random noise. The abundances themselves must sum to one, since the total contributions of all of the endmembers cannot exceed 100%. An additional equation is used to couple the values of the abundances and enforce this requirement:

$$\sum_{i=1}^{m} c_{ij} = 1 \tag{7.2}$$

Additionally, negative abundances are not physically possible, which implies an additional constraint,

$$c_{ij} > 0 \tag{7.3}$$

In the linear mixture model, one of the endmembers spectra represents the "shade point." This represents the response of the sensor to a completely dark substance on the ground. It is mathematically treated identically to the other material spectra within the scene. Ideally, this endmember would be represented by all zeros. However, due to sensor calibration and atmospheric backscatter, this endmember will generally have a nonzero radiance.

Care must be taken when applying the linear mixture model. This model can only be applied to images where the reflectance or radiance of mixed materials is proportional to the weighted sum of the constituent spectra's radiance or reflectance. For example, this model is not applicable in cases of intimate (fine grain) mixing of minerals in rock as internal reflections become important. Shallow water also mixes nonlinearly with the submerged sediment and similarly atmospheric effects interact nonlinearly with ground radiance. If the atmosphere is consistent over the entire image, this is not a concern since it simply acts to consistently scale the ground radiance as a function of wavelength received at the sensor. Varying atmosphere within a hyperspectral scene introduces a variable nonlinearity that must be corrected. Nonetheless, despite its shortcomings, in a large number of applications both terrestrial and planetary the linear mixing model has been used successfully and has become a standard part of hyperspectral data processing.

7.2.2. Convex Geometry and Hyperspectral Data

The combination of Eqs. (7.1) and (7.2) form a linear system [9,11,15]. An intuitive method for understanding this system is to visualize it as a geometric object. Viewing linear equations as a geometric system is a common means of understanding an otherwise abstract set of algebraic functions. Most operations with linear equations (including least-squares analysis) have direct analogs to the geometry of vectors (cf. Rawling [16], pp. 153–167). In this context, each n-banded pixel in a hyperspectral image is considered n point in an n-dimensional space (an n-tuple) where the axes are defined by the sensor bands. This relationship has been exploited frequently in the development of algorithms for data processing. Subspace projections such as the principal components transform act to rotate the hyperspectral data into a different axis. If the axis is more optimal for the description of the data, then data reduction can be achieved be truncating the dimensionality of the projected image to include only the most important components. In a similar way, linear unmixing of hyperspectral data acts as an affine transform, which scales, rotates, and translates the data to a physically based coordinate system.

Geometrically, Eqs. (7.1) and (7.2) form a simplex enclosing an m-dimensional space defined with each channel of the spectrometer acting as an axis. A simplex is the simplest geometric object that can contain a space of a given dimension. For example, in two-dimensional space the simplex is a triangle, which consists of three vertices. In three-dimensional space, the simplex is a tetrahedron. The vertices of the simplex are given by the spectra of the endmembers, and they correspond to pure pixels of a given constituent substance. Interior space within the simplex represents feasible mixtures of each material, where Eqs. (7.1), (7.2), and (7.3) are satisfied. This geometric relationship is invariant to rotation, scaling, and translation, and Eqs. (7.1), (7.2), and (7.3) are identical under a subspace transform such as the principal components transform.

A simplex represents the simplest form of a convex set. A convex set is a subset of points in a space which encompass the points of the larger set in such a way that a line connecting any two points in the set passes solely through the interior of the set (for more detail, see Lay [17], p. 11). The convex set can be thought of as a subset that spatially contains all points of the superset. For any given set of points in a space of arbitrary dimension, a convex set can be defined. In two dimensions, shapes such as a triangle or rectangle are convex, whereas shapes such as a star are not.

For linearly mixed hyperspectral data the endmember spectra represent a convex set. The specific geometric characteristics of linearly mixed hyperspectral data combined with an assumption as to the purity of the hyperspectral data set can be used to formulate an algorithm to find endmember spectra. The fact that the endmembers populate a convex set enclosing the data can be used to prove that the algorithm described subsequently will unconditionally converge to the correct solution for a limited range of data sets.

7.2.3. Maximizing the Volume of an Inscribed Simplex

The maximum volume transform is an extremely basic algorithm. The algorithm simply acts to maximize the volume of a simplex containing image pixels. The optimization procedure performs a stepwise maximization by sequentially evaluating each pixel in the image.

The mathematical definition of the volume of a simplex formed by the a set of endmember estimates (**E**) is proportional to the determinant of the set augmented by a row of ones:

$$V^0(\mathbf{E}) = \frac{1}{n!} abs \begin{vmatrix} 1 & 1 & \cdots & 1 \\ \vec{e}_1 & \vec{e}_2 & \cdots & \vec{e}_m \end{vmatrix} \tag{7.4}$$

where $\vec{e}_i$ represents an n-tuple vector containing the bands of the potential endmember set.

7.2.4. Pre-processing

It should be noted that the determinant in Eq. (7.4) is only defined in the case where $n = m - 1$. A transform that reduces the dimensionality (such as the principal components transform) must be performed in the case of hyperspectral data where n is usually much greater than m. This requires *a priori* an estimate of the number of endmembers within the image.

Any linear transform can be used for dimensionality reduction [2,14] with the caveat that the resulting endmembers will be solely composed of the subspace's basis. This transform must be reversed at the completion of this optimization process in order to have the endmembers natural form. Alternatively, the pixel locations of the solution endmember spectra can be retained, and the endmembers can be retrieved from the original (untransformed) image. Popular subspace projections used for dimensionality reduction of hyperspectral data include the principal components transform and the "minimum noise fraction" (MNF) transform [14]. The result of this transform is a new image where the bands represent linear combinations of the original sensor bands. The transform selects the linear transform based on some measure of error minimization. In the case of the principal components transform, the transform basis is chosen to minimize residual image error, while for the MNF transform the basis is chosen such that the signal-to-noise ratio is maximized on the produced component images.

Determining the number of endmembers in a hyperspectral image (and thus the number of transform bands to use) is difficult task. If available, an *a priori* estimate of the number of endmembers in a scene could be used. For example, in a geologic scene, a geologic map that defines the geologic units within the area could be used. This, however, requires a degree of knowledge of the scene that is rarely available. Traditional factor analysis cannot be used because this requires knowledge of the endmembers themselves.

The MNF transform has the benefit of determining the number of endmembers simultaneously with the transformation of the data into a lower-dimensional subspace. The MNF transform produces three output produces: the transformed image, the eigenvectors used to perform the transformation, and the eigenvalues that provide the signal-to-noise ratio of each transformed band. The point at which the transform eigenvalues fall below one provides an estimate of the true number of significant degrees of freedom in the image, and thus the total number of discernible endmembers within the hyperspectral image. The transform can then be truncated by eliminating transform bands with order greater than the total number of endmembers. This will produce a matrix of correct dimension for Eq. (7.4).

7.2.5. Stepwise Maximization of Endmember Volume

Since the calculated volumes are only used in comparison with each other, the dimensional factor in Eq. (7.4) can be dropped to give the factorial of the data dimension; for example,

$$V(\mathbf{E}) = abs\begin{vmatrix} 1 & 1 & \cdots & 1 \\ \vec{e}_1 & \vec{e}_2 & \cdots & \vec{e}_m \end{vmatrix} \propto V^0(\mathbf{E}) \tag{7.5}$$

An image pixel is evaluated by replacing one pixel in the present endmember set with the image pixel, producing a "trial endmember" set for endmember i and pixel j, $(\mathbf{E}^{'}_{ij})$.

$$\mathbf{E}^{'}_{ij} = \begin{bmatrix} 1 & \cdots & 1 & 1 & 1 & \cdots & 1 \\ \vec{e}_1 & \cdots & \vec{e}_{i-1} & \vec{p}_j & \vec{e}_{i+1} & \cdots & \vec{e}_m \end{bmatrix} \tag{7.6}$$

The MVT algorithm works by trying each pixel in the image in place of each potential endmember. If the replacement of the trial endmember results in an increase in the volume of the potential endmember set, the pixel is accepted as a potential endmember. It can be described more precisely as

```
function MVT(P, dim)
  Initialize endmember set E with random pixels from P
      for each pixel j in image P do:
          for each endmember i do:
              if (V(E'_ij) > V(E)) then:
                  E <- E'_ij
      return E
```

After each image pixel has been tried in place of each potential endmember, the members of the potential endmember represent the estimates for the true image endmembers. This algorithm amounts to a very simple nonlinear inversion for the largest simplex that can be inscribed within the data.

In some ways the simplex maximization algorithm described above is the direct opposite of the approach used in Orasis [5], which uses a "minimum volume transform." Both methods use the same underlying model and similar statistical assumptions to produce endmember spectra. It would seem counterintuitive that dramatically opposite approaches would produce compatible results, but test cases where the same data were processed with N-FINDR and Orasis produced similar looking material maps (see Figure 7.1 and Winter and Winter [18] for a detailed examination). The reason for these similar results is that the N-FINDR simplex inflation approach will find an interior estimate for the endmembers, and Orasis's exterior shrink-wrap estimate will find an exterior estimate. In data with low error rates and pure pixels, these results will be similar.

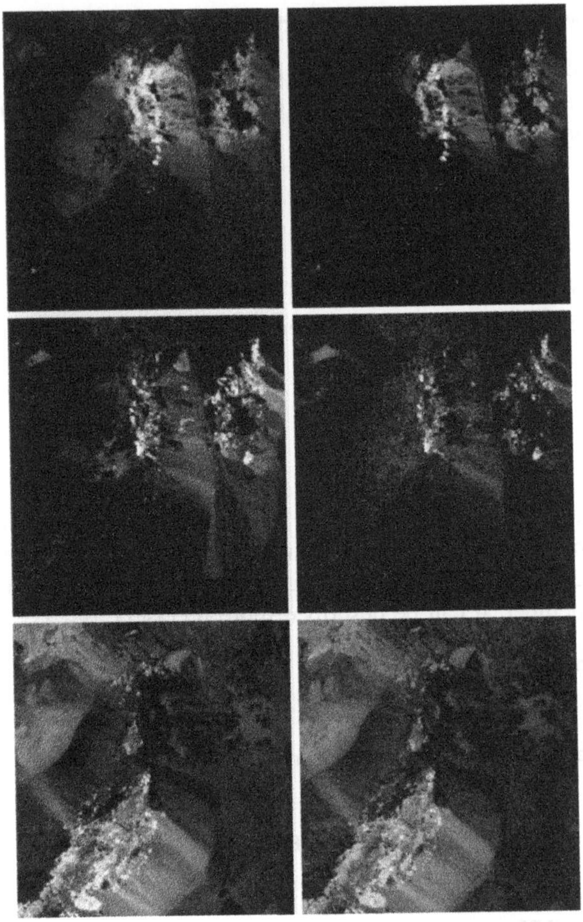

Figure 7.1. Fractional abundance planes for N-FINDR (left) and Orasis (right) derived automatically from analyzing the AVIRIS Cuprite, NV data. The top, middle, and bottom images show the prevalence of Alunite, Kaolinite, and Calcite, respectively.

7.2.6. Unmixing

Once the pure pixels are found, their spectra can be used to "unmix" the original image. This produces a set of images, each of which shows the abundance for each endmember for each pixel. These abundance planes are produced by inverting the mixture equation [Eq. (7.1)] using least-squares analysis. Determining optimal endmember contributions (c_k) for a given pixel (p_k) using least squares uses the solution of the normal equations for Eq. (7.1) [16]:

$$\vec{c}_j = (\mathbf{E}^{\mathrm{T}}\mathbf{E})^{-1}\mathbf{E}^{\mathrm{T}}\vec{p}_j \tag{7.7}$$

where **E** represents a matrix whose columns are the endmembers in the endmember set, and $\vec{c}_j$ represents a column vector containing the endmember contributions for pixel $\vec{p}_j$. Alternatively, nonnegatively constrained least-squares analysis can be used to enforce Eq. (7.3). This derives the solution for optimal mixtures subject to the physical constraint that no mixture coefficient can be negative. This algorithm is too complex to describe here. For more details on positively constrained regression, the reader is referred to Lawson and Hanson [19] for a full description of one approach.

7.3. CONVERGENCE OF THE ALGORITHM: MODEL DATA

7.3.1. Theoretically Perfect Data

Consider the behavior of this algorithm with perfectly linearly mixed hyperspectral data [i.e., data that behaves according to Eqs. (7.1), (7.2), and (7.3)]. In addition, it will be assumed that pure samples of each endmember are present in the data set. This algorithm can be shown to converge unconditionally to where the spectra in the potential endmember set equal the endmember spectra (pure materials) in the image. This model can be considered a less than realistic representation of hyperspectral data and is discussed solely to illustrate the convergent properties of this algorithm. Data sets that are more realistic will be examined in subsequent sections.

The volume of the endmember set is calculated by determinant [Eq. (7.5)]; if there is an increase in volume due to the substitution of a given pixel j in place of endmember i, then

$$\frac{V(\mathbf{E}'_{ij})}{V(\mathbf{E})} > 1 \tag{7.8}$$

The determinant of a matrix can be calculated through Laplacian development by minors ([20], p. 169). This is the method often used to calculate matrix determinants "by hand" and allows the development of equations describing the way a matrix changes with a linear perturbation. The expansion using development for the first column of matrix $\mathbf{E}$ is given by

$$|\mathbf{E}| = M_{11} + \sum_{i=1}^{m-1} (-1)^i e_{i1} M_{(i+1),1} \tag{7.9}$$

where M_{ij} is the determinant of the minor matrix formed by deleting row i and column j from the potential endmember matrix $\mathbf{E}$.

Consider the case where the space around the first potential endmember is probed for suitability by checking the increase in volume. The rate of increase of volume due to the small perturbation in the spectrum of potential endmember 1 (e_1) can be derived by taking the partial derivative of Eq. (7.9):

$$\frac{\partial V(\mathbf{E})}{\partial \vec{e}_1} = \sum_{i=1}^{m-1} (-1)^i M_{(i+1),1} \tag{7.10}$$

The rate of increase in volume of the potential endmember set is a linear function of the change in position between the evaluated pixel and the original

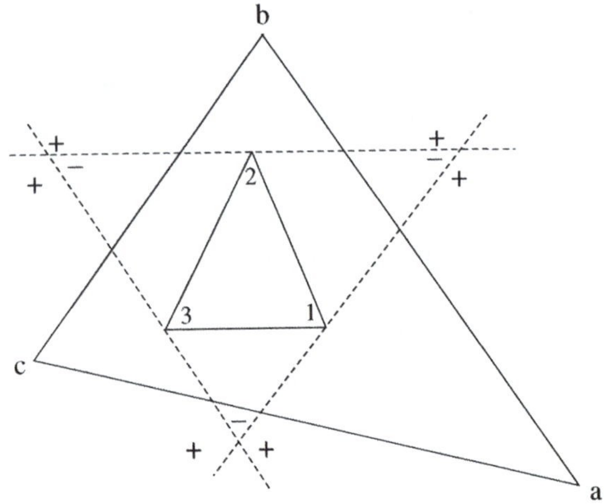

Figure 7.2. A simplex of potential endmembers (1, 2, 3) for three-component data sits within a simplex of actual endmembers (a, b, c). The lines intersecting the potential endmembers demark the half-spaces of volume increase for each potential endmember.

endmember. Perpendicular to this line and intersecting the current potential endmember is a hyperplane that defines the boundary between increasing and decreasing volume (Figure 7.2). Pixels evaluated "behind" this line (toward the center of the potential endmember set) will not increase the set's volume and will be rejected. Pixels in front of this plane (away from the center of the present endmember set) will be accepted as a potential endmember in place of the current estimate of endmember 1. The form of the Eq. (7.10) is identical for any given endmember in the endmember set, however, the vector of minors will change.

If a pixel results in an increase in volume, it is accepted as a member of the potential endmember set. The minor matrices used in Eqs. (7.9) and (7.10) will then need to be recalculated. Even with the changes in the minor matrices, the result will be of a similar linear form.

This progressive updating procedure for the endmember set is analogous to the use of the simplex method for the solution of linear programming problems (for a detailed discussion, see Ficken [21]). The simplex method is a widely used means of determining an optimal solution for linear function bounded by constraints. It can handle highly complex problems consisting of hundreds of independent variables. The constraints are formulated as inequalities that, when combined, form a simplex in parameter space. Using the simplex method to find the optimal solution reduces the complex multidimensional optimization to a systematic combinatorial process. The simplex method relies on the fact that the optimization of any linear function limited to a space bounded by a simplex always has an optimal solution at one or more of the vertices. More generally, an optimum solution of any linear function within any convex space occurs at a vertex ([17], p. 171).

In the case of the algorithm at hand, Eq. (7.10) defines the function to be optimized. The constraints [Eq. (7.3)] combined with the formulation of the problem

[Eqs. (7.1) and (7.2)] define the feasible region for the problem in a way similar to that of the bounds used in the simplex method. The fact that the optimization of a linear function on convexly constrained data sets has been exploited in hyperspectral data before with the "pixel purity index" [11]. The procedure used for in this method differs from the simplex method in that the procedure optimizes in a discrete, stepwise manner. Furthermore, the constraints defined by Eqs. (7.1), (7.2), and (7.3) require knowledge of the endmember fractions, and thus implicitly the endmember spectra themselves. This optimization process would appear circular with the determination of endmembers requiring knowledge of the endmembers themselves. However, the fact that the data forms a simplex spatially means that any endmember spectrum, which sits at a vertex, will be chosen by the algorithm. The optimization requires only that the image data distribution behave according to the linear mixture model [i.e., Eqs. (7.1), (7.2), and (7.3)] and that endmember spectra be in the scene.

7.2. Application to Perfect Synthetic Data

To show the algorithm working with perfect data, the "maximum volume transform" algorithm is now applied to a synthetic data set. The data set used here is constructed from mixtures of three synthetic saw-toothed spectra (Figure 7.3), reflecting the interaction of different materials. At each corner of the image is a pure spectrum of each endmember respectively, with concentration dropping as the reciprocal of distance from the corner. In addition, Eq. (7.2) is enforced by renormalizing the image such that the sum of the fractional contributions equals one. This is necessary to ensure that the linear mixture model is followed.

Prior to applying the algorithm, the MNF transform was applied to determine the number of image endmember spectra and reduce the image to the appropriate dimensionality. The "maximum volume transform" procedure was then applied to the reduced data set, and the vertices of the final simplex were recovered. These were then used to produce radiance unit endmembers by an inverse MNF transformation. The original image was subsequently unmixed using these endmembers using least-squares analysis, producing fractional abundance planes. The resulting fractional abundance planes are equal to the model abundance planes to machine precision.

7.4. CONVERGENCE WITH IMPERFECT DATA

Maurice Craig argued against the methodology described here, suggesting instead a minimum volume transform [22]. In his suggested methodology a simplex that contains the data is shrunk down to the smallest volume containing the data, commonly referred to as a "shrink-wrapping" process. This is similar to the process that Orasis [5] uses. As with the maximum volume transform described in this paper, at the end of the optimization process the vertices of the simplex contain the coordinates of the endmembers. Craig argued that the approach outlined here

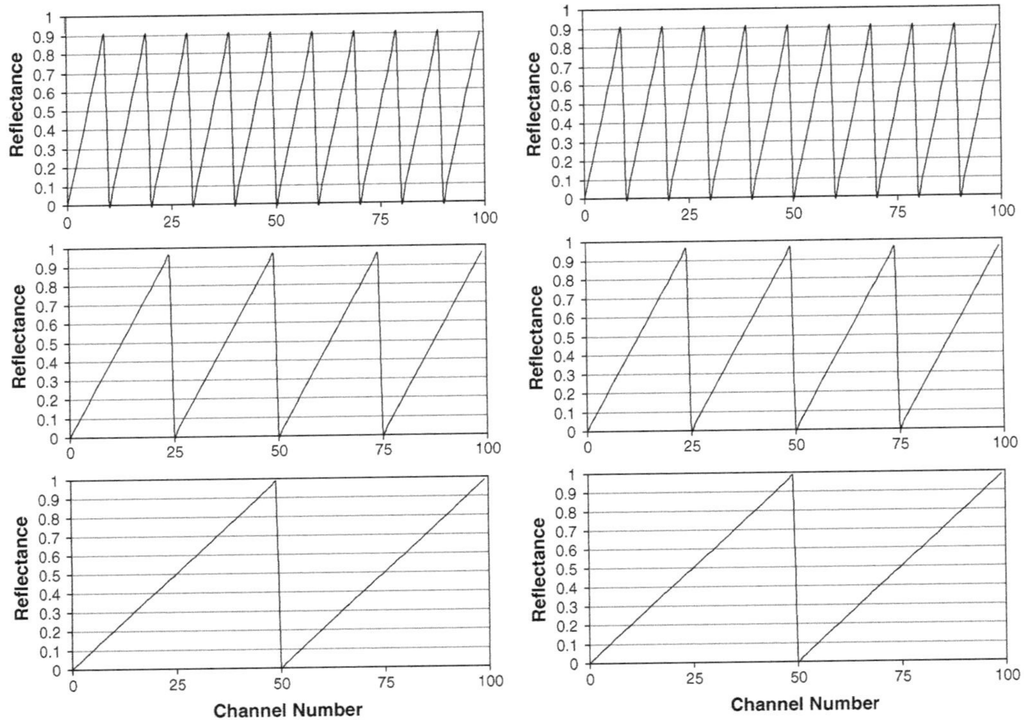

Figure 7.3. Synthetic endmember spectra used to construct test data (left) and algorithm derived spectra (right). Here 100-band spectra were constructed using saw-tooth functions of various periods. These spectra were mixed in various proportions to form a 100-band hyperspectral image. The derived endmembers exactly match the original endmembers.

produces "orientation ambiguity," nonsensible fractions (negative abundances and abundances that do not sum to one), in data sets without pure pixels. To illustrate Craig's misgivings, it is useful to consider the following practical example.

The collection of hyperspectral data does not occur under laboratory conditions. The collection of data occurs at spatial resolutions that may be larger than a generally occurring sample of constituent material. For example, the NASA JPL Airborne Visible Infrared Imaging Spectrometer (AVIRIS) sensor has an instantaneous field of view of 20 m at its operational altitude of 20 km [23]. In a given AVIRIS image, any object that occurs in the scene at a maximum spatial resolution of less than 20 m will never be seen in pure form. Small objects (in this case less than 20 m in size) will always be mixed with their neighbors. There exists a significant probability that no pure pixels will be present in a scene for a given endmember class. This problem increases with decreasing sensor spatial resolution.

As an example, consider an urban remote sensing application using AVIRIS at operational altitude where a classification based on roof type is desired. The hypothetical researcher does not know *a priori* the endmembers (in this case the types of roofs in the scene) and hopes to derive them using the maximum volume transform described here. If the largest roof in the scene is 2000 ft^2, then the pixel size will be approximately 13 m. This is less than the sensor resolution of AVIRIS. All roof pixels in the scene will therefore be mixed with whatever happens to be adjacent, probably grass, asphalt, or cement.

This algorithm will fail in this case because of the assumption that pure pixels exist within an image is not valid. It is therefore imperative that this algorithm behaves reasonably in the absence of such pure spectra. Reasonable behavior will be defined as selecting a pixel close to an actual (absent) endmember while still unconditionally converging to any pure pixels that do exist. In other words, reasonable behavior means that the algorithm will not catastrophically fail when its assumptions are not valid. The performance of this algorithm as the purity of one or more endmembers is reduced can be examined both theoretically and empirically. The theoretical discussion looks at the behavior of the algorithm based on linear optimization theory, which the empirical discussion looks ate the performance with a synthetic data set.

7.4.1. Theoretical Discussion of Imperfect Data

In the case where pure endmember spectra are not contained within the image, the mixture model remains the same; that is, Eqs. (7.1), (7.2), and (7.3) are still valid. The algorithm will fail to converge to the correct solution, however, due to the truncation of the data distribution, resulting in a lack of pure pixels to select.

What is the result of algorithms convergence in data sets without a pure pixel of one endmember class? Because the optimum solution of a linear function [Eq. (7.10)] in a convex set always occurs at a vertex of the set. Furthermore, for any arbitrary set of points in space, a convex set can be defined which encompasses the points.

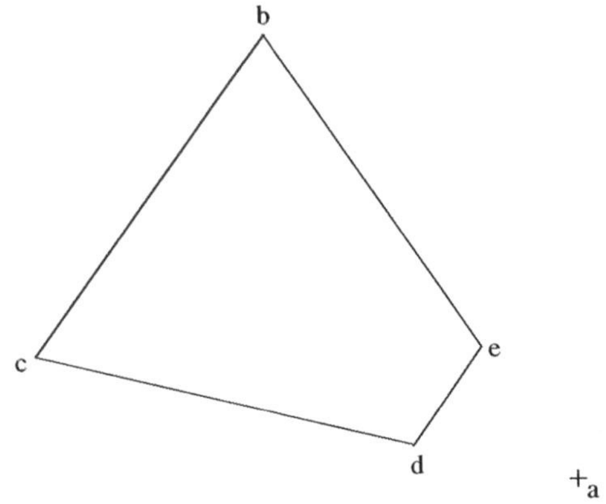

Figure 7.4. A convex hull for truncated two-dimensional image data.

Figure 7.4 shows the convex hull of a truncated set of two-dimensional image data. In a complete set of data, where pixels representing 100% of endmembers a, b, and c were present, the convex hull would be defined by the endmembers themselves (i.e., [a,b,c]). Figure 7.4 shows the case where the distribution of endmember a is not complete. There is no pixel in the data set with a concentration of endmember a above a certain threshold. In this case, the convex hull for the data is designated by points b, c, d, e.

How will maximum volume algorithm converge on this data set? The fact that the procedure follows a stepwise linear optimization requires that the algorithm select points on the convex hull. Since the algorithm updates each endmember position separately, convergence to the pure endmember spectra c and b will not be affected. Convergence toward endmember a will terminate at either of the convex hull points d or e. Which point the algorithm will converge to depends on the exact orientation of points d and e and on the exact form of the transient equation defining the optimization. The result is deterministic in the sense that the algorithm will always converge to the same incorrect endmember for class a; however, the result is not predicable without a detailed understanding of the exact nature of the data. As far as an analyst is concerned, the algorithm will converge to endmember d or e at random.

Figure 7.5 illustrates a slightly more complex distribution of incomplete data. Again, the distribution of endmember reflects a lack of complete purity. In this example, however, the distribution is has a less clean edge, resulting in the convex hull [b, c, d, e, f].

This simple experiment offers good insight as to the behavior of the maximum volume transform with respect to less-than-perfect data. In a sense, this algorithm fails because it does not converge to the correct endmembers (i.e., a, b, and c). However, the algorithm does not pathologically fail. It will correctly converge to whichever endmembers are present in the data.

This is borne out in a simple numerical experiment along the lines of that in Section 7.3.2. The data were prepared identically except for the fact that the first

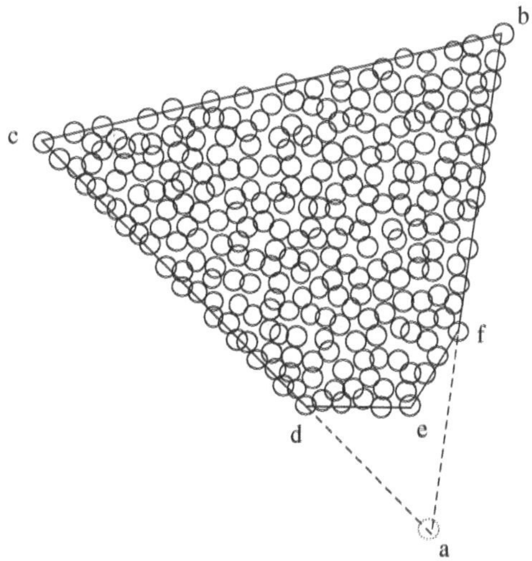

Figure 7.5. An example of two-dimensional linearly mixed data with no pure pixel for endmember a. The convex hull for these data is defined by the points b, c, d, e, f, because the line connecting these points contains all points in the data set.

endmember was limited to only 50% purity. In this case the distribution of data will resemble that shown in Figure 7.4, with one vertex of the triangular distribution cut off. In this case the maximum volume transform should converge to the second and third endmembers as before, but the first endmember spectra will be misidentified as a mixed pixel. The endmember spectra retrieved from this set of synthetic data are shown in Figure 7.6, and they match expectations. The endmember estimates for the second and third endmember are correct, but the first endmember estimate is a mixture of the first two endmember spectra.

7.5. EXAMPLE APPLICATION: HYPERSPECTRAL DATA

The region of Cuprite region in southwestern Nevada has provided a test bed for the remote sensing community since the mid-1970s. The geology of this specific region was discussed in detail by Abrams and Ashley [24]. The following geological context information is paraphrased from their work. The Cuprite region consists of an extensive exposure of Tertiary volcanics and Quaternary deposits. Sections of the volcanics underwent extensive hydrothermal alteration during the mid- to late Miocene. The alteration pattern studied here consists of concentric circles of alteration, consisting (from center outwards) of silicified, opalized, and argillized rocks. The silicified rocks show the highest degree of alteration and generally contain quartz, calcite, and some minor alunite and kaolinite. The opalized rocks consist generally of opaline silica with the addition of alunite and kaolinite and minor calcite. The argillized rocks occur sparsely at the edge of the deposit, and they consist of poorly exposed unaltered quartz and sandine, along with altered plagioclase.

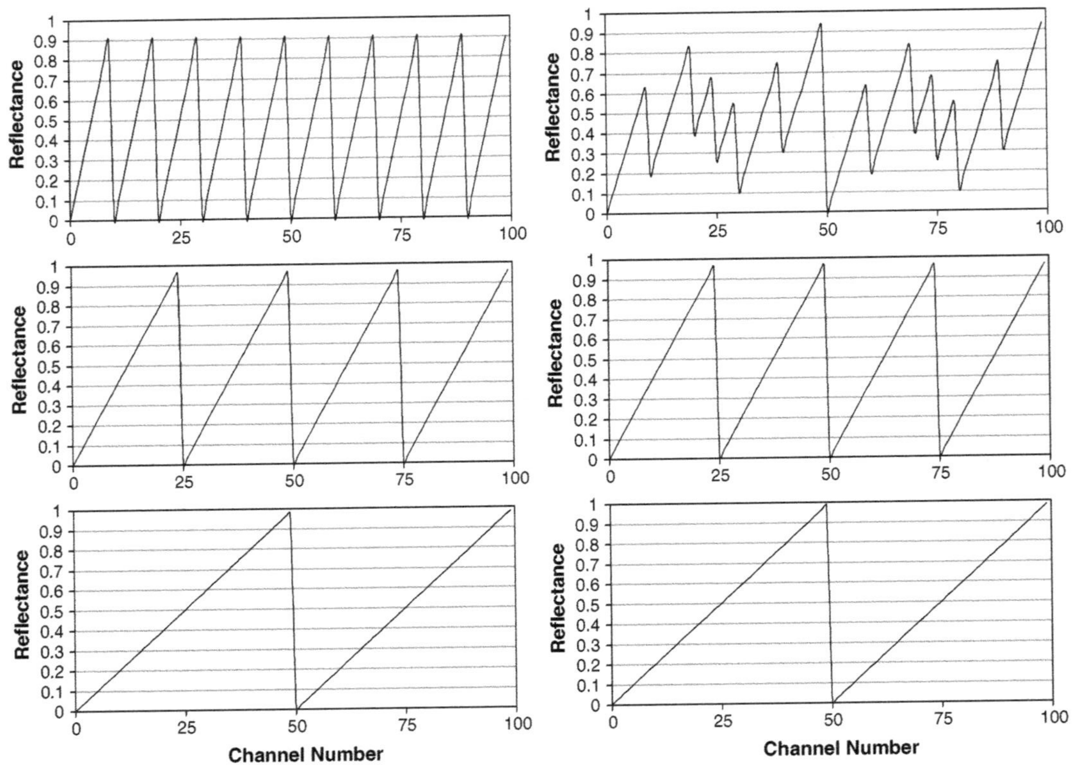

Figure 7.6. Synthetic endmember spectra used to construct test data without pure pixels for one endmember (left) and algorithm-derived spectra. As previously, 100-band spectra were constructed using saw-tooth functions of various periods. These spectra were mixed in various proportions to form a 100-band hyperspectral image. The top endmember has been incorrectly derived as the algorithm has converged on a mixed pixel. Results for the other endmembers are unaffected, however, and the correct result is returned.

In theory the different mineralogical units would represent endmembers within a hyperspectral scene. Each pixel within the image would represent a linear sum of a finite number of archetypical materials (the image endmember spectra). The application of the algorithm described here on hyperspectral data set taken over this region would be expected to produce endmember spectra that could be used to identify the mineralogical units in the area. Furthermore, subsequent unmixing of the derived endmembers should produce material maps that correlate well with published geological data.

7.5.1. Data Sets Used

Two Cuprite, Nevada data sets were processed using this algorithm. These were acquired using AVIRIS, NASA JPL's airborne visible/infrared imaging spectrometer [25], and HYMAP [26], HyVista Corporation's 126 spectral band airborne visible infrared imaging spectrometer. AVIRIS was designed and built by Jet Propulsion Laboratory, in Pasadena, CA. The sensor provides spectrographic data in 224 contiguous spectral bands from 400 to 2500 nm. Generally, AVIRIS flies aboard an ER-2 aircraft at an altitude of 20 km above sea level at approximately 730 km/h. HYMAP provides 126-band coverage across the reflective solar wavelength region of 0.45–2.5 mm, with contiguous spectral coverage (except in the atmospheric water absorption bands) with spectral bandwidths between 15 and 20 nm. The sensor is mounted on a three-axis gyro-stabilized platform, which reduces distortion due to aircraft motion. Hymap provides a signal-to-noise ratio of over 500:1. It has a 512-pixel swath width covering 61.3 degrees, giving a 2.5-mrad IFOV across track, and a 2.0-mrad IFOV along track. Typical ground spatial resolutions range from 3 to 10 m.

7.5.2. Comparison of Results with Geologic Field Data

The algorithm was applied to both scenes, assuming (based on an examination of geologic maps of the area) that there were 10 endmember spectra in the scene. The data were pre-processed using a standard principal components transform to the nine most statistically important bands as is necessary for the endmember determination algorithm. The maximum volume transform algorithm described above was then applied to the data set. Since endmember spectra are difficult to interpret in principal components space, the spectra in the original image corresponding to the endmember estimates were selected and were used instead. The resulting endmember spectra and unmixed fraction planes for AVIRIS are shown in Figures 7.7 and 7.8, respectively. The results of application of this algorithm for HYMAP are shown in Figures 7.9 and 7.10.

Figure 7.11 shows selected derived endmember spectra for both the AVIRIS and HYMAP Cuprite, NV data sets identified and compared with high-resolution USGS laboratory spectra [27]. There is a recognizable match between the absorption features of derived AVIRIS endmember spectra and the absorption features of laboratory spectra for five minerals (alunite, buddingtonite, kaolinite, calcite, and

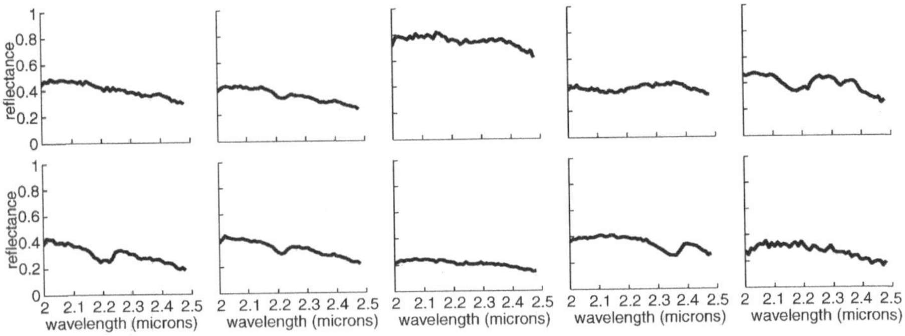

Figure 7.7. Endmember spectra derived using the maximum volume transform endmember spectra estimation algorithm for the AVIRIS Cuprite scene.

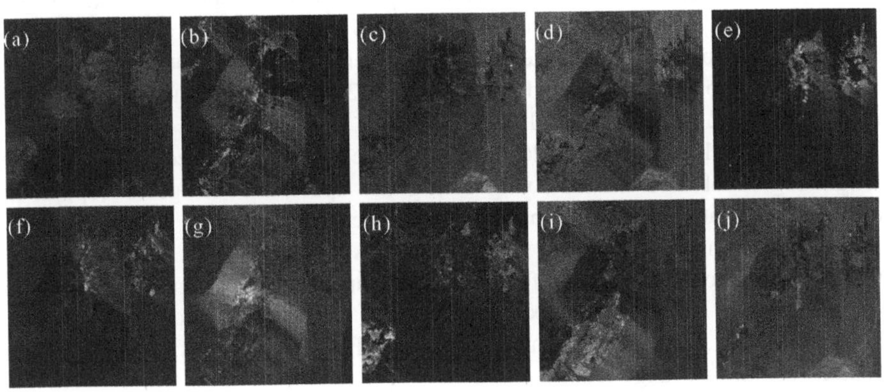

Figure 7.8. Unmixing the endmember spectra shown in Figure 7.7 from the AVIRIS Cuprite scene produces the abundance maps or fraction planes for the image. These images show the pixel-by-pixel abundance of each endmember component across the scene. The images are scaled so that zero abundance (the absence of a material) is black and full abundance (100% fill) is white. Shades of gray indicate varying levels of fractional abundance.

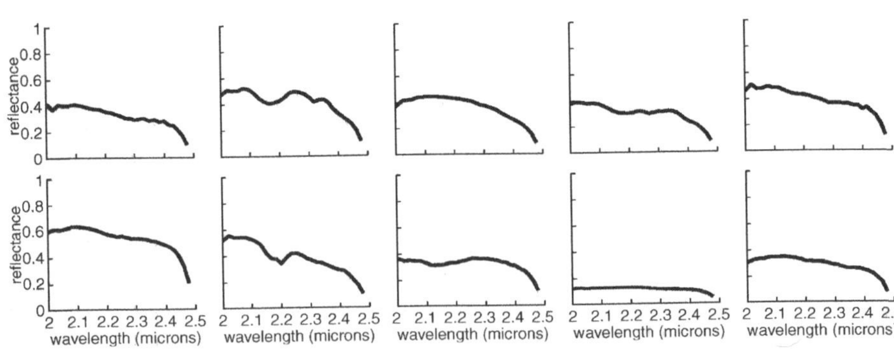

Figure 7.9. Endmember spectra derived using the maximum volume transform algorithm for the HYMAP Cuprite, NV scene.

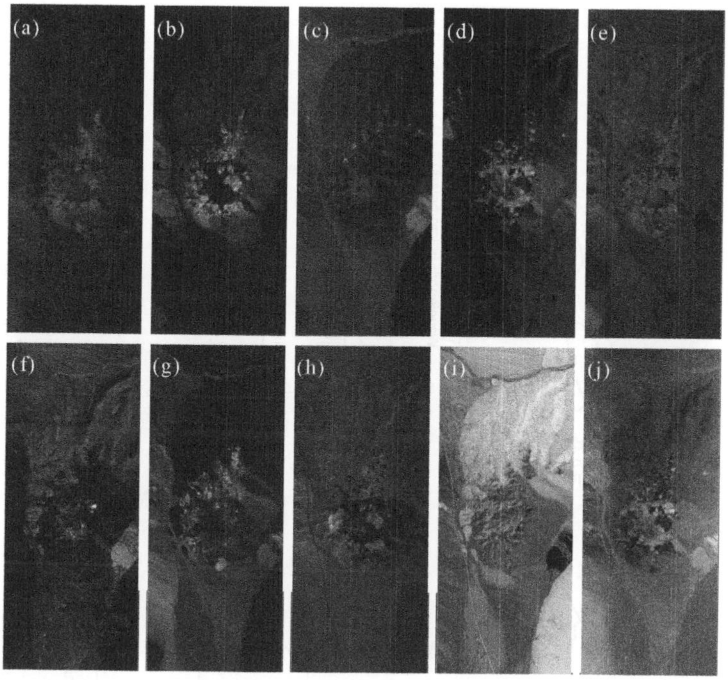

Figure 7.10. Unmixing the endmember spectra shown in Figure 7.7 from the HYMAP Cuprite scene produces the abundance maps or fraction planes for the image. These images show the pixel-by-pixel abundance of each endmember component across the scene. The images are scaled so that zero abundance (the absence of a material) is black and full abundance (100% fill) is white. Shades of gray indicate varying levels of fractional abundance.

muscovite). Looking at the unmixed abundance maps for both AVIRIS and HYMAP, neither the apparent calcite nor muscovite appear to be present in the HYMAP scene. For the remaining minerals, the results provide a recognizable match to laboratory spectra. Interestingly, buddingtonite was unknown in Cuprite until using airborne remote sensing data by Goetz and Srivastava [28].

Figure 7.7 shows the derived image endmember spectra and material abundance maps for the AVIRIS Cuprite scene. Not all endmember spectra are identifiable due to the lack of absorption spectra; however, spectra e, f, and i clearly resemble alunite, kaolinite, and calcite. Spectra b and g are difficult to identify, however, based on the examination of mineral maps of the region, they appear to represent muscovites. The material maps (shown in Figure 7.8) resulting from the application of the algorithm described here on these two hyperspectral data sets also compare favorably with geologic maps of the region developed using spectrographic methods by Clark et al. [1]. For example, the fraction planes derived from the AVIRIS Cuprite image in Figure 7.8, image b, e, f, h and i, correspond to Jarosite, Alunite, Kaolinite, Alunite/Kaolinite mix, and Calcite, respectively, in Figure 9B in Clark et al. [1].

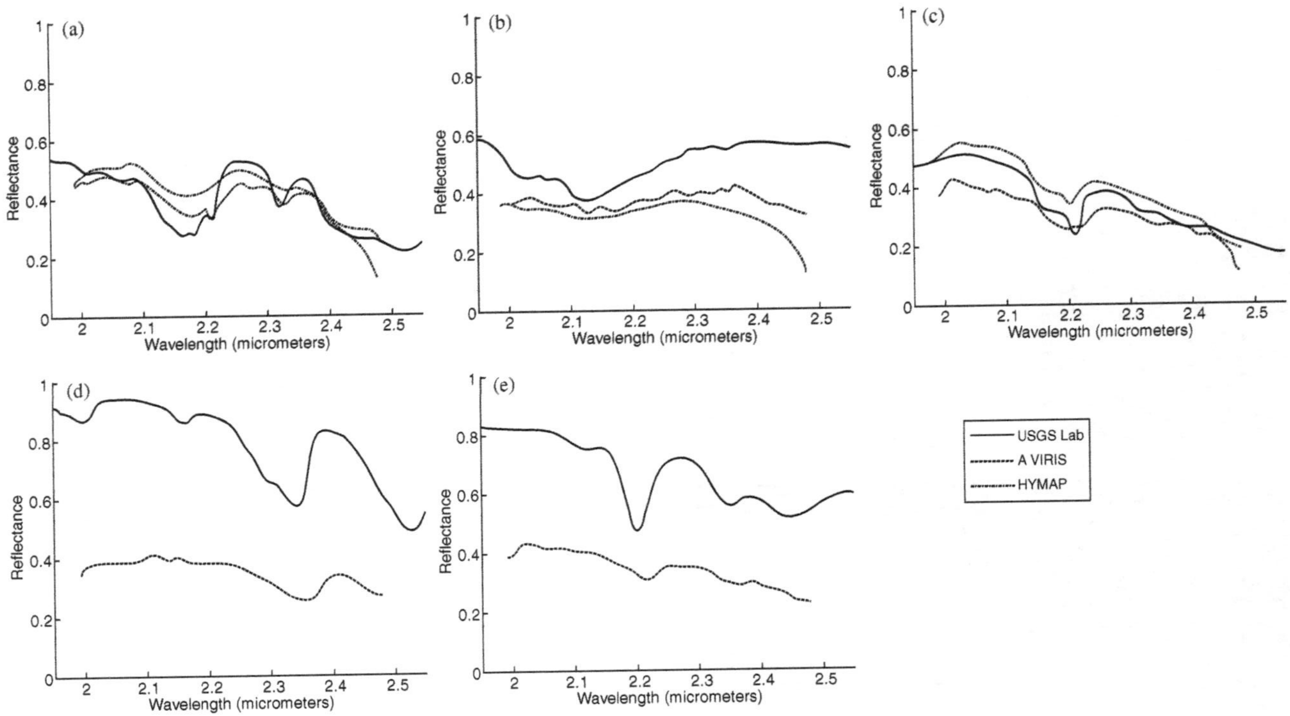

Figure 7.11. A comparison of derived endmember spectra for the AVIRIS (dashed) and HYMAP (dash-dotted) Cuprite scenes. Spectra correspond to the minerals alunite, buddingtonite, kaolinite, calcite, and muscovite (a–e, respectively). For comparison, USGS laboratory-derived spectra for these five minerals is shown (solid).

7.6. CONCLUSION

Endmember spectra determination algorithms offer good potential for the automated exploitation of hyperspectral imagery. With a handful of assumptions, they can reduce a multi-gigabyte hyperspectral image consisting of hundreds of bands into a set of archetypical spectra. These spectra can be used as a sort of spectral shorthand for the image. One of the simpler algorithms to accomplish this task, called a maximum volume transform (based on the commercial N-FINDR algorithm), consists of a simplex inflation procedure. This transform can be shown to converge to a correct solution in perfect data and to a reasonable solution in less-than-perfect data. Importantly, the algorithm appears to perform well with real-world hyperspectral data.

These algorithms require simplified models of the physical processes that produce hyperspectral data, as well as some assumptions as to the statistical distribution of endmember mixing process. While there is still progress to be made using these methods, the accuracy of these approaches will always be limited by the underlying accuracy of these assumptions. Future gains in automated hyperspectral image processing will require better, more accurate physical models and fewer assumptions.

ACKNOWLEDGMENTS

I would like to gratefully acknowledge the National Geospatial Intelligence Agency (grants NMA401-02-1-2013 and NMA201-00-1-2002) and the Intelligence Community Postdoctoral Research Fellowship Program for fiscal support vital to the preparation of this manuscript.

REFERENCES

1. R. N. Clark, G. A. Swayze, K. E. Livo, R. F. Kokaly, S. J. Sutley, J. B. Dalton, R. R. McDougal, and C. A. Gent, Imaging spectroscopy: Earth and planetary remote sensing with the USGS Tetracorder and expert systems, *Journal of Geophysics Research*, vol. 108(E12), no. 5131, pp. 5-1–5-44, 2003.
2. J. A. Richards, *Remote Sensing Digital Image Analysis: An Introduction*, Springer-Verlag, Berlin, 1999.
3. I. Reed, and X. Yu, Adaptive multiband CFAR detection of an optical pattern with unknown spectral distribution, *IEEE Transactions on Acoustics, Speech and Signal Processing*, vol. 38, pp. 1760–1770, October 1990.
4. J. S. Tyo, D. I. Diersen, A. Konsolakis, and R. C. Olsen, Principal components-based display strategy for spectral imagery, *IEEE Transactions on Geoscience and Remote Sensing*, vol. 41, 708–718, 2003.
5. J. Bowles, M. Daniel, J. Grossman, J. Antoniades, M. Baumback, and P. Palmadesso, Comparison of output from Orasis and pixel purity calculations, *Proceedings of SPIE*, vol. 3438, pp. 148–156, 1998.

6. M. E. Winter, Fast autonomous spectral endmember determination in hyperspectral data, *Proceedings of the Thirteenth International Conference on Applied Geologic Remote Sensing*, Vol. II, pp 337–344, Vancouver, B.C., Canada, 1999.
7. M. E. Winter, N-FINDR: An algorithm for fast autonomous spectral end-member determination in hyperspectral data, *Proceedings of SPIE*, vol. 3753, 1999; Imaging Spectrometry V., pp. 266–275.
8. M. Winter, A proof of the N-FINDR algorithm for the automated detection of endmembers in a hyperspectral image, *Proceedings of SPIE*, vol. 5425, pp. 31–41, 2004; Algorithms and Technologies for Multispectral, Hyperspectral, and Ultraspectral Imagery X, edited by Sylvia S. Shen and Paul E. Lewis.
9. J. W. Boardman, Analysis, understanding and visualization of hyperspectral data as convex sets in *n*-space, *Proceedings of SPIE*, vol. 2480. pp. 14–22, 1995.
10. J. B. Adams, M. O. Smith, and P. E. Johnson, Spectral mixture modelling: A new analysis of rock and Soil Types at the Viking Lander Site, *Journal of Geophysical Research*, vol. 91, pp. 8098–8112, 1986.
11. J. W. Boardman, K. A. Kruse, and R. O. Green, Mapping target signatures via partial unmixing of AVIRIS data, in *Summaries of the Fifth Annual JPL Airborne Geoscience Workshop*, Pasadena, CA, vol. 1, 1995.
12. J. Bowles, P. Palmadesso, J. Antoniades, M. Baumbeck, and L. J. Rickard, Use of filter vectors in hyperspectral data analysis, in *Infrared Spaceborne Remote Sensing III*, edited by M. S. Scholl and B. F. Andresen, *Proceedings of SPIE* 2553, 148–157, 1995.
13. R. A. Neville, Staenz, K, Szerdedi, T, Lebfebre, and J. Hauff, Automatic endmember extraction from hyperspectral data for mineral exploration, in *Proceedings of the Fourth International Airborne Remote Sensing Conference*, Vol. II, pp. 891–897, Ottawa, Ontario, Canada, 1999.
14. A. Green, M. Berman, P. Switzer, and M. D. Craig, A transformation for ordering multispectral data in terms of image quality with implications for noise removal, *IEEE Transactions on Geoscience and Remote Sensing*, vol. 26, no. 1, pp. 65–74, 1988.
15. J. W. Boardman, Automating spectral unmixing of AVIRIS data using convex geometry concepts, *Summaries of the Fourth Annual JPL Airborne Geoscience Workshop*, vol. 1, pp. 11–14, Jet Propulsion Laboratory, Pasadena, CA, 1994.
16. J. O. Rawlings, *Applied Regression Analysis: A Research Tool*, Wadsworth and Brooks/Cole Advanced Books & Software, Pacific Grove, CA, 1988.
17. S. R. Lay, *Convex Sets and Their Applications*, John Wiley & Sons, New York, 1982.
18. E. M. Winter and M. E. Winter, Autonomous hyperspectral endmember determination methods, *Proceedings of SPIE*, vol. 3870, pp. 150–158, 2000; *Sensors, Systems, and Next-Generation Satellites* III, edited by H. Fujisada and J. B. Lurie, 2000.
19. C. L. Lawson, and R. J. Hanson, *Solving Least Squares Problems*, Prentice Hall, Englewood Cliffs, NJ, 1974.
20. G. Arfken, *Mathematical Methods for Physicists*, Academic Press, New York, 1985.
21. F. A. Ficken, *The Simplex Method of Linear Programming*, Holt, Rinehart and Winston, New York, 1961.
22. M. D. Craig, Minimum volume transforms for remotely sensed data, *IEEE Transactions on Geoscience and Remote Sensing*, vol. 32, pp. 542–552, 1994.

23. W. M. Porter, and H. T. Enmark, A system overview of the airborne visible/infrared imaging spectrometer (AVIRIS), JPL Publication 87-38, Jet Propulsion Laboratory, Pasadena, CA, 1987.

24. M. J. Abrams, and R. P. Ashley, Alteration mapping using multispectral images—Cuprite Mining District, Esmeralda County, Nevada: U.S. Geological Survey Open File Report 80-367, 1980.

25. G. Vane, R. O. Green, T. G. Chrien, H. T. Enmark, E. G. Hansen, and W. M. Porter, The airborne visible/infrared imaging spectrometer (AVIRIS), *Remote Sensing of the Environment*, vol. 44, pp. 127–143, 1993.

26. T. Cocks, R. Jenssen, A. Stewart, I. Wilson, and T. Shields, The HYMAP(TM) airborne hyperspectral sensor: The system, calibration, and Performance, presented at *1st EARSEL Workshop on Imaging Spectroscopy, Zurich*, October 1998.

27. R. N. Clark, G. A. Swayze, A. Gallagher, T. V. V. King, and W. M. Calvin (1993) The U. S. Geological Survey, Digital Spectral Library: Version 1: 0.2 to 3.0 microns, *U.S. Geological Survey, Open File Report 93-592*, 1993.

28. A. F. H. Goetz, and V. Srivastava, Mineralogical mapping in the Cuprite Mining District, Nevada, in *Proceedings of the Airborne Imaging Spectrometer Data Analysis Workshop*, JPL Publication 85-41, pp. 22–29, Jet Propulsion Laboratory, Pasadena, CA, 1985.

CHAPTER 8

HYPERSPECTRAL DATA REPRESENTATION

XIUPING JIA
School of Information Technology and Electrical Engineering, University College, The University of New South Wales, Australian Defence Force Academy, Campbell ACT 2600, Australia

JOHN A. RICHARDS
College of Engineering and Computer Science, The Australian National University, Canberra ACT 0200, Australia

8.1. INTRODUCTION

The mainstay machine learning tool for thematic mapping in remote sensing has been maximum likelihood classification based on the assumption of Gaussian models for the distribution of data vectors in each class. It has been in constant use since first introduced in the late 1960s [1] and has been responsible for the success of many research-based and applied thematic mapping projects. Alternative classification methods have been introduced in the meantime, including neural networks [2] and support vector machines [3], but maximum likelihood classification remains a popular labelling tool because of its ease of use and the good results it generates if used properly, particularly when applied to multispectral data sets.

When used with hyperspectral data, maximum likelihood classification can suffer from the problem known as either the Hughes phenomenon [4] or the curse of dimensionality [5]. In essence, that refers to the failure of an ostensibly properly trained classifier to be able to generalize to unseen data with high accuracy. In maximum likelihood classification it is related, in particular, to not having sufficient training data available to form reliable estimates of the class conditional covariance matrices in the multivariate normal models. If N is the dimensionality of the data, then it is generally felt that a minimum of $10(N+1)$, with desirably as many as $100(N+1)$, training pixels per class is required; otherwise, unreliable statistics are likely and generalization will be poor.

Hyperspectral Data Exploitation: Theory and Applications, Edited by Chein-I Chang

A number of approaches has been proposed to reduce the impact of the dimensionality of the data when estimating class statistics. Each has its advantages and limitations. There has, however, been no systematic comparative evaluation of those methods designed particularly to make the maximum likelihood approach work effectively on high-dimensional remote sensing image data. (Nor has there been an evaluation of statistical versus nonparametric and other methods for effective thematic mapping from such hyperspectral data sets, but that is beyond the scope of this treatment.)

It is the purpose of this chapter to (a) provide an analysis of methods based upon supervised maximum likelihood classification and (b) present some comparative analyses. In essence, we are looking for an optimal dimensionality reduction or subspace projection schema that allows thematic mapping to be carried out effectively and efficiently with hyperspectral imagery.

8.2. THE MAXIMUM LIKELIHOOD METHODOLOGY

The fundamental assumption behind maximum likelihood classification is that each class can be represented by a multivariate normal model $N(\mathbf{m}, \Sigma)$. In general, this is a poor assumption in that the distribution of pixels in the classes of interest to the user—the so-called information classes—are not well-represented by normal models. However, the technique works particularly well if the information classes are resolved into sets of subclasses—often called spectral classes—each of which can be modelled acceptably by normal distributions [6, 7]. The mapping between spectral and information classes can be unique [8] or distributed [9]. If that multimodality is not resolved with the maximum likelihood classifier, then suboptimal results can be generated. Nevertheless, in the following we will assume that multimodality is not an issue with the fairly simple data sets tested.

The attraction of the maximum likelihood rule compared with more recent procedures such as the neural network and support vector machines (with kernels) is that the training process is relatively straightforward and standard software is widely available [10]. It is also theoretically appealing and, in the full Bayes' form, different penalties can be applied to different misclassifications, allowing the decisions to be biased in favor of more important outcomes. It also lends itself well to the incorporation of spatial context through the use of Markov random field [11] and relaxation labeling modifications [7].

As noted above, however, a major consideration in applying the maximum likelihood rule is to ensure that reliable sample estimates of $\mathbf{m}$ and Σ, the class mean and covariance, are generated. This requires sufficient training samples to be available for each (spectral) class. Estimation of the covariance matrix is the most demanding requirement, so we can assume that if the covariance is reliable, the mean vector is also reliable. For N spectral dimensions the covariance matrix is symmetric of size $N \times N$ with $\frac{1}{2}N(N+1)$ distinct elements. To avoid singularity, at least $N(N+1)$ independent samples is needed. Fortunately, each N-dimensional pixel vector contains N samples (one for each waveband) so that the minimum number of independent training pixels required is $(N + 1)$. Because of the difficulty in ensuring independence, Swain and Davis [1] recommend choosing more than this minimum, as noted earlier.

With the very large value of N for hyperspectral data sets, we can almost never guarantee that we will have sufficient training pixels per class to ensure good estimates of class statistics and thus good generalization. As a consequence, application of Gaussian maximum likelihood methods to hyperspectral imagery demands some form of dimensionality reduction. Standard feature reduction methods based on class separability measures such as divergence, JM distance, and transformed divergence [7] cannot be used effectively as a result of several considerations. First, the number of permutations of band subsets to be checked for hyperspectral data can be impractically large; second, they also depend on the estimation of class-specific covariance matrices and thus can suffer the same problem that we are trying to avoid, unless the feature subsets are small; finally, the resulting subsets may not be as information-rich as the features consisting of linear combinations of bands generated by transformation-based methods for feature reduction. Likewise, class-dependent transformations such as canonical analysis as procedures for selecting a smaller set of transformed features require class conditional covariance information. Suboptimal dimensionality reduction methods such as the principal components transformation perform better since the global covariance matrix needed for computing the transformation matrix is estimated from the aggregate of the training samples over all classes. Thus more reliable estimates should be obtained. The drawback of those methods, however, is that the transformations generated are not optimized for maximizing class separation in the reduced (transformed) band sets; instead, they are optimized for rank ordering global data variance by band.

There are better approaches. One is to devise feature selection methods that do not depend on covariance matrices for their operation; in other words, they are distribution-free. Another is to see whether the dependence of the class covariance matrix on the principal dimensionality of the problem can be modified. A third approach is to generate acceptable approximations to the class conditional covariances from poor sample estimates by adding in proportions of better estimated, yet perhaps less relevant, measures. This latter approach sometimes goes under the name of regularization.

8.3. OTHER CLASSIFICATION APPROACHES FOR USE WITH HYPERSPECTRAL DATA

Because they have a linear discriminant function basis, neural networks and support vector classifiers have been proposed as viable tools for handling hyperspectral data classification.

The neural network approach is attractive because it does not suffer from the same problem of dimensionality and, like the maximum likelihood rule, it is inherently multiclass. It has, however, two limitations that are particularly important with hyperspectral imagery. First, as in any neural network, the number and sizes of the hidden layers need to be set. Generally, the problem is overspecified and some form of pruning is used to generate a minimum network that will solve the problem at hand. Nevertheless, the issue is not straightforward. Second, a very large number of iterations is sometimes required to find a solution. Camps-Valles and Bruzzone [12] and Melgani and Bruzzone [13] show that accuracies as high as 87–94% can be obtained by mapping nine classes with 200-channel data. They used the AVIRIS Indian Pine data set, recorded in 1992, which covers an area of mixed agriculture and forestry in northwestern Indiana.

More recently, the support vector machine (SVM) has been used with hyperspectral data sets [3, 12]. It is the use of the kernel transformation employed in conjunction with the support vector machine that renders it useful for real remote sensing problems. It is based on two notions: First, the SVM finds the *optimal* hyperplane that separates a two-class problem. It is distribution-free and generates a hyperplane whose location depends only on those pixels from each of the two classes that are nearest to it. Second, having solved the linearly separable problem efficiently, the SVM seeks a data transformation $\boldsymbol{x} \rightarrow \phi(\boldsymbol{x})$, presumably to a higher-order space, within which inseparable data in the original domain becomes linearly separable and thus amenable to treatment by the SVM. On the nine-class, 200-channel Indian Pines data set, Camps-Valles and Bruzzone [12] have demonstrated that accuracies of around 95% are possible.

Limitations of the SVM approach include a training time that can be large and is quadratically dependent on the number of training samples, the need to find optimal kernel parameters (sometimes through a set of trial training runs), and the fact that the binary SVM has to be embedded in a binary decision tree, which can lead to large classification times.

Because of the high spectral definition provided by the fine spectral resolution in hyperspectral imagery, it is possible to adopt a scientific approach to pixel labeling based on spectroscopic knowledge. This approach is used in the Tetracorder technique [14]. It seeks to exploit the fact that the recorded reflectance spectrum is almost as good as that which would be recorded in the laboratory. Features known from spectroscopy (in particular applications) to be diagnostically informative are identified and used, through a knowledge-based reasoning methodology and library searching, to label pixels. This approach is essentially a transformation to a subspace that is defined by the characteristics of the diagnostic features. Most often they are absorption features, characterized by their spectral positions, widths, and depths.

8.4. CANDIDATE APPROACHES FOR HANDLING HYPERSPECTRAL DATA WITH MAXIMUM LIKELIHOOD METHODS

We review now those maximum likelihood-based methods that have been used successfully to date for handling hyperspectral data, in preparation for a detailed comparative analysis.

8.4.1. Feature Reduction Methods

Sometimes it is possible to reduce dimensionality by selecting a subset of the available channels; this might be guided by some foreknowledge of the ranges of wavelength appropriate to particular applications (for example, middle infrared data are of little value to water studies) or it may be simplistic, such as deleting every second band.

A more rigorous approach is to use standard feature assessment techniques such as divergence. However, they are not usually successful since, as noted earlier, they need class-conditional covariance data for their computation.

A better approach, therefore, is to seek feature reduction via transformation of the data, such that sets of transformed bands known to be less discriminating can be disregarded.

The essence of feature reduction via transformation is to establish a criterion that expresses the notional separability among classes, such as

$$J = \Sigma_w^{-1}\Sigma_a \tag{8.1}$$

in which Σ_w is an average measure (expressed as a covariance matrix) of the within class distribution of pixels about their respective means and Σ_a is a measure of the scatter among the classes themselves (often expressed as a covariance matrix of the class means about the global mean). Clearly, J will be large if the classes on the average are small and well-separated.

What we try to do by transformation is seek a coordinate rotation that maximizes J and coincidentally allows less discriminating features to be removed. Generally, the latter is achieved by finding the eigenvalues of J. The largest eigenvalues correspond to those axes along which the classes are maximally separated; the smallest eigenvalues indicate those axes in which separation is poor. By selecting the set of eigenvalues that account for, say, 99% of the variance we can discard axes (and thus features) that are minimally important for separation. The eigenvectors corresponding to the eigenvalues retained provide the actual transformation matrix required to go from the original features to the reduced set of transformed features.

The problem with Eq. (8.1) of course is that the average within class covariance matrix requires the class-conditional covariances to be available beforehand; this potentially defeats the value of feature reduction via transformation for hyperspectral data. However, Kuo and Landgrebe [15] have devised a distribution-free measure of data scatter, which can be used in place of the covariance matrices, to render this approach to feature reduction viable. Furthermore, their measure focuses attention on those pixels near class boundaries when deriving the required axis transformation. Known as nonparametric weighted feature extraction (NWFE), it has been shown to perform well in practice and is available in the MultiSpec image analysis system. The basis for NWFE is summarized in the following. Further details are available in Landgrebe [16].

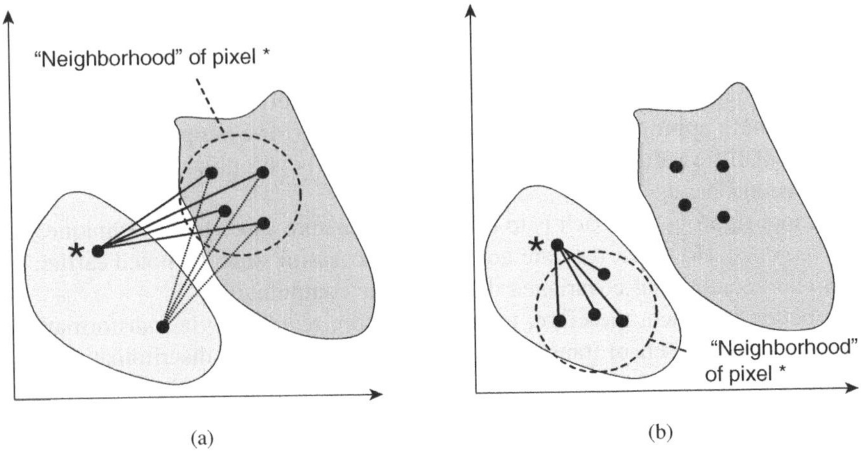

Figure 8.1. Developing a nonparametric measure of (a) among-class scatter and (b) within-class scatter.

Consider the two-dimensional, two-class example of Figure 8.1a, for simplicity. A *between-class* measure of scatter can be devised in the following manner. Consider the pixel marked by an asterisk in the figure. The scatter of all pixels in the other class about that pixel is computed, but their contributions are weighted according to how far away they are from that pixel. Those closest are given the higher weighting so that the pixels in the second class that are much further removed hardly contribute to the measure of scatter. In this manner it will be those pixels in the vicinity of the boundary that contribute most to the scatter measure and thus ultimately to the determination of the axis transformation that will be used for feature reduction.

Every pixel in the first class is used in turn as the centroid about which distance-weighted scatter is computed. Those measures are then averaged. The pixels in the other classes in turn are used for the centroids, with all the computed scatter measures then averaged. Thus we have a final measure that is the average over all pixels from each class as they scatter about all pixels from every other class.

In a similar manner we develop a *within-class* measure of scatter by examining the average distance-weighted measure of scatter about each pixel of the *same* class, as depicted in Figure 8.1b.

While computationally demanding, the average scatter matrices developed in this manner favor transformed features that give maximal separation across class boundaries. Features selected in this way have been shown to lead to good maximum likelihood classifier performance with hyperspectral data. Kuo and Landgrebe [17], for example, have shown that the original 200 channels of a Washington, DC Mall data set can be reduced to as few as five and still deliver accuracies in excess of 90% when mapping six classes with 40 training samples each.

8.4.2. Approximating the Class-Conditional Covariance Matrices

As noted earlier, for N dimensions at least $10(N+1)$ samples per class is seen to be necessary to estimate reliably the elements of the corresponding covariance matrix Σ_i.

If there are a total of K classes, then the total set of training pixels will be $10K(N+1)$. Thus, while it may be difficult to generate reliable estimates of the class conditional covariances, nevertheless, for any reasonable values of K it is likely that there might be sufficient data to obtain reasonably good estimates of global quantities, such as the class-independent covariance matrix Σ, computed using all the training pixels. Based on this, an approximation to the sample class conditional covariance matrix, which performs better than the poorly estimated sample class conditional covariance matrix itself, is

$$\hat{\Sigma}_i = \alpha_i\Sigma + (1-\alpha_i)\Sigma_i \tag{8.2}$$

in which Σ_i is the class-conditional estimate obtained from the class-specific training data and Σ is the global covariance estimate obtained using all the available training data. Values for the parameters α_i (which will differ from class to class) have to be found to give best results on generalization. That can be done through the repetitive application of the Leave-One-Out Classification (LOOC) approach to accuracy determination.

Better estimates are obtained when diagonal versions of the covariances are used in Eq. (8.2), such as in the more complex form:

$$\begin{aligned} \hat{\Sigma}_i &= (1-\alpha_i)\,\mathrm{diag}\,\Sigma_i + \alpha_i\Sigma_i & 0 \le \alpha_i \le 1 \\ &= (2-\alpha_i)\Sigma_i + (\alpha_i - 1)\Sigma & 1 < \alpha_i \le 2 \\ &= (3-\alpha_i)\Sigma + (\alpha_i - 2)\,\mathrm{diag}\,\Sigma, & 2 < \alpha_i \le 3 \end{aligned} \tag{8.3}$$

This has been shown by Landgrebe [16] to perform very well in practice because it implements various approximations to the discrimination problem dependent on the value of α_i.

8.4.3. Subspace Approximation

Another approach to hyperspectral thematic mapping using maximum likelihood methods is to exploit the substructure of the class covariance matrix so that it can be represented by a set of independent, smaller matrices [18, 19]. This method finds a set of independent feature subspaces by examining the block diagonal nature of the covariance matrix that results from ignoring low correlations between sets (blocks) of bands. In essence, it identifies regions of the reflectance spectrum within which adjacent bands are strongly correlated, because their respective covariances are important, and between which covariance can be ignored.

(a)

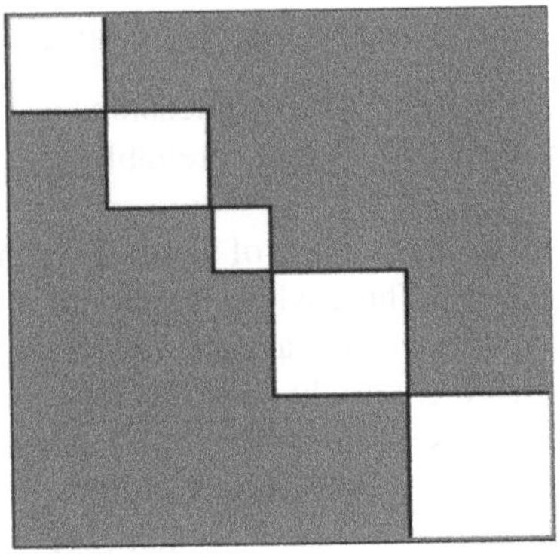

(b)

Figure 8.2. (a) Correlation matrix for 196 bands of the AVIRIS Jasper Ridge image, in which white indicates ± 1 correlation and black represent 0 correlation. (b) Selection of the highly correlated diagonal blocks.

Figure 8.2a shows the correlation matrix of a hyperspectral image data set, while Figure 8.2b shows a set of blocks that can be specified by ignoring those regions in the correlation matrix away from the highly correlated diagonal blocks.

The great benefit of this approximation is that it allows a decomposition of the covariance matrix, since only the interactions between those bands which are considered significant are retained. The covariance matrix itself, over all bands, is the sum of the blockwise covariance matrices. Thus, the number of training pixels required for reliable covariance estimation is established by the size of the largest block of bands. This can be as much as an order of magnitude less than required for the complete covariance matrix.

8.4.4. Data Space Transformation

The cluster space method [9] is another viable means for handling hyperspectral data. It uses nonparametric clustering to transform the hyperspectral measurements into a set of data classes. They can then be related to the information classes of interest via statistics learned from available training data. Since no second-order statistics are required, the method does not suffer from the high dimensionality of the training data—quite the contrary, the cluster-generated transformation obviates the problem, provided that clustering can be carried out effectively.

The technique is similar in principle to vector quantization for compression and shares its advantages. It is also, however, a generalization of the hybrid supervised–unsupervised thematic mapping methodology [8], long known to be a valuable schema for use with multispectral data, so that the multimodality of information classes is handled implicitly.

Although it overcomes the dimensionality problem, we have not chosen to incorporate it into the comparative analysis carried out for this investigation.

8.4.5. The Need for a Comparison

All of the candidate approaches just outlined can be made to work well, some more so in particular applications. However, despite significant experience with each, to date there has been no systematic comparative analysis undertaken to reveal which (if any) is near optimal and whether certain of the methods are better matched to particular application domains. That is surprising, given the importance of hyperspectral data and the magnitude of the problem to do with dimensionality.

While some comparative studies have been performed, many suffer the limitation that the techniques are often applied in a simplistic way without devising or exploiting schemas or methodologies that are matched to the characteristics of the particular algorithm to optimize its performance. Although this has been known since the early days of thematic mapping from remotely sensed data [4], many investigators still make the mistake of applying techniques without due regard to the complexities of the data and the limitations of the algorithms. This has been discussed recently, with the proposition that any reasonable algorithm can be made to work well, provided that its properties and limitations are well understood [10].

8.5. AN EXPERIMENTAL COMPARISON

We have undertaken an experimental comparative study of the maximum likelihood-based methods using two different images. The first is a Hyperion data set (geo_fremont198.BIL) recorded in 2001 over Fremont, California. Bands with zero values were removed, leaving 198 bands. The second is the Indiana Indian Pines data set with 220 bands.

Experiment 1. A simple classification was carried out with the Fremont data to gain an initial assessment of whether each of the techniques performs as expected in giving good results with high-dimensionality data. Only every second band was used for this exercise, giving 99 in total.

Four classes were selected: (1) wetland, (2) lawn, (3) new residential area, and (4) dry grass/bare soil. The numbers of training and testing pixels for each class, respectively, were 125, 92; 138, 93; 125, 75; 131, 105.

Seven different classifications were performed: One involved separating all (four) classes, and the other six involved separating the classes by pairs. The results are shown in Table 8.1, along with a classification performed using a simple maximum likelihood classifier, with no attention to dimensionality reduction.

In respect of each technique, the following should be noted:

- The number of features selected in NWFE were determined by making sure that those features contained 99% of the sum of the eigenvalues; in each case the number of features used is shown in parentheses in Table 8.1.

TABLE 8.1. Comparison of the Methods on the Testing Data (Fremont Data Set)

Classes	Standard Maximum Likelihood	Regularization	NWFE (Features)	Block-Based Maximum Likelihood
All four	64.9	90.4	84.4 (59)	92.6
1 and 2	58.4	98.4	80.0 (66)	90.3
1 and 3	98.2	100	99.4 (50)	100
1 and 4	100	100	100 (44)	100
2 and 3	92.3	100	99.4 (53)	100
2 and 4	81.3	100	93.4 (51)	100
3 and 4	82.2	82.2	90.6 (68)	95.0

- For the block-based method the blocks in the class conditional covariance matrices were defined by the band sets 1–15, 16–26, 27–48, 49–70, 71, 72, 73, 74, 75, 76–99.

From this simple comparison it is clear that all methods work acceptably, and certainly better than simple maximum likelihood classification overall.

Experiment 2. Several different trials were carried out with the Indian Pines data, covering a range of classes. The number of features was varied to see how the performance of each algorithm coped. Nine reduced feature subsets were created for the exercise, ranging from 220 bands (the original), on which the maximum likelihood rule is expected to perform poorly, to 44 features, on which maximum likelihood classification should perform well. The other methods should perform acceptably over the full range, given they have been devised to cope with the dimensionality problem. The various feature subsets were obtained by the processes described in Table 8.2.

TABLE 8.2. Defining the Reduced Feature Subsets

Number of Features	How Selected
220	Original set
198	Delete every 10th band
176	Delete every 5th band
165	Delete every 4th band
147	Delete every 3rd band
110	Use every 2nd band
74	Use every 3rd band
55	Use every 4th band
44	Use every 5th band

An interesting consideration arises in relation to using the NWFE approach. Two sets of results are shown. One applies the NWFE transform to each of the reduced feature data sets; then only those features that account for 99% of the original data variance are retained. This approach is referred to as NWFE (b). The other approach, referred to as NWFE (a), uses the NWFE itself to effect the feature reduction from the original 220 bands by doing a single transformation and then retaining the best (most discriminating) numbers of transformed features corresponding to the numbers of bands in the reduced feature data sets. While the comparison with the other techniques is not exact, in the sense that the original bands sets are different, this approach shows the strength of the NWFE approach in that it is losing more variance as we retain smaller numbers of transformed features. At the other extreme, it will suffer the same as full maximum likelihood when the dimensionality is high because the resulting covariance estimates are based on too many features.

Table 8.3 and Figure 8.3 show the results (average classification accuracy on testing data over all classes) of mapping the data into the five classes shown in Table 8.4, with varying numbers of features. Also shown in Table 8.3 is the number of features finally used (at the 99% variance level) for the NWFE (b) technique. The block-based maximum likelihood approach segmented the covariance matrix into three band ranges—namely, 1–102, 111–148, and 163–220—chosen by inspection of the global correlation matrix for the image, shown in Figure 8.4a. Only the diagonal elements were kept for those bands not in the blocks specified. Those same block definitions were kept as the number of bands was subsampled to provide the various reduced sets of bands.

Several aspects of these results are salient. First, the performance of the unmodified maximum likelihood classification is acceptable until the number of features reaches about 170, comparable to half the number of training samples per class. Clearly, beyond that number there are serious problems in the estimation of the class signatures, as expected. Second, all other methods perform well and

TABLE 8.3. Classification Accuracies for the Indian Pines Data Set (Percent) with Five Classes

Number of Bands	Standard Maximum Likelihood	Regularization	NWFE (a)	NWFE (b)	Block Based Maximum Likelihood	NWFE (b) features
220	54.7	82.5	58.3	86.4	85.43	113
198	69.3	83.5	71	87.4	85.3	99
176	78.9	82.5	80.1	87.3	85.56	86
165	83.7	84	82.9	87	85.17	79
147	83.1	82.3	84.1	87.1	84.65	69
110	88.3	84.9	86.5	88.1	85.17	47
74	86.5	82	88.1	88.7	84.65	30
55	87.5	84.5	87	88.2	84.78	20
44	88.1	83.7	87.3	87.9	83.99	16

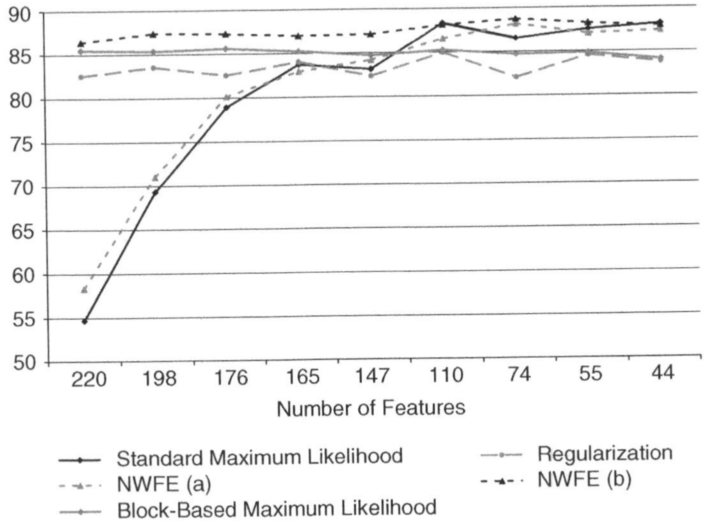

Figure 8.3. Classification results on the Indian Pines data set with five classes.

comparably over the range of a number of features apart from NWFE (a), which suffers in the same way as straight maximum likelihood, as expected.

A second test was carried out with a pair of difficult-to-separate classes (corn and soybean—no till), as summarized in Table 8.5. Classification performance as a function of numbers of features is seen in Table 8.6 and Figure 8.5. In this case it can be seen that none of the procedures performs very well, but the others are almost always better than standard maximum likelihood. Interestingly, the block-based method improves with fewer features.

It is interesting to note in Figure 8.5 how well the regularization approach works over the full range of feature sets; it exhibits a robustness not seen with the other techniques, and it performs better—with the exception of the block-based method, when the number of features is small. Again, the standard maximum likelihood method, and those classifications based on features selected through NWFE, behave relatively poorly for large feature sets, and they only show improvement

TABLE 8.4. Five Classes in the Indian Pines Data Set, with Numbers of Training and Testing Pixels

Class	Training Pixels	Testing Pixels
Grass	224	160
Hay	242	138
Woods	223	136
Corn—no till	252	170
Soybean—no till	234	158

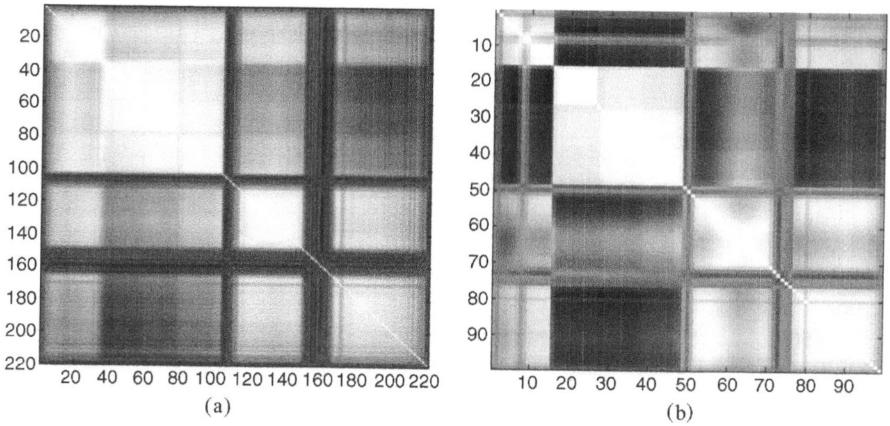

Figure 8.4. Correlation matrices for (a) the Indian Pines data set (blocks chosen defined by bands 1–102, 111–148, and 163–220) and (b) the Fremont data set (blocks chosen defined by bands 1–15, 16–26, 27–48, 49–70, and 76–99). For each data set, only the diagonal entries were kept for those bands not in the blocks.

TABLE 8.5. Two Classes, with Numbers of Training and Testing Pixels

Class	Training Pixels	Testing Pixels
Corn	233	196
Soybean—no till	234	158

TABLE 8.6. Classification Accuracies on the Indian Pines Data Set (Percent) with Two Difficult-to-Separate Classes

Number of Bands	Standard Maximum Likelihood	Regularization	NWFE (a)	NWFE (b)	Block-Based Maximum Likelihood	NWFE (b) Features
220	55.6	72.3	56.3	56.6	61.93	148
198	57.4	72.6	53	57.1	62.44	134
176	54.1	72.3	55.1	57.4	61.17	121
165	58.4	72.8	56.6	57.6	62.69	113
147	61.7	73.4	56.6	55.3	62.69	102
110	62.9	70.1	57.6	60.9	68.27	77
74	60.7	74.1	57.1	61.2	67.77	53
55	67	72.3	55.6	66.8	75.38	40
44	64.5	75.9	55.6	71.6	84.26	32

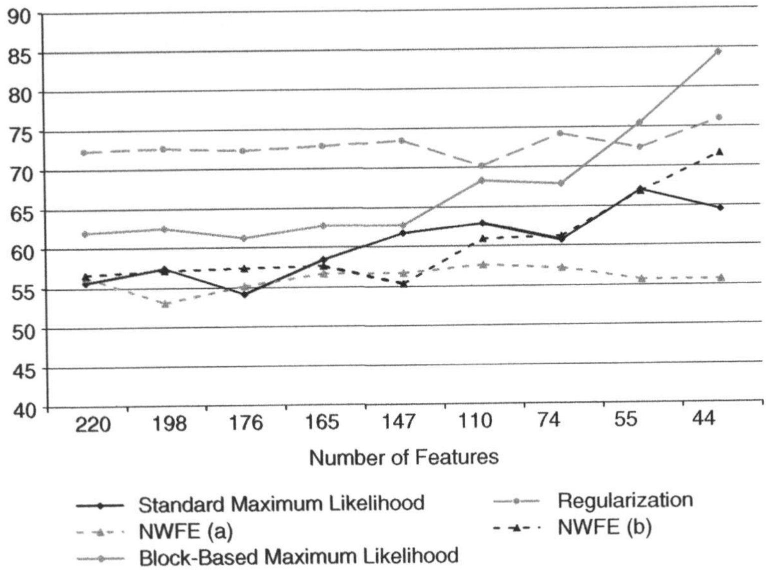

Figure 8.5. Classification results on Indian Pines data set with two difficult-to-separate classes.

when the number of features is approximately larger than half the number of training samples.

Experiment 3. A third test was performed using the Fremont data set involving three classes shown in Table 8.7. Table 8.8 and Figure 8.6 summarize the results obtained, essentially reflecting the results found with the five class Indian Pines exercise. The block boundaries for the block-based maximum likelihood approach are given in Figure 8.4b.

It is clear from these experiments that each of the candidate methods for rendering thematic mapping by maximum likelihood methods viable with hyperspectral data works well; they also yield comparable results on the data sets used. In endeavoring to choose among them, to see which may be the preferred approach, we turn our attention now to ease of training.

TABLE 8.7. Three Classes in the Fremont Data Set, with Numbers of Training and Testing Pixels

Class	Training Pixels	Testing Pixels
Oak woodland	205	206
Salt evaporation	215	161
Industrial	239	153

TABLE 8.8. Classification Accuracies on the Fremont Data Set (Percent) with Three Classes

Number of Bands	Standard Maximum Likelihood	Regularization	NWFE (a)	NWFE (b)	Block-Based Maximum Likelihood	NWFE (b) Features
198	57.1	99.4	57.7	98.7	99.42	74
179	67.5	99.4	73.8	99	99.23	68
159	83.1	99.6	80.6	99.2	99.62	62
149	90.2	99.6	86.3	99	99.62	58
132	92.9	99.8	90.4	99.4	99.81	52
99	98.1	99.6	96.3	99.4	99.81	42
66	99.6	100	98.7	99.4	99.81	28
50	100	100	99	99.8	100	22

Apart from the block-based simplification of maximum likelihood, the regularization and NWFE-based approaches require significant degrees of processing during training. In the case of regularization that amounts to means for finding the weights α_i in Eq. (8.3), while in the case of feature reduction via the NWFE transformation, compilation of the scattering matrices, eigenanalysis and transformation steps are needed. As an indication of the demand of these steps, we examine the time demand to finalize training. Figure 8.7 shows the total CPU time required (based on running MultiSpec on a 1.33-GHz PowerPC G4 processor)

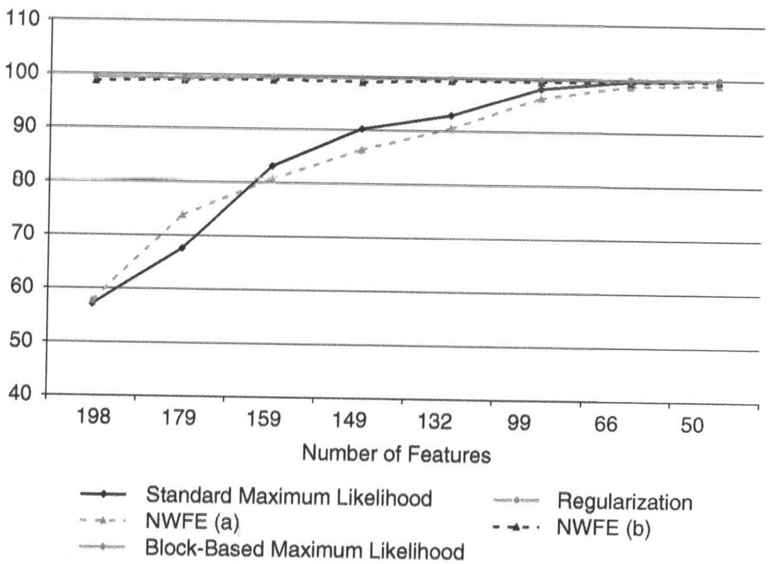

Figure 8.6. Classification results on the Fremont data set, with three classes.

for the regularization, NWFE, and standard maximum likelihood training phases as a function of number of bands and number of classes. As is to be expected, the NWFE technique is the most time-demanding, followed closely by regularization, both of which are substantially slower than the training step for standard maximum likelihood classification. While the training time for block-based maximum likelihood classification has not been shown explicitly, it will be comparable to that for the standard approach, the only additional effort required being that

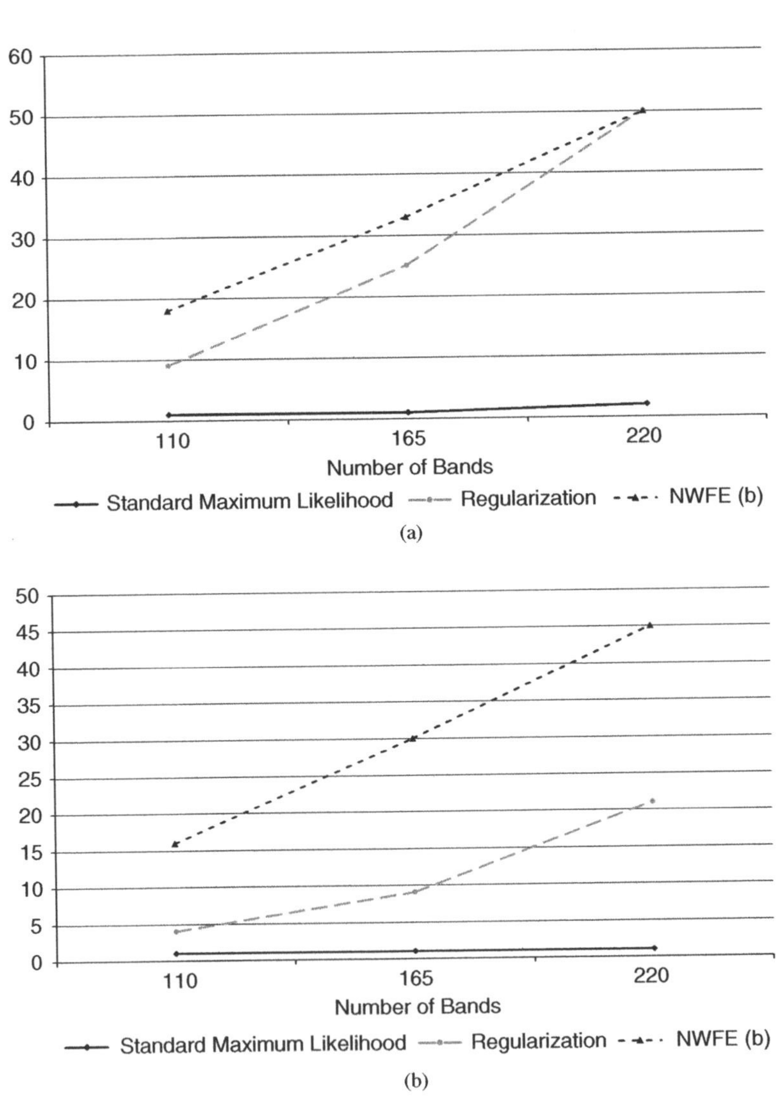

Figure 8.7. Classification time (as seconds of CPU usage) versus number of bands for (a) five classes and (b) two classes and (c) as a function of the number of classes for 220 bands.

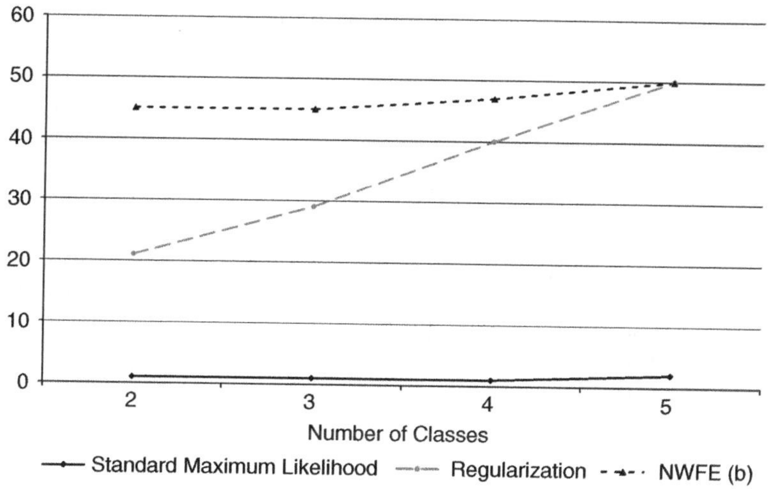

(c)

Figure 8.7. (*Continued*)

of preselecting the block boundaries to use. That is usually a simple manual task, although it could be automated by the application of edge detection methods to the correlation matrix shown in image form in Figure 8.2.

It is also important to consider a comparison of classification times. This will not be influenced by the size of the image or the number of classes, but will depend principally on the number of bands. Table 8.9 gives a simple comparison based on how each technique lines up against the classification time required for standard maximum likelihood classification. Some assumptions are made concerning the feature sets used to make such a comparison possible.

8.6. A QUALITATIVE EXAMINATION

It is possible to do a simple qualitative comparison among the methods, which provides some guidance to a future more rigorous analysis. Table 8.10 shows such an

TABLE 8.9. Relative Classification Times

Technique	Classification Time
Standard maximum likelihood	100%
Regularization	100%
NWFE (b)	44% with 2/3 features kept
Block-based maximum likelihood	50% with 2 equi-sized blocks
Block-based maximum likelihood	33% with 3 equi-sized blocks

TABLE 8.10. Qualitative Comparison of the Methods

Method	Description	Advantages	Disadvantages
Transforms (PCA, etc.)	Based on global covariance information; transformed axes with low variance can be ignored.	• Simple to apply • Global covariance matrix is generally well-conditioned.	• Best transformed axes usually don't align with class separability.
NWFE	Distance-weighted within-class and among-class scattering matrices are computed and used to define an axis transformation, from which the most significant subset of features is selected.	• Well-behaved • Distribution-free	• Requires moderately complex scattering matrices to be compiled. • Scattering matrix estimation may be affected by dimensionality problem, in which case regularization methods may be required as well.
Regularization	The class conditional covariance matrix is approximated by the weighted sum of global and class-conditional measures.	• Good results can be obtained with very simple approximations • Global covariance matrix is well-enough conditioned to render method viable. • Process can be iterated to provide improved results.	• Weighting coefficients need to be found, often involving a detailed LOOC approach. • Weighting coefficients need to be established for each new application. • Training process can be time-consuming.
Block-based	Block diagonal approximations to the class-conditional covariance matrices are found by looking for natural blocks of high correlation, globally.	• Dimensionality is reduced to that of the largest block. • Standard maximum likelihood algorithms can be used. • Simple to implement	• Block boundaries will be application-specific, and they may have to be identified with every new application. • Correlations among widely separated bands are ignored.
Cluster space	A set of clusters is found that act as the link between pixel vectors and information classes.	• Clustering is simple and does not require the use of second-order statistics. • The cluster conditional probabilities of pixel membership are one-dimensional.	• Clustering step may involve data of high dimensionality unless feature reduction through transformation is applied first. Thus, potentially suffers the same desadvantages as transformation approaches.

TABLE 8.10. (*Continued*)

Method	Description	Advantages	Disadvantages
		• It provides a statistical linkage between spectral and information classes.	• Cluster selection is important.
Diagnostic feature identification	The spectrum for a pixel is examined for absorption-like features that are felt to be diagnostic. The background spectrum is then removed.	• Dimensionality problem is obviated through the identification of diagnostic features. • Can handle mixtures. • Expert spectroscopic knowledge can be exploited.	• Requires careful identification and selection of diagnostic features. • Requires substantial expert knowledge and an extensive rule base.

examination and includes the principal components transformation, which is sometimes used to help overcome the dimensionality problem.

8.7. DISCUSSION

The poor generalization observed with most standard statistical classification procedures when applied to hyperspectral data is related directly to the sparseness of the data space and the unusual distribution of the data vectors within that space. It is straightforward to show, in principle, that the hyperspectral space is almost totally empty and that the concept of clusters and classes is dependent very much on whether the data are capable of being concentrated in such a sparse domain. Moreover, if the data are assumed to be uniformly distributed, then it can be shown that it is concentrated in the outer shell of the hyperspectral domain, while if it is assumed to be normally distributed, then it concentrates more toward the tails of the distribution functions [20]. As a consequence, retaining full dimensionality when seeking to apply standard thematic mapping techniques is not an option, and subspace transformations are essential. Thankfully, the very high degree of redundancy present usually means that substantial dimensionality reduction is possible, provided that care is taken. Alternatively, approximation techniques for the class-conditional covariances are viable.

In the study carried out here, it is clear that the candidate methods for handling hyperspectral data effectively based on maximum likelihood principles, in general, all work well in terms of achieving comparable classification accuracies. Where they differ, however, is in the complexity and timeliness of training. Depending on the number of bands and classes, the results above suggest that the successful approaches of regularization and feature reduction via the NWFE transformation can require an order-of-magnitude increase in training time compared with standard maximum likelihood classification and thus also the block-based variant of maximum likelihood.

An interesting aspect of the classification results presented in Figs 8.3 and 8.6 is that the ability of standard maximum likelihood classification to generalize breaks down when the number of features exceeds half the number of class-conditional training pixels; restated, this implies that reasonable results with standard maximum likelihood are being achieved whenever the number of training pixels per class(s) exceeds twice the number of bands, that is,

$$s \geq 2N$$

This is significantly fewer that the $10(N+1)$ generally considered necessary for reliable signature generation. While the less stringent result above has been derived on the basis of just a few trials involving only as many as five classes, it is nevertheless consistent with the observations of Van Neil et al. [20] based on four classes. However, more rigorous trials involving more complex data sets with a greater number of classes are needed before this smaller lower bound on the number of training samples is widely adopted.

Likewise, this early comparative analysis hyperspectral data analysis of techniques also needs more trials, and comparison to nonparametric methods, such as those based on neural networks and support vector machines.

ACKNOWLEDGMENT

The authors are grateful to David Landgrebe of Purdue University for making available the MultiSpec software used in the classification exercise presented here. They also wish to thank Bing Xu and Peng Gong of the University of California, Berkeley, for providing the Hyperion data set and the associated ground truth data.

REFERENCES

1. P. H. Swain and S. M. Davis, *Remote Sensing: The Quantitative Approach*, McGraw-Hill, New York, 1978.
2. J. A. Benediktsson, P. H. Swain, and O. K. Ersoy, Neural network approaches versus statistical methods in classification of multisource remote sensing data, *IEEE Transactions on Geoscience and Remote Sensing*, Vol. GE-28, pp. 540–552, 1990.
3. J. A. Gualtieri and R. F. Cromp, Support vector machines for hyperspectral remote sensing classification, *Proceedings of SPIE 27th AIPR Workshop on Advances in Computer Assisted Recognition*, Vol. 3584, 221–232, 1998.
4. G. F. Hughes, On the mean accuracy of statistical pattern recognizers, *IEEE Transactions on Information Theory*, Vol. IT-14, 56–63, 1968.
5. R. O. Duda, P. E. Hart, and D. G. Stork, *Pattern Classification*, 2nd edition, John Wiley & Sons, New York, 2001.

6. J. A. Richards and D. J. Kelly, On the concept of spectral class, *Remote Sensing Letters*, Vol. 5, 987–991, 1984.
7. J. A. Richards and X. Jia, *Remote Sensing Digital Image Analysis*, 3rd edition, Springer, Berlin, 1999.
8. M. D. Fleming, J. S. Berkebile, and R. M. Hoffer, Computer-aided analysis of Landsat-1 MSS data: A comparison of three approaches including a modified clustering approach, *Proceedings of Symposium on Machine Processing of Remotely Sensed Data*, LARS/ Purdue University, West Lafayette, IN, pp. 54–61, June 3–5, 1975.
9. X. Jia and J. A. Richards, Cluster space representation for hyperspectral classification, *IEEE Transactions on Geoscience and Remote Sensing*, Vol. 40, 593–598, 2002.
10. J. A. Richards, Is there a best classifier? Image and Signal Processing for Remote Sensing XI, *SPIE International Symposium on Remote Sensing Europe 2005*, Brugge, Belgium, September 19–22, 2005.
11. B. Jeon and D. A. Landgrebe, Classification with spatio-temporal interpixel class dependency contexts, *IEEE Transactions on Geoscience and Remote Sensing*, Vol. 30, 663–672,1992.
12. G. Camp-Valles and L. Bruzzone, Kernel-based methods for hyperspectral image classification, *IEEE Transactions on Geoscience and Remote Sensing*, Vol. 43, pp. 1351–1362, 2005.
13. F. Melgani and L. Bruzzone, Classification of hyperspectral remote sensing images with support vector machines, *IEEE Transactions on Geoscience and Remote Sensing*, Vol. 42, pp. 1778–1790, 2004.
14. R. N. Clark, G. A. Swayze, K. E. Livio, R. F. Kokaly, S. J. Sutley, J. B. Dalton, R. R. McDougal, and C. A. Gent, Imaging spectroscopy: Earth and planetary remote sensing with the USGS Tetracorder and expert systems, *Journal of Geophysical Research*, Vol. 108 (E12), pp. 5131–5175, 2003.
15. B-C. Kuo and D. A. Landgrebe, Nonparametric weighted feature extraction for classification, *IEEE Transactions on Geoscience and Remote Sensing*, Vol. 42, 1096–1105, 2004.
16. D. A. Landgrebe, *Signal Theory Methods in Multispectral Remote Sensing*, John Wiley & Sons, Hoboken, NJ, 2003.
17. B-C. Kuo and D. A. Landgrebe, A robust classification procedure based on mixture classifiers and nonparametric weighted feature extraction, *IEEE Transactions on Geoscience and Remote Sensing*, Vol. 40, pp. 2486–2494, 2002.
18. X. Jia, Classification of Hyperspectral Data Sets, Ph.D. thesis, The University of New South Wales, Kensington, Australia, 1996.
19. X. Jia and J. A. Richards, Efficient maximum likelihood classification for imaging spectrometer data sets, *IEEE Transactions on Geoscience and Remote Sensing*, Vol. 32, pp. 274–281, 1994.
20. T. G. Van Neil, T. R. McVicar, and B. Datt, On the relationship between sample size and data dimensionality: Monte Carlo analysis of broadband multi-temporal classification, *Remote Sensing of Environment*, Vol. 98, pp. 468–480, 2005.

CHAPTER 9

OPTIMAL BAND SELECTION AND UTILITY EVALUATION FOR SPECTRAL SYSTEMS

SYLVIA S. SHEN
The Aerospace Corporation, Chantilly, VA, USA

9.1. INTRODUCTION

Hyperspectral remote sensing technology has advanced rapidly in the last two decades. Sensors have been built that provide data with higher spectral fidelity than the traditional multispectral systems. While these fine spectral resolution sensors facilitate accurate detection and identification, the high volume and dimensionality of the data substantially increase the transmission bandwidth and the computational complexity of analysis. Additionally, redundancy in the hyperspectral data exists due to strong correlation between adjacent spectral bands. It is therefore desirable to have a method to optimize the selection of spectral bands to reduce the data dimensionality and, at the same time, maintain the distinct spectral features necessary for target discrimination. Optimization of spectral bands as described in this chapter is an important preliminary analysis tool in the evolution of sensor development, data extraction, and the development of spectral data analysis products for specific applications.

A commonly applied technique for dimensionality reduction is the principal component analysis or the Karhunen–Loeve transform of the scene data. The principal component analysis is an old tool in multivariate statistical data analysis. The principal components are the eigenvectors of the covariance matrix. The projection of data onto the principal components is called the principal component transform or the Karhunen–Loeve transform [1, 2] or the Hotelling transform [3]. The construction of the principal components is described below.

Let **X** denote the n-dimensional random vector representing observations from a population. (In spectral data analyses, n is the number of spectral bands.)

$$\mathbf{X} = (x_1, \ldots, x_n)^t \tag{9.1}$$

Hyperspectral Data Exploitation: Theory and Applications, Edited by Chein-I Chang

Let $\boldsymbol{\mu}_\mathbf{X}$ and $\Sigma_\mathbf{X}$ denote the mean vector and the covariance matrix of $\mathbf{X}$.

$$\boldsymbol{\mu}_\mathbf{X} = \mathbf{E}\{\mathbf{X}\} \tag{9.2}$$

$$\Sigma_\mathbf{X} = \mathbf{E}\{(\mathbf{X} - \boldsymbol{\mu}_\mathbf{X})(\mathbf{X} - \boldsymbol{\mu}_\mathbf{X})^t\} \tag{9.3}$$

From a sample of observation vectors $\mathbf{X}_1, \ldots, \mathbf{X}_M$, the sample mean vector $\mathbf{m}_\mathbf{X}$ and the sample covariance matrix $\mathbf{C}_\mathbf{X}$ can be calculated as estimates of the mean vector $\boldsymbol{\mu}_\mathbf{X}$ and the covariance matrix $\Sigma_\mathbf{X}$:

$$\mathbf{m}_\mathbf{X} = \frac{1}{\mathbf{M}} \sum_{i=1}^{\mathbf{M}} \mathbf{X}_i \tag{9.4}$$

$$\mathbf{C}_\mathbf{X} = \frac{1}{\mathbf{M} - 1} \sum_{i=1}^{\mathbf{M}} (\mathbf{X}_i - \mathbf{m}_\mathbf{X})(\mathbf{X}_i - \mathbf{m}_\mathbf{X})^t \tag{9.5}$$

The principal components are the eigenvectors of the sample covariance matrix $\mathbf{C}_\mathbf{X}$. The eigenvectors e_i and the corresponding eigenvalues λ_i are the solutions to the eigen-equation

$$\mathbf{C}_\mathbf{X} e_i = \lambda_i e_i \tag{9.6}$$

or the eigen-equation in matrix notation:

$$|\mathbf{C}_\mathbf{X} - \lambda \mathbf{I}| = 0 \tag{9.7}$$

In principal component analysis, the eigenvectors are usually ordered according to descending eigenvalues. These principal components have certain desirable properties by way of their construction. Among these properties is the fact that the first principal component is in the direction where the data have the largest spread and the second principal component is in the direction where the data have the second largest spread and are orthogonal to the first principal component. Therefore a set of a few leading principal components capturing most variation of the data are often used in data analysis. However, since these principal components are linear combinations of the original spectral bands, they lack any physical interpretation and they are difficult, if not nearly impossible, to realize in sensor system designs. Similar conclusions can be reached about their use and applicability to specific applications and spectral products. Therefore, an alternative approach is warranted, in which specific wavelength regions can be identified that correspond to observable features in objects of interest. The general term for this approach is band selection.

Several band selection methods have been investigated in the recent literature. A two-stage method that selects a subset of bands having the largest canonical correlation with the principal components has been suggested by Velez-Reyes et al. [4]. Gruninger et al. [5] have proposed a band selection method based on an endmember analysis technique. Withagen et al. [6] have selected bands for a multispectral 3CCD camera based on a Mahalanobis-distance-based metric and the classification results.

Target detection and material identification are two primary application areas for the use of hyperspectral data. A remote sensing instrument measures the interaction of electromagnetic radiation with a specific material (solid, liquid, or gas) or the self-emission of a material. Hence the performance and utility of spectral data in target detection and material identification clearly depends on whether the measured data captured the electromagnetic wavelengths where the target materials of interest have distinct characteristics. An optimal band selection technique was developed to address the needs for target detection and material identification [7], which selects spectral bands that permit the best material separation. A detailed description of this optimal band selection technique and the utility assessment of the selected bands constitute the remainder of this chapter.

9.2. BACKGROUND

Redundancy in the hyperspectral data exists due to strong correlation between adjacent spectral bands. More specifically, spectra of solids and liquids, even when viewed through a long atmospheric path, generally do not vary rapidly as a function of wavelength. For this reason, spectral resolution beyond a certain value may provide little or no extra information needed for target detection or material identification. The question then becomes, How many spectral bands do we really need and where should we place these bands? The answer depends on the application for which we are selecting and optimizing spectral bands. The types of applications that are of primary interest are target detection, discrimination, and material identification. For these types of applications, material separation is the key to success. If we can determine the spectral band wavelength locations that can separate a large collection of materials, then target detection, discrimination, and material identification can be achieved with high success rates. To this end, a study was conducted with the following two objectives. The first objective was to determine, for any given number of bands, where (wavelength locations) the spectra of target materials of interest differ from each other. The second objective was to assess the relative performance afforded by the various band sets to determine the minimum number of bands needed to achieve satisfactory target detection and material identification. An optimal band selection technique was developed to achieve the overall goal of determining the minimum number of bands, along with their placement, needed to separate target materials.

The basic principle behind this technique is to optimally select spectral bands that provide the highest material separability. As will be described in Section 9.3, this technique combines an information-theory-based criterion for band selection with a genetic algorithm to search for a near-optimal solution. This methodology was applied to 612 material spectra from a combined database to determine the band locations for 6-, 9-, 15-, 30-, and 60-band sets in the 0.42- to 2.5-μm spectral region that permit the best material separation.

The optimal band locations are given in Section 9.4.1. In the subsequent subsections, these optimal band locations for the 6-, 9-, and 15-band sets are compared to the bands of existing multiband systems such as Landsat 7, Multispectral Thermal Imager, Advanced Land Imager, Daedalus, and M7.

In Section 9.5, these optimal band sets are evaluated in terms of their utility related to anomaly/target detection and material identification using multiband data cubes generated from two HYDICE [8, 9] cubes. Comparisons are made between the exploitation results obtained from these optimal band sets and those obtained from the original 210 HYDICE bands. The assessment of the relative performance afforded by the various band sets allowed the determination of the minimum number of bands needed to produce satisfactory target detection, discrimination, and material identification.

Since this technique selects the actual spectral band wavelength locations for any given number of bands that permit the best material separation, it can be used in system design studies to provide an optimal sensor cost, data reduction, and data utility trade-off relative to a specific application.

9.3. THEORY/METHODOLOGY

This band selection technique uses an information-theory-based criterion for selecting the bands. The criterion determines the entropy-based information [10] contained in the selected bands and the degree of separation of the material spectra. The principle for selection is based on the following. For the multiple band data to have utility and be informative, measurements with the selected bands must be capable of separating different materials or classes of materials. The greater the separation, the more useful the bands are. Entropy is a measure of information contained in the selected bands. A higher entropy indicates that more information is contained in a particular band set and therefore a higher degree of separation.

Clearly, entropy is a function of the quantization setting of the spectral values. If one quantizes the spectral values, there are a finite number of possible values that each component of a spectral vector can assume. As the quantization is made coarser, numerically similar values will become indistinguishable as their quanta become equal (mapped into the same integer value). Each setting of quantization has associated with it a histogram of an n-dimensional vector (where n is the number of bands selected) from which the entropy can be computed for that quantization setting. Entropy is a measure of the amount of information contained within the histogram of the n-dimensional spectral vectors. By adjusting the coarseness of quantization and the choice of bands, the degree of separation between materials can be measured.

In order to find the band set (wavelength locations) for a specific number of bands having the highest entropy, a genetic algorithm was used for global search, combined with a terminal exhaustive local search. A genetic algorithm [11, 12] is an optimization technique that can search for the optimum case without evaluating all the candidate cases. It is an example of a more general class of methods called stochastic optimization [13]. These techniques are useful when the search space is too large and has too complicated a structure to permit the use of a method from the gradient descent family. Gradient descent optimization techniques search for a local maximum (or minimum) of a function by taking steps proportional to the positive (or negative) gradient of the function at the current point [14]. Stochastic optimization seeks to search more thoroughly the space without being trapped in a local optimum. Genetic algorithms can be particularly effective in finding solutions where the individual pieces of the solution are important in combination, or where a sequence is important. There is no guarantee of finding the global optimum, but if the genetic algorithm is well-designed, at least a number of important near-optimal solutions can be generated. Once the genetic algorithm has reduced the solution space, an exhaustive local search is employed to improve the final solution.

9.3.1. Information-Theory-Based Criterion for Band Selection

The selection criterion or the fitness function/objective function used in the genetic algorithm optimization is the entropy measure relative to some quantization bin width. If a particular choice of n band wavelengths $\lambda = (\lambda_1, \lambda_2, \lambda_3, \ldots, \lambda_n)$ is made and if Q is the quantization binwidth, then the reflectance spectrum $\mathrm{Ref}_k(\lambda)$ of the kth material in the signature database associated with band set λ and quantization Q is represented by the following n-dimensional discrete vector:

$$V[\mathrm{Ref}_k(\lambda)] = \left[\mathrm{Int}\left(\frac{\mathrm{Ref}_k(\lambda_1)}{Q}\right), \mathrm{Int}\left(\frac{\mathrm{Ref}_k(\lambda_2)}{Q}\right), \ldots, \mathrm{Int}\left(\frac{\mathrm{Ref}_k(\lambda_n)}{Q}\right)\right] \quad (9.8)$$

where $\mathrm{Int}(\cdot)$ represents the operation of taking the nearest integer value. The entropy H associated with the band set λ is

$$H(\lambda) = -\sum p(V) \log_2(p(V)) \quad (9.9)$$

where $p(V) = \#(V)/N$, and the summation is taken over all discrete vectors V in n-dimensional space where $\#(V) > 0, \#(V)$ is the number of histogram counts associated with the vector V, and N is the total number of counts (i.e., the total number of material reflectance spectra in the database). Clearly, H is a function of the band set λ, quantization Q, and the database used.

This entropy calculation was applied to a collection of 612 spectra representing man-made and natural materials. A material signature database was first generated by combining the NEF (nonconventional exploitation factors) database [15] and the ground measurements taken by the Topographic Engineering Center (TEC) and other organizations during the various HYDICE collection campaigns (e.g., desert radiance, forest radiance, urban radiance, island radiance, alpine radiance, etc.). The combined database was then pruned to eliminate duplicates to arrive at the final set of 612 spectra.

The material spectra in the NEF database cover wavelengths from 0.3 to 15.0 μm. The database is wavelength sampled differently in three regions. Reflectance spectra are sampled at 2-nm resolution from 0.3 to 0.8 μm, at 20-nm resolution from 0.8 to 5.0 μm, and at 100-nm resolution from 5.0 to 15.0 μm.

The TEC's ground measurements taken during the various HYDICE collection campaigns cover wavelengths from 0.35 to 2.5 μm. These measurements are sampled at 5-nm resolution.

Since the different wavelength sampling resolution will cause the band selection procedure to put more emphasis in the finer sampled spectral region, we should ideally resample all spectra in the combined database to a common resolution to avoid biasing the selection. For example, 20 nm, the least common multiple of 2, 5, and 20 nm (the three sampling resolutions over the 0.42- to 2.5-μm spectral region originally used in the combined database), would be the ideal resolution at which all spectra should be resampled. However, in order to have sufficient number of bands to select from, all spectra in the combined database were resampled to 10-nm resolution in the 0.42- to 0.8-μm region and to 20-nm resolution in the 0.8- to 2.5-μm region, yielding 124 spectral bands to choose from.

The entropies were calculated for band sets $\lambda = (\lambda_1, \lambda_2, \lambda_3, \ldots, \lambda_n)$ with bands λ_j selected from the 124 spectral bands.

9.3.2. Radiative Transfer Considerations

It is important to consider that airborne or spaceborne systems sense radiance energy propagated through an atmosphere at the system's entrance aperture. Material spectral libraries in the 0.4- to 2.5-μm region, either laboratory- or field-collected, are typically in reflectance units. For this reason, the 612 material reflectance spectra in the combined database were first converted to total spectral radiances at the aperture. The conversion was accomplished based on an interpolation of 20 MODTRAN [16] runs for a particular time of day and sun position. The entropy calculations were then performed on the 612 adjusted material radiance spectra. More detail on this reflectance to radiance conversion is described below.

Recognizing that spectra of remotely sensed targets are altered by the effects of the atmosphere and the characteristics of the sun, reflectance measurements were adjusted to account for the effects of solar illumination, atmospheric down-welling, reflectance, and propagation up through the atmosphere to simulate the radiance spectra seen at the sensor. The adjustment made to the 612 spectra in the combined database is as follows. The spectral radiance seen at the sensor can be expressed by the following equation:

$$\begin{aligned}&\text{Total spectral radiance at aperture } (\lambda)\\&\quad = \text{Atmospheric transmission } (\lambda) \times \text{Ground-reflected radiance}(\lambda, \text{albedo})\\&\qquad + \text{Path-scattered radiance}(\lambda)\\&\qquad + \text{Atmospheric radiance}(\lambda)\end{aligned} \tag{9.10}$$

where λ represents wavelength.

Since ground-reflected radiance is a nonlinear function of albedo, 20 MODTRAN runs were made for a range of albedos (0.01–0.95 at a spacing of 0.05) and for the wavelength region chosen for this analysis (0.42–2.5 μm). To adjust a material reflectance from the combined database for a given wavelength, the ground-reflected radiance term in the above equation was interpolated using the ground-reflected radiance of the two albedos that are closest to that material reflectance at that wavelength. The interpolated value was multiplied by the atmospheric transmission, and the corresponding path scattered radiance and the atmospheric radiance were added. This procedure simulates what the database reflectance measurements would look like at an entrance aperture outside the atmosphere. The entropy-based band selection procedure was then performed on the 612 adjusted material radiance spectra.

9.3.3. The Genetic Algorithm

The genetic algorithm works on an individual where an individual in this study is a band set. A band set is a subset of the 124 bands.

From this point, the optimization proceeds as a basic genetic algorithm as described in Goldberg [12]. Each individual is selected for "mating" with others in the population based on the fitness function/objective function related to its entropy-based information content.

Each parent in the mating contributes the lower portion of its set of band wavelengths to one of a pair of "offspring" and contributes the upper to another. This process mimics the biological process of genetic crossover during cell meiosis. The point at which each parent's band set is separated is a random variable. However, the crossover point is set to be the same for both parents in order to keep the length of the band sets constant. This rule tends to propagate strings of "successful" bands so that observations at these wavelengths will separate the material spectra in the database. New wavelengths may enter the population in a step that mimics the biological process of mutation. The mutation point is once again a random variable.

A diagram that illustrates the crossover and mutation processes using example band sets of 7 bands is given below. (Numbers indicated are band wavelength locations in microns.)

PARENTS (MATING)

A	0.56	0.65	0.70	1.00 ×	1.14	1.26	2.24
B	0.44	0.52	0.74	0.90 ×	1.31	1.56	1.78

(x denotes crossover point)

OFFSPRING (FROM CROSSOVER)

C	0.56	0.65	0.70	1.00	1.31	1.56	1.78
D	0.44	0.52	0.74	0.90	1.14	[1.26]	2.24

([ ] denotes site of a mutation)

OFFSPRING (FROM MUTATION)

C	0.56	0.65	0.70	1.00	1.31	1.56	1.78
D'	0.44	0.52	0.74	0.90	1.14	1.72	2.24

The genetic algorithm optimization procedure was executed as follows. First, for a specific number of bands (6, 9, 15, etc.), a set of 40 individuals was randomly selected. An individual is a band set (of 6, 9, 15, etc., bands). Two individuals were then randomly selected to form a parent pair. Forty pairs were selected. Each pair went through the mating process described above to produce two offspring. The better offspring of the two (i.e., the offspring with the higher entropy) was retained. After all 40 pairs went through the mating process, 40 offspring were retained. This process is termed "one generation." After the first generation, these 40 retained offspring went through the mating process to produce another set of 40 offspring (i.e., second generation). The process evolved for 40 generations. The best offspring (i.e., the offspring with the highest entropy from all generations) was retained. The entire 40-generation process is repeated eight times, each time starting with 40 randomly selected individuals and ending with the best offspring. The best five offspring from executing the 40-generation evolution process eight times were introduced to the terminal local search process to arrive at the final solution. This local search involved moving each solution wavelength by the smallest feasible increment until no improvements in entropy could be obtained. The best band set emerging from the terminal local search was the final solution.

For this study, the band optimization procedure described above was run separately to obtain optimal band sets of 6, 9, 15, 30, and 60 bands.

9.4. OPTIMAL BAND SET RESULTS

The first objective of this optimal band selection study was to determine, for any given number of bands, where (wavelength locations) the spectra of target materials of interest differ from each other. The second objective was to assess the relative exploitation performance afforded by the derived band sets to determine the minimum number of bands needed. To achieve these objectives, the band optimization procedure described in Section 9.3 was run to obtain optimal band sets comprised of 6, 9, 15, 30, and 60 bands out of a total of 124 bands. The band locations of these optimal band sets are shown in Section 9.4.1. Comparisons of the band locations of these optimal band sets with those of several existing multispectral systems are given in Section 9.4.2. These comparisons should provide insight into performance of these systems against scenes containing materials like those in the database used for this study.

9.4.1. Information-Theory-Based Optimal Band Sets

When executing the band optimization procedure separately to obtain optimal band sets of 6, 9, 15, 30, and 60 bands, a quantization binwidth of 10^{-3} W/ $(\text{cm}^2 \cdot \text{ster} \cdot \mu\text{m})$ was chosen. This quantization setting resulted in entropies ranging from 6.057 to 8.713 for these optimal band sets, out of a maximum of 9.26. (*Note*: The maximum entropy is achieved when the 612 atmospherically adjusted material radiance spectra are distinct. The entropy in this case is $H(\lambda) = \log_2 612 = 9.26$.)

The band locations and entropies of the optimal band sets are given Table 9.1.

TABLE 9.1. Optimal Band Sets and Their Entropies

Number of Bands	Entropy	Band Locations (in Microns)									
60	8.713	0.43	0.44	0.47	0.48	0.49	0.50	0.51	0.55	0.56	0.57
		0.58	0.59	0.63	0.64	0.65	0.66	0.68	0.69	0.71	0.72
		0.73	0.74	0.75	0.76	0.77	0.78	0.79	0.80	0.82	0.84
		0.86	0.90	0.98	1.04	1.06	1.08	1.10	1.22	1.24	1.26
		1.28	1.46	1.54	1.56	1.58	1.60	1.62	1.64	1.80	1.82
		1.86	1.88	1.92	1.94	1.96	1.98	2.08	2.14	2.18	2.38
30	8.429	0.43	0.48	0.49	0.51	0.55	0.56	0.57	0.67	0.72	0.74
		0.75	0.76	0.78	0.79	0.80	0.82	0.84	0.86	1.04	1.06
		1.08	1.24	1.28	1.56	1.62	2.16	2.38	2.40	2.44	2.48
15	7.521	0.48	0.49	0.56	0.68	0.72	0.75	0.78	0.79	0.84	1.02
		1.08	1.24	1.96	2.16	2.40					
9	6.633	0.49	0.56	0.68	0.75	0.79	0.88	1.02	1.24	2.22	
6	6.057	0.49	0.56	0.67	0.75	0.88	1.06				

9.4.2. Comparison of Optimal Band Sets to Existing Multispectral Systems

In this section, the band locations of the various information-theory-based optimal band sets are compared to the band locations of several existing multiband systems. Table 9.2 shows the bands for Landsat-7 ETM+, Multispectral Thermal Imager (MTI), Advanced Land Imager (ALI), Daedalus AADS 1268, and M7. The first three are space-based systems. The latter two are airborne systems. All of these systems are wide-band multispectral systems. All but ALI have spectral bands outside of the study spectral region of 0.42–2.5 μm. Those bands are not compared. The bands that are in the study spectral region of 0.42–2.5 μm are shown in bold in Table 9.2.

In the following three subsections, optimal band sets of 6, 9, and 15 bands are each compared to existing multispectral systems that have comparable number of bands.

Note: In comparing the optimal band sets to existing multispectral systems, it is important to be aware of the functional criteria behind the development and use of a particular multispectral sensor and the criteria used in deriving the optimal bands sets.

9.4.2.1. Optimal Band Set of 6 Bands. Six out of the seven Landsat-7 bands are within our study spectral region of 0.42–2.5 μm. Comparing the band locations of the information-theory-based optimal band set of six bands with the six Landsat-7 bands, Figure 9.1 shows that four out of our six optimal band locations fall within the Landsat bands.

TABLE 9.2. Band Locations in Microns for Landsat-7, MTI, ALI, Daedalus, and M7

Landsat-7		MTI		ALI		Daedalus		M7	
1	**0.45–0.52**	A	**0.45–0.52**	1′	**0.433–0.453**	1	**0.42–0.45**	1	**0.45–0.47**
2	**0.53–0.61**	B	**0.52–0.60**	1	**0.450–0.515**	2	**0.45–0.51**	2	**0.48–0.50**
3	**0.63–0.69**	C	**0.62–0.68**	2	**0.525–0.605**	3	**0.51–0.59**	3	**0.51–0.55**
4	**0.78–0.90**	D	**0.76–0.86**	3	**0.630–0.690**	4	**0.58–0.62**	4	**0.55–0.60**
5	**1.55–1.75**	E	**0.86–0.90**	4	**0.775–0.805**	5	**0.61–0.66**	5	**0.60–0.64**
6	10.40–12.50	F	**0.91–0.97**	4′	**0.845–0.890**	6	**0.63–0.73**	6	**0.63–0.68**
7	**2.09–2.35**	G	**0.99–1.04**	5′	**1.200–1.300**	7	**0.71–0.82**	7	**0.68–0.75**
		H	**1.36–1.39**	5	**1.550–1.750**	8	**0.81–0.95**	8	**0.79–0.81**
		I	**1.55–1.75**	7	**2.080–2.350**	9	**1.60–1.80**	9	**0.81–0.92**
		J	3.50–4.10			10	**2.10–2.40**	10	**1.02–1.11**
		K	4.87–5.07			11	8.20–10.5	11	**1.21–1.30**
		L	8.00–8.40			12	8.20–10.5	12	**1.53–1.64**
		M	8.40–8.85					13	**1.54–1.75**
		N	10.20–10.70					14	**2.08–2.20**
		O	**2.08–2.35**					15	**2.08–2.37**
								16	10.40–12.50

9.4.2.2. Optimal Band Set of 9 Bands. Ten out of the 15 MTI bands, all of the nine ALI bands, and 10 out of the 12 Daedalus bands are within our study spectral region of 0.42–2.5 μm.

Figure 9.1 shows that the band locations of seven out of the optimal band set of nine bands fall within the MTI bands. A slightly different group of seven out of the optimal set of nine bands fall within the ALI bands. Another slightly different group of seven out of the optimal nine bands fall within the Daedalus bands.

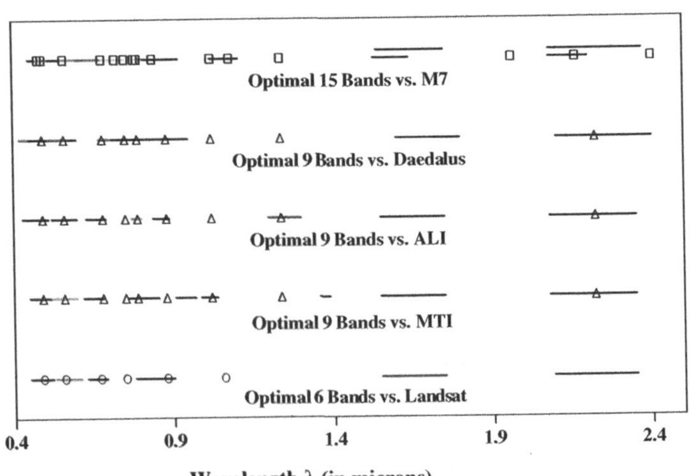

Figure 9.1. Comparisons of optimal band sets to existing multispectral systems.

TABLE 9.3. Comparisons Between Optimal Band Sets and Existing Multispectral Systems

Number of Bands	Band Locations (in Microns)										System
15	**0.48**	**0.49**	**0.56**	**0.68**	**0.72**	**0.75**	0.78	**0.79**	**0.84**	**1.02**	M7
	1.08	**1.24**	1.96	**2.16**	2.40						
9	**0.49**	**0.56**	**0.68**	**0.75**	**0.79**	**0.88**	1.02	1.24	**2.22**		Daedalus
9	**0.49**	**0.56**	**0.68**	0.75	**0.79**	**0.88**	1.02	**1.24**	**2.22**		ALI
9	**0.49**	**0.56**	**0.68**	0.75	**0.79**	**0.88**	**1.02**	1.24	**2.22**		MTI
6	**0.49**	**0.56**	**0.67**	0.75	**0.88**	1.06					Landsat 7

9.4.2.3. Optimal Band Set of 15 Bands. Fifteen out of the 16 M7 bands are within our study spectral region of 0.42–2.5 μm. Significant overlap exists between M7 Bands 12 and 13, as well as between M7 Bands 14 and 15. As a result, only 13 of the M7 bands are distinct bands. Figure 9.1 compares these 13 M7 bands with the band locations of our optimal band set for 15 bands, and we see that 12 out of the 15 optimal band locations fall within the M7 bands.

Table 9.3 summarizes the comparisons between optimal band sets of 6, 9, and 15 bands and the existing multispectral systems. For each optimal band set, a band location is denoted in boldface type if it falls within a spectral band of the multispectral system listed on the right-hand side of Table 9.3. It can easily be seen that there is a high correlation between the band locations of the various optimal band sets and the bands of these five existing multiband systems.

9.5. UTILITY ASSESSMENT OF OPTIMAL BAND SETS

To address the second objective, the utility of these information-theory-based optimal band sets was evaluated in terms of their exploitation performance. Two HYDICE (hyperspectral digital imagery collection experiment) scenes were selected for this evaluation. The HYDICE sensor [8, 9] is a nadir-viewing, push-broom imaging spectrometer. Reflected solar energy is measured along a ground swath approximately 1 km wide based on a design flight altitude of 6 km (20,000 ft). The ground sampling distance (GSD) varies from 0.75 to 3 m depending on the aircraft altitude above ground level. The spectral range of the instrument extends from the visible into the short wave infrared (0.4–2.5 μm). The spectrum is sampled in 210 contiguous channels, nominally 10 nm wide across the spectral range.

The two HYDICE cubes selected for use in this study were collected at 5000 ft with a GSD of 0.75 m. The first hyperspectral cube is a scene with a desert background; the second data cube is a forest scene. Both scenes contain a variety of targets: vehicles, fabric, calibration panels, and a large assortment of material panels (metals, painted plastic, painted wood, painted rubber, fabrics, building materials, and roofing materials) in five different sizes: 3 m × 3 m, 2 m × 2 m, 1 m × 1 m, 2.75 m × 2.75 m, and 2.4 m × 2.4 m.

To assess the relative utility of the optimal band sets shown in Table 9.1, the two HYDICE data cubes were used to generate multiband data cubes using the information-theory-based optimal band sets. Two nonliteral exploitation functions (i.e., anomaly detection and material identification) were then performed on these multiband data cubes. Performance was measured in terms of the anomaly detection and material identification success rates when the false alarm rate was held constant for all multiband data cubes.

9.5.1. Anomaly Detection

Anomaly detection is considered to be a type of surveillance problem that does not require the use of reference signatures of target materials of interest. Anomaly detection was performed by applying the linear unmixing algorithm to each multiband data cube, using background materials as endmembers [17–19]. A threshold was applied to the resulting residual image to produce a detection map. The thresholds were chosen empirically so that the false alarm rate (FAR) was constant for all multiband data cubes. FAR is defined as the number of false alarm pixels divided by the total number of pixels in the scene. FAR is kept constant so that a fair comparison can be made of the detection success rates across all multiband data sets. In this study, the thresholds for anomaly detection were chosen so that the FAR is 0.0004 for all cases.

Table 9.4 shows the anomaly detection results obtained from the multiband data cubes generated for the optimal band sets of 6, 9, 15, 30, and 60 bands as well as the original 210-band data cube for the desert scene. Table 9.5 shows the results from the forest scene.

TABLE 9.4. Anomaly Detection Results for the Desert Scene

		Number of Correctly Detected Targets					
Target Type	Number of Targets	Original 210-Band	60-Band	30-Band	15-Band	9-Band	6-Band
Vehicle	10	10	9	10	10	10	5
Fabric	7	7	6	6	5	4	4
Material panel							
3 m × 3 m	37	32	31	31	30	31	18
2 m × 2 m	10	10	10	10	8	9	6
1 m × 1 m	10	7	7	6	6	8	3
Total	74	66	63	63	59	62	36
Total number of pixels		288,000	288,000	288,000	288,000	288,000	288,000
Total number of FAs		130	132	131	133	118	129
FAR		0.0004	0.0004	0.0004	0.0004	0.0004	0.0004

TABLE 9.5. Anomaly Detection Results for the Forest Scene

Target Type	Number of Targets	Number of Correctly Detected Targets					
		Original 210-Band	60-Band	30-Band	15-Band	9-Band	6-Band
Vehicle	14	9	9	8	9	8	8
Fabric	3	3	3	3	3	3	2
Rubber object	3	2	2	2	2	2	2
Material panel							
3 m × 3 m	10	10	10	9	10	10	8
2 m × 2 m	10	9	9	9	10	9	7
1 m × 1 m	10	8	8	8	8	6	5
2.75 m × 2.75 m	3	3	3	3	2	3	2
2.4 m × 2.4 m	3	2	2	3	2	3	2
Total	56	46	46	45	46	44	36
Total number of pixels		288,000	288,000	288,000	288,000	288,000	288,000
Total number of FAs		136	119	123	120	127	134
FAR		0.0004	0.0004	0.0004	0.0004	0.0004	0.0004

The results in Tables 9.4 and 9.5 show that either no or some slight degradation in anomaly detection performance was observed between the 60-, 30-, 15-, or 9-band data and the original 210-band data for both the desert and forest scenes, with only one exception. The exception is the 15-band data for the desert scene where moderate degradation was observed. Significant degradation was observed for the 6-band data for both scenes. The results also show that there was no degradation in anomaly detection performance between the 60-band data and the original 210-band data for the forest scene.

9.5.2. Material Identification

Material identification is a type of spectral reconnaissance problem that uses known target signatures. Material identification was performed by first converting each multiband data set to apparent reflectances using the empirical line method, with coefficients derived from reflectance panels located in the scene. A background suppression in conjunction with a spectral matching technique [17–19] was used to identify each target material of interest using TEC's ground reflectance spectra as reference signatures. A threshold was used along with a filtered vector to produce a material identification map. In this study, the thresholds were chosen so that the FAR is 0.0004 for all materials of interest and for all multiband data sets.

Table 9.6 shows the material identification results obtained from the multiband data cubes generated for band sets of 6, 9, 15, 30, and 60 bands as well as the original 210-band data cube for the desert scene. Table 9.7 shows the results from the forest scene.

TABLE 9.6. Material Identification Results for the Desert Scene

		Number of Correctly Identified Targets					
Target Type	Number of Targets	Original 210-Band	60-Band	30-Band	15-Band	9-Band	6-Band
Vehicle	6	3	6	1	3	0	0
Fabric	1	1	1	1	0	0	0
Material panel							
3 m × 3 m	7	7	7	7	4	0	0
2 m × 2 m	4	4	4	4	2	0	0
1 m × 1 m	4	4	4	4	2	0	0
Total	22	19	22	17	11	0	0
Total number of pixels		288,000	288,000	288,000	288,000	288,000	288,000
Total number of FAs		129	127	127	124	128	129
FAR		0.0004	0.0004	0.0004	0.0004	0.0004	0.0004

TABLE 9.7. Material Identification Results for the Forest Scene

		Number of Correctly Identified Targets					
Target Type	Number of Targets	Original 210-Band	60-Band	30-Band	15-Band	9-Band	6-Band
Vehicle 1	4	4	4	4	4	4	0
Vehicle 2	3	3	3	3	1	0	0
Vehicle 3	4	3	3	2	0	2	0
Vehicle 4	3	3	3	1	0	0	0
Fabric	2	2	2	2	1	1	0
Material panel							
3 m × 3 m	3	3	3	3	1	2	0
2 m × 2 m	3	3	3	3	0	2	0
1 m × 1 m	3	3	3	3	0	0	0
2.75 m × 2.75 m	3	3	3	3	1	2	0
Total	28	27	27	24	8	13	0
Total number of pixels		288,000	288,000	288,000	288,000	288,000	288,000
Total number of FAs		128	128	127	129	124	129
FAR		0.0004	0.0004	0.0004	0.0004	0.0004	0.0004

Tables 9.6 and 9.7 show that no degradation in material identification performance was observed between the 60-band data and the original 210-band data for both scenes. Slight degradation was observed between the 30-band data and the original 210-band data for both scenes. Significant degradation was observed for 15-, 9-, and 6-band data sets. No targets were correctly identified using the 9- or 6-band data for the desert scene or using the 6-band data for the forest scene.

9.6. SUMMARY/CONCLUSIONS

The information-theory-based band selection methodology described in this chapter is a valuable tool for the development of both spectral sensor systems and their associated information related spectral products.

This band selection technique was applied to the 612 adjusted material spectra (adjusted to account for atmospheric effects) in a combined database to determine, for band sets of 6, 9, 15, 30, and 60 bands, the band locations that permit the best material separation.

Section 9.4 illustrated the high correlation between the band locations of the various optimal band sets and the bands of five existing multiband systems. This correlation is a very useful tool to predict the performance of any one of these existing multispectral sensor systems in extracting information from scenes containing materials like those in the material spectral database used for this study.

The optimal band sets were also evaluated in terms of their utility related to anomaly detection and material identification against well documented and "ground truthed" data collections.

The good anomaly detection performance shown in Section 9.5 indicates that the 60-, 30-, 15-, and 9-band locations, selected by our information-theory-based methodology, contain sufficient discriminating power to separate a large assortment of man-made materials from both the desert and forest background materials. Even with 9 bands, there is still enough distinct spectral information to separate man-made materials from natural background materials. However, the discriminating power diminishes significantly when the number of bands decreases to 6.

The material identification results shown in Section 9.5 indicate that a larger number of bands are needed to positively identify different material types than to separate man-made materials from natural background materials. Band sets of 15 bands or less do not contain sufficient spectral information to allow correct identification of different material types.

The material identification results also indicate that the 60-band locations, selected by our information-theory-based methodology, contain sufficient diagnostic spectral features to produce the same material identification performance as the original 210-band data. The result reconfirms the findings from an earlier study conducted by the author on spatial resolution, spectral resolution, and signal-to-noise trade-off [18, 19].

The optimal band location results determined by this method clearly depend on the nature of the material database used for measuring the entropy. The database should be well chosen with both materials to be identified and potential backgrounds present, and in appropriate abundances. As databases with more varied material types become available, the band locations for the different band sets can be updated. Nevertheless, this study showed that a subset of well-chosen 9 or more spectral bands performed exceedingly well in separating man-made materials from natural background materials. The study also showed that well-chosen 30 or more bands performed extremely well in positively identifying different material types. These results can provide significant insight into the development and optimization of multiband spectral sensors and algorithms for the preparation of specific information related products derived from spectral data.

More importantly, unlike the principal components, which are linear combinations of the original spectral bands and have no physical meaning, the optimal bands selected by this information-theory-based band selection technique are the actual wavelength bands that permit the best material separation. These optimal bands can be used in system design studies to provide an optimal sensor cost, data reduction, and data utility trade-off relative to a specific application or mission.

Multispectral sensor systems can be built using these optimal band locations. Hyperspectral sensor systems can selectively activate and transmit a subset of bands optimized for a specific scenario using this band selection technique. As the scenario changes, this technique can select a different set of spectral bands optimized for the new scenario and upload the new set of bands for the hyperspectral sensor system to activate and transmit. By using this technique to select a reduced number of optimal spectral bands, the data transmission bandwidth is reduced and the computational complexity is reduced, but the distinct spectral features needed for target discrimination is maintained.

REFERENCES

1. K. Karhunen, Uber Lineare Methoden in der Wahrsccheilichkeitsrechnung, *Annales Academiae Scientiarum Fennicae, Seried A1: Mathematica-Physica*, vol. 37, 1947.
2. M. Loéve, Probability Theory, Van Nostrand, New York, 1963.
3. H. Hotelling, Analysis of a complex of statistical variables into principal components, *Journal of Educational Psychology*, vol. 24, 1933.
4. M. Velez-Reyes, D. M. Linares, and L. O. Jimenez, Two-stage band selection algorithm for hyperspectral imagery, *Proceedings of SPIE on Algorithms and Technologies for Multispectral, Hyperspectral, and Ultraspectral Imagery* VIII, vol. 4725, 2002.
5. J. Gruninger, R. Sundberg, M. Fox, R. Levine, W. Mundkowsky, M. S. Salisbury, and A. H. Ratcliff, Automated optimal channel selection for spectral imaging sensors, *Proceedings of SPIE on Algorithms and Technologies for Multispectral, Hyperspectral, and Ultraspectral Imagery VII*, vol. 4381, 2001.
6. P. J. Withagen, E. den Breejen, E. M. Franken, A. N. de Jong, and H. Winkel, Band selection from a hyperspectral data cube for a real-time multispectral 3CCD camera, *Proceedings of SPIE on Algorithms and Technologies for Multispectral, Hyperspectral, and Ultraspectral Imagery VII*, vol. 4381, 2001.
7. S. S. Shen, and E. M. Bassett, Information theory based band selection and utility evaluation for reflective spectral systems, *Proceedings of SPIE on Algorithms and Technologies for Multispectral, Hyperspectral, and Ultraspectral Imagery VIII*, vol. 4725, 2002.
8. R. Basedow, P. Silverglate, W. Rappoport, R. Rockwell, D. Rosenburg, K. Shu, R. Whittlesey, and E. Zalewski, The HYDICE instrument design, *Proceedings of the International Symposium on Spectral Sensing Research*, vol. 1, 1992.
9. R. Basedow, HYDICE system: Implementation and performance, *Proceedings of SPIE*, vol. 2480, 1995.

10. A. Papoulis, *Probability, Random Variables and Stochastic Processes*, 2nd edition, McGraw-Hill, New York, 1984.
11. J. H. Holland, Genetic algorithms, *Scientific American*, vol. 267, no. 1, 1992.
12. D. E. Goldberg, *Genetic Algorithms*, Addison-Wesley, Reading, MA, 1989.
13. J. C. Spall, *Introduction to Stochastic Search and Optimization: Estimation, Simulation, and Control*, John Wiley & Sons, New York, 2003.
14. J. Nocedal, and S. Wright, *Numerical Optimization*, Springer-Verlag, New York, 1999.
15. Nonconventional Exploitation Factors Data System (NEFDS) Specifications, ORD-312-92.
16. A. Berk, L. S. Bernstein, and D. C. Robertson, MODTRAN: A Moderate Resolution Model for LOWTRAN 7, Geophysics Laboratory, GL-TR-89-0122, 1989.
17. S. .S Shen, Relative utility of HYDICE and multispectral data for object detection, identification, and abundance estimation, *Proceedings of SPIE on Hyperspectral Remote Sensing and Applications*, vol. 2821, 1996.
18. S. S. Shen, Multiband sensor system design tradeoffs and their effects on remote sensing and exploitation, *Proceedings of SPIE on Imaging Spectrometry*, vol. 3118, 1997.
19. S. S. Shen, Spectral/Spatial/SNR trade study, *Proceedings of Spectroradiometric Science Symposium*, 1997.

CHAPTER 10

FEATURE REDUCTION FOR CLASSIFICATION PURPOSE

SEBASTIANO B. SERPICO, GABRIELE MOSER, AND ANDREA F. CATTONI
Department of Biophysical and Electronic Engineering, University of Genoa, I-16145 Genoa, Italy

10.1. INTRODUCTION

The spectral signatures of different land-cover types in hyperspectral remote-sensing images represent a very rich source of information that may allow an accurate separation of cover classes to be obtained by means of suitable pattern-classification techniques. This is one of the main reasons why interest is devoted to hyperspectral sensors, which provide an accurate sampling of such signatures, based on a huge number of channels (e.g., some hundreds) with narrow spectral bands [1, 2]. However, dealing with hundreds of narrow-band channels involves problems in the acquisition phase (noise), storage and transmission phases (data size), and processing phase (complexity) [1].

In the context of supervised classification, an additional problem is represented by the so-called "Hughes phenomenon" that appears when the training-set size is not large enough to ensure a reliable estimation of the classifier parameters. As a result, a significant reduction of the classification accuracy can be observed [3–5]. Specifically, as the number of features n employed for classification increases, the number of parameters p of a given supervised classifier also grows (e.g., p grows linearly with n for a linear classifier, quadratically for a Bayesian Gaussian classifier, and even exponentially for some nonparametric classifiers [3]). However, the size of the training set to be used to estimate such parameters is typically fixed. Hence, an increase in n can be expected to yield an improvement in the classification accuracy only if the training-set size is large enough to estimate all the parameters of the classifier. Therefore, as n increases, the classification accuracy is expected to increase until a maximum value is reached and then to decrease

Hyperspectral Data Exploitation: Theory and Applications, Edited by Chein-I Chang

monotonically, due to the worsening in the quality of the classifier-parameter estimates [4].

In order to overcome the Hughes phenomenon, a pattern-recognition approach can be applied: The original hyperspectral bands ("h-bands") are considered as features and feature-reduction algorithms are applied [6]. In particular, feature-selection algorithms have been proposed in the literature [7] to select a (sub-)optimal subset of the complete set of h-bands. As an alternative, a more general approach based on transformations (feature extraction) can be adopted [3]. Usually, linear transformations are applied to project the original feature space onto a lower-dimensional subspace that preserves most of the information [2, 3].

The purpose of this chapter is to propose a procedure to extract nonoverlapping spectral bands ("s-bands") of variable bandwidths and spectral positions from hyperspectral images, in such a way as to optimize the accuracy for a specific classification problem. The interest of this kind of procedure can lie in the reduction of the number of bands drawn from a hyperspectral image or in a case-based design of the spectral bands of a programmable sensor (e.g., the MERIS sensor on board of the ENVISAT satellite). The proposed procedure represents a special case of feature transformation. With respect to other approaches, the kind of transformation employed (i.e., the averaging of contiguous h-bands to generate s-bands) shows the advantage of allowing the interpretability of the new features (i.e., the s-bands) to be saved. Therefore it can be considered as a compromise between the two requirements of generality and interpretability. We recall that methods for spectral band design were proposed in De Backer et al. [8] and in Wiersma and Landgrebe [9], where completely different approaches were adopted, which were based on (a) the minimization of the mean-square representation error by the application of the Karhunen–Loeve expansion [10] to the spectral response function [9] and (b) the numerical optimization of the parameters of a set of Gaussian optical filters according to an inter-class distance measure [8], respectively.

The chapter is organized as follows. Section 10.2 provides a review of the literature concerning feature-reduction algorithms for hyperspectral data classification. Then, the proposed band-extraction approach is described in Section 10.3 and the results of its application to real hyperspectral data are presented in Section 10.4. Finally, conclusions are drawn in Section 10.5.

10.2. PREVIOUS WORK ON FEATURE REDUCTION FOR HYPERSPECTRAL DATA CLASSIFICATION

10.2.1. Feature Selection

Feature-selection techniques generally involve both a search algorithm and a criterion function [11, 12]. The search algorithm generates possible "solutions" of the feature-selection problem (i.e., subsets of features), which are then compared by applying the criterion function as a measure of the effectiveness of each solution. An exhaustive search for the optimal solution turns out to be intractable from a computational viewpoint, even for moderate values of the number of features [11]. The "branch-and-bound" approach has been applied to feature selection as a nonexhaustive strategy to find the globally optimum solution [13]. However, the reduction of the computation amount is not enough to make its application feasible for problems with hundreds of features [14]. Therefore, several suboptimal approaches to feature selection have been proposed in the literature [12, 15, 16]. The simplest suboptimal search strategies are the sequential forward selection (SFS) and the sequential backward selection (SBS) techniques [11, 16] that identify the best feature subset that can be obtained by adding to (SFS), or removing from (SBS), the current feature subset, one feature at a time, until the desired number of features is achieved. A serious drawback of both methods is that they do not allow backtracking (e.g., in the case of SFS, once a feature is selected at a given iteration, it cannot be removed in any successive iteration). The sequential forward floating selection (SFFS) and the sequential backward floating selection (SBFS) methods improve the standard SFS and SBS techniques by dynamically changing the number of features included (SFFS) or removed (SBFS) at each step and by allowing the reconsideration of the features included or removed at the previous steps [16, 17]. The two suboptimal search algorithms presented in Bruzzone and Serpico [7] (namely, the "Steepest Ascent" and the "Fast Constrained Search") are based on a formalization of the feature-selection problem as a discrete optimization problem in a suitable binary multidimensional space and allow different trade-offs between the effectiveness of the selected features and the computational time required to find a solution. Several other methods based on attractive concepts like feature similarity measures [18], graph-searching algorithms [19], neural networks [20], genetic methods [12, 21–23], simulated annealing [24], finite mixture models [14, 25], "tabu search" metaheuristics [26], spectral distance metrics [27], and parametric feature weighting [28] were also explored in the literature.

10.2.2. Feature Extraction

The main target of a feature-extraction technique is the reduction of the data dimensionality by mapping the feature space onto a lower-dimensional space. Usually, linear transformations, for which the transformation matrix is optimized in order to minimize the information loss, are adopted. A basic parametric method is the Discriminant Analysis Feature Extraction (DAFE), which is based on the maximization of a functional (namely, the Rayleigh coefficient), expressed as a ratio of a between-class scatter matrix to an average within-class scatter matrix [2, 3]. DAFE allows a simple closed-form computation of the transformation matrix, but suffers from a serious drawback—that is, the possibility to extract at most $(M - 1)$ features, M being the number of classes. When still operating in a parametric context, extensions of the DAFE approach are proposed in Landgrebe [29] by allowing different weights to be assigned to distinct couples of classes according to the distance of the respective class means; in Kuo and Landgrebe [30] by integrating regularized leave-one-out covariance estimators in the computation of scatter matrices; or in Du and Chang [31] by integrating *a priori* knowledge in order to align the class means with predefined target directions in the transformed feature space. Nonparametric generalizations of DAFE have also been developed. For instance, the "Nonparametric Discriminant Analysis" (NDA) extends the DAFE approach by introducing a modified nonparametric definition of the between-class scatter matrix, based on a "K-nearest neighbors" approach [3]. Nonparametric expressions for the within-class scatter matrix are also integrated in the DAFE framework by the modified NDA method proposed in Bressan and Vitri [32] and by the "Nonparametric Weighted Feature Extraction" technique developed in Kuo and Landgrebe [33]. A "Penalized Discriminant Analysis," which modifies DAFE in order to improve its effectiveness also when many highly correlated features are present, has been proposed in Hastie et al. [34] and applied to hyperspectral data analysis in Yu et al. [35]. A kernel-based strategy [36] is employed in Baudat and Anouar [37] and in Mika et al. [38] to generalize DAFE and to formulate a nonparametric and nonlinear "Kernel Fisher Discriminant" technique.

The Decision Boundary Feature Extraction (DBFE) [39] employs information about the decision hypersurfaces associated to a given parametric Bayesian classifier, in order to define an intrinsic dimensionality for the classification problem and a corresponding optimal linear mapping. The "Projection Pursuit" is a technique based on the numerical maximization of a functional (named "projection index") that is directly computed in the transformed lower-dimensional space [40]. A limitation of the method lies in the fact that, in general, only a local maximum of the projection index can be found, and consequently suboptimal feature extraction is obtained [40, 41].

A further feature-extraction approach consists in grouping the original features in subsets of highly correlated features, in order to transform separately the features in each subset [42]. This procedure allows one to work in lower-dimensional spaces and to apply, for example, classical techniques based on the estimation of data-covariance matrices (the estimation of this matrix in a hyperdimensional space may be problematic [2]). Feature grouping has been proposed in conjunction with the Principal Component Analysis (PCA) transform [42], with the Fisher transformation [43], with the classification approaches based on binary hierarchical classifiers and on error correcting output codes [44], and with the "tabu-search" metaheuristic approach [45, 46].

In Bruce et al. [47] the integration of DAFE and of feature extraction based on discrete wavelet transforms [48] is proposed for hyperspectral image classification. The combination of PCA and of morphological transformations with structuring elements of varying sizes is proposed in Benediktsson et al. [49] in order to address the problem of classifying hyperspectral high-resolution imagery of urban areas. Multichannel generalizations of the classical morphological operators are introduced in Plaza et al. [50] and are exploited to perform the feature reduction and the classification of hyperspectral images acquired over urban and agricultural areas.

10.3. THE PROPOSED BAND-EXTRACTION METHOD

10.3.1. Problem Formulation

We assume that a hyperspectral image of a given site is available and that it contains n original h-bands, where a set $\Omega = \{\omega_1, \omega_2, \ldots, \omega_M\}$ of information classes (defined by the user) appears. A set of labeled pixels for all such classes should also be available, divided into a training set and a test set.

An s-band can be obtained by averaging a group of contiguous channels of the hyperspectral image, therefore it can be identified by the starting and the ending h-bands of such a group. Specifically denoting by $\mathcal{H} = \{x_1, x_2, \ldots, x_n\}$ the set of the available n h-bands, we aim at partitionating $\mathcal{H}$ into a set $\mathcal{S} = \{y_1, y_2, \ldots, y_m\}$ of m nonoverlapping and contiguous s-bands. Therefore, the indexes of the ending h-band of the rth s-band and of the starting h-band of the $(r+1)$th s-band $(r = 1, 2, \ldots, m-1)$ coincide. By denoting by τ_r $(\tau_r \in \{1, 2, \ldots, n\})$ this threshold index $(r = 1, 2, \ldots, m-1)$, the collection $\mathcal{S}$ of the extracted s-bands is unambiguosly determined by the set $\{\tau_1, \tau_2, \ldots, \tau_{m-1}\}$ of the thresholds between consecutive s-bands. After introducing, for simplicity, two further dummy thresholds $\tau_0 = 0$ and $\tau_m = n$, the rth s-band y_r is computed as follows:

$$y_r = \frac{1}{\tau_r - \tau_{r-1}} \sum_{\ell=\tau_{r-1}+1}^{\tau_r} x_\ell, \qquad r = 1, 2, \ldots, m \tag{10.1}$$

Note that $\tau_r < \tau_{r+1}$ for $r = 0, 1, \ldots, m$. Therefore, the problem of the extraction of the m s-bands can be formulated as the optimization of $(m-1)$ thresholds between consecutive s-bands. To this end, as in the feature-selection context, an optimization procedure can be developed, based on a functional measuring the quality of each admissible configuration of thresholds and endowed with a suitable search strategy.

Focusing first on the former issue, we adopt an inter-class distance measure as a functional measuring the quality of each configuration of extracted s-bands. Such a functional represents a quantitative measure of the separability of the classes in the transformed m-dimensional feature space. Several inter-class distance measures have been proposed in the literature, such as (a) the Bhattacharyya distance, related by the Chernoff bound to the error probability of a binary Bayesian classifier [3, 51], (b) the divergence [52] and the normalized divergence [1], based on information theory, or (c) the Jeffries–Matusita distance [1, 53], strictly related to the Bhattacharyya distance. We assume to operate in a Bayesian classification framework and we assume a Gaussian model for the class-conditional probability density function (PDF) of each class (which is a usual approach when dealing with hyperspectral data classification [2]). Under these assumptions, we adopt the average Jeffries–Matusita distance as an inter-class distance measure, since it turns out to be related to the performances of the Gaussian "maximum *a posteriori*" classifier (GMAP, for short).

Thanks to the linearity of the simple band-extraction mapping (10.1), the class-conditional distribution $p_i^S(\cdot)$ of the transformed feature vector* $y^S = [y_1, y_2, \ldots, y_m]^T$ conditioned to each class ω_i turns out to be a Gaussian $\mathcal{N}(m_i^S, \Sigma_i^S)$ where m_i^S and Σ_i^S are the ω_i-conditional mean and covariance of y^S, respectively ($i = 1, 2, \ldots, M$). Under this Gaussianity assumption, the well-known average Jeffries–Matusita distance is adopted, and is defined as follows [1, 53]:

$$J(\mathcal{S}) = \sum_{i=1}^{M-1} \sum_{j=i+1}^{M} P_i P_j \cdot \sqrt{2 - 2\exp[-B_{ij}(\mathcal{S})]} \tag{10.2}$$

where P_i is the prior probability of ω_i ($i = 1, 2, \ldots, M$), and

$$B_{ij}(\mathcal{S}) = \frac{1}{8}(m_i^S - m_j^S)^T \left(\frac{\Sigma_i^S + \Sigma_j^S}{2}\right)^{-1} (m_i^S - m_j^S) + \frac{1}{2}\ln \frac{\det\left(\frac{\Sigma_i^S + \Sigma_j^S}{2}\right)}{\sqrt{\det \Sigma_i^S \det \Sigma_j^S}} \tag{10.3}$$

is the Bhattacharyya distance betweeen ω_i and ω_j ($i, j = 1, 2, \ldots, M$; $i \neq j$). Therefore, in order to minimize the probability of classification error, we adopt $J(\cdot)$ as a functional to guide the band-extraction process; that is, we search for a set of s-bands maximizing $J(\cdot)$.

Four discrete search algorithms have been developed in order to perform this maximization task, whose description is reported in the following subsections.

10.3.2. Sequential Forward Band Partitioning

The first considered strategy is the "Sequential Forward Band Partitioning" (SFBP), which is conceptually similar to the SFS strategy for feature selection. The key idea of the method lies in increasing iteratively the number of extracted s-bands, by saving at each iteration the current set of thresholds and by adding the new threshold that yields the largest increase in the functional $J(\cdot)$.

More specifically, first we apply an exhaustive search among all possible thresholds that optimizes the adopted criterion. As a result, the best single threshold, which splits the set of h-bands in two s-bands, is obtained. Then we add a further threshold, maintaining the first threshold fixed and exploring all the possibilities by applying an exhaustive search again. Hence, this second threshold will split one of the previously obtained two s-bands, thus generating a set of three s-bands. The process is iterated, by adding a new threshold at a time to the previously introduced ones. An exhaustive search, which is similar to the previous one, is applied at each step.

The procedure is stopped when the desired number m of s-bands is obtained (i.e., when $(m-1)$ non-dummy thresholds have been fixed). When we denote by $\mathcal{S}$ the

*All the vectors in the chapter are implicitly assumed to be column vectors, and the superscript "T" stands for the matrix transpose operator.

current configuration of s-bands, formally identified with the related collection $\{\tau_0, \tau_1, \ldots, \tau_m\}$ of thresholds, the following pseudo-code summarizes the SFBP algorithm:

SFBP Algorithm

```
{
    initialize S = {1, n} and J_max = -∞;
    for r ∈ {1,2,...,m-1}
    {
        for each possible threshold τ ∈ {1,2,...,n-1}
        {
            if τ ∉ S and J(S ∪ {τ}) > J_max,
            then set S* = S ∪ {τ} and update J_max = J(S*);
        }
    }
    update S = S*;
}
```

Like SFS, the proposed SFBP procedure has the advantage of conceptual simplicity, but it also has the drawback of not allowing backtracking; that is, once a threshold has been fixed at a given iteration, it will never be removed at any successive iteration.

10.3.3. Steepest Ascent Band Partitioning

The second search strategy, the "Steepest Ascent Band Partitioning" (SABP), is initialized with a solution containing m s-bands, which can be randomly generated or obtained by another search algorithm (e.g., the previously described one) or provided by the user (e.g., based on *a priori* knowledge of the problem) and aims at improving iteratively such solution by identifying the "direction of the highest local increase" of the functional $J(\cdot)$. Specifically, such optimization strategy, which is similar to the one adopted by the Steepest Ascent (SA) algorithm proposed in Serpico and Bruzzone [7] for feature selection, aims at performing a complete local exploration in the space of the configurations of $(m-1)$ non-dummy thresholds. By defining a local move as a change in the configuration of the thresholds that involves the modification of only one threshold, at each SABP iteration, the local move providing the highest increase in $J(\cdot)$ is searched exhaustively.

Operatively, at each iteration, this exhaustive local exploration is performed by removing the current threshold, one at a time, thus obtaining a temporary set of $(m-1)$ s-bands. The removed threshold is replaced with a new one in order to have m thresholds again. The new optimal threshold is exhaustively searched for inside the $(m-1)$ temporary s-bands, by splitting one of them into two s-bands. All the solutions generated by performing such local moves are evaluated on the basis of the adopted criterion $J(\cdot)$. If the best new solution is better than the initial one, then it is selected as the new solution; otherwise, the method is stopped. The local search is performed again by starting from the new solution, and so on.

Hence, the iterative procedure terminates when no increase in the criterion function can be obtained by any of the above-defined local moves. As for SA [7], it can be proven that this procedure converges in a finite number of iterations (see Appendix), although the number of iterations required to reach convergence is not known in advance. A pseudo-code for the method is the following one, where $\mathcal{S}$ and $\mathcal{S}^0$ stand for the current and the initial sets of extracted s-bands (identified, as usual, with the corresponding sets of thresholds), respectively:

SABP Algorithm

```
{
    initialize S = S^0 = {τ_r}_{r=0}^m and J_max = J(S^0);
    set StopFlag = false;
    do
    {
        set J* = −∞;
        for r ∈ {1, 2, ..., m − 1}
        {
            for each possible threshold τ ∈ {1, 2, ..., n − 1}
            {
                if τ ∉ S − {τ_r} and J(S − {τ_r} ∪ {τ}) > J*,
                then set S* = S − {τ_r} ∪ {τ} and update J* = J(S*);
            }
        }
        if J* > J_max,
        then update S = S* and J_max = J*;
        else StopFlag = true;
    }
    while StopFlag = false.
}
```

The concepts of "local exploration" and "local move," introduced here with an intuitive meaning, can be rigorously formalized within a metric-theoretic framework. As detailed in the Appendix, the set of all possible configurations of $(m - 1)$ thresholds can be endowed with a metric-space structure [54, 55], by mapping the set of all the possible configurations of s-bands onto a suitable binary multidimensional space and by defining a suitable distance function in this space. In this framework, each and every of the above-mentioned local moves generates a configuration of s-bands having distance 2 from the current configuration, thus allowing SABP to explore exhaustively, at each iteration, the radius-2 neighborhood of the current set of thresholds. In addition, according to this interpretation, the final configuration selected by SABP can be proven to be a local maximum point of $J(\cdot)$ in this metric space (see Appendix A).

10.3.4. Fast Constrained Band Partitioning

The third proposed search strategy—that is, the "Fast Constrained Band Partitioning" (FCBP)—starts from an initial solution (like SABP) and then progressively improves it with an optimization strategy which is similar to the one adopted by the Fast Constrained Search (FCS) algorithm proposed in [7] for feature selection. The key idea of the method lies in simplifying the SABP procedure in order to make the computation time shorter and deterministic, although it gives up the possibility to perform an exhaustive analysis of all the possible local moves. In particular, FCBP removes the first threshold and tries all possible ways to replace it in the same way as SABP. If the best solution obtained is better than the starting one, then it is immediately considered as the new solution. Then the second threshold is removed and all replacements are explored and evaluated in order to see if a better solution can be found. The procedure is iterated only once, until the replacement of each of the original m thresholds has been tried. A pseudo-code for the method is the following one, where $\mathcal{S}$ and $\mathcal{S}^0$ stand for the same sets defined for SABP in the previous section:

FCBP Algorithm

```
{
    initialize S = S^0 = {τ_r}_{r=0}^m and J_max = J(S^0);
    for r ∈ {1,2,...,m−1}
    {
        set J* = −∞;
        for each possible threshold τ ∈ {1,2,...,n−1}
        {
            if τ ∉ S − {τ_r} and J(S − {τ_r} ∪ {τ}) > J*,
            then set S* = S − {τ_r} ∪ {τ} and update J* = J(S*);
        }
        if J* > J_max,
        then update S = S* and J_max = J*;
    }
}
```

From an algorithmic viewpoint, the difference between SABP and FCBP lies in the fact that, at each step, SABP tries to replace each and every threshold before comparing all the generated solutions and accepting a new one; on the contrary, FCBP can update the solution at each attempt to replace a threshold. In particular, differently from SABP, FCBP does not perform a complete exploration of all the possible local moves (i.e., of a whole neighborhood centered on the current solution), but it modifies the configuration of the thresholds, once a local move yielding an increase in $J(\cdot)$ is found in the subset of local moves related to the removal of a threshold. In addition, the stop criteria are different: SABP continues until its exploration stops giving improvements, while FCBP iterates only once over the thresholds to explore the related replacements, which, in general, does not allow a local maximum point of $J(\cdot)$ to be reached. From the viewpoint of performances, SABP is expected to be more powerful, because each change of solution is based on a more extensive search; on the other hand, FCBP is faster and its number of iterations is deterministic. In particular, the total number of moves explored by FCBP is equal to the number of moves explored by SABP in each single iteration.

10.3.5. Convergent Constrained Band Partitioning

The fourth proposed technique does not derive directly from a similar approach employed in the feature selection context, but is specifically introduced in the present band-synthesis context. Specifically, the goal of this method is to reach a trade-off between the SABP and FCBP approaches, by combining both the local optimality, which is typical of SABP, and the shorter computation time of FCBP. The key idea of the algorithm, named "Convergent Constrained Band Partitioning" (CCBP), lies in iterating the FCBP procedure not once, but several times, until convergence is reached.

Specifically, CCBP is an iterative procedure which is initialized (like SABP and FCBP) with an initial configuration of m s-bands. At each step, CCBP performs a complete FCBP procedure and assesses the quality of the configuration of s-bands obtained at the end of each iteration by the adopted functional. The method stops when no more increase in the functional can be obtained, like SABP. As described in further detail in the Appendix, a convergence theorem can be proved also for CCBP, which guarantees that the method will stop in a finite number of iterations and that the final set of s-bands is a local maximum point of the functional in the multidimensional binary metric space representing the collection of all configurations of s-bands. Therefore, the same good analitycal properties hold for SABP and CCBP.

On the other hand, CCBP, as compared with SABP, is expected to take a shorter computation time to reach convergence, because each CCBP iteration does not perform an exhaustive local exploration of all possible local moves (like SABP) but it modifies the threshold configuration, once a threshold replacement yielding an increase in $J(\cdot)$ is found (like FCBP). However, the number of iterations needed to reach convergence is not known in advance for CCBP as well, which makes its computation time nondeterministic (as in the case of SABP). A pseudo-code for CCBP (with the same notations and conventions adopted for SABP and FCBP) is presented as follows:

CCBP Algorithm

{

  initialize $\mathcal{S} = \mathcal{S}^0 = \{\tau_r\}_{r=0}^{m}$ and $J_{max} = J(\mathcal{S}^0)$;
  StopFlag = false.

```
    do
    {
        for r ∈ {1,2,...,m − 1}
        {
            set J* = −∞;
            for each possible threshold τ ∈ {1,2,...,n − 1}
            {
  if τ∉S − {τ_r} and J(S − {τ_r} ∪ {τ}) > J*,
  then set S* = S − {τ_r} ∪ {τ} and update J* = J(S*);
            }
            if J* > J_max,
            then update S = S* and J_max = J*;
            else StopFlag = true;
        }
    }
    while StopFlag = false.
}
```

10.4. EXPERIMENTAL RESULTS

10.4.1. Data Set for Experiments

The proposed band-partitioning methodology was tested on the well-known hyperspectral "Indian Pine" data set, consisting of a 145 × 145-pixel portion of an AVIRIS image acquired on NW Indian Pine in June 1992 [2]. Not all of the 220 original bands were employed in the experiments, since 18 bands were affected by atmosphere-absorption phenomena and consequently were discarded. Hence, the considered data dimensionality is $n = 202$. As an example, two bands from the adopted data set are shown in Figure 10.1. Nine classes have been selected and are represented by ground truth data (Table 10.1). The subdivision of the ground-truth data into training and test data was not performed randomly but by defining spatially disjoint training and test fields for each class, in order to reduce as much as possible the correlation between the samples used to train the system and the ones employed to test its performances.

10.4.2. Classification Results

The effectiveness of the band-partitioning approach to feature reduction for hyperspectral data classification was evaluated by employing the developed SFBP, SABP, FCBP, and CCBP algorithms in order to generate m s-bands, $2 \leq m \leq 30$, and by applying a parametric GMAP classifier in the transformed feature space, the class parameters being estimated as sample-means and sample-covariance matrices [52].

Figure 10.1. "Indian Pine" data set employed for experiments: (a) band 16 (central wavelength: 547.60 nm); (b) band 193 (central wavelength: 2232.07 nm).

For each value of m, SABP, FCBP, and CCBP were initialized with the m s-bands generated by SFBP.

As shown in Figure 10.2, the behaviors of the classification accuracies (overall accuracy, OA; and average accuracy, AA) provided by SFBP, SABP, FCBP, and CCBP, as m varies in [2, 30] were quite similar, which suggests an overall comparable effectiveness of the four techniques from the viewpoint of the classification-map quality. In particular, the best result, in terms of overall accuracy, was provided by FCBP with 22 features (OA = 81.73%); however, the peak overall accuracies given by the four algorithms were close to each other, namely, 81.06% for SFBP with $m = 24$; 81.38% for SABP with $m = 25$; 81.73% for FCBP with $m = 22$; 81.38% for CCBP with $m = 25$ (the same as SABP).

This similarity is further confirmed by an analysis of the s-band configurations yielding such peak accuracies (see Table 10.2), since the optimal configurations identified by SABP, FCBP, and CCBP (initialized by the same SFBP configuration)

TABLE 10.1. Number of Training and Test Samples in the "Indian Pine" Data Set Used for Experiments

Class	Training Samples	Test Samples
Corn—no till	762	575
Corn—min	435	326
Grass/pasture	232	225
Grass/trees	394	283
Hay—windrowed	235	227
Soybean—no till	470	443
Soybean—min	1428	936
Soybean—clean till	328	226
Woods	728	487

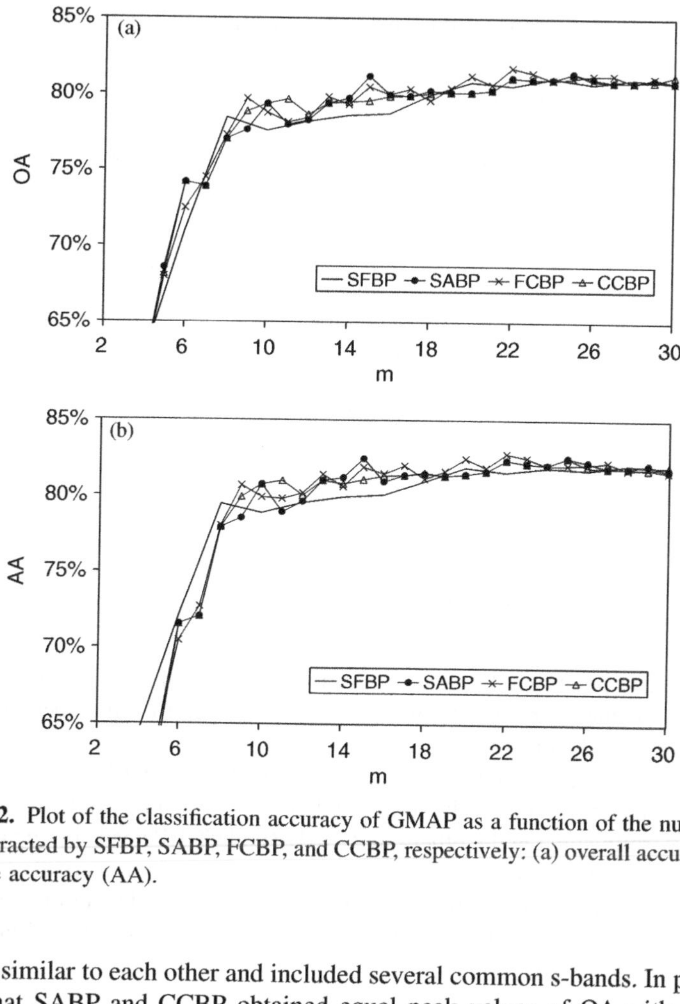

Figure 10.2. Plot of the classification accuracy of GMAP as a function of the number m of s-bands extracted by SFBP, SABP, FCBP, and CCBP, respectively: (a) overall accuracy (OA); (b) average accuracy (AA).

were very similar to each other and included several common s-bands. In particular, the fact that SABP and CCBP obtained equal peak values of OA with the same number of extracted s-bands (namely, $m = 25$) is explained by the fact that both methods converged to the same configuration of thresholds (see Table 10.2). This confirms the good convergent properties proved in the Appendix for such methods, and it points out that, in the present experiment, they converged to the same local maximum point of the functional $J(\cdot)$.

On the contrary, different behaviors can be noted from the viewpoint of the computational burden. In Figure 10.3 we show the numbers of exhaustive threshold searches performed by SABP, FCBP, and CCBP before the termination of the methods. All four methods modify and/or enlarge the number of extracted thresholds by repeatedly performing exhaustive searches of a single threshold location in the set of the available h-bands. In fact, an exhaustive search for a threshold value $\tau \in \{0, 1, \ldots, n-1\}$ is performed in the inner loop of the pseudo-code of each of the four proposed techniques. The overall number of such searches is a meaningful

TABLE 10.2. Configurations of the Thresholds Yielding the Highest Test-Set Overall Accuracies for SFBP ($m = 24$), SABP ($m = 25$), FCBP ($m = 22$), and CCBP ($m = 25$)[a]

	Threshold τ_r					Threshold τ_r			
r	SFBP	SABP	FCBP	CCBP	r	SFBP	SABP	FCBP	CCBP
1	12	13	12	13	13	101	94	101	94
2	20	18	18	18	14	111	101	112	101
3	26	27	27	27	15	123	112	122	112
4	30	30	30	30	16	132	122	141	122
5	31	31	31	31	17	143	132	147	132
6	33	33	33	33	18	148	141	157	141
7	35	37	37	37	19	151	148	168	148
8	37	51	52	51	20	157	151	175	151
9	44	61	61	61	21	168	157	184	157
10	54	66	66	66	22	176	168	—	168
11	81	75	75	75	23	186	175	—	175
12	94	84	94	84	24	—	184	—	184

[a]For each method the ordered list of the non-dummy thresholds τ_r ($r = 1, 2, \ldots, m - 1$) yielding the maximum test-set overall accuracy is reported.

measure of computational burden. In addition, this measure is directly related to the complexity of the four algorithms and does not depend on the specific hardware configuration used to test them.

Note that for each value of m in [2, 30], SFBP and FCBP exactly perform m exhaustive searches, so that, the total numbers of search operations for such methods are deterministic. On the other hand, SABP and CCBP need significantly higher numbers of searches (see Figure 10.3). SABP requires a comparatively much larger number of exhaustive searches to explore the space of the threshold configurations before reaching a local maximum point of the Jeffries–Matusita functional. In

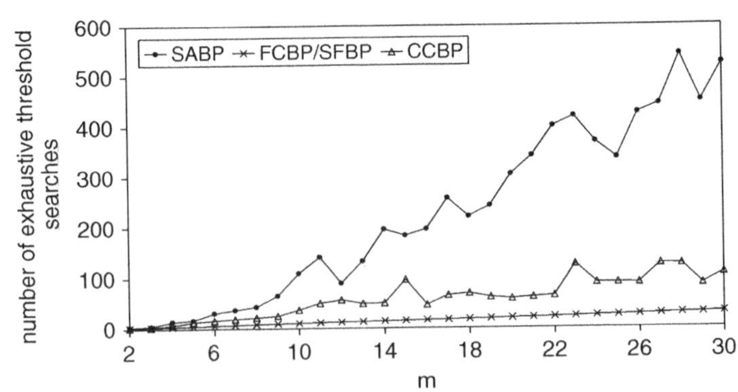

Figure 10.3. Behavior of the number of exhaustive threshold searches required to extract m s-bands by using SFBP, SABP, FCBP, and CCBP, respectively, as a function of m.

TABLE 10.3. Classification Accuracies Obtained by FCBP for $m = 22$ Before (Left) and After (Right) Grouping the Classes Corresponding to the Same Vegetation Type

Class	Accuracy (%)	Class	Accuracy (%)
Corn—no till	81.57	Corn	77.36
Corn—min	67.79	Grass	94.88
Grass/pasture	89.78	Hay—windrowed	100.00
Grass/trees	98.94	Soybean	92.09
Hay—windrowed	100.00	Woods	98.36
Soybean—no till	43.12		
Soybean—min	85.04	Overall accuracy	90.21
Soybean—clean till	80.53	Average accuracy	92.54
Woods	98.36		
Overall accuracy	81.73		
Average accuracy	82.79		

addition, the number of exhaustive threshold searches performed by SABP is nondeterministic, because the number of SABP iterations needed to reach convergence is not known in advance. The same conclusion holds for CCBP, although, as expected, the experiments pointed out that CCBP needs a much lower number of iterations than SABP to reach a local maximum point of $J(\cdot)$. Note, in particular, that, as mentioned above, SABP and CCBP converged, for $m = 25$, to the same local maximum, but the numbers of exhaustive threshold searches required by SABP and CCBP to reach this point were 337 and 89, respectively; that is, a much shorter time was needed by CCBP.

We stress that the accuracies above were, in general, not very good both due to the above-mentioned choice of the training and test fields and to the large spectral overlapping among several classes. In particular, most classification errors were due to the confusion between "corn—no till" and "corn—min," between "grass/pasture" and "grass/trees," and among "soybean—no till," "soybean—min," and "soybean—clean till," with these groups of classes representing very similar vegetated covers. This is confirmed by Table 10.3, which shows, as an example, the accuracy obtained for each class by FCBP with $m = 22$, as well as the accuracies resulting from grouping the above-mentioned critical classes in three "higher level" classes: "corn," "grass," and "soybean." A sharp accuracy increase results from this grouping operation, and it yields a 90.67% overall accuracy and a 92.69% average accuracy. The corresponding classification map is shown in Figure 10.4a.

10.4.3. Comparison with Previously Proposed Feature-Reduction Methods

In order to further assess the capabilities of the proposed approach, a comparison was made with the performances of the well-known SFS feature-selection algorithm and of the DBFE feature-transformation method. SFS was applied to select

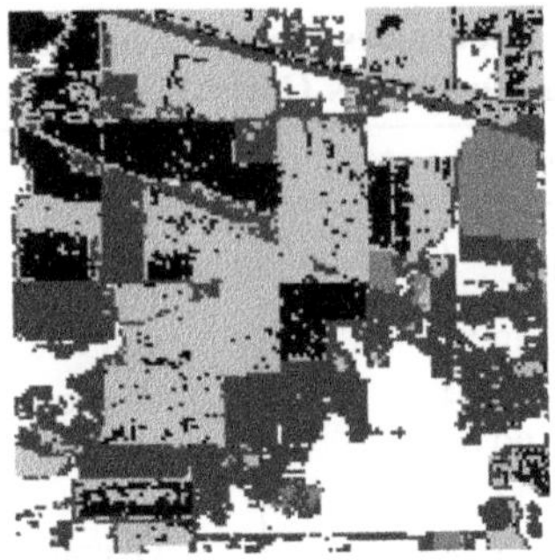

(a)

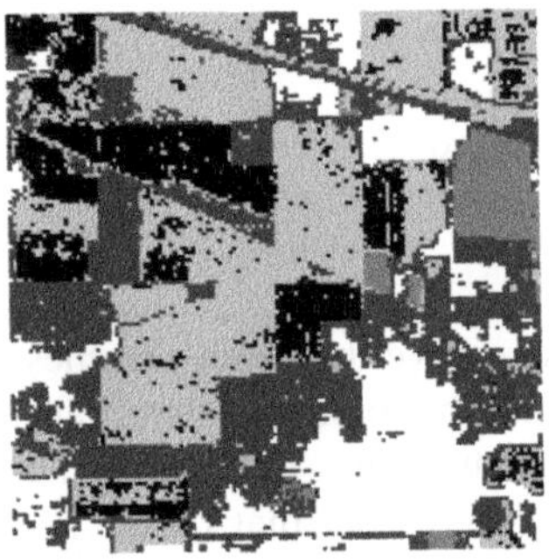

(b)

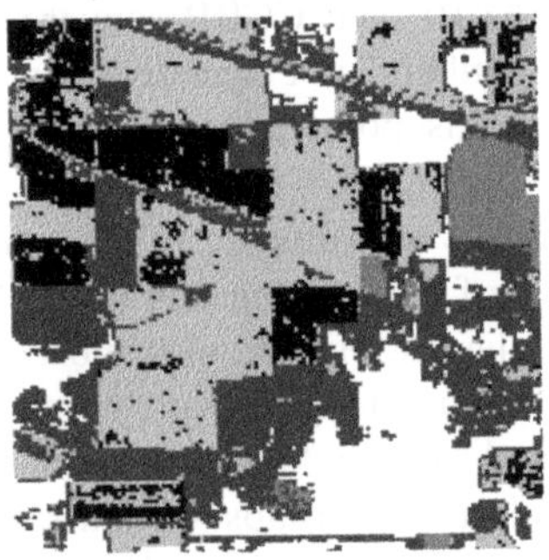

(c)

Figure 10.4. Classification maps generated by GMAP applied to the sets of features extracted by: (a) FCBP ($m = 22$); (b) DBFE with $m = 14$ and pre-reduction performed by SFS; (c) DBFE with $m = 16$ and pre-reduction performed by CCBP. Color legend: black represents "corn," dark grey represents "grass," middle gray represents "hay—windrowed," light gray represents "soybean," white represents "woods."

a suboptimal subset of features which aims at maximizing the average Jeffries–Matusita functional. DBFE is known as an effective parametric feature-extraction methodology, computing a linear feature transform according to an analysis of the decision boundaries separating the decision regions corresponding to distinct classes in the original hyperdimensional space [39]. DBFE is known from the literature to provide high accuracies [39–41] and was adopted here as a benchmark parametric extraction approach. However, as stated in Jimenez and Landgrebe [39], a preliminary reduction stage is usually necessary in order to apply DBFE efficiently, since this method involves the estimation of the class-conditional covariance matrices in the original hyper-dimensional space, which can be a critical operation, due to the Hughes phenomenon. In particular, according to the results of the comparative study [41], SFS was employed here in this pre-reduction role (from 202 to 30 features). Moreover, the choice of SFS is also explained by its simplicity.

An accuracy comparison between SFS and FCBP is shown in Figure 10.5 (thanks to the above-mentioned similarity among the classification results of the proposed methods, focusing here on FCBP involves no loss of generality). For almost all values of $m \in [2, 30]$, the band-partitioning algorithms achieved better accuracies (both OA and AA) than SFS, thus suggesting a higher class discrimination capability. On the other hand, a comparison between FCBP and DBFE

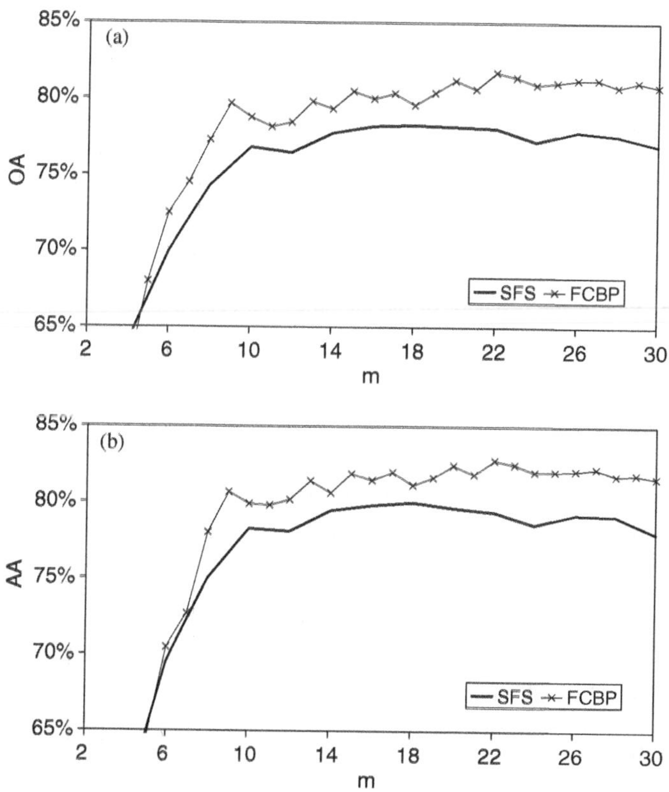

Figure 10.5. Plot of the classification accuracy of GMAP as a function of the number m of selected/extracted features, for FCBP and SFS: (a) overall accuracy (OA); (b) average accuracy (AA).

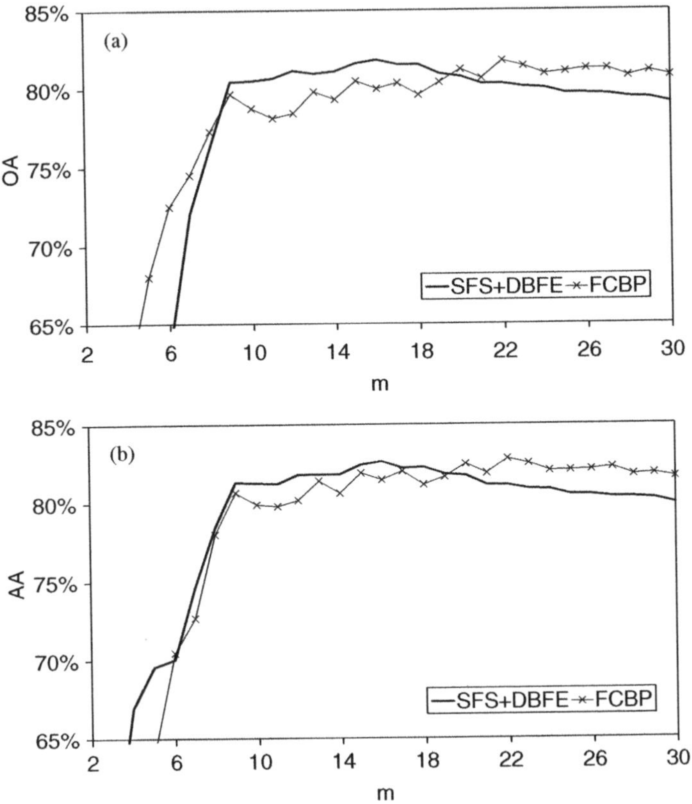

Figure 10.6. Plot of the behavior of the classification accuracy of GMAP as a function of the number m of extracted features, for FCBP and DBFE; the latter is applied with a pre-reduction step performed by SFS: (a) overall accuracy (OA); (b) average accuracy (AA).

(the latter with preprocessing based on SFS; see Figure 10.6) points out that DBFE provided higher accuracy values than FCBP for $9 \leq m \leq 19$, whereas FCBP obtained better classification performances than DBFE for $m \geq 20$ (for $m \leq 8$, both methods exhibited values of OA and AA below 80%; that is, both classification results were poor). Furthermore, the peak overall accuracy achieved by DBFE was 81.71% ($m = 16$), which was slightly better than the peak accuracies obtained by the band-partitioning approach. On the other hand, by grouping together the classes corresponding to the same vegetation type (see Table 10.4), DBFE provided an 89.27% overall accuracy and a 91.71% average accuracy, which were slightly lower than the corresponding values given by FCBP (see Table 10.3). These results globally suggest a similar effectiveness for the proposed approach and for DBFE from the viewpoint of the quality of the resulting classification maps. A visual comparison between the maps given by FCBP and DBFE (see Figure 10.4) confirms this conclusion.

TABLE 10.4. Classification Accuracies Obtained by DBFE, with $m = 14$ and Pre-reduction Performed by SFS, Before (left) and After (right) Grouping the Classes Corresponding to the Same Vegetation Type

Class	Accuracy (%)	Class	Accuracy (%)
Corn—no till	80.87	Corn	75.47
Corn—min	61.04	Grass	94.88
Grass/pasture	88.89	Hay—windrowed	100.00
Grass/trees	98.94	Soybean	91.46
Hay—windrowed	100.00	Woods	96.71
Soybean—no till	60.27		
Soybean—min	82.59	Overall accuracy	89.27
Soybean—clean till	74.34	Average accuracy	91.71
Woods	96.71		
Overall accuracy	81.81		
Average accuracy	82.63		

However, we note that such performances of DBFE are intrinsically limited by the information loss due to the use of SFS in the pre-reduction step, whose choice may in general affect the final results. In the following section this aspect will be further investigated, by suitably combining DBFE with the proposed algorithms.

10.4.4. Combination of the Band-Partitioning Approach and of DBFE

An interesting operational characteristic of the band-partitioning approach consists in the fact that the method computes covariance-matrix estimates only in the transformed lower-dimensional space, thus avoiding the critical process of the covariance-parameter estimation in the original space. In particular, this property suggests that the band-partitioning procedures can be operationally suitable to be employed also as pre-reduction tools for DBFE. In this section, this specific application of the proposed methods is experimentally assessed.

Specifically, the classification accuracy achieved by DBFE was evaluated by applying the method in the 30-dimensional feature space obtained by extracting 30 s-bands by using the proposed band-partitioning approaches. In particular, DBFE was employed to perform a further reduction to m features, with $m \in [2, 30]$. As shown in Figure 10.7, the experiments suggested a very high effectiveness of this combined feature-reduction strategy, because DBFE increased the accuracies obtained by all the proposed band-partitioning algorithms. In particular, the best performances were obtained, in this case, by performing a pre-reduction with CCBP, which yielded an overall accuracy equal to 83.23% with $m = 15$ features (see Table 10.5). As compared with the peak accuracies obtained by the band-partitioning methods, we can also note that this combined approach allows both (a) an increase in the classification accuracy to be achieved and (b) a further

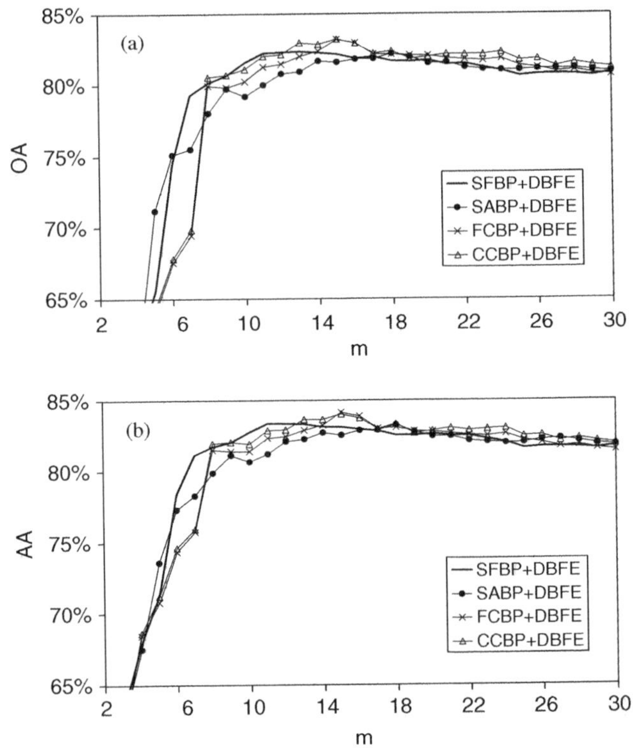

Figure 10.7. Plot of the behavior of the classification accuracy of GMAP as a function of the number m of extracted features, for DBFE with pre-reduction of the original set of h-bands with SABP, FCBP, and CCBP: (a) overall accuracy (OA); (b) average accuracy (AA).

TABLE 10.5. Classification Accuracies Obtained by DBFE, with $m = 15$ and Pre-reduction Performed by CCBP, Before (Left) and After (Right) Grouping the Classes Corresponding to the Same Vegetation Type

Class	Accuracy (%)	Class	Accuracy (%)
Corn—no till	84.00	Corn	80.24
Corn—min	67.48	Grass	94.69
Grass/pasture	88.89	Hay—windrowed	100.00
Grass/trees	98.94	Soybean	93.27
Hay—windrowed	100.00	Woods	97.54
Soybean—no till	54.18		
Soybean—min	85.15	Overall accuracy	91.28
Soybean—clean till	80.09	Average accuracy	93.15
Woods	97.54		
Overall accuracy	83.23		
Average accuracy	84.03		

reduction in the number of features to be obtained. The corresponding classification map is shown in Figure 10.4c.

These results confirm the great potential of the DBFE methodology (which allows optimizing effectively the classification accuracy of the Bayesian classifiers), but also suggest a further application of the proposed band-partitioning approach as an efficient pre-processor for DBFE. This is further confirmed by the overall comparison shown in Figure 10.8, which summarizes the behaviors of the classification accuracies of the proposed/considered approaches (i.e., band partitioning, selection, and transformation) as functions of m. Globally, the best results (from the viewpoints of both (a) the overall behavior of the accuracy versus m and (b) the peak accuracy values) were obtained by combining DBFE with the band-partitioning strategy (as a reference, the combination of CCBP and DBFE is considered in Figure 10.8), because this approach achieved higher values of OA and AA than SFS and DBFE for almost all values of $m \geq 8$ (for $m \leq 7$ classification maps

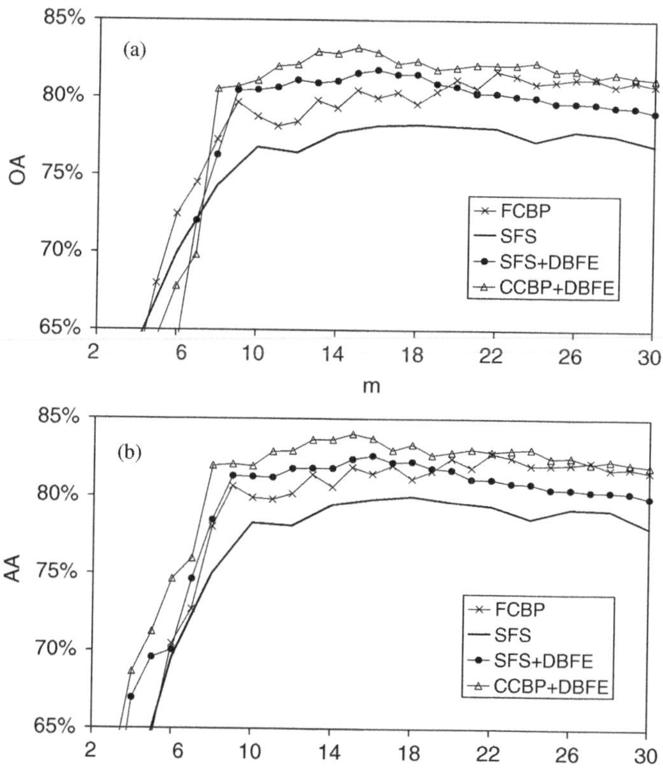

Figure 10.8. Overall comparison among the classification results obtained by the proposed/considered methods. Plot of the classification accuracy of GMAP as a function of the number m of selected/extracted features, for FCBP, SFS, and DBFE (applied with both SFS and CCBP as pre-processors): (a) overall accuracy (OA); (b) average accuracy (AA).

with OA $< 80\%$ were generated by all considered methods). Good classification results were also granted by the band-partitioning methodology (in particular, FCBP is considered in Figure 10.8) and by DBFE with SFS-based pre-processing, whereas worse accuracies were obtained by the simple SFS selection technique.

10.5. CONCLUSIONS

In the present chapter an innovative feature-transformation methodology exploiting numerical discrete search strategies in order to extract a set of synthetic nonoverlapping contiguous multispectral bands from a given hyperspectral image has been proposed. Specifically, three search strategies originally developed in the feature-selection context have been reformulated and extended to the present band-synthesis context, and a fourth innovative technique has also been introduced here.

The numerical experiments on real data assess the effectiveness of the band-partitioning approach as a feature-reduction tool, and they highlight the fact that the proposed methods achieve better accuracies than SFS, which is adopted as a reference feature-selection algorithm. The results are similar to the ones of DBFE chosen as a benchmark feature extraction method (and applied after a dimensionality pre-reduction is performed by SFS). On the other hand, very good classification results are obtained by combining together the proposed band-synthesis algorithms and the DBFE transformation method, by using the proposed techniques in order to perform the initial pre-reduction stage required by DBFE. Such results suggest the proposed techniques to be effective in two different feature-reduction contexts—that is, both as independent feature-reduction tools and as pre-processors for DBFE. In particular, the good accuracies granted by DBFE with this type of pre-processing stage further confirm the effectiveness of this well-known parametric approach and the need for an accurate preliminary choice of the algorithm applied in the pre-reduction stage.

As far as a comparison among the proposed techniques is concerned, we can note that quite similar accuracy results are provided by the four proposed methods, with the best peak accuracy values being achieved by FCBP and CCBP. However, the four methods exhibit different theoretical and computational properties. Specifically, a fast and deterministic computation time is guaranteed for SFBP and FCBP, but for such methods the resulting configurations of s-bands are not expected, in general, to be maximum points for the adopted functional. On the contrary, (local) optimality theorems are proved with regard to SABP and CCBP, which guarantee that these techniques converge to local maximum points of the functional. However, both methods exhibit a nondeterministic overall execution time, because the numbers of iterations needed to reach such local maxima are not known in advance. In particular, the specific search strategies adopted by CCBP and SABP suggest that CCBP is significantly faster than SABP in reaching convergence. This theoretical conjecture is also confirmed by the experiments, which pointed out a large difference in the number of iterations needed by the two methods to reach local maxima of the functional.

We note that, differently from the usual feature transformation techniques (such as DBFE itself), the developed band synthesis approach allows saving a physical meaning for the transformed features, which represent the bands acquired by a synthetic multispectral sensor. From this viewpoint, the proposed method aims at combining the flexibility of the extraction approach to feature reduction with the availability of a physical meaning for the features in the lower-dimensional space, which is typical of the selection approaches. In addition, the band-extraction process involves the estimation of class-conditional means and covariances only in the transformed lower-dimensional space, thus limiting the possible impact of the Hughes phenomenon on the estimation accuracy and on the resulting effectiveness of the feature-transformation process. Furthermore, this band-synthesis methodology may be applied in order to provide useful information for the design of multispectral sensors, since it automatically identifies sets of multispectral bands which are optimized as far as given land-cover classification problems are concerned.

An interesting development of this activity would be a further validation of SFBP, SABP, FCBP, and CCBP, in conjunction with different functionals (e.g., the divergence or directly the overall accuracy over a validation set), in order to assess both the effectiveness of such functionals from a classification viewpoint and the flexibility of the proposed optimization techniques.

ACKNOWLEDGMENTS

The authors would like to thank Professor David A. Landgrebe from Purdue University (USA) for providing the "Indian Pine" data set freeware at the website: ftp://ftp.ecn.purdue.edu/biehl/MultiSpec/92AV3C. We would also like to thank Massimo D'Incà for his support to the implementation of the methods.

APPENDIX: METRIC-THEORETIC INTERPRETATION OF THE SABP AND CCBP METHODS

In the present Appendix a suitable metric-space structure, which is employed to formalize the SABP and CCBP methods and to state their convergence properties, is introduced. Specifically, according to Section 10.3, we shall identify the set $\mathcal{S}$ of m s-bands with the corresponding set $\{\tau_0, \tau_1, \tau_2, \ldots, \tau_{m-1}, \tau_m\}$ of thresholds between consecutive s-bands, as follows:

$$0 = \tau_0 < \tau_1 < \tau_2 < \cdots < \tau_{m-1} < \tau_m = n \tag{10.4}$$

In order to formalize the threshold selection problem in a metric space perspective, an alternative representation of the possible configurations of s-bands is introduced. Specifically, $\mathcal{S}$ and the related collection $\{\tau_r\}_{r=0}^{m}$ of thresholds can be equivalently represented by introducing a binary n-dimensional string* $B = (B_1, B_2, \ldots, B_n)$

*Note that we denote by B_ℓ the ℓth component of a binary n-dimensional string B ($\ell = 1, 2, \ldots, n$).

($B_\ell \in \{0,1\}$ for $\ell = 1,2,\ldots,n$) such that $B_\ell = 1$ if ℓ is one of the threshold values (i.e., if $\ell = \tau_r$ for some $r = 0,1,2,\ldots,m$) and $B_\ell = 0$ otherwise. In particular, a binary string B with $B_n = 1$, m bits equal to 1 (including B_n), and $(n-m)$ bits equal to 0, which biunivocally identifies the collection $\{\tau_r\}_{r=0}^{m}$ of thresholds and the configuration $\mathcal{S}$ of s-bands, is obtained. More formally, by marking by $w(B)$ the "weight" of the string B—that is, the number of unitary bits in B [55]—the set of all the binary n-dimensional strings representing the configurations of s-bands is given by

$$\mathcal{B}_m^n = \{B \in \{0,1\}^n : B_n = 1,\ w(B) = m\} \tag{10.5}$$

It is well known that the set $\{0,1\}^n$ of binary n-dimensional strings can be endowed with a metric-space structure, by introducing the so-called Hamming distance, that is, the following function $d : \{0,1\}^n \times \{0,1\}^n \to \mathbb{R}$[56]:

$$d(B,B') = w(B \oplus B'), \qquad B, B' \in \{0,1\}^n \tag{10.6}$$

where "$\oplus$" stands for the usual exclusive-or operator [55]. Operatively, $d(B,B')$ is the number of bits of B, which are different from the corresponding bits in B' ($B, B' \in \{0,1\}^n$) and can be proved to satisfy the axioms of a metric function [56]:

- Positivity: $d(B,B') \geq 0$ for all $B, B' \in \{0,1\}^n$ with equality if and only if $B = B'$.
- Symmetry: $d(B,B') = d(B',B)$ for all $B, B' \in \{0,1\}^n$.
- Triangle inequality: $d(B,B') \leq d(B,B'') + d(B',B'')$ for all $B, B', B'' \in \{0,1\}^n$.

Such a metric-space structure is naturally inherited by the subset $\mathcal{B}_m^n$, which represents all the configurations of s-bands, thus allowing one to quantify the difference between two distinct configurations of s-bands and to reconsider the functional $J(\cdot)$ measuring the quality of each set of s-bands as a real function $J: \mathcal{B}_m^n \to \mathbb{R}$ defined over $\mathcal{B}_m^n$.

In addition, the metric definition implicitly allows the concepts of neighborhood and of local maximum of a function to be introduced. Specifically, the neighborhood of center $\bar{B} \in \mathcal{B}_m^n$ and radius $\delta > 0$ is defined as the set $\{B \in \mathcal{B}_m^n: d(B,\bar{B}) \leq \delta\}$ consisting of the configurations of s-bands whose distance from $\bar{B}$ is not higher than δ. Similarly, $\bar{B} \in \mathcal{B}_m^n$ is a local maximum point for $J(\cdot)$ if there exists $\delta > 0$ such that $J(B) \geq J(\bar{B})$ for all $B \in \mathcal{B}_m^n$ with $d(B,\bar{B}) \leq \delta$, that is, if $J(\bar{B})$ is the maximum value of $J(\cdot)$ over a neighborhood centered on $\bar{B}$.

The existence of such notions on $\mathcal{B}_m^n$ allows studying analytically the behaviors of the SABP and CCBP procedures. The following lemma about the specific behavior of the metric d over $\mathcal{B}_m^n$ (and not over the whole space $\{0,1\}^n$) turns out to be useful in this study.

Lemma. The minimum nonzero value of the Hamming distance $d(\cdot,\cdot)$ on $\mathcal{B}_m^n$ is 2; that is, $d(B,B') \geq 2$ for all $B,B' \in \mathcal{B}_m^n$ such that $B \neq B'$.

Proof. According to Eq. (10.6), d takes on only nonnegative integer values. Hence, if $B,B' \in \mathcal{B}_m^n$ and $B \neq B'$, then $d(B,B') \geq 1$. Let us suppose that $d(B,B') = 1$; hence, there is just one bit (say, the ℓth bit) in which B and B' differ; that is, $B_\ell = 1$ and $B'_\ell = 0$ or $B_\ell = 0$ and $B'_\ell = 1$. In the former case, $w(B) = w(B') + 1$, but this cannot occur because $w(B) = w(B') = m$ for $B,B' \in \mathcal{B}_m^n$. Similarly, also the case $B_\ell = 0$ and $B'_\ell = 1$ is not allowed, which completes the proof by contradiction.

Theorem 1. Denoting by $B^t \in \mathcal{B}_m^n$ the binary string corresponding to the set of m s-bands generated at the tth SABP iteration ($t = 1, 2, \ldots$), the set of local moves explored by SABP at the tth step is the radius-2 neighborhood centered on B^t (i.e., B^{t+1} is obtained by SABP by exhaustively exploring the whole radius-2 neighborhood centered on B^t). In addition, SABP converges in a finite number of iterations to a local maximum point of $J(\cdot)$.

Proof. Each local move tested by SABP in order to compute B^{t+1} generates a configuration of s-bands corresponding to a binary string $B \in \mathcal{B}_m^n$ obtained from B^t by removing a unitary bit (i.e., a threshold) and by adding another unitary bit in a different position, previously occupied by a zero bit (see Section 10.3.3). In both cases, Eq. (10.6) gives $d(B,B^t) = 2$. Conversely, if $B \in \mathcal{B}_m^n$ has distance 2 from B^t, then B and B^t differ in just two bits (say, the ℓth and the hth bits). This occurs in one of the following four cases:

1. $B_\ell = 0, B_\ell^t = 1, B_h = 0$, and $B_h^t = 1$.
2. $B_\ell = 1, B_\ell^t = 0, B_h = 0$, and $B_h^t = 1$.
3. $B_\ell = 0, B_\ell^t = 1, B_h = 1$, and $B_h^t = 0$.
4. $B_\ell = 1, B_\ell^t = 0, B_h = 1$, and $B_h^t = 0$.

Cases 1 and 4 are not allowed, because they would imply $w(B) = w(B^t) - 2$ and $w(B) = w(B^t) + 2$, respectively, whereas $w(B) = w(B^t) = m$ for all $B, B^t \in \mathcal{B}_m^n$. Case 2 means that a unitary bit was present in the hth position of B^t while it has been moved to the ℓth position in B; that is, a threshold was in the hth h-band in B^t and it has been moved to the ℓ-th h-band in B. This is one of the local moves performed by SABP. Similarly, also Case 3 is a well-defined local move made by SABP. Therefore, at the tth SABP iteration the whole radius-2 neighborhood of the current binary string B^t is exhaustively explored.

If in the first iteration of SABP no configuration allows an increase in $J(\cdot)$ to be obtained with respect to the initial configuration, then SABP stops immediately. Otherwise, each SABP iteration increases the value of $J(\cdot)$, which does not allow the algorithm to return to a configuration of s-bands generated in a previous iteration (i.e., if s, t are iteration numbers with $s > t$, we have $J(B^s) > J(B^t)$, and therefore $B^s \neq B^t$). Since SABP stops when no local increase is possible and since $\mathcal{B}_m^n$ is a finite set, SABP will stop in a finite number of iterations. In particular, according to the SABP stop condition, the final configuration B^* will be such that $J(B^*) \geq J(B)$ for all $B \in \mathcal{B}_m^n$ with $d(B, B^*) \leq 2$. We note that, since 2 is the minimum nonzero value of $d(\cdot, \cdot)$, B^* satisfies the above definition of local maximum point of $J(\cdot)$ with $\delta = 2$, which completes the proof.

Thus, SABP allows escaping from the local maxima when a higher value of $J(\cdot)$ is present in a neighborhood. According to the present metric interpretation, the procedure could be further generalized, by exploring larger neighborhoods of the current configuration of s-bands (i.e., neighborhoods of radius $\delta > 2$) at each iteration. This would improve the capability of the method to escape from the local maxima, but would also increase the complexity of the method and the resulting computation time.

Theorem 2. CCBP converges in a finite number of iterations to a local maximum point of $J(\cdot)$.

Proof. If in the first iteration of CCBP no configuration allows an increase in $J(\cdot)$ to be obtained with respect to the initial configuration, then CCBP stops immediately. Otherwise, each CCBP iteration (like each SABE iteration) increases the value of $J(\cdot)$; therefore, the proof of finite-time convergence reported above for SABE also holds for CCBP.

In order to complete the current proof, we have to demonstrate that the convergence point $B^* \in \mathcal{B}_m^n$ is a local maximum point for $J(\cdot)$. We cannot directly extend the proof presented above for SABE, because it relies on the fact that SABE explores a whole radius-2 neighborhood at each iteration, whereas CCBP does not exhibit this property.

According to Section 10.3.5, CCBP stops because $J(B) \leq J(B^*)$ for all $B \in \mathcal{B}_m^n$ which are explored by CCBP during the last iteration. Since this iteration is a full run of FCBP, CCBP first explores each solution $B \in \mathcal{B}_m^n$ which can be obtained by removing the first non-dummy unitary bit in B^* and by replacing it with another unitary bit in B, which is placed in a different position (not already occupied by a unitary bit). Therefore $d(B, B^*) = 2$. When marking by $B^{\#}$ the best one among such explored solutions, if the inequality $J(B^{\#}) > J(B^*)$ was satisfied, CCBP would immediately update the current binary string by accepting $B^{\#}$ as the new solution. However, B^* is, by definition, the convergence point of CCBP. Therefore, the inequality $J(B^{\#}) > J(B^*)$ is false by contradiction; that is, $J(B^*) \geq J(B^{\#}) \geq J(B)$ for all $B \in \mathcal{B}_m^n$ which can be obtained by replacing the first non-dummy unitary bit in B^*.

By the inductive repetition of this argument, we conclude that $J(B^*) \geq J(B)$ also for all $B \in \mathcal{B}_m^n$ which can be obtained by replacing each of the other $(m-2)$ non-dummy unitary bits. This means that the CCBP last iteration is also identical to an SABP iteration. Hence, according to Theorem 1, the whole radius-2 neighborhood of B^* is explored at this iteration and B^* is a local maximum point for $J(\cdot)$ with $\delta = 2$, which completes the proof.

A comparison between the proofs of Theorems 1 and 2 suggests a further insight into the differences between SABP and CCBP. At each iteration, SABP explores a whole neighborhood of the current solution, by making only distance-2 moves and by choosing the direction of the maximum increase in $J(\cdot)$. CCBP explores a less regular subset of the space $\mathcal{B}_m^n$, making more than one distance-2 moves at each iteration. This approach globally allows more distant solutions to be reached in a single step but does not identify, in general, the direction of the maximum increase in $J(\cdot)$. Hence, the local maximum point reached by SABP may be expected to be better than the one obtained by CCBP. On the other hand, CCBP could reach a local maximum point faster than SABP.

REFERENCES

1. J. A. Richards and X. Jia, *Remote Sensing Digital Image Analysis*, 3rd edition, Springer-Verlag, Berlin, 1999.
2. D. A. Landgrebe, *Signal Theory Methods in Multispectral Remote Sensing*, Wiley-Interscience, New York, 2003.
3. K. Fukunaga, *Introduction to Statistical Pattern Recognition*, 2nd edition, Academic Press, New York, 1990.
4. G. F. Hughes, On the mean accuracy of statistical pattern recognizers, *IEEE Transactions on Information Theory*, vol. 14, no. 1, pp. 55–63, 1968.
5. L. O. Jimenez and D. A. Landgrebe, Supervised classification in high-dimensional space: Geometrical, statistical, and asymptotical properties of multivariate data, *IEEE Transactions on Systems, Man, and Cybernetics—Part C*, vol. 28, no. 1, 39–54, 1998.
6. G. Shaw and D. Manolakis, Signal Processing for hyperspectral image explotation, *IEEE Signal Processing Magazine*, vol. 19, no. 1, p. 12, 2002.
7. S. B. Serpico and L. Bruzzone, A new search algorithm for feature selection in hyperspectral remote sensing images, *IEEE Transactions on Geoscience and Remote Sensing, Special Issue on Analysis of Hyperspectral Image Data*, vol. 39, no.7, pp. 1360–1367, 1994.
8. S. De Backer, P. Kempeneers, W. Debruyn, and P. Scheunders, A band selection technique for spectral classification, *IEEE Geoscience and Remote Sensing Letters, in print, available at http://ieeexplore.ieee.org/,* 2005.
9. J. Wiersma and D. A. Landgrebe, Analytical design of multispectral sensors. *IEEE Transactions on Geoscience and Remote Sensing*, vol. 18, no. 2, pp. 180–189, 1980.
10. W. K. Pratt, *Digital Image Processing*, 2nd edition, Wiley Interscience, New York, 1991.
11. A. Jain and D. Zongker, Feature selection: Evaluation, application, and small sample performance, *IEEE Transactions on Pattern Analysis and Machine Intelligence*, vol. 19, pp. 153–189, 1997.
12. M. Kudo and J. Sklansky, Comparison of algorithms that select features for pattern classifiers, *Pattern Recognition*, vol. 33, pp. 25–41, 2000.
13. P. M. Narendra and K. Fukunaga, A branch and bound algorithm for feature subset selection, *IEEE Transactions on Computers*, vol. 26, pp. 917–922, 1977.
14. P. Pudil, Feature selection toolbox software package, *Pattern Recognition Letters*, vol. 23, pp. 487–492, 2002.

15. L. Bruzzone and S. B. Serpico, A technique for feature selection in multiclass cases, *International Journal of Remote Sensing*, vol. 21, pp. 549–563, 2000.
16. P. Pudil, J. Novovicova, and J. Kittler, Floating search methods in feature selection, *International Journal of Remote Sensing*, 15, pp. 1119–1125, 1994.
17. P. Somol, P. Pudil, J. Novovicova, and P. Paclik, Adaptive floating search methods in feature selection, *Pattern Recognition Letters*, vol. 20, pp. 1157–1163, 1999.
18. P. Mitra, C. A. Murthy, and S. K. Pal, Unsupervised feature selection using feature similarity. *IEEE Transactions on Pattern Analysis and Machine Intelligence*, vol. 24, no. 3, pp. 301–312, 2002.
19. M. Ichino and J. Sklansky, Optimum feature selection by zero-one integer programming, *IEEE Transactions on Systems, Man and Cybernetics*, vol. 14, pp. 737–746, 1984.
20. A. Verikas and M. Bacauskiene, Feature selection with neural networks, *Pattern Recognition Letters*, vol. 23, pp. 1323–1335, 2002.
21. W. Siedlecki and J. Sklansky, A note on genetic algorithms for large-scale feature selection, *Pattern Recognition Letters*, vol. 10, pp. 335–347, 1989.
22. B. Yu, S. De Backer, and P. Scheunders, Genetic feature selection combined with composite fuzzy nearest neighbor classifiers for hyperspectral sattelite imagery, *Pattern Recognition Letters*, vol. 23, pp. 183–190, 2002.
23. H. Yao and L. Tian, A genetic-algorithm-based selective principal component analysis (ga-spca) method for high-dimensional data feature extraction. *IEEE Transactions on Geoscience and Remote Sensing*, vol. 41, no. 6, pp. 1469–1478, 2003.
24. W. Siedlecki and J. Sklansky, On automatic feature selection, *International Journal of Pattern Recognition and Artificial Intelligence*, vol. 2, pp. 197–210, 1988.
25. P. Pudil, J. Novovicova, N. Choakjarenwanit, and J. Kittler, Feature selection based on the approximation of class densities by finite mixtures of special type, *Pattern Recognition*, vol. 28, no. 9, pp. 1389–1398, 1994.
26. H. Zhang and G. Sun, Feature selection using tabu search method, *Pattern Recognition*, vol. 35, pp. 701–711, 2002.
27. N. Keshava, Distance metrics and band selection in hyperspectral processing with applications to material identification and spectral libraries, *IEEE Transactions on Geoscience and Remote Sensing*, vol. 42, no. 7, pp. 1552–1565, 2004.
28. R. Huang and M. He, Band selection based on feature weighting for classification of hyperspectral data. *IEEE Geoscience and Remote Sensing Letters*, vol. 2, no. 2, pp. 156–159, 2005.
29. M. Loog, R. P. W. Duin, and R. Haeb-Umbach, Multiclass linear dimension reduction by weighted pairwise fisher criteria, *IEEE Transactions on Pattern Analysis and Machine Intelligence*, vol. 23, pp. 762–766, 2001.
30. B.-C. Kuo and D. A. Landgrebe, A covariance estimator for small sample size classification problems and its application to feature extraction, *IEEE Transactions on Geoscience and Remote Sensing*, vol. 40, no. 4, pp. 814–819, 2002.
31. Q. Du and C.-I Chang, A linear constrained distance-based discriminant analysis for hyperspectral image classification, *Pattern Recognition*, vol. 34, pp. 361–373, 2001.
32. M. Bressan and J. Vitri, Nonparametric discriminant analysis and nearest neighbor classification, *Pattern Recognition Letters*, vol. 24, pp. 2743–2749, 2003.

33. B.-C. Kuo and D. A. Landgrebe, Nonparametric weighted feature extraction for classification, *IEEE Transactions on Geoscience and Remote Sensing*, vol. 42, no. 5, pp.1096–1105, 2004.
34. T. Hastie, A .Buja, and R. Tibshirani, Penalized discriminant analysis. *Annals of Statistics*, vol. 23, no. 1, pp. 73–102, 1995.
35. B. Yu, I. M. Ostland, P. Gong, and R. Pu, Penalized discriminant analysis of *in situ* hyperspectral data for conifer species recognition, *IEEE Transactions on Geoscience and Remote Sensing*, vol. 37, no. 5, pp. 2569–2577, 1999.
36. K.-R. Muller, S. Mika, G. Ratsch, K. Tsuda, and B. Scholkopf, An introduction to kernel-based learning algorithms, *IEEE Transactions on Neural Networks*, vol. 12, no. 2, pp. 181–201, 2001.
37. G. Baudat and F. Anouar, Generalized discriminant analysis using a kernel approach, *Neural Computation*, vol. 12, pp. 2385–2404, 2000.
38. S. Mika, G. Ratsch, B. Scholkopf, A. Smola, J. Weston, and K.-R. Muller, Invariant feature extraction and classification in kernel spaces, in *Advances in Neural Information Processing Systems*, vol. 12, MIT Press, Cambridge, MA, 1999.
39. C. Lee and D. A. Landgrebe, Feature extraction based on decision boundaries, *IEEE Transactions on Pattern Analysis and Machine Intelligence*, vol.15, no. 4, pp. 388–400, 1993.
40. L. O. Jimenez and D. A. Landgrebe, Hyperspectral data analysis and feature reduction via projection pursuit, *IEEE Transactions on Geoscience and Remote Sensing*, vol. 37, no. 6, pp. 2653–2667, 1999.
41. S. B. Serpico, M. D'Incà, F. Melgani, and G. Moser, A comparison of feature reduction techniques for classification of hyperspectral remote-sensing data, in *Proceedings of the SPIE Conference on Image and Signal Processing for Remote Sensing VIII, Crete, Greece, 22–27 September*, pp. 347–358, 2002.
42. X. Jia and J. A. Richards, Segmented principal components transformation for efficient hyperspectral remote-sensing image display and classification, *IEEE Transactions on Geoscience and Remote Sensing*, vol. 37, no. 1, pp. 538–542, 1999.
43. S. Kumar, J. Ghosh, and M. M. Crawford, Best-bases feature extraction algorithms for classification of hyperspectral data. *IEEE Transactions on Geoscience and Remote Sensing*, vol. 39, no.7, pp. 1368–1379, 2001.
44. J. T. Morgan, J. Ham, M. M. Crawford, A. Henneguelle, and J. Ghosh, Adaptive feature spaces for land cover classification with limited ground truth data, *International Journal of Pattern Recognition and Artificial Intelligence*, vol. 18, no. 5, pp. 777–799, 2004.
45. D. Korycinski, M. M. Crawford, J. W. Barnes, and J. Ghosh, Adaptive feature selection for hyperspectral data analysis using a binary hierarchical classifier and tabu search, in *Proceedings of the 2003 IEEE International Geoscience and Remote Sensing Symposium, Toulouse, France, 21–25 July*, Vol. 1, pp. 297–299, 2003.
46. D. Korycinski, M. M. Crawford, and J. W. Barnes, Adaptive feature selection for hyperspectral data analysis, in *Proceedings of the SPIE Conference on Image and Signal Processing for Remote Sensing IX, Barcelona*, Spain, pp. 213–225, 2003.
47. L. M. Bruce, C. H. Koger, and J. Li, Dimensionality reduction of hyperspectral data using discrete wavelet transform feature extraction, *IEEE Transactions on Geoscience and Remote Sensing*, vol. 40, no. 10, 2002.
48. S. Mallat, *A Wavelet Tour of Signal Processing*, Academic Press, New York, 1999.

49. J. A. Benediktsson, J. A. Palmason, and J. R. Sveinsson, Classification of hyperspectral data from urban areas based on extended morphological profiles, *IEEE Transactions on Geoscience and Remote Sensing*, vol. 43, no. 3, pp. 480–491, 2005.
50. A. Plaza, P. Martínez, J. Plaza, and R. Perez, Dimensionality reduction and classification of hyperspectral image data using sequences of extended morphological transformations. *IEEE Transactions on Geoscience and Remote Sensing*, vol. 43, no. 3, pp. 466–479, 2005.
51. H. L. Van Trees, *Detection, Estimation and Modulation Theory*, Vol. 1, John Wiley & Sons, New York, 1968.
52. R. O. Duda, P. E. Hart, and D. G. Stork, *Pattern Classification*, 2nd edition, John Wiley & Sons, New York, 2001.
53. L. Bruzzone, F. Roli, and S. B. Serpico, An extension to multiclass cases of the Jeffreys–Matusita distance, *IEEE Transactions on Pattern Analysis and Machine Intelligence*, vol. 33, pp. 1318–1321, 1995.
54. E. Cech, *Topological Spaces*, John Wiley & Sons, New York, 1966.
55. M. B. Smyth and R. Tsaur, Hyperconvex semi-metric spaces, *Topology Proceedings*, 26, pp. 791–810, 2002.
56. A. B. Carlson, P. B. Crilly, and J. C. Rutledge, *Communication Systems*, McGraw-Hill, New York, 2001.
57. R. B. Ash, *Information Theory*, Dover, New York, 1965.

CHAPTER 11

SEMISUPERVISED SUPPORT VECTOR MACHINES FOR CLASSIFICATION OF HYPERSPECTRAL REMOTE SENSING IMAGES

LORENZO BRUZZONE, MINGMIN CHI, AND MATTIA MARCONCINI
Department of Information and Communication Technology, University of Trento, I-38050 Trento, Italy

11.1. INTRODUCTION

The recent development of sensor technology resulted in the possibility to design hyperspectral sensors that can acquire remote sensing images in hundreds of spectral channels. Hyperspectral sensors are able to sample the reflective portion of the electromagnetic spectrum ranging from the visible region (0.4–0.7 m) through the near-infrared (about 2.4 m) in contiguous bands about 10 nm wide. Therefore, they represent an important technological evolution from earlier multispectral sensors, which typically collect spectral information in only a few wide noncontiguous bands. The high spectral resolution of hyperspectral sensors allows a detailed analysis of the spectral signature of land-cover classes (e.g., shape of narrow absorption bands), thus permitting to discriminate even species with very similar spectral behaviors (e.g., different types of forest). This results in the possibility to increase the classification accuracy with respect to the use of multispectral sensors.

From a methodological perspective, given the complexity of the classification process, usually supervised techniques are preferred to unsupervised techniques for the analysis of hyperspectral images. However, a critical issue in supervised classification of hyperspectral images is the definition of a proper training set for the learning of the classification algorithm. In this context, two main problems should be addressed relating to (i) the quantity of the available training patterns and (ii) the quality of the available training samples.

Hyperspectral Data Exploitation: Theory and Applications, Edited by Chein-I Chang

As regards the quantity of training patterns, in most applications the number of available training samples is not sufficient for a proper learning of the classifier, since gathering reliable prior information is often too expensive both in terms of economic costs and time. In particular, if the number of training samples is relatively small compared to the number of features (and thus of classifier parameters to be estimated), the problem of the curse of dimensionality (i.e., the Hughes phenomenon [1]) arises. This results in the risk of overfitting the training data and may involve poor generalization capabilities of the classifier. This problem is very critical with hyperspectral data.

Concerning the quality of training data, there are two important issues to consider in relation to hyperspectral images, which result in unrepresentative training sets that affect the accuracy of the classification process: (a) the correlation among training patterns taken from the same area and (b) the nonstationary behavior of the spectral signature of each land-cover class in the spatial domain of the scene.

As regards the first issue, in real applications the training samples are usually taken from the same site and often appear as different spatial clusters composed of neighboring pixels in the remote sensing images. As the autocorrelation function of an image is not impulsive in the spatial domain, this violates the required assumption of independence among samples included in the training set, thus reducing the information conveyed to the classification algorithm by the considered training patterns. Concerning the second issue, physical factors related to both ground and atmospheric conditions affect the spectral signatures in the spatial domain of the image. From a theoretical point of view, in order to obtain a good characterization of the nonstationary behavior of the spectral signatures of information classes, for each investigated land-cover category, training samples should be collected from different areas of the scene. Nevertheless, such a requirement is seldom satisfied in practical applications, thus degrading the generalization ability of the classification system in the learning of the classifier.

The aforementioned critical points result in *ill-posed* classification problems (see Baraldi et al. [2] for a detailed description of *ill-posed* problems), which cannot be properly solved with standard supervised classification techniques. Although some work has been carried out in the remote sensing literature to address *ill-posed* problems in relation to the analysis of hyperspectral images, at the present no general techniques capable to face both the quantity and quality problems of training data have been proposed. For this reason, it is very important to investigate and develop new methodologies that are able to exploit all the potentialities of such kind of data in real applications.

From an analysis of the literature, one can observe that the two most promising approaches to the classification of hyperspectral images consist in (i) the employment of semisupervised learning methods that take into account both labeled and unlabeled samples in the learning of the classifier and (ii) the use of supervised kernel-based methods, like Support Vector Machines (SVMs), which are intrinsically robust to high-dimensional problems.

As concerns the first issue, in the last few years there has been increasing interest in the use of semisupervised learning methods that exploit both labeled and unlabeled samples for addressing *ill-posed* problems with hyperspectral data [3, 4]. In remote sensing literature, this type of problem has been mainly addressed by semisupervised classifiers based on parametric or semiparametric techniques that approximate class distributions by a specific statistical model [4–6]. A possible approach to the learning problem is to use the Expectation-Maximization (EM) algorithm [7] for a maximum likelihood estimation of the parameters of information classes (the EM algorithm is typically defined under the assumption that the data are generated according to some simple known parametric models). In terms of Fisher information, Shahshahani and Landgrebe [4] proved that additional unlabeled samples are helpful for semisupervised classification in the context of a Gaussian Maximum Likelihood (GML) classifier under a zero-bias assumption. By assuming a Gaussian Mixture Model (GMM), the authors used the iterative EM algorithm to estimate model parameters with both labeled and unlabeled samples to better estimate the parameters of the GMM. In order to limit the negative influence of semilabeled samples (which are originally unlabeled samples that obtain labels during the learning process) on the estimation of the parameters of a GML classifier, in Tadjudin and Landgrebe [8] a weighting strategy was introduced (i.e., full weights are assigned to training samples, while reduced weights are defined for semilabeled samples during the estimation phase of the EM algorithm). However, the covariance matrices are highly variable when the size of the training set is small. To overcome this problem, in Jackson and Landgrebe [6] an adaptive covariance estimator was proposed to deal with *ill-posed* problems in the classification of hyperspectral data. In the adaptive quadratic process, semilabeled samples are incorporated in the training set to estimate regularized covariance matrices so that the variance of these matrices can be smaller compared to the conventional counterparts [8].

As concerns kernel-based methods, recently they have been promisingly applied to classification of hyperspectral remote sensing images. Kernel-based classifiers map data from the original input space to a kernel feature space of higher dimensionality, and then they solve a linear problem in that space. The most widely used kernel-based classifiers in the analysis of hyperspectral images are: Support Vector Machines (SVMs) [9], Kernel Fisher Discriminant Analysis (KFDA) [10, 11], and regularized AdaBoost [12]. Among the others, SVMs have been most extensively studied in the analysis of hyperdimensional data and proved to be very effective, outperforming many other systems in a wide variety of applications. SVMs exploit the principles of the statistical learning theory proposed by Vapnik [9] and attempt to separate samples belonging to different classes by tracing maximum margin hyperplanes in the kernel space where samples are analyzed. The success of SVMs in classification of hyperspectral data is justified by three main general reasons:

1. Their intrinsic effectiveness, with respect to traditional classifiers, which results in high classification accuracies and very good generalization capabilities.
2. The convexity of the objective function used in the learning of the classifier, which results in a unique solution (i.e., the system cannot fall into suboptimal solutions associated with local minima).

3. The possibility of representing the convex optimization problem in a dual formulation, where only nonzero Lagrange multipliers are necessary for defining the separation hyperplane (which is a very important advantage in the case of large datasets).

Despite these properties, for small-size training sets (i.e., in *ill-posed* problems), large deviations are possible for the empirical risk. In addition, small sample size can force the overfitting or underfitting of supervised learning. This may result in low classification accuracy as well as poor generalization capabilities.

To solve this kind of problem, transductive SVMs (TSVMs) [9, 13], which exploit both labeled and unlabeled samples in the learning phase, have been recently proposed in the machine learning community [14]. Bennett and Demiriz [15] implemented linear semisupervised SVMs,† and the results obtained on some of the standard UCI data sets showed little improvement when insufficient training information is available. Joachims [16] solved the quadratic optimization problem for the implementation of TSVMs with an application to text classification. The effectiveness of TSVMs for text classification (in a high-dimensional feature space) was supported by theoretical and experimental findings. A limitation of this implementation is that it requires an estimation of the ratio between unlabeled positive and negative samples for transductive learning at the beginning of the learning algorithm; nevertheless, in real applications, this prior knowledge is usually not available. Hence, by prefixing the number of expected positive patterns, the algorithm may lead to underfitting. Accordingly, Chen et al. [17] proposed a progressive TSVM (PTSVM) algorithm that can overcome such drawback by labeling in a pairwise-like way positive and negative samples. Though there is still some debate about whether the transductive inference can be successful in semisupervised classification [18], it has been proved both empirically and theoretically [9, 13, 15, 16] that TSVMs can be effective in handling problems where few labeled data are available (small-size labeled data sets), while unlabeled data are easy to obtain (e.g., text categorization, biological recognition, etc.).

In this chapter we address classification of hyperspectral data by introducing some semisupervised techniques based on SVMs. In particular, we focus the attention on two different kinds of semisupervised approaches, which exploit optimization algorithms for minimizing the cost function in the dual and in the primal formulation of the learning problem.

As regards the former approach, we address hyperspectral classification by a novel progressive semisupervised SVM classifier (referred as PS^{3}VM), which is an improvement of a transductive SVM technique recently presented in Bruzzone

†In the literature, *semisupervised learning* commonly refers to the employment of both labeled and unlabeled data for training and contrasts *supervised learning* (in which all available data are labeled) or *unsupervised learning* (in which all available data are unlabeled). *Transductive learning*, instead, is used to contrast *inductive learning*. A classifier is transductive if it only works on the labeled and unlabeled training data and cannot handle unseen data. Nevertheless, under this convention, TSVMs are actually inductive classifiers. The name TSVM originates from the intention to work only on the observed data, according to Vapnik [14].

et al. [19]. From a theoretical point of view, this technique presents two main methodological novelties with respect to Joachims [16] and Zhang and Oles [17]: (i) It exploits an original iterative semisupervised procedure that adopts a weighting strategy for the unlabeled patterns based on a time-dependent criterion and (ii) it exploits an adaptive convergence criterion able to fit the specific investigated problem, such that it is not necessary to estimate the expected number of iterations (which might be a difficult task in cases where few prior knowledge about the examined data is available).

Concerning the approach developed in the primal formulation, we initially introduce a semisupervised SVM that exploits a gradient descent algorithm to minimize the cost function (∇S^3VM). Then, we propose the use of a Low-Density Separation (LDS) algorithm based on the cluster assumption—that is, on the idea that the true decision boundary between the considered classes should be in low-density regions in the feature space.

In order to assess the effectiveness of the aforementioned techniques, many *ill-posed* classification problems have been defined using an hyperspectral Hyperion image acquired on the Okavango Delta area (Botswana).

The rest of the chapter is organized as follows. The next section introduces the basic principles of standard supervised SVMs both in the primal and in the dual formulations. Sections 11.3 and 11.4 describe the investigated semisupervised methodologies developed in the dual and the primal formulation, respectively, for binary classification problems. Section 11.5 presents the strategy adopted for the extension to the multiclass case of the aforementioned binary classifiers, whereas Section 11.6 deals with possible model selection strategies. Experimental results are presented in Section 11.7. Finally, Section 11.8 draws the conclusions of this chapter.

11.2. BACKGROUND: SVM METHODS FOR CLASSIFICATION OF HYPERSPECTRAL IMAGES

SVMs are large margin classifiers that exploit the principles of the statistical learning theory [14]. If an L2-norm regularizer is used, the optimization problem related to the learning of SVMs can be represented as a quadratic convex optimization problem with inequality constraints. For such optimization problems in nonlinear optimization theory, duality is preferred; thus, SVMs are often solved in dual representation by introducing Lagrange multipliers. However, this is not mandatory since one can also implement SVMs in the primal representation [20, 21]. In this section, we briefly review the basics for SVMs both in the dual and in the primal formulations.

11.2.1. SVMs In The Dual Formulation

Let $X = \{\mathbf{x}_l\}_{l=1}^n$ be the set of n available training samples and $Y = \{y_l\}_{l=1}^n$ be the set of associated labels. Standard SVMs are linear binary inductive learning classifiers where data in input space are linearly separated by the hyperplane:

$$h : f(\mathbf{x}) = \mathbf{w} \cdot \mathbf{x} + b = 0 \tag{11.1}$$

with a maximum geometric margin:

$$\frac{2}{\| \mathbf{w} \|^2} \tag{11.2}$$

where $\mathbf{x}$ represents a generic sample, $\mathbf{w}$ is a vector normal to the hyperplane, and b is a constant such that $\frac{b}{\|\mathbf{w}\|^2}$ represents the distance of the hyperplane from the origin. The objective of SVMs is to solve the following quadratic optimization problem with proper inequality constraints:

$$\begin{cases} \min_{\mathbf{w},b} \left\{ \frac{1}{2} \| \mathbf{w} \|^2 \right\} \\ y_l(\mathbf{w} \cdot \mathbf{x}_l + b) \geq 1 \qquad \forall l = 1, \ldots, n \end{cases} \tag{11.3}$$

Since direct handling of inequality constraints is difficult, Lagrange theory is usually exploited by introducing Lagrange multipliers $\alpha_{l=1}^{n}$ for the quadratic optimization problem. This leads to an alternative dual representation:

$$\begin{cases} \max_{\alpha} \left\{ \sum_{l=1}^{n} \alpha_l - \frac{1}{2} \sum_{l=1}^{n} \sum_{i=1}^{n} y_l y_i \alpha_l \alpha_i \langle \mathbf{x}_l, \mathbf{x}_i \rangle \right\} \\ \alpha_l \geq 0, \quad 1 \leq l \leq n \\ \sum_{l=1}^{n} y_l \alpha_l = 0 \end{cases} \tag{11.4}$$

According to the following KKT conditions (which are necessary and sufficient conditions for solving (11.4) with respect to $\boldsymbol{\alpha}$),

$$\begin{cases} \dfrac{\partial L(\mathbf{w}, b, \boldsymbol{\alpha})}{\partial \mathbf{w}} = 0 \\ \dfrac{\partial L(\mathbf{w}, b, \boldsymbol{\alpha})}{\partial b} = 0 \\ \alpha_l \geq 0, \qquad 1 \leq l \leq n \\ y_l(\mathbf{w} \cdot \mathbf{x}_l + b) - 1 \geq 0 \\ \alpha_l [y_l(\mathbf{w} \cdot \mathbf{x}_l + b) - 1] = 0 \end{cases} \tag{11.5}$$

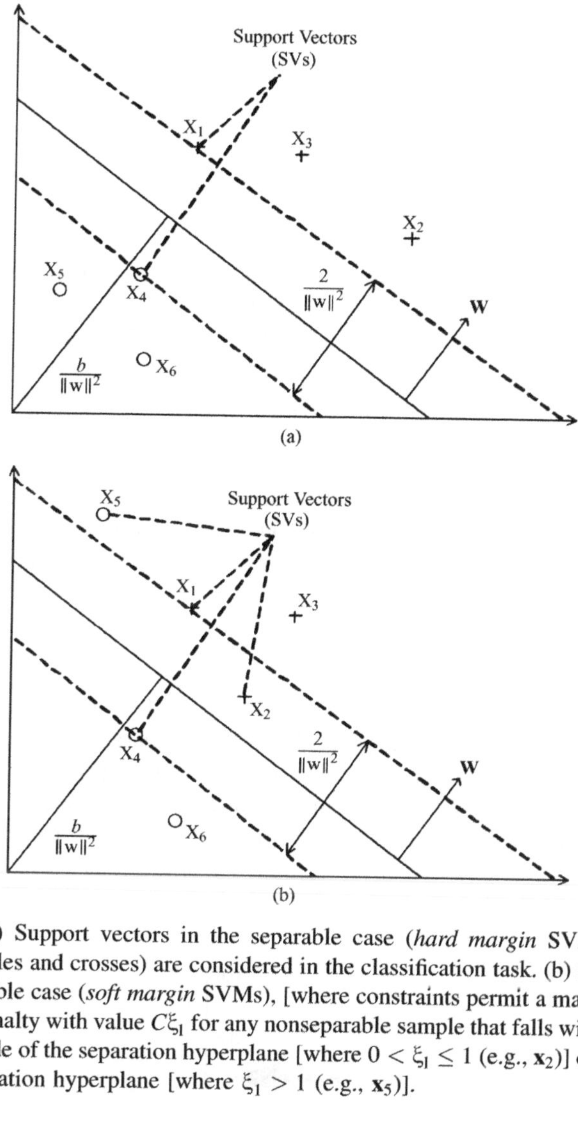

Figure 11.1. (a) Support vectors in the separable case (*hard margin* SVM), where two classes (i.e., circles and crosses) are considered in the classification task. (b) Support vectors in the nonseparable case (*soft margin* SVMs), [where constraints permit a margin less than 1 and contain a penalty with value $C\xi_l$ for any nonseparable sample that falls within the margin on the correct side of the separation hyperplane [where $0 < \xi_l \leq 1$ (e.g., $\mathbf{x}_2$)] or on the wrong side of the separation hyperplane [where $\xi_l > 1$ (e.g., $\mathbf{x}_5$)].

it is possible to demonstrate that only the training samples associated with nonzero dual variables (Lagrangian multipliers) contribute to define the separation hyperplane. These training samples are called *support vectors* (SVs) (e.g., $\mathbf{x}_1$ and $\mathbf{x}_4$ in Figure 11.1 a). The SVs lie on the margin bounds, while the remaining training samples are irrelevant for the classification.

To allow the possibility that in a nonseparable case some training samples violate (11.4) for increasing the generalization ability of the classifier, the constraints are softened by introducing the slack variables ξ_l and the associated *penalization parameter* C (also called *regularization parameter*). Accordingly, the new well-posed optimization problem (using L2 regularization) becomes

$$\begin{cases} \min_{\mathbf{w},b,\xi} \left\{ \frac{1}{2} \| \mathbf{w} \|^2 + C \sum_{l=1}^{n} \xi_l \right\} \\ y_l(\mathbf{w} \cdot \mathbf{x}_l + b) \geq 1 - \xi_l \qquad \forall l = 1, \ldots, n \\ \xi_l > 0 \end{cases} \tag{11.6}$$

The SVMs with the above-described soft constraints are called *soft margin* SVMs. To emphasize the difference with this nonseparable case, the optimization problem in (11.3) is called *hard margin* SVM. In soft margin SVMs, a set of SVs consists of training samples in the upper and lower margins and outlier samples (e.g., $\mathbf{x}_1$, $\mathbf{x}_2$ $\mathbf{x}_4$, and $\mathbf{x}_5$ in Figure 11.1 b). A classifier characterized by good generalization ability is designed by controlling both the classifier capacity and the sum of the slack variables $\sum_l \xi_l$. The latter can be shown to provide an upper bound to the number of training errors. After carrying out optimization on (11.6) with some quadratic optimization techniques (e.g., *Chunking, Decomposition methods*, etc.), one can obtain the dual variables α_l and thus $\mathbf{w}$. Hence, it is possible to predict the label for a given sample $\mathbf{x}$ according to

$$\hat{y} = \text{sgn}[f(\mathbf{x})] \tag{11.7}$$

that is, the labeling is "+1" when $f(\mathbf{x}) \geq 0$ and "−1" otherwise.

If the data in the input space cannot be linearly separated, they can be projected into a higher-dimensional feature space (e.g., a Hilbert space H) with a nonlinear mapping function $\Phi(\cdot)$, where the inner product between the two mapped feature vectors $\mathbf{x}_l$ and $\mathbf{x}_i$ becomes $\langle \Phi(\mathbf{x}_l), \Phi(\mathbf{x}_i) \rangle$ (Figure 11.2). In this case, due to Mercer's

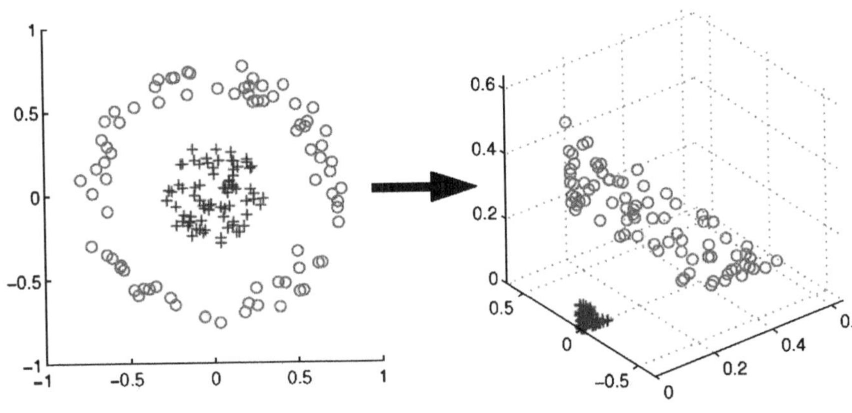

(a) (b)

Figure 11.2. Example of nonlinear transformation. When a set of samples cannot be linearly separated in the input space (a), their separation hyperplane can be constructed in another space (e.g., a Hilbert space) by nonlinear mappling Φ (b).

theorem, if we replace the inner product in (11.6) with a kernel function $k(\mathbf{x}_l, \mathbf{x}_i) = \langle \Phi(\mathbf{x}_l), \Phi(\mathbf{x}_i) \rangle$ (i.e., "*kernel trick*"), we can avoid representing the feature vectors explicitly. Thus, the dual representation (11.6), with the constraint $0 \leq \alpha_l \leq C$ in the nonseparable case, can be expressed in terms of the inner product with a kernel function as follows:

$$\max_{\alpha} \left\{ \sum_{l=1}^{n} \alpha_l - \frac{1}{2} \sum_{l=1}^{n} \sum_{i=1}^{n} y_l y_i \alpha_l \alpha_i k(\mathbf{x}_l, \mathbf{x}_i) \right\}$$
$$0 \leq \alpha_l \leq C, \qquad 1 \leq l \leq n \tag{11.8}$$
$$\sum_{l=1}^{n} y_l \alpha_l = 0$$

where $\mathbf{K}$ is called *kernel matrix* (or *Gram matrix*) and denotes the n squared positive definite matrix whose elements are $K_{li} = k(\mathbf{x}_l, \mathbf{x}_i)$. $\mathbf{K}$ is symmetric (i.e., $k(\mathbf{x}_l, \mathbf{x}_i) = k(\mathbf{x}_i, \mathbf{x}_l)$) and satisfies the following condition:

$$\sum_{i,l} \alpha_i \alpha_l K_{il} \geq 0 \tag{11.9}$$

Therefore, it represents a measure of the similarity among the data. Unlike in other classification techniques such as Multilayer Perceptron Neural Networks [22], the kernel $k(\cdot, \cdot)$ ensures that the objective function is convex; hence, there are no local maxima in the cost function in (11.8) for standard supervised SVMs. Due to the "*kernel trick*," one can see that the number of operations required to compute the inner product in the nonlinear SVMs by evaluating the kernel function is not necessarily proportional to the number of features [23]. Hence, the use of kernels in sparse representation potentially circumvents the high-dimensional feature problem inherent in hyperspectral remote sensing data.

11.2.2. SVMs In The Primal Formulation

The quadratic optimization problem (with inequality constraints in L2-norm SVMs) makes most of the literature to focus on the employment of Lagrange theory. However, in references 20 and 21 the authors proved that optimization problems in SVMs can be also solved directly in the primal formulation. From a theoretical point of view, primal and dual optimizations are equivalent in terms of both solution quality and time complexity. Nevertheless, primal optimization can be superior when it comes to approximate the solution, because it is focused on directly determining the best discriminant function in the original input space [20]. A possible implementation of primal optimization of SVMs is to use a local minimization technique (such as gradient descent [24]) on the original representation. In this

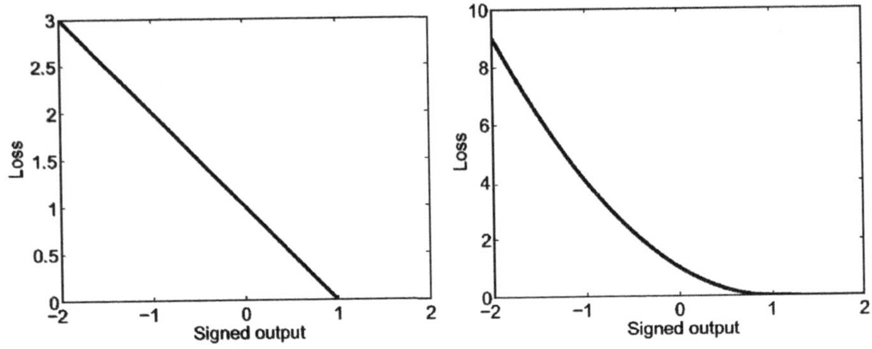

Figure 11.3. Loss for the labeled samples in (11.10) when (a) $p = 1$, hinge loss $H(t) = \max(0, 1 - t)$ and (b) $p = 2$, quadratic loss $H(t) = \max(0, 1 - t)^2$.

case, the minimization problem (11.3) can be rewritten without explicit constraints as follows:

$$\min_{\mathbf{w},b}\left\{\frac{1}{2}\|\mathbf{w}\|^2 + C\sum_{l=1}^{n} H(y_l(\mathbf{w}\cdot\mathbf{x}_l + b))\right\} \tag{11.10}$$

where $H(y_l(\mathbf{w}\cdot\mathbf{x}_l + b))$ is the loss for the training patterns $\mathbf{x}_l \in X$, defined by $H(t) = \max(0, 1 - t)^p$. When $p = 1$ a hinge loss is used (cf. Figure 11.3 a), and when $p = 2$ a quadratic loss is considered (cf. Figure 11.3 b). In this chapter we only take into account the quadratic loss for labeled samples, since the hinge loss is nondifferentiable.

With respect to (11.10), we define a labeled sample $\mathbf{x}_l$ given the vector $\mathbf{w}$, as a support vector if $y_l(\mathbf{w}\cdot\mathbf{x}_l + b) < 1$; that is, the loss on such sample is not equal to zero [20]. Note that (11.10) is an unconstrained optimization problem. For simplicity, in the following discussion we ignore the offset b, since all the algebra presented below can be extended easily for taking it into account. If gradient descent is used, provided that $H(\cdot)$ is differentiable, the gradient of (11.10) with respect to $\mathbf{w}$ is given by

$$\nabla = \mathbf{w} + C\sum_{l=1}^{n}\frac{\partial H(y_l\mathbf{w}\cdot\mathbf{x}_l)}{\partial \mathbf{w}}\mathbf{x}_l y_l \tag{11.11}$$

where $\frac{\partial H(y_l\mathbf{w}\cdot\mathbf{x}_l)}{\partial \mathbf{w}}$ is the partial derivative of $H(y_l\mathbf{w}\cdot\mathbf{x}_l)$ with respect to $\mathbf{w}$. For the optimal solution $\mathbf{w}^*$, the gradient vanishes such that $\nabla_{\mathbf{w}^*} = 0$. Hence, the solution is a linear combination of input data:

$$\mathbf{w}^* = \sum_{l=1}^{n}\beta_l\mathbf{x}_l \tag{11.12}$$

where

$$\beta_l = -C\frac{\partial H(y_l \mathbf{w} \cdot \mathbf{x}_l)}{\partial \mathbf{w}} y_l \tag{11.13}$$

This result is also known as *Representer Theorem* [25]. Then, it is possible to replace $\mathbf{w}$ in (11.10) with (11.12) as follows:

$$\min_{\beta}\left\{\frac{1}{2}\sum_{l=1}^{n}\sum_{i=1}^{n}\beta_l\beta_i\langle \mathbf{x}_l, \mathbf{x}_i\rangle + C\sum_{l=1}^{n}H\left(y_l\sum_{i=1}^{n}\beta_i\langle \mathbf{x}_l, \mathbf{x}_i\rangle\right)\right\} \tag{11.14}$$

Any optimization technique (e.g., Newton method) can be used to solve (11.14) with respect to $\boldsymbol{\beta}$ (for the details the reader is referred to Chapelle [20]). For nonlinear SVMs in the primal, again it is possible to exploit the "*kernel trick*" used for nonlinear SVMs in the dual, where the inner product $\langle \mathbf{x}_l, \mathbf{x}_i\rangle$ is replaced by a kernel function.

11.3. PROPOSED PS^{3}VM IN DUAL FORMULATION

In this section, we introduce the PS^{3}VM algorithm (Table 11.1), which is an improvement of the transductive SVM technique recently presented in the literature in Bruzzone et al. [19]. This algorithm is specifically designed to tackle the problem of hyperspectral image classification. The attention will be focused on the two-class case; for the generalization to the multiclass case the reader is referred to Section 11.5.

The PS^{3}VM algorithm exploits the standard theoretical approach of supervised SVMs developed in the dual formulation presented in Section 11.2. According to Section 11.2, let $X = \{\mathbf{x}_l\}_{l=1}^{n}$ be the set of n available training samples and let $Y = \{y_l\}_{l=1}^{n}$ be the set of associated labels. Let $X^* = \{\mathbf{x}_u^*\}_{u=1}^{m}$ be the unlabeled set consisting of m unlabeled samples, and let $Y^* = \{y_u^*\}_{u=1}^{m}$ be the corresponding predicted labels obtained according to the classification model after learning with the training set. $\mathbf{x}_l, \mathbf{x}_u^* \in R^N$ and $y_l, y_u^* \in \{-1, +1\}$. Similarly to supervised SVMs, the nonlinear mapping $\phi : R^N \to F$ to a higher (possibly infinite) dimensional (Hilbert) feature space is defined.

The PS^{3}VM technique is based on an iterative algorithm. From a theoretical point of view, it is composed of three main phases, defined on the basis of the type of samples employed in the training process and of their weights in the cost function: (i) *initialization* (only original training samples with a single regularization term), (ii) *semisupervised learning* (original training samples with a single regularization term plus originally unlabeled patterns with a regularization term based on a temporal criterion), and (iii) *convergence* (original training samples with a single regularization term plus originally unlabeled patterns with a single regularization term).

TABLE 11.1. Learning Procedure for the Proposed Binary PS^{3}VM

Begin

- $i = 0$
- $X^{(0)} = \{\mathbf{x}_1, \ldots, \mathbf{x}_n\}$, $X^{*(0)} = \{\mathbf{x}_1^*, \ldots, \mathbf{x}_m^*\}$, $S^{(0)} = \emptyset$
- Solve : $\begin{cases} \min\limits_{\mathbf{w}^{(0)} b^{(0)}, \xi^{(0)}} \left\{ \frac{1}{2} \| \mathbf{w}^{(0)} \|^2 + C \sum\limits_{l=1}^{n} \xi_l^{(0)} \right\} \\ y_l^{(0)} (\langle \phi(\mathbf{x}_l), \mathbf{w}^{(0)} \rangle + b^{(0)}) \geq 1 - \xi_l^{(0)} \quad \forall l = 1, \ldots, n \\ \xi_l^{(0)} \geq 0 \end{cases}$

Repeat

- $\forall \mathbf{x}_u^* \in X^{*(i)}, \forall \mathbf{x}_u^* \in X^{*(i)}$ calculate $f^{(i)}(\mathbf{x}_u^*) = \mathbf{w}^{(i)} \cdot \mathbf{x}_u^* + b^{(i)}$
- Update the set containing the pseudo-labeled patterns in the upper side of the margin band:

$$H_{\text{up}}^{(i)} = \{\mathbf{x}_u^* | \mathbf{x}_u^* \in X^{*(i)},\ 0 \leq f^{(i)}(\mathbf{x}_u^*) < 1\}$$

- Update the set containing the pseudo-labeled patterns in the lower side of the margin band:

$$H_{\text{low}}^{(i)} = \{\mathbf{x}_u^* | \mathbf{x}_u^* \in X^{*(i)}, -1 < f^{(i)}(\mathbf{x}_u^*) < 0\}$$

- Sort $H_{\text{up}}^{(i)}$ according to

$$f^{(i)}(\mathbf{x}_u^{*\text{up}}) \geq f^{(i)}(\mathbf{x}_{u+1}^{*\text{up}}),\ \forall u = 1, \ldots, |H_{\text{up}}^{(i)}| - 1,\ \mathbf{x}_u^{*\text{up}} \in H_{\text{up}}^{(i)}$$

- $H_{\text{low}}^{(i)}$ according to

$$f^{(i)}(\mathbf{x}_u^{*\text{low}}) \leq f^{(i)}(\mathbf{x}_{u+1}^{*\text{low}}),\ \forall u = 1, \ldots, |H_{\text{low}}^{(i)}| - 1,\ \mathbf{x}_u^{*\text{low}} \in H_{\text{low}}^{(i)}$$

- Compute $\lambda^{(i)} = \min(|H_{\text{up}}^{(i)}|, \rho)$, $\delta^{(i)} = \min(|H_{\text{low}}^{(i)}|, \rho)$
- Update the set containing the semilabeled patterns selected at the ith iteration:

$$H^{(i)} = \{\mathbf{x}_1^{*\text{up}}, \ldots, \mathbf{x}_{\lambda^{(i)}}^{*\text{up}}, \mathbf{x}_1^{*\text{low}}, \ldots, \mathbf{x}_{\delta^{(i)}}^{*\text{low}}\},\ \mathbf{x}_u^{*\text{up}} \in H_{\text{up}}^{(i)},\ \mathbf{x}_u^{*\text{low}}, \in H_{\text{low}}^{(i)}$$

- Update the set containing all the semilabeled patterns:

$$J^{(i)} = J_1^{(i)} \cup J_2^{(i)} \cup \ldots \cup J_\gamma^{(i)} \begin{cases} J_1^{(i)} = H^{(i)} \\ J_k^{(i)} = J_{k-1}^{(i-1)} - S^{(i)},\ \forall k = 2, \ldots, \gamma - 1 \\ J_\gamma^{(i)} = (J_\gamma^{(i-1)} \cup J_{\gamma-1}^{(i-1)}) - S^{(i)} \end{cases}$$

- $\eta^{(i)} = |J^{(i)}|$
- $i = i + 1$
- Update the training set: $X^{(i)} = (X^{(i-1)} \cup H^{(i-1)}) - S^{(i-1)}$
- Update the unlabeled set: $X^{*(i)} = (X^{*(i-1)} - H^{(i-1)}) - S^{(i-1)}$
- Define the regularization parameter for each semilabeled pattern:

for $\forall j = 1, \ldots, \eta^{(i-1)}$

$$C_j^* = \left(\frac{C^{*\max} - C^{*(0)}}{(\gamma - 1)^2} \right) (k-1)^2 + C^{*(0)} \Leftrightarrow (\mathbf{x}_j^* \in J_k^{(i-1)}), \qquad \gamma, k \in_0$$

TABLE 11.1. (*Continued*)

- Solve : $$\begin{cases} \min\limits_{\mathbf{w}^{(i)},b^{(i)},\xi^{(i)},\xi^{*(i)}} \left\{ \frac{1}{2} \| \mathbf{w}^{(i)} \|^2 + C\sum_{i=1}^{n} \xi_i^{(i)} + \sum_{j=1}^{\eta^{(i-1)}} C_j^* \xi_j^{*(i)} \right\} \\ y_l^{(i)} \cdot (\mathbf{w}^{(i)} \cdot \mathbf{x}_l + b^{(i)}) \geq 1 - \xi_l^{(i)}, \quad \forall l = 1, \ldots, n, \ \mathbf{x}_l \in X^{(i)} \\ y_l^{*(i)} \cdot (\mathbf{w}^{(i)} \cdot \mathbf{x}_j^* + b^{(i)}) \geq 1 - \xi_j^{*(i)}, \quad \forall j = 1, \ldots, \eta^{(i-1)}, \ \mathbf{x}_j^* \in J^{(i-1)} \\ \xi_l^{(i)}, \xi_j^{*(i)} \geq 0 \end{cases}$$
- Update the set containing the mislabeled patterns at the ith iteration:

$$S^{(i)} = \begin{cases} \emptyset, & i = 0, 1 \\ \{\mathbf{x}_u^* | (\mathbf{x}_u^* \in X^{(i)}, \mathbf{x}_u^* \in X^{(i-1)}), y^{*(i)} \neq y^{*(i-1)}\}, & i \geq 2 \end{cases}$$

Until $\begin{cases} |M^{(i)}| \leq \lceil \beta \cdot m \rceil \\ |S^{(i)}| \leq \lceil \beta \cdot m \rceil \end{cases}$ where $M^{(i)} = \{\mathbf{x}_u^* | \mathbf{x}_u^* \in X^{*(i)}, -1 < f^{(i)}(\mathbf{x}_u^*) < 1\}$

- end = i
- Fix : $\begin{cases} y_l = y_l^{\text{end}}, & y_j^* = y_j^{*(\text{end})} \\ J = J^{(\text{end}-1)}, & \eta = \eta^{(\text{end}-1)}, \ X = X^{(0)} \end{cases}$
- Solve : $$\begin{cases} \min\limits_{\mathbf{w},b,\xi,\xi^*} \left\{ \frac{1}{2} \| \mathbf{w} \|^2 + C\sum_{i=1}^{n} \xi_i + C^{*\max} \sum_{j=1}^{n} \xi_j^* \right\} \\ y_l \cdot (\mathbf{w} \cdot \mathbf{x}_l + b) \geq 1 - \xi_l, \quad \forall l = 1, \ldots, n, \ \mathbf{x}_l \in X \\ y_j^* \cdot (\mathbf{w} \cdot \mathbf{x}_j^* + b) \geq 1 - \xi_j^*, \quad \forall j = 1, \ldots, \eta, \ \mathbf{x}_j^* \in J \\ \xi_l, \xi_l^* \geq 0 \end{cases}$$

End

Phase 1: Initialization. The first phase corresponds to the initial step of the entire process ($i = 0$).† We have that $X^{(0)} = \{\mathbf{x}_1, \ldots, \mathbf{x}_n\}$ and $X^{*(0)} = \{\mathbf{x}_1^*, \ldots, \mathbf{x}_m^*\}$. As for both the TSVM [16] and the PTSVM [17] algorithms, a standard supervised SVM is used to obtain an initial separation hyperplane based on training data alone $\{\mathbf{x}_l\}_{l=1}^{n} \in X^{(0)}$. Let $\boldsymbol{\xi}^{(0)} = \{\xi_1^{(0)}, \ldots, \xi_n^{(0)}\}$ be the vector of the slack variables associated with the patterns of $X^{(0)}$.

The bound cost function to minimize is the following:

$$\begin{cases} \min\limits_{\mathbf{w}^{(0)},b^{(0)},\xi^{(0)}} \left\{ \frac{1}{2} \| \mathbf{w}^{(0)} \|^2 + C\sum_{l=1}^{n} \xi_l^{(0)} \right\} \\ y_l^{(0)}(\phi(\mathbf{x}_l) \cdot \mathbf{w}^{(0)} + b^{(0)}) \geq 1 - \xi_l^{(0)} \quad \forall l = 1, \ldots, n \\ \xi_l^{(0)} \geq 0 \end{cases} \tag{11.15}$$

†The superscript (i), $i \in N$, refers to the values of the parameters at the ith iteration.

According to the resulting decision function $f^{(0)}(\mathbf{x}) = \mathbf{w}^{(0)} \cdot \mathbf{x} + b^{(0)}$, "pseudo" labels $\{y_u^*\}_{u=1}^m$ are given to the unlabeled samples $\{\mathbf{x}_u^*\}_{u=1}^m$, which are therefore called *pseudo-labeled* patterns.

Phase 2: Semisupervised Learning. The second phase starts with iteration $i = 1$ and represents the core of the PS^{3}VM. At the generic iteration i, each pattern of $X^{*(i)}$ is analyzed and the value of the decision function determined at iteration $i-1$ is computed. Because support vectors bear the richest information (i.e., they are the only patterns that affect the position of the separation hyperplane), among the "informative" samples (i.e., the ones in the margin band) unlabeled samples closest to the margin bounds have the highest probability to be correctly classified.

Let us define the two following subsets:

$$H_{\text{up}}^{(i)} = \{\mathbf{x}_u^* | \mathbf{x}_u^* \in X^{*(i)}, \quad 0 \leq f^{(i)}(\mathbf{x}_u^*) < 1\} \tag{11.16}$$

$$H_{\text{low}}^{(i)} = \{\mathbf{x}_u^* | \mathbf{x}_u^* \in X^{*(i)}, \quad -1 < f^{(i)}(\mathbf{x}_u^*) < 0\} \tag{11.17}$$

where $f^{(i)}(\mathbf{x}) = \mathbf{w}^{(i)} \cdot \mathbf{x} + b^{(i)}$. $H_{\text{up}}^{(i)}$ is made up of the patterns of the unlabeled set that at the iteration i lie between the two hyperplanes $h : f^{(i)}(\mathbf{x}) = 0$ and $h_1 : f^{(i)}(\mathbf{x}) = +1$, with the lower bound included. $H_{\text{low}}^{(i)}$ is made up of the patterns of the unlabeled set that at the iteration i lie in the space between the two hyperplanes $h : f^{(i)}(\mathbf{x}) = 0$ and $h_2 : f^{(i)}(\mathbf{x}) = -1$. Without loss of generality, we can sort the aforementioned sets according to

$$f^{(i)}(\mathbf{x}_u^{*\text{up}}) \geq f^{(i)}(\mathbf{x}_{u+1}^{*\text{up}}), \qquad \forall u = 1, \ldots, |H_{\text{up}}^{(i)}| - 1, \mathbf{x}_u^{*\text{up}} \in H_{\text{up}}^{(i)} \tag{11.18}$$

$$f^{(i)}(\mathbf{x}_u^{*\text{low}}) \leq f^{(i)}(\mathbf{x}_{u+1}^{*\text{low}}), \qquad \forall u = 1, \ldots, |H_{\text{low}}^{(i)}| - 1, \mathbf{x}_u^{*\text{low}} \in H_{\text{low}}^{(i)} \tag{11.19}$$

This means that:

- The first element of $H_{\text{up}}^{(i)}$ has the smallest Euclidean distance from h_1, whereas the last element of $H_{\text{up}}^{(i)}$ has the smallest Euclidean distance from h.
- The first element of $H_{\text{low}}^{(i)}$ has the smallest Euclidian distance from h_2, whereas the last element of $H_{\text{low}}^{(i)}$ has the smallest Euclidean distance from h.

The proposed approach is inspired by the PTSVM algorithm in which, at each iteration of the learning process, one positive and one negative example are labeled simultaneously. In particular, the samples in the margin having the maximum and the minimum values of the decision function, respectively, are associated with the sign of their decision function. Nevertheless, as two patterns may not be sufficiently representative for tuning the position of the hyperplane, in the algorithm an alternative strategy is adopted. In greater detail, at each iteration, both the first ρ patterns belonging to $H_{\text{up}}^{(i)}$ and to $H_{\text{low}}^{(i)}$ (where the value $\rho \geq 1$ is defined *a priori* by the user), whose pseudo-labels are "+1 " and "−1," respectively, are selected and inserted into the training set (cf. Figure 11.4 and Figure 11.5). Such samples are defined as

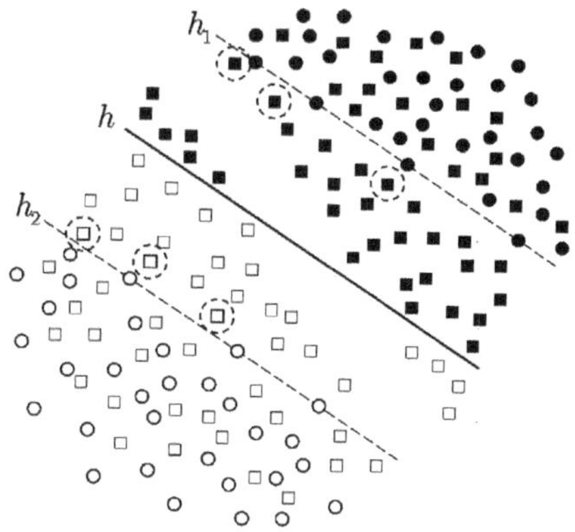

Figure 11.4. Margin and separation hyperplane resulting after the first iteration of the PS^{3}VM algorithm for a simulated data set. Training patterns are shown as white (class "−1") and black (class "+1") circles. Pseudolabeled samples belonging to class "−1" and "+1" are shown as white and black squares, respectively. The separation hyperplane is shown as a solid line, whereas the dashed lines define the margin. The dashed circles highlight the ρ (ρ = 3) semilabeled patterns selected from both the upper and the lower side of the margin at the first iteration.

"*semilabeled*" patterns. On the one hand, this strategy has the advantages of (i) speeding-up the learning process, thus reducing the number of iterations required to reach the convergence, and (ii) capturing in a more reliable way the information present in the unlabeled samples. On the other hand, high values of ρ might result in an unstable learning process (too many labeling errors may be considered at a given iteration).

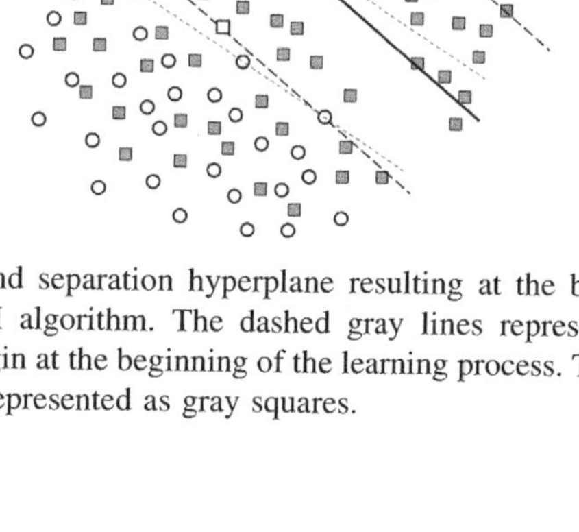

Figure 11.5. Margin and separation hyperplane resulting at the beginning of the second iteration of the PS^{3}VM algorithm. The dashed gray lines represent both the separation hyperplane and the margin at the beginning of the learning process. The remaining originally unlabeled patterns are represented as gray squares.

As the cardinality of $H_{\text{up}}^{(i)}$ and $H_{\text{low}}^{(i)}$ may become lower than ρ , the number of patterns selected from the upper and lower side of the margin band at the generic iteration i are given by

$$\lambda^{(i)} = \min(|H_{\text{up}}^{(i)}|, \rho) \tag{11.20}$$

$$\delta^{(i)} = \min(|H_{\text{low}}^{(i)}|, \rho) \tag{11.21}$$

The new set containing the semilabeled samples at step i is defined as

$$H^{(i)} = \{\mathbf{x}_1^{*\text{up}}, \ldots, \mathbf{x}_{\lambda^{(i)}}^{*\text{up}}, \mathbf{x}_1^{*\text{low}}, \ldots, \mathbf{x}_{\delta^{(i)}}^{*\text{low}}\}, \qquad \mathbf{x}_u^{*\text{up}} \in H_{\text{up}}^{(i)}, \mathbf{x}_u^{*\text{low}} \in H_{\text{low}}^{(i)} \tag{11.22}$$

Note that $H^{(0)} = \emptyset$ and $|H^{(i)}| \leq 2\rho$.

Let $J^{(i)}$ represent the set containing all the patterns selected from $X^{*(i)}$ that have been assigned always the same label until iteration i, therefore $J^{(i)} \cup X^{*(i)} = X^{*(0)}$. A dynamical adjustment is necessary for taking into account that the position of the separation hyperplane changes at each iteration. Let

$$S^{(i)} = \begin{cases} \emptyset, & i = 0, 1 \\ \{\mathbf{x}_u^* | (\mathbf{x}_u^* \in X^{(i)}, \mathbf{x}_u^* \in X^{(i-1)}), y^{*(i)} \neq y^{*(i-1)}\}, & i > 1 \end{cases} \tag{11.23}$$

represent the set of samples belonging to $J^{(i)}$ whose labels at iteration i are different than those at iteration $i - 1$. Note that $S^{(i)} \subseteq J^{(i-1)}$ and $S^{(0)} = S^{(1)} = \emptyset$. If the label of a semilabeled pattern at iteration i, $y^{*(i)}$, is different from the one at iteration $i - 1$, $y^{*(i-1)}$ (label inconsistency), such a label is erased, and the semilabeled pattern is reset to the unlabeled state and moved into $X^{*(i)}$. In this way, it is possible to reconsider this pattern at the following iterations of the semisupervised learning procedure. Therefore, we have

$$J^{(i)} = (J^{(i-1)} - S^{(i)}) \cup H^{(i)} \tag{11.24}$$

$$X^{*(i)} = (X^{*(i-1)} - H^{(i-1)}) \cup S^{(i-1)} \tag{11.25}$$

As will be underlined in the following, the PS^{3}VM algorithm aims at gradually increasing the regularization parameter for the semilabeled patterns that have been given a label, according to a time-dependent criterion. The set $J^{(i)}$ is partitioned into a finite number of subsets γ,

$$J^{(i)} = J_1^{(i)} \cup J_2^{(i)} \cup \ldots \cup J_\gamma^{(i)} \tag{11.26}$$

where γ is a free parameter called *growth rate* and represents the maximum number of iterations for which the user allows the regularization parameter to increase:

$$\begin{cases} J_1^{(i)} = H^{(i)} \\ J_k^{(i)} = J_{k-1}^{(i-1)} - S^{(i)}, & \forall k = 2, \ldots, \gamma - 1 \\ J_\gamma^{(i)} = (J_\gamma^{(i-1)} \cup J_{\gamma-1}^{(i-1)}) - S^{(i)} \end{cases} \tag{11.27}$$

The subset $J_1^{(i)}$ includes all the patterns belonging to X^* which obtained a label at iteration i. Each subset $J_k^{(i)}$, $k = 2, \ldots, \gamma - 1$, includes all the samples that belong to the subset with index $k-1$ at iteration $i-1$ and are labeled in the same way after the tuning of the discriminant hyperplane. At iteration i, the subset with index γ includes all the samples of the subsets J_γ and $J_{\gamma-1}$ which do not belong to the subset S at step i. Note that $J_k^{(i)} = \emptyset, \forall i < k$. Let $\eta^{(i-1)} = |J_k^{(i-1)}|$. The bound minimization problem can be written as

$$\begin{cases} \min\limits_{\mathbf{w}^{(i)}, b^{(i)}, \xi^{(i)}, \xi^{*(i)}} \left\{ \frac{1}{2} \| \mathbf{w}^{(i)} \|^2 + C \sum\limits_{i=1}^{n} \xi_i^{(i)} + \sum\limits_{j=1}^{\eta^{(i-1)}} C_j^* \xi_j^{*(i)} \right\} \\ y_l^{(i)} \cdot (\mathbf{w}^{(i)} \cdot \mathbf{x}_l + b^{(i)}) \geq 1 - \xi_l^{(i)}, \qquad \forall l = 1, \ldots, n, \; \mathbf{x}_l \in X^{(i)} \\ y_j^{*(i)} \cdot (\mathbf{w}^{(i)} \cdot \mathbf{x}_j^* + b^{(i)}) \geq 1 - \xi_j^{*(i)}, \qquad \forall j = 1, \ldots, \eta^{(i-1)}, \; \mathbf{x}_j^* \in J^{(i-1)} \\ \xi_l^{(i)}, \xi_j^{*(i)} \geq 0 \end{cases} \tag{11.28}$$

The semilabeled samples in the training set $\mathbf{x}_j^* \in J^{(i-1)} \subseteq X^{(i)}$ are associated with a regularization parameter $C_j^* = C_j^*(k)$ that depends on the kth subset $J_k^{(i-1)}$ which they belong to at iteration $i-1$. In the learning process of PS³VMs a proper choice for both the regularization parameters C and C^* is very important. The purpose of C and C^* is to control the number of misclassified samples that originally belong to the training set and the unlabeled set, respectively. On increasing their values the penalty associated with errors on the training and semilabeled samples increases. In other words, the larger the regularization parameter, the higher the influence of the associated samples on the selection of the discriminant hyperplane. As regards the semisupervised procedure, it has to be taken into account that the statistical distribution of semilabeled patterns could be rather different compared to that of the original training data. Thus, they should be considered gradually in the semisupervised process so as to avoid instabilities in the learning process. For this reason, the algorithm adopts a weighting strategy based on a temporal criterion. The regularization parameter for the semilabeled patterns C^* increases in a quadratic way, depending on the number of iterations they last inside the set $J^{(i)}$. $\forall j = 1, \ldots, \eta^{(i-1)}$ we have

$$C_j^* = \left(\frac{C^{*\max} - C^{*(0)}}{(\gamma - 1)^2} \right) (k-1)^2 + C^{*(0)} \Leftrightarrow (\mathbf{x}_j^* \in J_k^{(i-1)}) \tag{11.29}$$

where $C^{*(0)}$ is the initial regularization value for semilabeled samples (this is a user defined parameter), and $C^{*\max}$ is the maximum cost value of semilabeled samples and is related to that of training patterns (i.e., $C^{*\max} = \tau \cdot C$, $\tau \leq 1$ being a constant; a reasonable choice has proved to be $\tau = 0.5$). Based on (11.29), it is possible to define an indexing table so as to identify regularization values of semilabeled samples easily according to the number of iterations included in the training set.

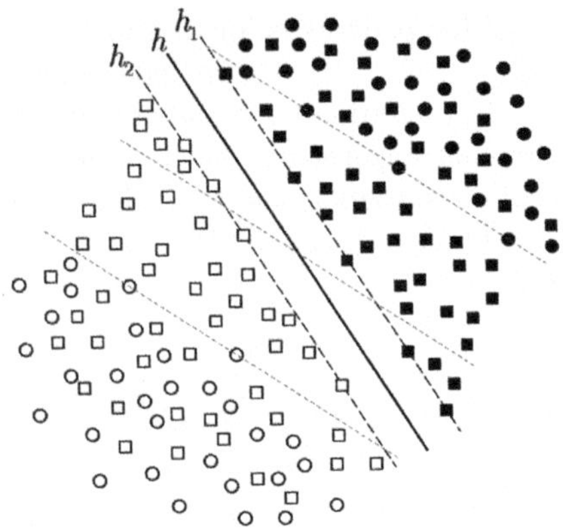

Figure 11.6. Final margin and separation hyperplane resulting after the last iteration of the PS^{3}VM algorithm in an ideal situation. The dashed gray lines represent both the separation hyperplane and the margin at the beginning of the learning process.

Phase 3: Convergence. From a theoretical viewpoint, it can be assumed that the convergence is reached if any of the originally unlabeled samples lies into the margin band (cf. Figure 11.6). Nevertheless, such a choice might result in a high computational load. Moreover, it may happen that even if the margin band is empty, the number of inconsistent patterns at the current iteration is not negligible. For these reasons, the following empirical stopping criteria has been defined:

$$\begin{cases} |M^{(i)}| \leq \lceil \beta \cdot m \rceil \\ \\ |S^{(i)}| \leq \lceil \beta \cdot m \rceil \end{cases} \quad \text{where } M^{(i)} = \{\mathbf{x}_u^* | \mathbf{x}_u^* \in X^{*(i)}, -1 < f^{(i)}(\mathbf{x}_u^*) < 1\} \quad (11.30)$$

where m is the number of originally unlabeled samples and β is a constant fixed *a priori* that tunes the sensitivity of the learning process to unlabeled patterns. This means that convergence is reached if both the number of mislabeled samples and the number of pseudo-labeled patterns which lie into the margin band at the current iteration are lower than or equal to $\lceil \beta \cdot m \rceil$. A reasonable empirical choice for β has proved to be $\beta = 0.03$.

As soon as the convergence is reached, the corresponding iteration is denoted as $i = \text{end}$. Moreover, in order to simplify the notation, the following reductions are introduced:

$$\begin{cases} y_l = y_l^{(\text{end})}, & y_j^* = y_j^{*(\text{end})} \\ J = J^{(\text{end}-1)}, & \eta = \eta^{(\text{end}-1)}, \quad X = X^{(0)} \end{cases} \quad (11.31)$$

The final minimization problem is defined as

$$\begin{cases} \min\limits_{(\mathbf{w},b,\xi,\xi^*)} \left\{ \frac{1}{2} \| \mathbf{w} \|^2 + C \sum\limits_{i=1}^{n} \xi_i + C^{*\max} \sum\limits_{j=1}^{\eta} \xi_j^* \right\} \\ y_l \cdot (\mathbf{w} \cdot \mathbf{x}_l + b) \geq 1 - \xi_l, \qquad \forall_l = 1, \ldots, n, \ \mathbf{x}_l \in X \\ y_j^* \cdot (\mathbf{w} \cdot \mathbf{x}_j^* + b) \geq 1 - \xi_j^*, \qquad \forall j = 1, \ldots, \eta, \ \mathbf{x}_j^* \in J \\ \xi_l, \xi_j^* \geq 0 \end{cases} \tag{11.32}$$

where the entire set of semilabeled samples J is associated with the same regularization parameter, $C^{*\max}$. Finally, all the originally unlabeled patterns $\mathbf{x}_u^* \in X^{*(0)}$ are labeled according to the resulting separation hyperplane.

11.4. SEMISUPERVISED SVMs IN PRIMAL FORMULATION

In this section, we introduce two semisupervised SVM techniques recently presented in the machine learning literature that address the hyperspectral classification problem in the primal formulation. In particular, we present an S^{3}VM with optimization based on gradient descent (∇S^{3}VM) [24] and a Low-Density Separation (LDS) algorithm [26]. As in the previous section, the attention will be focused on the two-class case; for the generalization to the multiclass case the reader is referred to Section 11.5.

11.4.1. General Framework of S^{3}VM in the Primal Formulation

In order to simplify the notation, let us consider the following definition:

$$\tilde{X} = \{\tilde{\mathbf{x}}_1, \ldots, \tilde{\mathbf{x}}_{n+m}\} = \{\mathbf{x}_1, \ldots, \mathbf{x}_n, \mathbf{x}_1^*, \ldots, \mathbf{x}_m^*\} \tag{11.33}$$

where the first n elements $\tilde{\mathbf{x}}_1, \ldots, \tilde{\mathbf{x}}_n$ correspond to the available labeled samples (i.e., $\{\mathbf{x}_l\}_{l=1}^n \in X$), whereas the remaining m elements $\tilde{\mathbf{x}}_{n+1}, \ldots, \tilde{\mathbf{x}}_{n+m}$ correspond to the available unlabeled samples (i.e., $\{\mathbf{x}_u^*\}_{u=1}^m \in X^*$). As in practical applications, data are not usually linearly separable, only the soft margin implementation will be taken into account in the following. In order to consider the available unlabeled samples in the learning process, an additional term is added to the cost function in (11.10), which becomes

$$\min_{\mathbf{w},b} \left\{ \frac{1}{2} \| \mathbf{w} \|^2 + C \sum_{i=1}^{n} H(y_i(\mathbf{w} \cdot \tilde{\mathbf{x}}_i + b)) + C^* \sum_{i=n+1}^{n+m} H(|\mathbf{w} \cdot \tilde{\mathbf{x}}_i|) \right\} \tag{11.34}$$

The symmetric hinge loss for unlabeled samples is nonconvex (cf. Figure 11.7 a); hence the resulting objective function is also nonconvex [16, 26]. Accordingly, the minimization problem becomes more complex with respect to standard inductive SVMs and different implementation techniques can yield significantly different results.

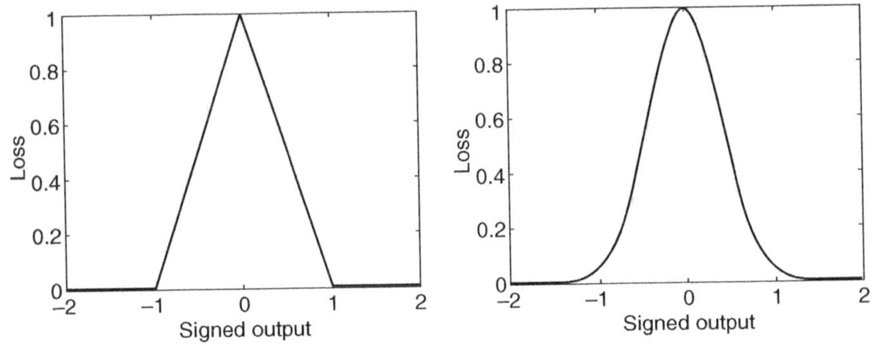

(a) (b)

Figure 11.7. Losses for unlabeled samples in the S^{3}VM objective functions (11.34) and (11.35): (a) symmetric hinge loss for unlabeled samples [$H(t) = \max\ (0,1 - |t|)$]; (b) Gaussian approximated hinge loss for unlabeled samples [$\tilde{H}(t) = \exp(-3t^2)$].

11.4.2. S^{3}VM with Gradient-Descent-Based Optimization (∇S^{3}VM)

In this section, gradient descent [24] is used to solve (11.34) in the primal formulation. As in Section 11.2, for simplicity we ignore the offset b. Since the last term in (11.34) is not differentiable, the minimization problem can be reasonably approximated by

$$\min_{\mathbf{w},b}\left\{\frac{1}{2}\parallel \mathbf{w} \parallel^2 + C\sum_{i=1}^{n} H(y_i\mathbf{w}\cdot\tilde{\mathbf{x}}_i) + C^*\sum_{i=n+1}^{n+m}\tilde{H}(\mathbf{w}\cdot\tilde{\mathbf{x}}_i)\right\} \tag{11.35}$$

where $\tilde{H}(t)$ is an approximation of the symmetric hinge loss for unlabeled samples. This approximation is defined by $\tilde{H}(t) = \exp(-st^2)$, where s is a constant. For instance, when $s = 3$, the Gaussian approximated symmetric hinge loss for unlabeled data is shown in Figure 11.7b. The loss for labeled samples is quadratic when $p = 2$ in (11.10) as depicted in Figure 11.3b.

Linear Case: Let us firstly consider a linear S^{3}VM. The gradient of (11.35) with respect to $\mathbf{w}$ is given by

$$\nabla = \mathbf{w} + 2C\sum_{i=1}^{n}\frac{\partial H(y_i\mathbf{w}\cdot\tilde{\mathbf{x}}_i)}{\partial\mathbf{w}}\tilde{\mathbf{x}}_i y_i - 2sC^*\sum_{i=n+1}^{n+m}\frac{\partial H(\mathbf{w}\cdot\tilde{\mathbf{x}}_i)}{\partial\mathbf{w}}(\mathbf{w}\cdot\tilde{\mathbf{x}}_i)\cdot\tilde{\mathbf{x}}_i \tag{11.36}$$

At the optimal solution $\mathbf{w}^*$, the gradient vanishes such that $\nabla_{\mathbf{w}^*} = 0$; hence, the optimum value is a linear combination of all the available samples as follows:

$$\mathbf{w}^* = \sum_{i=1}^{n+m}\beta_i\tilde{\mathbf{x}}_i \tag{11.37}$$

Replacing (11.37) in (11.35), we have

$$\min_{\beta}\left\{\begin{array}{l}\frac{1}{2}\sum_{i=1}^{n+m}\sum_{j=1}^{n+m}\beta_i\beta_j\ \tilde{\mathbf{x}}_i\cdot\tilde{\mathbf{x}}_j\\ +C\sum_{i=1}^{n}H\left(y_i\sum_{j=1}^{n+m}\beta_j\langle\tilde{\mathbf{x}}_i\cdot\tilde{\mathbf{x}}_j\rangle\right)\\ +C^*\sum_{i=n+1}^{n+m}\tilde{H}\left(\sum_{j=1}^{n+m}\beta_j\langle\tilde{\mathbf{x}}_i\cdot\tilde{\mathbf{x}}_j\rangle\right)\end{array}\right\} \tag{11.38}$$

Nonlinear Case: Let us now consider a nonlinear S^3VM with a kernel function $k(\cdot,\cdot)$ and an associated Reproducing Kernel Hilbert Space. According to the Representer Theorem [25], we have

$$\mathbf{w}^* = \sum_{i=1}^{n+m}\beta_i k(\tilde{\mathbf{x}}_i,\cdot) \tag{11.39}$$

In terms of β_i, (11.38) can be rewritten as

$$\begin{aligned}&\min_{\boldsymbol{\beta}}\left\{\begin{array}{l}\frac{1}{2}\sum_{i=1}^{n+m}\beta_i k(\tilde{\mathbf{x}}_i,\cdot)\sum_{j=1}^{n+m}\beta_j k(\tilde{\mathbf{x}}_j,\cdot)\\ +C\sum_{i=1}^{n}H\left(y_i\sum_{j=1}^{n+m}\beta_j k(\tilde{\mathbf{x}}_i,\tilde{\mathbf{x}}_j)\right)\\ +C^*\sum_{i=n+1}^{n+m}\tilde{H}\left(\sum_{j=1}^{n+m}\beta_j k(\tilde{\mathbf{x}}_i,\tilde{\mathbf{x}}_j)\right)\end{array}\right\}\\ &=\min_{\boldsymbol{\beta}}\left\{\frac{1}{2}\boldsymbol{\beta}^T\tilde{\mathbf{K}}\boldsymbol{\beta}+C\sum_{i=1}^{n}H(y_i\tilde{\mathbf{K}}_i^T\boldsymbol{\beta})+C^*\sum_{i=n+1}^{n+m}\tilde{H}(\tilde{\mathbf{K}}_i^T\boldsymbol{\beta})\right\}\end{aligned} \tag{11.40}$$

where $\tilde{\mathbf{K}} = [k(\tilde{\mathbf{x}}_i,\tilde{\mathbf{x}}_j)]_{i,j=1}^{n+m}$ is the *kernel matrix* and $\tilde{\mathbf{K}}_i$ is the ith column of $\tilde{\mathbf{K}}$. Since H and $\tilde{H}$ are first-order differentiable, we can optimize (11.40) by gradient descent.

This S^3VM optimized by gradient descent is called ∇S^3VM.

11.4.3. S^3VM Low-Density Separation (LDS) Algorithm

11.4.3.1. Graph Kernel. Due to the nonconvexity of the S^3VM objective function, the representation of data can be changed for simplifying the solution to the learning problem. The LDS algorithm assumes that the decision boundary between the considered classes should be in low-density regions in the feature space (*cluster assumption*). A possible strategy for applying this idea is to consider the density between a pair of patterns along a path in the whole data set [26]. Such path problems can be represented with a graph.

Let the undirected graph $G = (V, E)$ be derived from both the labeled and unlabeled sets such that the vertices V are the samples, and the generic edge $(i, j) \in E$ (weighted by W_{ij}) connects a pair of vertices. If a fully connected weighted graph is considered, the edges connect each vertex to all the remaining ones. If sparsity is desired, one possibility is to put edges only between the vertices that are nearest neighbors (e.g., by exploiting the thresholding degree (k-NN)† or the distance (ε-NN)‡).

The edge weight W_{ij} is a measure for the similarity between two vertices $\tilde{\mathbf{x}}_i$ and $\tilde{\mathbf{x}}_j$. For instance, if a Gaussian kernel is used and if the Euclidean distance $d_{ij} = \| \tilde{\mathbf{x}}_i - \tilde{\mathbf{x}}_j \|^2$ between $\tilde{\mathbf{x}}_i$ and $\tilde{\mathbf{x}}_j$ is considered, the weight value becomes

$$W_{ij} = \exp\left(-\frac{d_{ij}}{2\sigma^2} \right) \tag{11.41}$$

Let us assume that, on the one hand, if a pair of vertices are in the same cluster, then there exists a path connecting them such that the data density along the path is high. On the other hand, if two points are in different clusters, then there exists a low density area somewhere along the path. If the minimum density along a path q is assigned with a score marked as $S(q)$, then the path q connecting the vertices $\tilde{\mathbf{x}}_i$ and $\tilde{\mathbf{x}}_j$ in the same cluster has a high score; otherwise, if the path goes between clusters, it does not exist such a path with a high score. Let P_{ij} denote the set of the shortest paths§ with respect to the density connecting the two vertices $\tilde{\mathbf{x}}_i$ and $\tilde{\mathbf{x}}_j$ on a graph $G = (V, E)$ and $p \in V^l$ be a set of l-tuples of vertices along one path q, which is one of the paths P_{ij}. Consequently, we can define the similarity between a pair of vertices to maximize the scores in all paths, that is, $\max_{q \in P_{ij}} \{S(q)\}$. This path-based similarity measure is described in Fischer et al. [27]. The length of the path is represented as $|q|$. A path q connects the vertices $\tilde{\mathbf{x}}_{p_1}$ and $\tilde{\mathbf{x}}_{p_{|q|}}$ with $(\tilde{\mathbf{x}}_{p_k}, \tilde{\mathbf{x}}_{p_{k+1}}) \in E$ for $1 \leq k < |q|$. Fischer et al. [27] defined the dissimilarity between vertices $\tilde{\mathbf{x}}_i$ and $\tilde{\mathbf{x}}_j$ in a way that the maximum distance is estimated between a pair of vertices along one path q, that is, $d_q = \max_{k<|q|} \{d_{\tilde{\mathbf{x}}_{p_k} \tilde{\mathbf{x}}_{p_{k+1}}}\}$. The minimum distance among the maximum ones in all the paths is the final measure of the dissimilarity between vertices $\tilde{\mathbf{x}}_i$ and $\tilde{\mathbf{x}}_j$. Hence, we can write

$$d_{ij} = \max_{q \in P_{ij}} \{S(q)\} = \exp\left[-\frac{1}{2\sigma^2} \left(\min_{q \in P_{ij}} \{d_q\} \right) \right] \tag{11.42}$$

This is called *connectivity kernel*, which is positive definite [27]. However, from (11.42) it is possible to observe that the kernel values do not depend on the length

†Vertices $\tilde{\mathbf{x}}_i$ and $\tilde{\mathbf{x}}_j$ are connected by an edge if $\tilde{\mathbf{x}}_i$ is in the k-nearest-neighborhood of $\tilde{\mathbf{x}}_i$, or vice versa. k is an hyperparameter that controls the density of the graph. k-NN has the nice property of "*adaptive scales*," because the neighborhood radius is different in low- and high-data-density regions.

‡Vertices $\tilde{\mathbf{x}}_i$ and $\tilde{\mathbf{x}}_j$ are connected by an edge if the distance $d_{ij} \leq \varepsilon$. The hyperparameter ε controls the neighborhood radius. Although ò is continuous, the search for the optimal value is discrete.

§It can be computed by the Dijkstra' algorithm [28].

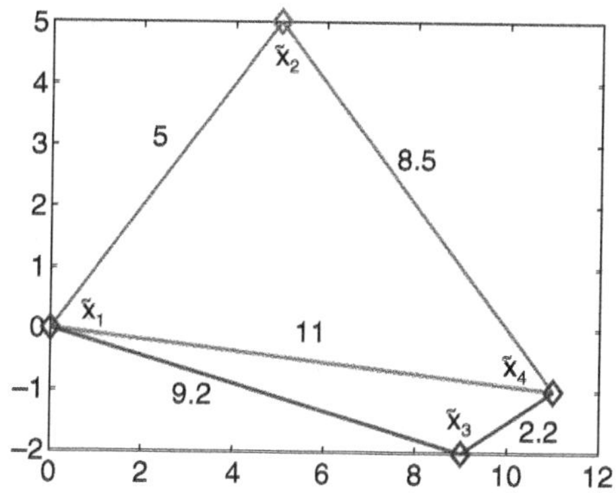

Figure 11.8. Examples of a simple graph with three paths connecting the vertices $\tilde{\mathbf{x}}_1$ and $\tilde{\mathbf{x}}_4$.

of the paths. If a path connects two vertices in two clusters, like a bridge that connects two clusters, the similarity might be taken from this path. To avoid this problem, d_q is "softened," that is,

$$d_q^{\rho} = \frac{1}{\rho} \ln\left(1 + \sum_{k=1}^{|q|-1} (e^{\rho d_{\tilde{\mathbf{x}}_{p_k} \tilde{\mathbf{x}}_{p_{k+1}}}} - 1)\right) \tag{11.43}$$

Thus

$$d_{ij}^{\rho} = \min_{q \in P_{ij}} \{(d_q^{\rho})^2\} \tag{11.44}$$

If $\rho \to 0$, d_q^{ρ} becomes the sum of the original distances along the path q, whereas if $\rho \to \infty$, (11.42) is recovered. If ρ is between 0 and ∞, a value between the maximum and the minimum is obtained for d_{ij}^{ρ}. For example, Figure 11.8 shows a simple graph with four vertices, where there exist three paths that connect the vertices $\tilde{\mathbf{x}}_1$ and $\tilde{\mathbf{x}}_4$:

$$q_r : \tilde{\mathbf{x}}_1 \to \tilde{\mathbf{x}}_2 \to \tilde{\mathbf{x}}_4, \qquad q_b : \mathbf{x}_1 \to \tilde{\mathbf{x}}_3 \to \tilde{\mathbf{x}}_4, \qquad q_g : \tilde{\mathbf{x}}_1 \to \tilde{\mathbf{x}}_4$$

The distance between the pairs along a path is assigned in the figure. The final distance d_{14}^{ρ} between the vertices $\tilde{\mathbf{x}}_1$ and $\tilde{\mathbf{x}}_4$ according to different ρ is shown in Table 11.2. From the experimental analysis, the value of ρ turns out to be crucial for obtaining good results.

11.4.3.2. Balancing Constraint. In real applications, the prior probabilities of information classes in hyperspectral remote sensing problems are unbalanced. In addition to the local minima, one of the problems of the objective function (11.35) is that it is inclined to give unbalanced solutions, classifying all the unlabeled samples in the same class. In Joachims [16], a balancing constraint was proposed in the sense that the class ratio in the unlabeled set should be the same as that in the labeled set. However, if this class ratio is not well estimated

TABLE 11.2. Distances of All the Paths Connecting the Vertices $\tilde{x}_1$ and $\tilde{x}_4$ (cf. Figure 11.8) According to Different ρ Values

ρ	d^ρ_{qr}	d^ρ_{qb}	d^ρ_{q8}	d^ρ_{q14}
0	13.5	11.4	11	11
0.1	10.93	10.16	11	10.16
1	8.52	9.22	11	8.52
∞	8.5	9.2	11	8.5

from the labeled set (this is a usual case in the presence of small-size training data problems), this constraint can be harmful. Nevertheless, from Chapelle and Zien [26] the following "soft" version of such a constraint can be adopted:

$$\frac{1}{m}\sum_{i=n+1}^{n+m} \mathbf{w}\cdot\tilde{\mathbf{x}}_i + b = \frac{1}{n}\sum_{i=1}^{n} y_i \tag{11.45}$$

This is in analogy to the treatment of the min-cut problem in spectral clustering, which is usually replaced by the normalized cut to enforce balanced solutions [29]. An easy way to enforce the constraint (11.45) is to translate all the unlabeled samples such that the mean of those samples is the origin, that is, $\sum_{i=n+1}^{n+m}\tilde{\mathbf{x}}_i = 0$. Then, by fixing $b = \frac{1}{n}\sum_{i=1}^{n} y_i$, an unconstrained optimization on $\mathbf{w}$ can be performed. In the feature space, this corresponds to a modified kernel [30],

$$\tilde{k}(\mathbf{x},\mathbf{y}) = k(\mathbf{x},\mathbf{y}) - \frac{1}{m}\sum_{i=n+1}^{n+m} k(\mathbf{x},\tilde{\mathbf{x}}_i) - \frac{1}{m}\sum_{i=n+1}^{n+m} k(\mathbf{y},\tilde{\mathbf{x}}_i) + \frac{1}{m^2}\sum_{i=n+1}^{n+m}\sum_{j=n+1}^{n+m} k(\tilde{\mathbf{x}}_i,\tilde{\mathbf{x}}_j) \tag{11.46}$$

11.4.3.3. LDS Algorithm Formulation. Since the distance between a pair of vertices is softened, it turns out that the kernel matrix $\tilde{\mathbf{K}}$ is not positive definite, except for two extreme cases: $\rho = 0$ and $\rho = \infty$. A possible solution to this problem is to use the Multidimensional Scaling (MDS) [31] in order to determine the matrix of the minimum distances from all n labeled points to all the $(n+m)$ points, which is given by

$$\boldsymbol{D}^\rho = [d^\rho_{ij}], \qquad i = 1,\ldots,n, \quad j = 1,\ldots,n+m \tag{11.47}$$

A nonlinear Gaussian transformation is then applied to $\boldsymbol{D}^\rho$ in order to obtain the kernel matrix $\tilde{\mathbf{K}}$, whose elements are

$$\tilde{\mathbf{K}}_{ij} = \exp\left(-\frac{d^\rho_{ij}}{2\sigma^2}\right) \tag{11.48}$$

TABLE 11.3. Low-Density Separation Algorithm

Begin

- Build the nearest neighbor graph G from all labeled and unlabeled samples.
- Compute the $n \cdot (n+m)$ distance matrix $\boldsymbol{D}^{\rho}$ of minimum ρ-path distances from all n labeled points to all the $(n+m)$ points according to

$$d_{ij}^{\rho} = \min_{q \in p_{ij}} \{(d_q^{\rho})^2\}$$

where

$$d_q^{\rho} = \frac{1}{\rho} \ln\left(1 + \sum_{k=1}^{|q|-1} (e^{\rho d_{\tilde{\mathbf{x}}_{pk}\tilde{\mathbf{x}}_{pk+1}}} - 1)\right)$$

- Perform a nonlinear transformation on $\boldsymbol{D}^{\rho}$ to get $\tilde{\mathbf{K}}$ using the Gaussian function:

$$\tilde{\mathbf{K}}_{ij} = \exp\left(-\frac{d_{ij}^{\rho}}{2\sigma^2}\right)$$

- Apply MDS to $\tilde{\mathbf{K}}$.
- Select the first q components to define a new kernel matrix $\hat{\mathbf{K}}$.
- Determine $\boldsymbol{\beta}^*$ by applying the gradient descent technique to the following objective function:

$$\min_{\boldsymbol{\beta}} \left\{ \frac{1}{2} \boldsymbol{\beta}^T \hat{\mathbf{K}} \boldsymbol{\beta} + C \sum_{i=1}^{n} H(y_i \hat{\mathbf{K}}_i^T \boldsymbol{\beta}) + C^* \sum_{i=n+1}^{n+m} \tilde{H}(\hat{\mathbf{K}}_i^T \boldsymbol{\beta}) \right\}$$

End

For details, the reader is referred to Chapelle and Zien [26]. The final LDS algorithm used for the learning of the classifier is summarized in Table 11.3.

11.5. PROPOSED STRATEGY FOR MULTICLASS PROBLEMS

Let us consider a multiclass problem defined by a set $\Omega = \{\omega_1, \ldots, \omega_S\}$ made up of S information classes. As in standard SVMs, all the semisupervised SVMs inherit the multiclass problem, and thus the learning process has to be based on a structured architecture made up of binary classifiers. Nevertheless, there is an important difference between SVMs and S^{3}VMs, which leads to a fundamental constraint when considering multiclass architectures: in the learning procedure of binary S^{3}VMs, it must be possible to give a proper classification label to all unlabeled samples. Let each binary S^{3}VM of the multiclass architecture solve a subproblem, where each pattern must belong to one of the two classes Ω_A and Ω_B, defined as proper subsets of the original set of labels Ω (i.e., $\Omega_A \subset \Omega$ is associated with patterns whose output is "+1", $\Omega_B \subset \Omega$ is associated with patterns whose output is "−1"). The semisupervised approach imposes that for each binary S^{3}VM of the multiclass architecture there must be an exhaustive representation of all the possible labels. In other words, the following simple but important constraint must be fulfilled:

$$\Omega_A \cup \Omega_B = \Omega \tag{11.49}$$

If (11.49) is not satisfied, it means that there are unlabeled patterns that the system is not capable of representing and/or classifying correctly. In order to take this constraint into account, a One-Against-All (OAA) multiclass strategy that involves a parallel architecture made up of S different S^{3}VMs (one for each class) can be adopted, as shown in Figure 11.9. The sth S^{3}VM solves a binary problem defined by one information class (e.g., $\{\omega_s\} \in \Omega$) against all the others (e.g., $\Omega - \{\omega_s\}$). In other words we have that

$$\Omega_A = \{\omega_s\}, \; \Omega_B = \Omega - \{\omega_s\} \tag{11.50}$$

It is clear that with this strategy all the binary S^{3}VMs of the multiclass architecture satisfy (11.49).

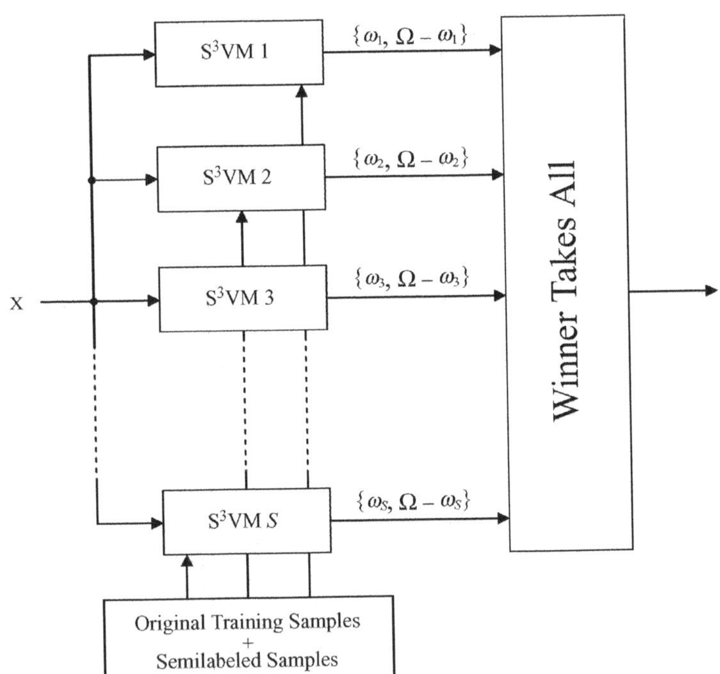

Figure 11.9. One-Against-All architecture for addressing the multiclass problems with the proposed semisupervised SVM approach.

The standard "Winner-Takes-All" (WTA) rule is used to make the final decision: for a generic pattern $\mathbf{x}$ the winning class is the one that corresponds to the SVM with the highest output, that is,

$$\mathbf{x} \in \omega_s \Leftrightarrow \omega_s = \arg \max_{i=1,\ldots,S} \{f_i(\mathbf{x})\} \tag{11.51}$$

where $f_i(\mathbf{x})$ represents the output of the ith S^3VM.

It is worth nothing that in the literature there are also other multiclass combination schemes that are adopted with standard supervised SVMs. For example, the One-Against-One (OAO) strategy is widely used and has proved to be more effective than the OAA strategy in many classification problems. However, the OAO scheme cannot be used with S^3VMs. In fact, this scheme involves $S \cdot (S-1)/2$ binary classifiers, which model all possible pairwise classification problems: each element of the multiclass architecture carries out a classification in which two information classes $\omega_s \in \Omega$ and $\omega_j \in \Omega$, $(s \neq j)$, are analyzed against each other. Consequently, for the generic binary classifier we have that

$$\Omega_A = \{\omega_s\}, \qquad \Omega_B = \{\omega_j\} \; s \neq j \tag{11.52}$$

It is clear that all the members of this multiclass architecture violate the constraint in (11.49) (i.e., $\Omega_A \cup \Omega_B \neq \Omega$); therefore, the OAO strategy cannot be used in the semisupervised framework.

11.6. MODEL SELECTION STRATEGY

As in standard supervised SVMs, also for semisupervised S^3VMs a key factor is the choice of the kernel function. When no prior knowledge is available (or prior information is not reliable, as in *ill-posed* problems), the best option is to use spherical kernels [32] which proved to be good general-purpose kernels. A widely used spherical kernel is the Gaussian radial basis function (RBF) defined as

$$k(\mathbf{x}, \mathbf{x}_l) = \exp\left(-\sum_f \left[\frac{\| \mathbf{x}^f - \mathbf{x}_l^f \|^2}{2\sigma^2}\right]\right) \tag{11.53}$$

where f is a feature index. After choosing the kind of kernel function and the values of kernel parameters (e.g., the spread σ in RBF kernels), the regularization parameter penalizing the training errors should be estimated in the training phase. These parameters are called hyperparameters, and choosing their best values (i.e., those that minimize the expected test error) is called model selection.

Concerning the selection of the parameters for a semisupervised SVM (e.g., the kernel parameters, the growth rate γ and the regularization parameters C and C^* for a PS^3VM), in general we suggest three main strategies depending on the number of originally available labeled patterns.

(i) From a theoretical viewpoint a very reliable strategy involves the generalization of the k-fold cross-validation to the semisupervised framework. $k+1$ disjoints labeled sets must be defined: k sets, $T_1, \ldots, T_k$, of (approximately) equal size and a validation set, V. For any given set of parameters, the considered semisupervised SVM must be trained k times. Each time $k-1$ subsets are put together to form the training set $X = \{T_1 \cup T_2 \cup \ldots \cup T_k\} - T_i$, $i = 1, \ldots, k$, whereas the remaining set T_i and the validation set V, both considered without their labels, are put together to generate the unlabeled set $X^* = T_i \cup V$. The performances are evaluated on T_i. The set of parameters with the lowest average error across all k trials is chosen. Finally, the learning is performed setting $X = T_1 \cup T_2 \cup \ldots \cup T_k$ and $X^* = V$ and the accuracy is evaluated on the validation set V. It is worth nothing that, in order to obtain reliable results, this strategy requires a reasonable number of labeled data; nevertheless, such a requirement is seldom satisfied in real applications. From a computational point of view, this strategy is quite expensive, because the learning has to be carried out k times; however, reasonable performances can be obtained even if $k = 2$.

(ii) When the small number of labeled patterns does not permit to define a reasonable validation set, it is possible to employ a simpler strategy. Two disjoints labeled sets, T_1 and T_2, can be defined: T_1 coincides with the training set X, whereas T_2, considered without its labels, corresponds to the unlabeled set X^*. The set of parameters that permits to obtain the lowest average of error on T_2 (considered with its labels in the validation) is selected. The computational burden is lower with respect to that of the strategy at point i), but this at the price of a lower reliability of the results (they depend on the definition of T_1 and T_2).

(iii) If the number of labeled samples is very small, the only two possible choices are leave-one-out and resubstitution methods. However, leave-one-out can become particularly critical in the presence of strongly minoritary classes, especially when information classes are represented by very few patterns (in the limit case in which only one pattern is available for a class, it cannot be applied). In such extreme cases the parameters should be selected on the basis of the training error (resubstitution error). From a theoretical viewpoint, this can lead to poor generalization capability in inductive learning. However, it is worth nothing that even if the distribution of the originally labeled data set and the distribution of unlabeled data set are slightly different, it is reasonable to expect that they represent a similar problem. Hence, even if the semisupervised approach cannot optimize the model on the unlabeled patterns (the model selection is carried out on training samples), it is able to adapt the model to all the available data. In other words, in the semisupervised process the support vectors that define the discriminant function change with respect to those identified in the inductive supervised learning carried out on training samples, thus fitting the model to all the available samples.

11.7. EXPERIMENTAL RESULTS

11.7.1. Data Set Description

The data set used in our experiments refers to the Okavango Delta area (Botswana). In particular, we used an hyperspectral image acquired in May 2001 by the Hyperion sensor of the EO-1 satellite at 30-m/pixel resolution (Figure 11.10). An area of 44.3×7.7 km was selected for the study. In order to reduce the effects of miscalibration, bad detectors, and atmospheric anomalies, a proper pre-processing of the data was performed. Water absorption bands were removed, therefore we considered 145 of the original 220 bands [10–55, 82–97, 102–119, 134–164, 187–220]. We chose 14 information classes that reflect the impact of flooding on vegetation in the investigated site and represent the land cover types in seasonal swamps, occasional swamps, and drier woodlands (see Table 11.4). We refer the reader to Ham et al. [33] for greater details on this data set.

From the set of available labeled samples, we defined 10 small-size training sets (keeping the proportions among the different classes) made up of only 5% (i.e., 156 samples) of randomly sampled original training patterns. Then we generated two different kinds of test sets: (i) a kind of test set composed of samples taken from

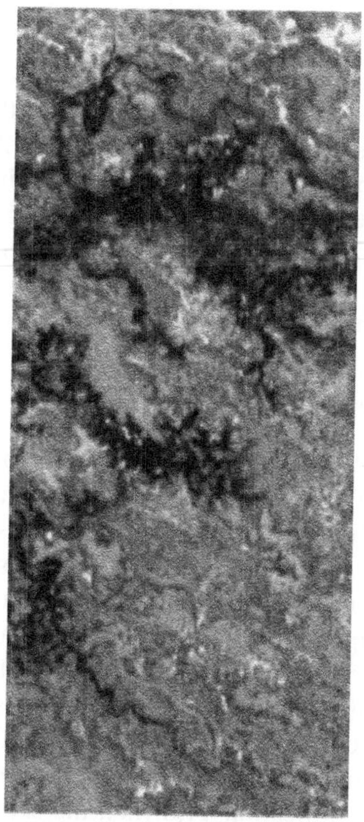

Figure 11.10. Detail of Band 217 of the Hyperion image used in the experiments.

TABLE 11.4. Number of Patterns for the Original Training Set, the Spatially Correlated (SC) Test Set, and the Spatially Uncorrelated (SU) Test Set[a]

Information Classes	Training Set	SC Test Set	SU Test Set
Water	270 (8.31%)	68 (8.31%)	126 (5.05%)
Hyppo grass	101 (3.09%)	26 (3.18%)	162 (6.5%)
Floodplain grasses 1	251 (7.74%)	63 (7.7%)	158 (6.34%)
Floodplain grasses 2	215 (6.63%)	54 (6.6%)	165 (6.62%)
Reeds	269 (8.27%)	68 (8.31%)	168 (6.74%)
Riparian	269 (8.27%)	68 (8.31%)	211 (8.46%)
Firescar	259 (7.98%)	65 (7.95%)	176 (7.06%)
Island interior	203 (6.26%)	51 (6.23%)	154 (6.17%)
Acacia woodlands	314 (9.67%)	79 (9.66%)	151 (6.05%)
Acacia shrublands	248 (7.65%)	62 (7.58%)	190 (7.65%)
Acacia grasslands	305 (9.38%)	77 (9.41%)	358 (14.35%)
Short mopane	181 (5.56%)	46 (5.62%)	153 (6.13%)
Mixed mopane	268 (8.27%)	67 (8.19%)	233 (9.34%)
Exposed soil	95 (2.92%)	24 (2.93%)	89 (3.57%)

[a](Classes 3 and 4 are both floodplain grasses that are seasonally inundated, but differ in their hydroperiod—that is, the amount of time inundated; classes 9, 10, and 11 represent different mixtures of acacia woodlands, shrublands, and grasslands and are named accorcing to the dominant class).

areas surrounding those of training patterns (referred as "spatially correlated (SC) test set"), which contains a 25% (i.e., 818 samples) of the aforementioned labeled training patterns, and (ii) a kind of test set composed of 2494 samples taken from a location geographically separate from that of the training data (referred as "spatially uncorrelated (SU) test set"). It is worth noting that the second data set (SU) properly models the spatial variability of the spectral signatures of classes.

We carried out several trials on the aforementioned data. In the following we present the most interesting results.

11.7.2. Results With PS^{3}VM

As regards the PS^{3}VM technique, we used in turn each of the 10 randomly sampled splits as the training set X. In all the experiments, we used Gaussian RBF kernel functions. According to the theoretical formulation presented in Section 11.3, we expect that the higher the number of unlabeled patterns employed in the learning phase, the greater should be the final classification accuracy. Nevertheless, in order to overcome the high computational burden due to the large size of the investigated test sets (in particular the SU test set), in our experiments we included in X^* only a subset made up of 25% of the considered original test samples.

As pointed out in Section 11.5, for addressing the multiclass problem, we used a One-Against-All architecture. Concerning the model selection, we employed the second strategy described in Section 11.6 (i.e., the set of parameters that permits to obtain the lowest average of error on X^* was selected). Usually, given a set of

TABLE 11.5. Average Overall Accuracy (%) and Kappa Coefficient of Accuracy Over 10 Splits for Both the Spatially Correlated (SC) and the Spatially Uncorrelated (SU) Test Sets Obtained by Exploiting 25% of the Test Samples for Model Selection

		Overall Accuracy (%)		Kappa Coefficient	
Test Set	Classifier	Mean	Standard Deviation	Mean	Standard Deviation
SC	SVM	88.28	1.76	0.873	0.019
	PS^{3}VM	89.55[a]	1.54	0.887	0.017
SU	SVM	73.51	2.41	0.652	0.026
	PS^{3}VM	79.23	1.53	0.713	0.017

[a]Boldface numbers represent

hyperparameters, the corresponding classification accuracy obtained by the selected multiclass architecture over the test set is computed. Then the set of hyperparameters that provides the highest classification accuracy over X^* is chosen. In our experiments, we used a slightly different strategy. In particular, we considered separately each *s*th binary subproblem: in each experiment for the *s*th binary PS^{3}VM we selected the set of hyperparameters which provided the lowest error rate on X^*. The final OAA multiclass architecture is made up by the S binary PS^{3}VMs that provide the best performances. Given the complexity of the investigated problem, we expect that in this way the classifier is able to better discriminate each single class from all the others, thus improving the final classification accuracy.

In the above-described strategy we assume that in the model selection phase the real labels associated with patterns belonging to X^* are known. Even if in practical applications no information is available about X^*, with this kind of experiment we aimed at determining an "upper bound" for the performances of the PS^{3}VM technique. In Table 11.5, we compare the results provided by the proposed PS^{3}VM with those obtained by supervised SVMs (also in this case we selected the best model according to the aforementioned model selection strategy). For both the investigated test sets, we computed the mean and the standard deviation over the 10 splits for both the overall accuracy and the kappa coefficient of accuracy. As expected, the PS^{3}VMs exhibited better accuracies than those obtained by the supervised SVMs.

As regards the SC test set, the average increase in the overall accuracy and in the kappa coefficient are +1.27% and +0.014, respectively. These results confirm the effectiveness of the proposed technique, which proved also more stable than the standard supervised SVMs as pointed out by the lower values of standard deviations (i.e., −0.22 for the overall accuracy and −0.002 for the kappa coefficient). It is worth nothing that in this experiment, training and test sets are correlated and thus also supervised SVMs can provide good results.

Concerning the SU test set, it should be underlined that, due to the differences in the spectral signatures of the information classes with respect to the training set, the investigated *ill-posed* problem was particularly complex. In the light of this aspect, the obtained results are particularly interesting. In all the trials, the PS^{3}VMs outperformed the supervised SVMs. In greater detail, the average increase in the overall accuracy and the kappa coefficient are +5.72% and +0.061, respectively. In addition, unlike the supervised SVMs, the PS^{3}VMs proved very stable also in this case providing lower values of standard deviation (i.e., −0.88 for the overall accuracy and −0.09 for the kappa coefficient), thus confirming their effectiveness.

11.7.3. Results With ∇S^{3}VM and LDS

As regards the approaches developed in the primal formulation, we performed two different kinds of experiments, based on different model selection strategies. In order to assess the effectiveness of the ∇S^{3}VM and LDS algorithms, also in this case we compared the results with those obtained by standard SVMs, whose model selection was the same adopted for the proposed semisupervised classifiers.

In the first set of experiments, for both the ∇S^{3}VM and LDS techniques we carried out a model selection based on a fivefold cross-validation for each of the 10 splits, according to what was reported in point (i) of Section 11.6. In this way, in contrast with the results reported in the previous section, we simulated a real problem in which only labels associated with training patterns were assumed available.

Table 11.6 reports the average Overall Accuracy over the 10 trials for both the considered test sets (SC and SU). As one can see, the accuracies obtained by the supervised SVMs for both the SC and SU test sets are lower than the ones reported in Table 11.5. Actually, this is only because in this case any prior information about the validation set X^* is taken into account during the model selection phase.

As pointed out in the previous subsection, the overall accuracy provided by the supervised SVM classifiers is already good for the SC test dataset. However, both ∇S^{3}VM and LDS provided a classification accuracy 1.62% higher than supervised

TABLE 11.6. Average Overall Accuracy (%) Over the 10 Trials for Both the Considered Test Sets (SC and SU) Obtained by Supervised SVM, ∇S^{3}VM, and LDS Techniques with a Fivefold Cross-Validation Model Selection Strategy

	Average Overall Accuracy (%)		
Test Set	SVM	∇S^{3}VM	LDS
SC	88.26	89.88[a]	89.88
SU	70.72	74.00	73.51

[a]Boldface numbers represent

TABLE 11.7. Average Overall Accuracy (%) Over the 10 Trials for the SU Test Set Obtained by Supervised SVM, ∇S^3VM and LDS Techniques[a]

Classifier	Average Overall Accuracy (%)
SVM	73.89
∇S^3VM	75.45
LDS	77.19[b]

[a]For each classifier the model was selected according to the lowest error rate on the test set.
[b]Boldface number represents

SVMs, thus confirming our expectations. It is worth noting that, as discussed in Section 11.4, when $\rho = \infty$, LDS is equivalent to ∇S^3VM. As in the SC test set, the best model for LDS detected $\rho = \infty$, and the two accuracies are the same.

Concerning the SU test set, both the semisupervised approaches provided very good performances, outperforming supervised SVMs. In greater detail, we can see an increase in the overall accuracy of +2.79% and +3.28% for the LDS algorithm and the ∇S^3VMs, respectively.

In the light of the above results, in the second set of experiments we focused the attention only on the spatially uncorrelated classification problem, because it proved more challenging. In this case, we adopted the model selection strategy described in point (ii) of Section 11.6. In particular, we used the entire investigated test set as X^* in the learning phase and then selected the classifier whose hyperparameters minimized the error rate on X^*. It is worth nothing that with this strategy we were able to determine an empirical limit for both the proposed semisupervised methodologies. Results are reported in Table 11.7. With respect to Section 11.7.2, in this case the information associated with the whole test set was exploited for the model selection, and therefore there is a little improvement for the average overall accuracy obtained by the supervised SVMs. Moreover, also the accuracies obtained by ∇S^3VM and LDS reasonably increased with respect to the previous set of experiments and confirmed again the validity of both the proposed approaches. On average, the overall accuracy provided by ∇S^3VMs increased of +1.56%, while the one obtained by the LDS algorithm increased of +3.3%.

In order to further analyze the effectiveness of the presented algorithms, we also carried out experiments with supervised SVMs with different proportions of labeled samples. In particular, trials with 15%, 30%, 50%, and 75% of original 3248 training patterns were conducted. The lowest error with 5% original labeled set obtained by ∇S^3VM is reported as baseline in Figure 11.11. From the figure, one can see that in the semisupervised setting, the average test error obtained with 5% of labeled samples included in the training set is almost equal to the error yielded with 50% of labeled samples included in the training set in the supervised setting. This further confirms the effectiveness of the proposed approaches.

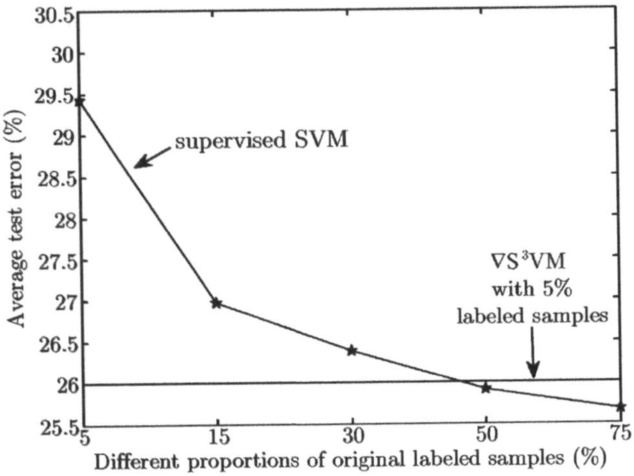

Figure 11.11. Average test errors over 10 splits for the SU test set provided by supervised SVMs for different proportion of original labeled set (e.g., 15%, 30%, 50%, and 75%) and lowest test error obtained by ∇S^3VM with 5% of original labeled set.

11.8. DISCUSSION AND CONCLUSION

In this chapter, we addressed classification of hyperspectral data by introducing some semisupervised techniques based on SVMs. Usually, SVMs are solved in dual representation with Lagrange theory; nevertheless, they can also be implemented in the primal representation [20, 21]. Accordingly, we focused the attention on two different kinds of semisupervised approaches, which exploit optimization algorithms for minimizing the cost function both in the dual and in the primal formulation of the learning problem.

Concerning the former approach, we introduced a novel Progressive Semi-supervised SVM classifier (PS^3VM), which improves the properties of the technique presented in Bruzzone et al. [19]. The proposed PS^3VM is based on an iterative self-labeling strategy: at each iteration the unlabeled samples in the margin band furthest from the separation hyperplane are labeled according to the current discriminant function and then included into the training set. With respect to the transductive SVMs already presented in the literature, PS^3VMs exhibit two main methodological novelties: (i) an original weighting strategy for the unlabeled patterns based on a time-dependent criterion and (ii) an adaptive convergence criterion able to fit the specific investigated problem without requiring any estimate of the number of iterations.

As regards the approaches developed in the primal formulation, we introduced (i) a semisupervised SVM that exploits a gradient descent algorithm for minimizing the cost function (∇S^3VM) and (ii) a Low-Density Separation (LDS) algorithm based on the idea that the true decision boundary between the considered classes should be in low-density regions in the feature space (i.e., *cluster assumption*).

The ∇S^3VM solves an unconstrained optimization problem in which hinge loss functions are associated with both labeled and unlabeled patterns. In the LDS algorithm, at the beginning the input data are represented on a graph to enforce the cluster assumption, and a distance measure between a pair of vertices is computed according to a path-based kernel. In order to consider the length of a path, the distance is softened in a flexible way and, finally, a gradient descent is employed on the newly derived input data.

In order to assess the effectiveness of the proposed approaches, we simulated many *ill-posed* problems. In particular, from a real hyperspectral image acquired on the Okavango Delta area (Botswana) we defined 10 small-size training sets and considered 14 information classes able to reflect the impact of flooding on vegetation in the investigated site. We generated two different kinds of test sets: (i) a test set made up of patterns taken from areas surrounding those of training patterns and (ii) a test set composed of patterns taken from completely disjoint areas with respect to the training patterns, which models the spatial variability of the spectral signatures of the considered information classes. In all cases, the proposed semisupervised techniques resulted in high and satisfactory classification accuracy, outperforming supervised SVMs. Thus, they represent a very promising approach to the classification of hyperspectral data.

It is worth nothing that, as in all semisupervised methods, also for PS^3VM it is not possible to guarantee an increase of accuracy in all cases. If the initial accuracy is particularly low (i.e., the most of the semilabeled samples are incorrectly classified), it is not possible to obtain good performances. In other words, the correct convergence of the learning depends on the definition of the unlabeled samples considered and, implicitly, on the "similarity" between the problems represented by the training patterns and the unlabeled samples. Nevertheless, this effect is common to all semisupervised classification approaches and is not peculiar of the proposed methods.

ACKNOWLEDGMENT

The authors wish to thank Professor M. Crawford (Purdue University, USA) for providing the Hyperion data used in the experiments.

REFERENCES

1. G. F. Hughes, On the mean accuracy of statistical pattern recognition, *IEEE Transactions on Information Theory*, vol. IT-14, pp. 55–63, 1968.
2. A. Baraldi, L. Bruzzone, and P. Blonda, Quality assessment of classification and cluster maps without ground truth knowledge, *IEEE Transactions on Geoscience and Remote Sensing*, vol. 43, no. 4, pp. 857–873, 2005.
3. V. N. Vapnik, *The Nature of Statistical Learning Theory*, 2nd edition, Springer-Verlag, Berlin, 1999.

4. B. M. Shahshahani and D. A. Landgrebe, The effect of unlabeled samples in reducing the small sample size problem and mitigating the Hughes phenomenon, *IEEE Transactions on Geoscience and Remote Sensing*, vol. 32, no. 5, pp. 1087–1095, 1994.
5. Q. Jackson and D. A. Landgrebe, Adaptive Bayesian contextual classification based on Markov random fields, *IEEE Transactions on Geoscience and Remote Sensing*, vol. 40, no. 11, pp. 2454–2463, 2002.
6. Q. Jackson and D. A. Landgrebe, An adaptive method for combined covariance estimation and classification, *IEEE Transactions on Geoscience and Remote Sensing*, vol. 40, no. 5, pp. 1082–1087, 2002.
7. A. Dempster, N. Laird, and D. Rubin, Maximum likelihood from incomplete data via the EM algorithm, *The Royal Statistical Society*, Series B, 1977.
8. S. Tadjudin and D. A. Landgrebe, Robust parameter estimation for mixture model, *IEEE Transactions on Geoscience and Remote Sensing*, vol. 38, no. 1, pp. 439–445, 2000.
9. V. N. Vapnik, *The Nature of Statistical Learning Theory*, 2nd edition, Springer-Verlag, Berlin, 1999.
10. S. Mika, G. Rätsch, B. Schölkopf, A. Smola, J. Weston, and K.-R. Müller, Invariant feature extraction and classification in kernel spaces, in *Advances in Neural Information Processing Systems*, Vol. 12, MIT Press, Cambridge, MA, 1999.
11. G. Camps-Valls and L. Bruzzone, Kernel-based methods for hyperspectral image classification, *IEEE Transactions on Geoscience and Remote Sensing*, vol. 43, no. 6, pp. 1351–1362, 2005.
12. G. Rätsch, B. Schökopf, A. Smola, S. Mika, T. Onoda and K.-R. Müller, Robust ensemble learning, in *Advances in Large Margin Classifiers*, pp. 207–219, MIT Press, Cambridge, MA, 1999.
13. A. Gammerman, V. Vapnik, and V. Vowk, Learning by transduction, in *Uncertainty in Artificial Intelligence*, Madison, WI, pp. 148–156, 1998.
14. V. N. Vapnik, *Statistical Learning Theory*, pp. 339–371, 434–437, 518–520, John Wiley & Sons, New York, 1998.
15. K. Bennett and A. Demiriz, Semisupervised support vector machines, in *Advances in Neural Information Processing Systems*, Vol. 10, pp. 368–374, MIT Press, Cambridge, MA, 1998.
16. T. Joachims, Transductive inference for text classification using support vector machines. International Conference on Machine Learning (ICML), 1999.
17. T. Zhang and F. Oles, A probability analysis on the value of unlabeled data for classification problems, International Joint Conference on Machine Learning, June 2000.
18. Y. Chen, G. Wang, and D. S., Learning with progressive transductive support vector machine, *Pattern Recognition Letters*, vol. 24, no. 12, pp. 1845–1855, 2003.
19. L. Bruzzone, M. Chi, and M. Marconcini, A novel transductive SVM for the semisupervised classification of remote sensing images, *IEEE Transactions on Geoscience and Remote Sensing*, in press.
20. O. Chapelle, Training a support vector machine in the primal, *Journal of Machine Learning Research*, submitted, 2006.
21. S. Keerthi and D. DeCoste, A modified finite Newton method for fast solution of large scale linear SVMs, *Journal of Machine Learning Research*, vol. 6, pp. 341–361, 2005.
22. C. M. Bishop, *Neural Networks for Pattern Recognition*, Clarendon, Oxford, UK, 1995.

23. N. Cristianini and J. Shawe-Taylor, *An Introduction to Support Vector Machines and Other Kernel-Based Learning Methods*, Cambridge University Press, Cambridge, UK, 1995.

24. S. Boyd and L. Vandenberghe, *Convex Optimization*, Cambridge University Press, Cambridge, UK, 2002.

25. G. Kimeldorf and G. Wahba, Some results on tchebycheffian spline functions, *Journal of Mathematical and Analytical Applications*, vol. 33, pp. 82–95, 1971.

26. O. Chapelle and A. Zien, Semisupervised classification by low density separation, in Tenth International Workshop on Artificial Intelligence and Statistics, 2005.

27. V. Fischer, B. Roth, and J. M. Buhmann, Clustering with the connectivity kernel, in *Advances in Neural Information processing Systems*, vol. 16, 2003.

28. E. W. Dijkstra, A note on two problems in connection with graphs, *Numerische Mathematik*, vol. 1, pp. 269–271, 1959.

29. T. Joachims, Transductive learning via spectral graph partitioning, presented at International Conference on Machine Learning (ICML), 2003.

30. B. Schölkopf and A.J. Smola. *Learning with Kernels*, p. 431, MIT Press, Cambridge, MA, 2002.

31. R. Duda, P. Hart, and D. Stork, *Pattern Classification*, 2nd edition, John Wiley & Sons, New York, 2000.

32. O. Chapelle, V. Vapnik, O. Bousquet, and S. Mukherjee, Choosing multiple parameters for support vector machines, *Machine Learning*, vol. 46, no. 1, pp. 131–159, 2002.

33. J. Ham, Y. Chen, M. Crawford, and J. Ghosh, Investigation of the Random Forest Framework for classification of hyperspectral data, *IEEE Transactions on Geoscience and Remote Sensing*, vol. 43, no. 16, 2005.

PART III

APPLICATIONS

CHAPTER 12

DECISION FUSION FOR HYPERSPECTRAL CLASSIFICATION

MATHIEU FAUVEL
Laboratoire des Images et des Signaux, 38402 Saint Martin d'Hères, France; and Department of Electrical and Computer Engineering, University of Iceland, 107 Reykjavik, Iceland

JOCELYN CHANUSSOT
Laboratoire des Images et des Signaux, 38402 Saint Martin d'Heres, France

JON ATLI BENEDIKTSSON
Department of Electrical and Computer Engineering, University of Iceland, 107 Reykjavik, Iceland

12.1. INTRODUCTION

With the development of remote sensing sensors, hyperspectral remote sensing images are now widely available. They are characterized by hundreds of spectral bands. For a classification task, the increased dimensionality of the data increases the capability to detect various classes with a better accuracy. But at the same time, classical classification techniques are facing the problem of statistical estimation in high-dimensional space. Due to the high number of features and small number of training samples, reliable estimation of statistical parameters is difficult [1]. Furthermore, it is proved that, with a limited training set, beyond a certain limit, the classification accuracy decreases as the number of features increases (Hughes phenomenon [2]). However, several classification algorithms have been proposed in the past few years.

Recently, *support vector machines* (SVMs) have shown to be well-suited for high-dimensional classification problems [3, 4]. With SVMs, classes are not characterized by statistical criteria but by a geometrical criterion. SVMs seek a separating hyperplane maximizing the distance with the closest training samples for two classes. This approach allows SVMs to have a very high capability of generalization

Hyperspectral Data Exploitation: Theory and Applications, Edited by Chein-I Chang

and, as a consequence, only require a few training samples. In addition, for nonlinearly separable data, SVMs use the kernel trick to map the data onto a higher-dimensional space where they are linearly separable [5]. Early work in classification of remotely sensed images by SVMs showed promising results [6, 7]. In Melgani and Bruzzone [8], several SVM-based classifiers are compared with other classical classifiers such as a K-nearest neighbors classifier and a neural network classifier. The SVMs using the kernel trick outperformed the other classifiers in terms of accuracy. Multiclass SVMs performances were also positively compared with a discriminant analysis classifier, a decision tree classifier, and a feedforward neural network classifier with a limited training set [9]. Though these experiments highlight the good generalization capability of SVMs, the data used were pre-processed; that is, three selected bands were used for the classification and thus, performances in high-dimensional space were not investigated. In both articles [8,9], the *Gaussian radial basis* kernels were shown to produce the best results. In Mercier [10], several spectral-based kernels were tested on hyperspectral data. These kernels were designed to handle spectral meaning, and, in particular, various non-Euclidean metrics were considered to characterize similarity between vectors. In Fauvel et al. [11], two kernels are considered and compared to assess the generalization capability of SVMs as well as the ability of SVMs to deal with high-dimensional feature spaces in the situation of very limited training set.

Another strategy consists in performing a classification with some feature extraction based on mathematical morphology [12, 13]. Such an approach was initially designed to classify panchromatic data from urban areas: Pesaresi and Benediktsson [14] proposed to construct a morphological profile by a composition of geodesic opening and closing operations with increasing sizes. A neural network approach was used for the pixel-wised classification of the extracted profile. The profile consists of multiple opening and closing transformations of the same base image and should be more effective in discrimination of different urban features. On the other hand, since the images are all transformation of the same image, there may be a lot of redundancy evident in the feature set. Therefore, feature extraction can be desirable in finding the most important features in the feature space. In Benediktsson [15], the method in Pesaresi and Benediktsson [15] is extended by including decision boundary feature extraction (DBFE) to reduce the redundancy in the morphological profile. Regarding the extension of this method to hyperspectral images, a simple approach was suggested in Dell'Acqua et al. [16], consisting in using only the first principal component (PC) of hyperspectral image data to build a morphological profile. In Benediktsson et al. [17], this method is extended. This issue is also addressed by Plaza et al. [18].

All these methods have their own characteristics and advantages, with none of them strictly outperforming all the others.

Usually, for a given data set, performances in terms of *global* and *by class* classification accuracies depend on the considered classes—that is, on their spectral and spatial characteristics. For instance, methods based on morphological filtering are well-suited to classify structures with a typical spatial shape, like man-made constructions. On the contrary, algorithms based on statistical approaches perform better for the classification of vegetation and soils. As a consequence, we propose to use several approaches and try to take advantage of the strengths of each algorithm. This concept is called *decision fusion* [19]. Decision fusion can be defined as the process of fusing information from several individual data sources after each data source has undergone a preliminary classification. For instance, Benediktsson and Kanellopoulous [19] proposed a multisource classifier based on a combination of several neural/statistical classifiers. The samples are first classified by two classifiers (a neural network and a multisource classifier); every sample with agreeing results is assigned to the corresponding class. In the case of a conflicting situation, a second neural network is used to classify the remaining samples. The main limitation of this method is the need of large training sets to train the different classifiers. Jeon et al. [20] used two decision fusion rules to classify multitemporal Thematic Mapper data. Recently, Lisini et al. [21] proposed to combine sources according to their class accuracies. In this chapter, the decision fusion rule is modeled with fuzzy data fusion rules.

Data fusion has been used successfully in many classification problems. Tupin et al. [22] combined several structure detectors to classify SAR images using Dempster–Shafer theory. Tupin and Roux [23], aggregated information extracted from SAR and optical images for building detection. In a first step, potential buildings edge were extracted from the SAR image and, in a second step, the buildings shape was extracted from the optical image using the SAR image information. Chanussot et al. [24] proposed several strategies to combine the output of a line detector applied to multitemporal images. Here, we consider a general framework to aggregate the results of different classifiers. Conflicting situations, where the different classifiers disagree, are solved by estimating the pointwise accuracy and modeling the global reliability for each algorithm [25]. This leads to the definition of an adaptive fusion scheme ruled by these reliability measures. The proposed algorithm is based on fuzzy sets and possibility theory. The framework of the addressed problem is modeled as follows: For a given data set, n classes are considered, and m classifiers are assumed to be available. For an individual pixel, each algorithm provides as an output a membership degree for each of the considered classes. The set of these membership values is then modeled as a fuzzy set, and the corresponding degree of fuzziness determines the local reliability of the algorithm. The global accuracy is manually defined for each class after a statistical study of the results obtained with each separately used classifier. Hence, the fusion is performed by aggregating the different fuzzy sets provided by the different classifiers. It is adaptively ruled by the reliability information and does not require any further training. The decision is postponed to the end of the fusion process in order to take advantage of each algorithm and enable more accurate results in conflicting situations. In previous studies, this method has been successfully applied to panchromatic data. Taking material from Fauvel et al. [11], this chapter presents the general framework and applies it to the classification of hyperspectral data, using the two previously mentioned algorithms as information sources.

This chapter is organized as follows. Fuzzy set theory and measures of fuzziness are briefly presented in Section 12.2. Section 12.2.2 presents the model for each classifier's output in terms of a fuzzy set. Then, the problem of information fusion is discussed in Section 12.3. The proposed fusion scheme is detailed in Section 12.4, and experimental results are presented in Section 12.5. Finally, conclusions are drawn.

12.2. FUZZY SET THEORY

Traditional mathematics assigns a membership value of 1 to elements that are members of a set, and 0 to those which are not, thus defining *crisp sets.* On the contrary, *fuzzy set* theory handles the concept of partial membership to a set, with real-valued membership degrees ranging from 0 to 1. Fuzzy set theory was introduced in 1965 by Zadeh as a way to model the vagueness and ambiguity in complex systems [26]. It is now widely used to process unprecise or uncertain data [27, 28]. In particular, it is an appropriate framework to handle the output of one given classifier for further processing since it usually does not come in a binary form and includes some ambiguity. In this section, we first recall general definitions and properties of fuzzy sets. Then, we go in detail through the model used for the representation of the classifiers' output.

12.2.1. Fuzzy Set Theory

Definitions

Definition 1 (Fuzzy Subset). A fuzzy subset* F of a reference set U is a set of ordered pairs $F = \{(x, \mu_F(x))|x \in U\}$, where $\mu_F : U \rightarrow [0, 1]$ is the membership function of F in U.

Definition 2 (Normality). A fuzzy set is said to be normal if and only if $\max \mu_F(x) = 1$.

Definition 3 (Support). The support of a fuzzy set F is defined as

$$\mathrm{Supp(F)} = \{\mathrm{x} \in \mathrm{U}|\mu_\mathrm{F}(\mathrm{x}) > 0\}$$

Definition 4 (Core). The core of a fuzzy set is the (crisp) set containing the points with the largest membership value (1). It is empty if the set is non-normal.

Logical Operations. Classical Boolean operations extend to fuzzy sets [26]. With F and G two fuzzy sets, classical extensions are defined as follows:

Equality. The equality between two fuzzy sets is defined as the equality of their membership functions:

$$\mu_F = \mu_G \Leftrightarrow \forall x \in U, \quad \mu_F(x) = \mu_G(x) \tag{12.1}$$

* For convenience, we will use the term "fuzzy set" instead of "fuzzy subset" in the following, where a fuzzy set F is described by its membership function μ_F.

Inclusion. The inclusion of a set in another one is defined by the inequality of their membership functions:

$$\mu_F \subset \mu_G \Leftrightarrow \forall x \in U, \quad \mu_F(x) \leq \mu_G(x) \tag{12.2}$$

Union. The union of two fuzzy sets is defined by the maximum of their membership functions:

$$\forall x \in U, (\mu_F \cup \mu_G)(x) = \max\{\mu_F(x), \mu_G(x)\} \tag{12.3}$$

Intersection. The intersection of two fuzzy sets is defined by the minimum of their membership functions:

$$\forall x \in U, (\mu_F \cap \mu_G)(x) = \min\{\mu_F(x), \mu_G(x)\} \tag{12.4}$$

Complement. The complement of a fuzzy set F is defined by

$$\forall x \in U, \mu_{\bar{F}}(x) = 1 - \mu_F(x) \tag{12.5}$$

Measures of Fuzziness. Fuzziness is an intrinsic property of fuzzy sets. To measure to what degree a set is fuzzy, and thus estimate the corresponding ambiguity, several definitions have been proposed [29, 30]. Ebanks [31] proposed to define the degree of fuzziness as a function f with the following properties:

1. $\forall F \subset U$, if $f(\mu_F) = 0$, then F is a crisp set.
2. $f(\mu_F)$ is maximum if and only if $\forall x \in U, \mu_F(x) = 0.5$.
3. $\forall(\mu_F, \mu_G) \in U^2, \quad f(\mu_F) \geq f(\mu_G)$ if $\forall x \in U \begin{cases} \mu_G(x) \geq \mu_F(x) & \text{if } \mu_F(x) \geq 0.5, \\ \mu_G(x) \leq \mu_F(x) & \text{if } \mu_F(x) \leq 0.5. \end{cases}$
4. $\forall F \in U, \quad f(\mu_F) = f(\mu_{\bar{F}})$. A set and its complement have the same degree of fuzziness.
5. $\forall(\mu_F, \mu_G) \in U^2, \quad f(\mu_F \cup \mu_G) + f(\mu_F \cap \mu_G) = f(\mu_F) + f(\mu_G)$.

Derived from the probability theory and the classical Shannon entropy, De Luca and Termini [29] defined a fuzzy entropy satisfying the above five properties:

$$H_{DTE}(\mu_F) = -K \sum_{i=1}^{n} \Big(\mu_F(x_i)\log_2(\mu_F(x_i)) + (1 - \mu_F(x_i))\log_2(1 - \mu_F(x_i))\Big) \tag{12.6}$$

Bezdeck [32] proposed an alternative measure of fuzziness based on a multiplicative class.

Definition 5 (Multiplicative Class). A multiplicative class is defined as

$$H_*(\mu_F) = K \sum_{i=1}^{n} g(\mu_F(x_i)), \quad K \in R^+ \tag{12.7}$$

where $g(\mu_F)$ is defined as

$$\begin{aligned} g(t) &= \tilde{g}(t) - \min_{0 \le t \le 1} \tilde{g}(t) \\ \tilde{g}(t) &= h(t)h(1-t) \end{aligned} \tag{12.8}$$

and h is a concave increasing function on [0,1]:

$$h : [0,1] \rightarrow R^2, \quad \forall x \in [0,1] h'(x) > 0 \quad \text{and} \quad h''(x) < 0 \tag{12.9}$$

The multiplicative class allows the definition of various fuzziness measures, where different choices of g lead to different behaviors. For instance, let $h : [0,1] \rightarrow R^+$ be $h(t) = t^\alpha, 0 < \alpha < 1$. The function h satisfies the required conditions for the multiplicative class, and the function

$$H_{\alpha QE}(\mu_F) = \frac{1}{n 2^{-2\alpha}} \sum_{i=1}^{n} \mu_F(x_i)^\alpha (1 - \mu_F(x_i))^\alpha \tag{12.10}$$

is a measure of fuzziness, namely the α-*Quadratic entropy.* Rewriting (12.10) as

$$\begin{aligned} H_{\alpha QE}(\mu_F) &= \frac{1}{n} \sum_{i=1}^{n} S_{\alpha QE}(\mu_F(x_i)) \\ S_{\alpha QE}(\mu_F(x_i)) &= \frac{\mu_F(x_i)^\alpha (1 - \mu_F(x_i))^\alpha}{2^{-2\alpha}} \end{aligned} \tag{12.11}$$

we can analyze the influence of parameter α (see Figure 12.1): The measure becomes more and more selective as α increases from 0 to 1. With α close to 0, all the fuzzy sets have approximately the same degree of fuzziness and the measure is not sensitive to changes in μ_F, whereas with α close to 1, the measure is highly selective with the degree of fuzziness quickly decreasing when the fuzzy set differs from $\mu_F = 0.5$. As a consequence, an intermediate value such as $\alpha = 0.5$ usually provides a good trade-off [32].

Examples of fuzzy sets and their fuzziness values are given in Figure 12.2 and Table 12.1, respectively. For the binary set, fuzziness is null with respect to property 1 above. Property 2 is fulfilled since the fuzziness is maximum for $\mu_F(x) = 0.5$.

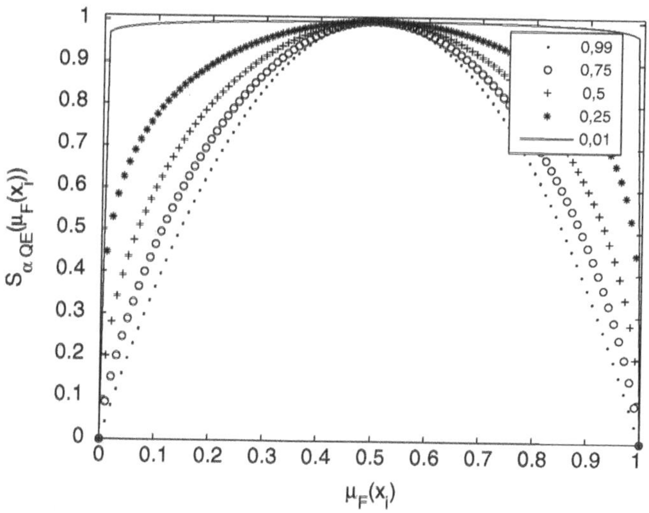

Figure 12.1. Influence of parameter α on $S_{\alpha QE}$. © 2006 IEEE.

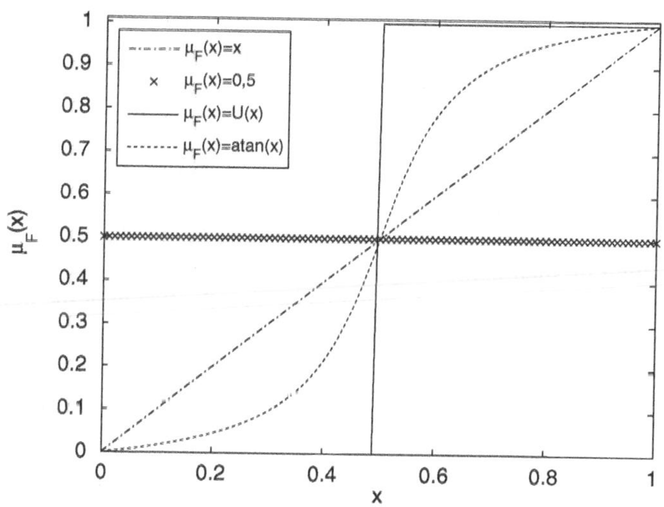

Figure 12.2. Example of four fuzzy sets with different degrees of fuzziness.

TABLE 12.1. Degree of Fuzziness for Different Fuzzy Sets Computed with α–*Quadratic Entropy*

$\alpha =$	0.01	0.25	0.5	0.75	0.99
$H_{\alpha QE}(\mu_F(x)) = a\tan(x)$	0.959	0.600	0.394	0.271	0.196
$H_{\alpha QE}(\mu_F(x)) = x$	0.967	0.725	0.549	0.423	0.333
$H_{\alpha QE}(\mu_F(x)) = U(x)$	0	0	0	0	0
$H_{\alpha QE}(\mu_F(x)) = 0.5$	0.993	0.840	0.707	0.594	0.503

Following condition 3, the fuzzy set with the arctan membership function has a lower fuzziness than the fuzzy set with the linear membership function.

12.2.2. Class Representation

An n-classes classification problem is considered, for which m different classifiers are available. For a given pixel x, the output of classifier i is the set of numerical values:

$$\{\mu_i^1(x), \mu_i^2, (x) \ldots, \mu_i^j(x), \ldots, \mu_i^n(x)\} \tag{12.12}$$

where $\mu_i^j(x) \in [0, 1]$ (after a normalization, if required) is the membership degree of pixel x to class j according to classifier i. The higher this value, the more likely it is that the pixel belongs to class j (if one single classifier is used, the decision is taken by selecting the class j maximizing $\mu_i^j(x)$: $\text{class}_{\text{selected}}(x) = \text{argmax}_j(\mu_i^j(x))$). Depending on the classifier, $\mu_i^j(x)$ can be of different nature: probability, posterior probability at the output of a neural network, membership degree at the output of a fuzzy classifier, and so on. In any case, the set $\pi_i(x) = \{\mu_i^j(x), j = 1, \ldots, n\}$ provided by every classifier i can be considered as a fuzzy set.

As a conclusion, for every pixel x, m fuzzy sets are computed, one by each classifier. This set of fuzzy sets constitutes the input for the fusion process:

$$\{\pi_1(x), \pi_2(x), \ldots, \pi_i(x), \ldots, \pi_m(x)\} \tag{12.13}$$

In Figure 12.3, two conflicting sets are represented: For this pixel, if one trusts the first classifier (on the left), then class number 4 would be selected whereas if one trusts the second classifier (on the right), then he would select class number 5. The handling of such conflicting situations is the central issue that needs to be addressed by the fusion system. As a matter of fact, the fusion of the nonconflicting results is of little interest in our case: Though it might increase our belief in the

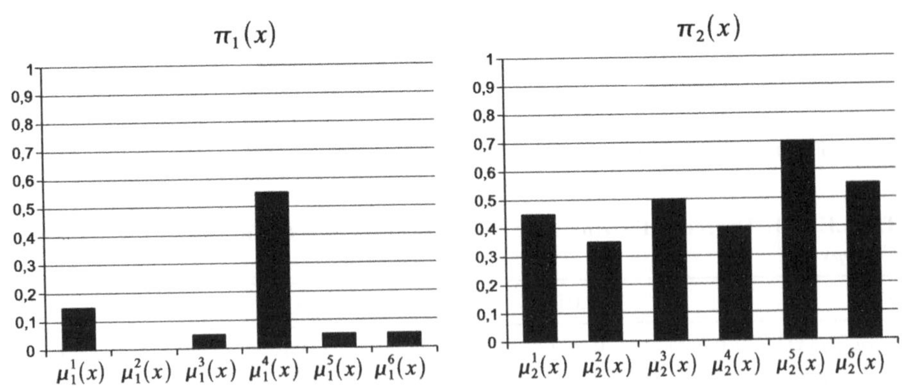

Figure 12.3. Example of two conflicting sets π for a given pixel x. © 2006 IEEE.

corresponding result, it certainly won't change the final decision and thus won't increase the classification performances. On the contrary, in the case of conflicting results, at least one classifier is wrong and the fusion gives a chance to correct this and increase the classification performances. Fuzzy set theory provides various combination operators to aggregate these fuzzy sets. Such combination operators are discussed in the next section.

12.3. INFORMATION FUSION

After briefly recalling the basics of data fusion, we discuss in this Section the problem of measuring the confidence of individual classifiers. We finally propose an adaptive fusion operator. In the following we denote the fuzzy set i by π_i and m the number of sources.

12.3.1. Introduction

Data fusion consists of combining information from several sources in order to improve the decision [33]. As previously mentioned, the most challenging issue is to solve conflicting situations where some of the sources disagree. Numerous combination operators have been proposed in the literature. They can be classified into three different kinds, depending on their behavior [34]:

- *Conjunctive Combination.* This corresponds to a *severe* behavior. The resulting fuzzy set is necessarily smaller than the initial sets and the core is included in the initial cores (it can only decrease). The largest conjunctive operator is the fuzzy intersection (12.4) leading to the following fuzzy set: $\pi_\wedge(x) = \bigcap_{i=1}^{N} \pi_i(x)$. *T-norms* are conjunctive operators. They are commutative, associative, increasing and with $\pi_i(x) = 1$ as a neutral element (i.e., if $\pi_2(x) = 1$ then $\pi_\wedge(x) = \pi_1(x) \cap \pi_2(x) = \pi_1(x)$). They satisfy the following property:

$$\pi_\wedge(x) \leq \min_{i \in [1,m]} \pi_i(x) \tag{12.14}$$

- *Disjunctive Combination.* This corresponds to an *indulgent* behavior. The resulting fuzzy set is necessarily larger than the initial sets and the core contains the initial cores (it can only increase). The smallest disjunctive operator is the fuzzy union (12.3), leading to the following fuzzy set: $\pi_\vee(x) = \bigcup_{i=1}^{N} \pi_i(x)$. *T-conorms* are disjunctive operators. They are commutative, associative, increasing, and with $\pi_i(x) = 0$ as neutral element. They satisfy the following property:

$$\pi_\vee(x) \geq \max_{i \in [1,m]} \pi_i(x) \tag{12.15}$$

- *Compromise Combination.* this corresponds to intermediate *cautious* behaviors. $T(a, b)$ is a compromise combination if it satisfies

$$\min(a,b) < T(a,b) < \max(a,b) \tag{12.16}$$

On illustrative purpose, we can consider the following toy problem. To estimate how old a person is, two estimates are available, each one modeled by a fuzzy set. These fuzzy sets are represented in Figure 12.4a; note that they are highly conflicting. From these two information sources, we want to classify a person into one of the three following classes: young (under 30), middle age (between 30 and 65), or

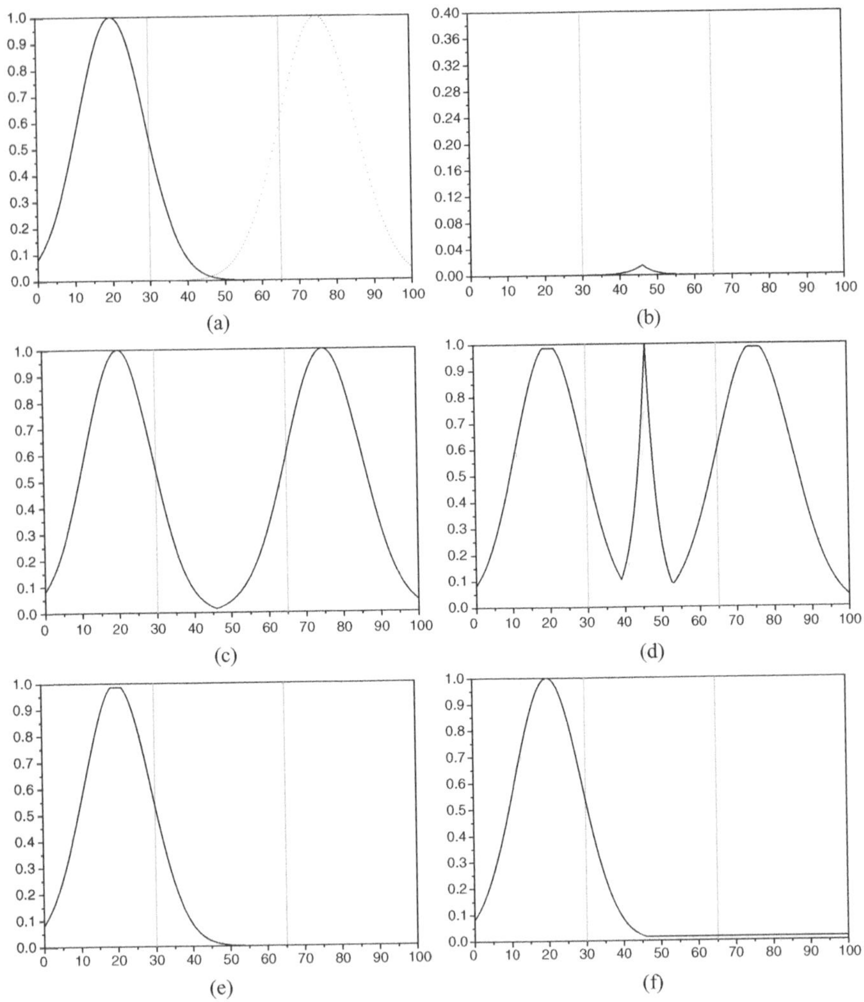

Figure 12.4. Examples of combination operator. (a) Two possibilities distribution. (b and c) The result of the min and the max operators, respectively. (d, e, and f) Results of the three compromise operators presented in (12.18), (12.19), and (12.20), respectively. © 2006 IEEE.

old (above 65). To illustrate the three possible modes of combination, we aggregate the information with the min operator (T-norm), the max operator (T-conorm), and the three different compromise operators. Results are presented in Figure 12.4. The decision is taken by selecting the class corresponding to the maximum membership.

Conjunctive Combination. Figure 12.4b presents the result obtained with the min operator—that is, the less severe conjunctive operator. It is a unimodal fuzzy set. This fuzzy set is subnormalized, but this problem could be solved using $\pi'_\wedge(x) = \frac{\pi_\wedge(x)}{\sup_x(\pi_\wedge(x))}$, but this would not change the shape of the result. In this case, the decision would be *middle age*, which is not compatible with any of the initial sources. In this case, the sources strongly disagree and the conjunctive fusion does not help in the classification. As a conclusion, conjunctive operators are not suited for conflicting situations.

Disjunctive Combination. Figure 12.4c presents the result obtained with the max operator—that is, the less indulgent disjunctive operator. The resulting membership function is multimodal and each maximum is of equal amplitude. Again, no satisfactory decision can be made.

Compromise Combination. Three different such operators are discussed. They are all based on the measure of the conflict between sources, defined as $1 - C$ with

$$C(\pi_1, \pi_2) = \sup_x \min(\pi_1(x), \pi_2(x)) \tag{12.17}$$

The compromise combination operators have been proposed by Dubois and Prade [35]. Bloch has classified these operators as *contextual-dependent (CD) operators* [36], where context can be, for example, a conflict between the sources, knowledge about reliability of a source, and some spatial information. These operators have been proposed in possibility theory [37], but they can also be used in fuzzy set theory for combining membership functions [36]. Being able to adapt to the context, these operators are more flexible and thus provide interesting results. The first considered operator (12.18):

$$\pi(x) = \begin{cases} \max\left(\frac{\min(\pi_1(x),\pi_2(x))}{C(\pi_1,\pi_2)}, \min(\max(\pi_1(x),\pi_2(x)), 1-C(\pi_1,\pi_2))\right) & \text{if } C(\pi_1,\pi_2) \neq 0 \\ \max(\pi_1(x),\pi_2(x)) & \text{if } C(\pi_1,\pi_2) = 0 \end{cases} \tag{12.18}$$

adapts its behavior as a function of the conflict between the sources:

- It is conjunctive if the sources have low conflict.
- It is disjunctive if the sources have high conflict.
- It behaves in a compromise way in case of partial conflict.

Figure 12.4d presents the obtained result using operator (12.18). Corresponding decision (middle age) is still not satisfactory.

In this case, some information on source reliability must be included, and the most reliable source(s) should be privileged in the fusion process. Different situations can be considered:

- It is possible to assign a numerical degree of reliability to each source.
- A subset of sources is reliable, but we do not know which one(s).
- The relative reliability of the sources are known, but with no quantitative values. However, priorities can be defined between the sources.

The two following adaptive operators are examples of *prioritized fusion operator* [35].

$$\pi(x) = \min(\pi_1(x), \max(\pi_2(x), 1 - C(\pi_1, \pi_2))) \tag{12.19}$$

$$\pi(x) = \max(\pi_1(x), \min(\pi_2(x), C(\pi_1, \pi_2))) \tag{12.20}$$

For both operators, when $C(\pi_1, \pi_2) = 0$, π_2 contradicts π_1 and only the information provided by π_1 is retained. In this case, π_2 is considered as a specific piece of information while π_1 is viewed as a fuzzy default value. Assuming that π_1 is more accurate than π_2, we get the result presented in Figures 12.4e and 12.4f, enabling a satisfactory decision.

As a conclusion, conjunctive and disjunctive combination operators are ill-suited to handle conflicting situations. These situations should be solved by CD operators, incorporating reliability information.

12.3.2. Measure of Confidence

Pointwise Accuracy. For a given pixel and a given classifier, we propose to interpret the degree of fuzziness of the fuzzy set $\pi_i(x)$ defined in (12.12) as a pointwise measure of the accuracy of the method. We intuitively consider that the classifier is *reliable* if one class has a high membership value while all the others have a membership value close to zero. On the contrary, when no membership value is significantly higher than the others, the classifier is *unreliable* and the results it provides should not be taken too much into account in the final decision. In other words, uncertain results are obtained when the fuzzy set $\pi_i(x)$ has a high fuzziness degree, the highest degree being reached for uniformly distributed membership values.

To reduce the influence of unreliable information and thus enhance the relative weight of reliable information, we weight each fuzzy set by

$$w_i = \frac{\sum_{k=0, k\neq i}^{m} H_{\alpha QE}(\pi_k)}{(m-1)\sum_{k=0}^{m} H_{\alpha QE}(\pi_k)} \tag{12.21}$$

$$\sum_{i=0}^{m} w_i = 1$$

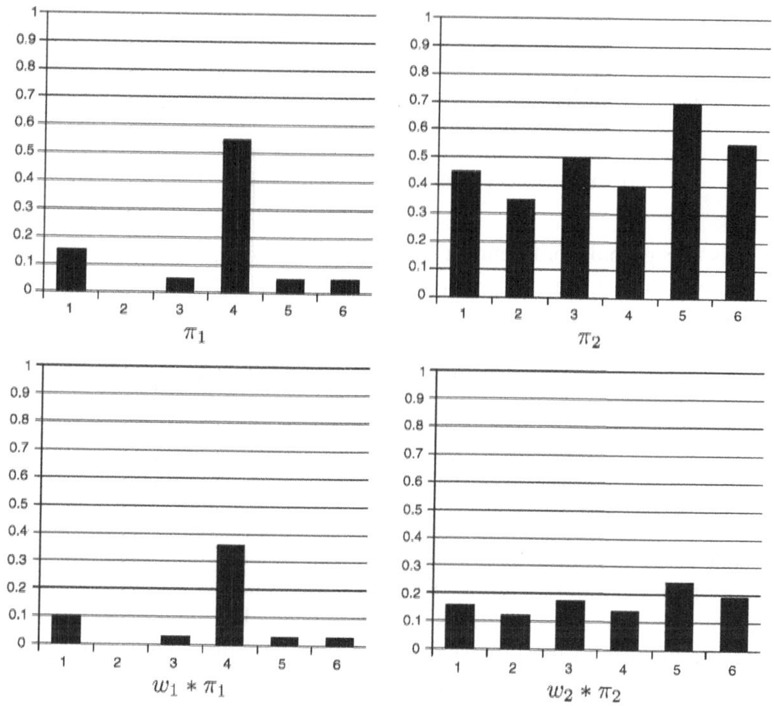

Figure 12.5. Normalization effects. This figure shows two fuzzy sets (π_1 and π_2) with different fuzziness ($H_{\alpha QE}(\pi_1) = 0.51, H_{\alpha QE}(\pi_2) = 0.97, w_1 = 0.65$, and $w_2 = 0.35$). The normalization effect is shown on the right-hand side. Influence of classifier 2 is more reduced by w_2 than classifier 1 is reduced by w_1. © 2006 IEEE.

where $\alpha = 0.5$, $H_{\alpha QE}(\pi_k)$ is the fuzziness degree of source k, and m is the number of sources. When a source has a low fuzziness degree, w_i is close to 1 and it only slightly affects corresponding fuzzy set. Figure 12.5 illustrates the effects of this normalization.

Global Accuracy. Beyond the adaptation to the local context described in the previous paragraph, we can also use prior knowledge regarding the performances of each classifier. This knowledge is modeled for each classifier i and for each class j by a parameter f_i^j. Such global accuracy can be determined by a separate statistical study on each of the used classifiers. If, for a given class j, the user considers that the results provided by classifier i are satisfactory, parameter f_i^j is set to one. Otherwise, it is set to zero. Since this decision is binary, we assume that for each class there is at least one method ensuring a satisfactory global reliability.

12.3.3. Combination Operator

Numerous combination rules have been proposed in the literature, from simple conjunctive or disjunctive rules, such as min or max operators, to more elaborated

CD operators, such as defined by (12.19) and (12.20) where the relative reliability of each source is used. However, with these operators, sources always have the same hierarchy and the fusion scheme does not adapt to the local context. In Fauvel et al. [11], we consider the following extension:

$$\mu_f^j(x) = \max\left(\min\left(w_i\mu_i^j(x), f_i^j(x)\right), \; i \in [1, m]\right) \tag{12.22}$$

where f_i^j is the global confidence of source i for class j, w_i is the normalization factor defined in (12.21), and μ_i^j is an element of the fuzzy set π_i defined in (12.12). This combination rule ensures that only reliable sources are taken into account for each class (predefined coefficients f_i^j) and that the fusion also automatically adapts to the local context by favoring the source that is locally the most reliable (weighting coefficients w_i).

12.4. THE FUSION SCHEME

We present here the complete considered fusion scheme. In a first step, each classifier is applied separately (but no decision is taken). In a second step, the results provided by the different algorithms are aggregated. The final decision is taken by selecting the class with the largest resulting membership value.

The fusion step is organized as follows:

For each pixel:

1. Separately build the fuzzy set $\pi_i(x) = \{\mu_i^1(x), \mu_i^2(x), \ldots, \mu_i^j(x), \ldots, \mu_i^n(x)\}$ for each classifier i, with n classes.
2. Compute the fuzziness degree $H_{\alpha QE}(\pi_i)$ of each fuzzy set $\pi_i(x)$.
3. Normalize data with w_i defined in (12.21).
4. Apply operator (12.22).
5. Select the class corresponding to the highest resulting membership degree.

The block diagram of the fusion process is given in Figure 12.6.

Note that in Figure 12.6, the range of the fuzzy sets is rescaled before the fusion step in order to combine data with the same range. This is achieved with the following range stretching algorithm:

- For all $\pi_i(x) = \{\mu_i^1(x), \ldots, \mu_i^j(x), \ldots, \mu_i^n(x)\}$, compute

$$M = \max_{j,x}[\mu_i^j(x)]$$

$$m = \min_{j,x}[\mu_i^j(x)]$$

- For all $\mu_i^j(x)$, compute

$$* \; \mu_i^j(x) = \frac{\mu_i^j(x) - m}{M - m}$$

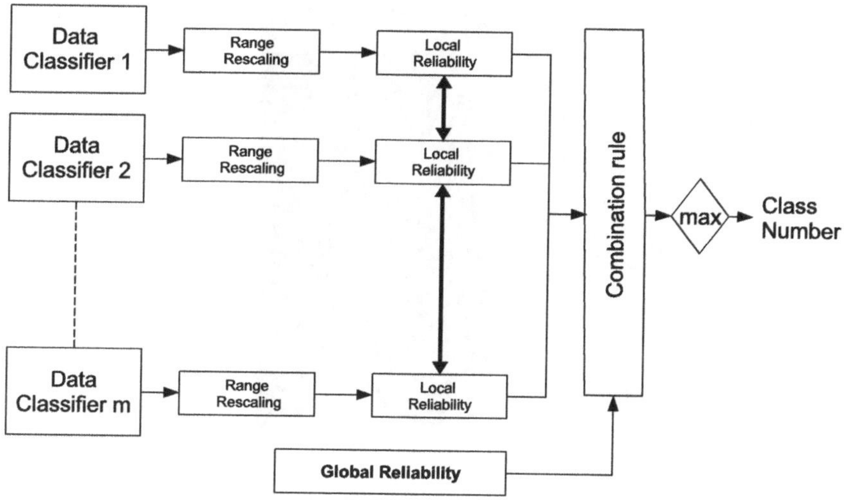

Figure 12.6. Block diagram of the fusion method. © 2006 IEEE.

12.5. EXPERIMENTAL RESULTS

12.5.1. Test Image

Airborne data from the ROSIS-03 (Reflective Optics System Imaging Spectrometer) optical sensor are used for the experiments. The flight over the city of Pavia, Italy, was operated by the Deutschen Zentrum fur Luft- und Raumfahrt (DLR, the German Aerospace Agency) in the framework of the HySens project, managed and sponsored by the European Union. According to specifications, the number of bands of the ROSIS-03 sensor is 115 with a spectral coverage ranging from 0.43 to 0.86 μm. The spatial resolution is 1.3 m per pixel. The original data set is 610 × 340 pixels. Some channels (12) have been removed due to noise. The remaining 103 spectral dimensions are processed.

Nine classes of interest are considered, namely: trees, asphalt, bitumen, gravel, metal sheet, shadow, bricks, meadow, and soil.

Figure 12.7a presents a three-channel color composite of the original data, where channels 80, 45, and 10 of the original data are used for red, green, and blue, respectively. The available reference data is shown in Figure 12.7b and the number of training and test samples is given in Table 12.2.

12.5.2. Classifier Based on Mathematical Morphology and Neural Network

In this section, we present a classifier based on a morphological feature extraction, the classification being performed with an artificial neural network. We first briefly recall the concept of granulometry. We then explain how the feature extraction is extended to the case of hyperspectral images.

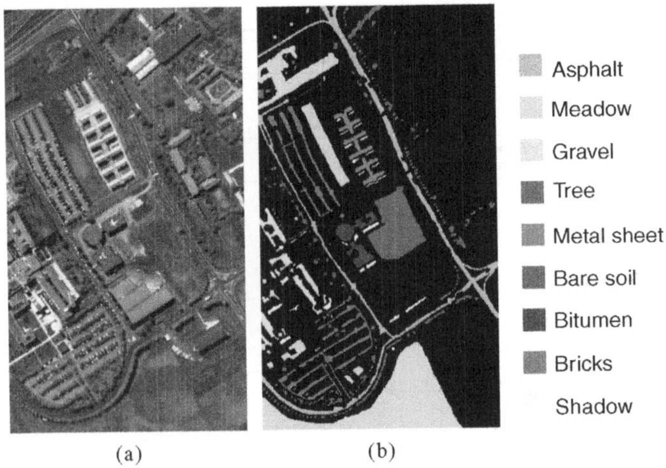

Figure 12.7. ROSIS University area. (a) Three-channel color composite, channels 80 (red), 45 (green) and 10 (blue). (b) Available reference data and information classes.

Feature Extraction and Granulometry. Granulometries are popular and powerful tools derived from the mathematical morphology theory. They are classically used for the analysis of the size distribution of particles in an image. They can be applied in various applications, ranging from the study of porous media to texture segmentation [38]. More information on mathematical morphology can be found in references 39 and 40. Reference 13 is an application-oriented book on morphological image analysis, from the principles to recent developments, and reference 12 is a survey paper investigating the use of advanced morphological operators in the general frame of satellite remote sensing.

Granulometries have recently been introduced in remote sensing image processing for the classification of urban areas [14, 15]. As a matter of fact, traditional pixel classification techniques used for remotely sensed rural areas turned to be

TABLE 12.2. Information Classes and Samples

Class		Samples	
No.	Name	Train	Test
1	Asphalt	548	6,304
2	Meadows	540	18,146
3	Gravel	392	1,815
4	Tree	524	2,912
5	Metal sheet	265	1,113
6	Bare soil	532	4,572
7	Bitumen	375	981
8	Brick	514	3,364
9	Shadow	231	795
	Total	3,921	40,002

ill-suited for urban areas, especially at high resolution. Beyond the spectral signature, there is actually a strong need for incorporating spatial information in the classification process. That can be achieved using granulometries.

The classical granulometry by opening is obtained by applying morphological opening operations with structuring elements (SE) of increasing size. The consequence is a progressive simplification of the image with a gradual disappearance of the features that are brighter than their immediate neighborhood. Each structure is removed when it becomes smaller than the SE. Using connected operators, such as geodesic reconstruction, no shape noise is introduced: At one given step, every structure is either totally removed or exactly preserved [41]. The opening being an anti-extensive operation, the evolution of the gray level for each pixel can be plotted as a monotonously decreasing curve, from its initial value to a lower bound corresponding to the smallest gray level value in the initial image as the size of the SE increases. Noting $I(x,y)$ as the original image and γ_s the morphological opening by reconstruction using a disk SE of radius s, the morphological profile $MP_\gamma(i)$ is defined for each pixel by

$$\begin{aligned} MP_\gamma(0) &= I(x,y) \\ MP_\gamma(1) &= \gamma_s[I(x,y)] \\ MP_\gamma(2) &= \gamma_{2s}[I(x,y)] \\ &\cdots\downarrow\cdots \\ MP_\gamma(p) &= \gamma_{ps}[I(x,y)] \end{aligned} \tag{12.23}$$

When the size of the SE reaches the characteristic size of one given structure—that is, when the SE does not fit inside the structure any longer—the structure is removed. Corresponding pixels are then assigned the gray level value of the darker region surrounding the structure. Opening operations only affect structures that are brighter than their immediate neighborhood. Other structures are left unchanged and thus lead to a constant MP_γ.

Similarly, a granulometry by closing is obtained using morphological closing operations, with the same set of structuring elements. Noting Φ_s, the morphological closing, by reconstruction with a structuring element of size s, it is given by

$$MP_\Phi(i) = \Phi_{is}[I(x,y)], \qquad i = 1,\ldots,p \tag{12.24}$$

The closing operation being dual to the opening operation, the closing-based profile provides information regarding the structures that are darker than their immediate surrounding and does not affect structures that are brighter than their surroundings.

Finally, in order to process simultaneously the bright and the dark structures of the image, the two MPs are concatenated:

$$\begin{aligned} MP(i) &= MP_\gamma(i) && \text{for } i = 1,\ldots,p \\ &= I(x) && \text{for } i = 0 \\ &= MP_\Phi(-i) && \text{for } i = -1,\ldots,-p \end{aligned} \tag{12.25}$$

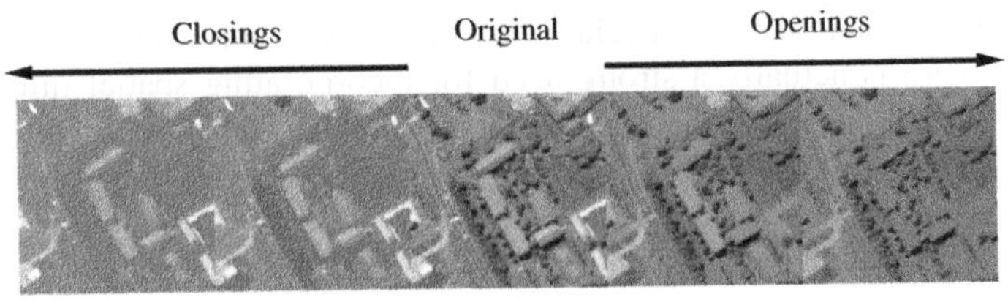

Figure 12.8. Simple morphological profile with 2 opening and 2 closings. In the shown profile, circular structuring elements are used with radius increment 4 ($r = 4, 8$pixels).

For illustration, Figure 12.8 presents one component of the original image (part of Fig. 12.9) and one granulometry with two openings and two closings.

The morphological profile (or its derivate) extracted for every pixel can be used as the input for an artificial network performing the classification [14], potentially by including decision boundary feature extraction (DBFE) to reduce the redundancy in the morphological profile [15].

Extension to Hyperspectral Data. Similar feature extractions have been proposed to deal with hyperspectral images [16, 17]. One solution consists in first decomposing the data into principal components (PC). Figure 12.9 presents the four first components in the decomposition of the ROSIS original data used in this study. It results in the concentration of the information over a few uncorrelated components. The proposed algorithm is then

- Perform a principal component analysis of the original hyperspectral data.
- Select the first PCs cumulating 90% of the whole information (the sum of the corresponding eigen values represents 90% of the sum of all the eigen values). In our experiment, this is achieved by selecting the three first PCs.
- Compute the morphological profile for each pixel, on each component separately. In our experiment, 10 openings and 10 closings are computed,

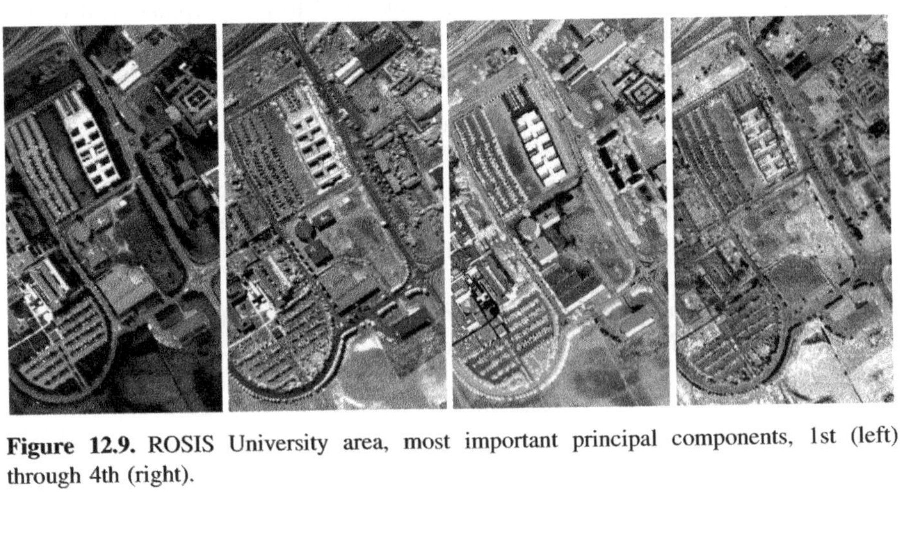

Figure 12.9. ROSIS University area, most important principal components, 1st (left) through 4th (right).

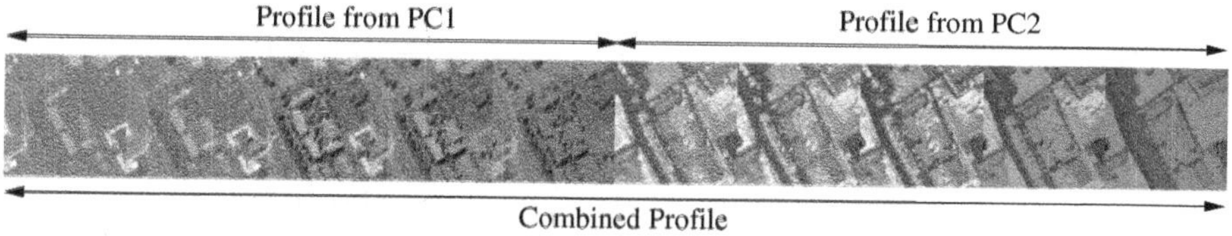

Figure 12.10. Extended morphological profile of two images. Each of the original profile has 2 opening and 2 closings. Circular structuring element with radius increment 4 was used ($r = 4, 8$).

with circular structuring elements with a 2-pixel increment. For one given component and for every pixel, the extracted vector has a dimension equal to $2 \times 10 + 1 = 21$.

- For every pixel, concatenate the profiles obtained with each component. This leads to a 3 components $\times$ 21 values = 63-dimensional feature vector. This is illustrated by Figure 12.10 in a reduced case.
- This vector is used as the input for the neural network for classification.

Results. Figure 12.12a presents the thematic classification map obtained with this neural-network-based classifier. The corresponding confusion matrix is given by Table 12.3. The classification accuracy is fairly good. However, one class, namely "gravel," has poor results with the corresponding pixels being spread over various classes and with a confusion with the "metal sheet" class.

Assessing the Pointwise Accuracy. For every pixel, the output of the neural network gives the posterior probability corresponding to each class. Consequently, the fuzzy set $\pi_i(x)$ is directly derived and is used as the input to the previously described fusion scheme for this neural network based classifier.

12.5.3. Classifier Based on Support Vector Machines

In this section, we present a second classifier which is detailed and discussed in Fauvel et al. [42]. It is based on support vector machines (SVMs), which are known to be well-suited for high-dimensional classification problems [3, 4]. As stated in the introduction, SVMs characterize classes using a geometrical criterion rather than statistical criteria. They seek a separating hyperplane maximizing the distance with the closest training samples for two classes. This approach allows SVMs to have a very high capability of generalization and, as a consequence, only require a few training samples. For nonlinearly separable data, SVMs use the kernel trick to map the data onto a higher-dimensional space where they are linearly separable [5].

Here, we consider multiclass SVMs without any feature reduction of the original hyperspectral data. The standard Gaussian radial basis kernel with L2-norm distance is used. This algorithm was proved to provide interesting classification accuracy, even in the case of a limited training set [42]. Note that other kernels

TABLE 12.3. Confusion Matrix for the Neural Network-Based Classifier

	Asphalt Ref.	Meadow Ref.	Gravel Ref.	Tree Ref.	Metal Sheet Ref.	Bare Soil Ref.	Bitumen Ref.	Brick Ref.	Shadow Ref.	Average Accuracy	Overall Accuracy
Asphalt	6157	1	86	1	1	52	58	36	13		
Meadow	44	18,476	17	22	1	154	0	49	0		
Gravel	0	19	31	4	0	0	0	9	0		
Tree	15	23	48	2968	4	9	1	70	8		
Metal sheet	0	3	993	6	1334	0	0	1200	1		
Bare soil	406	127	0	61	0	4791	17	1	1		
Bitumen	8	0	23	0	0	22	1254	0	0		
Brick	1	0	901	1	4	1	0	2317	0		
Shadow	0	0	0	1	1	0	0	0	924		
%	92.85	99.07	1.48	96.87	99.18	95.27	94.29	62.93	97.57	82.17	89.42

could be considered, such as the spectral angle mapper which basically computes the angle between two vectors in the vector space [10,42,43].

In the following, we present the used classifier and start by briefly recalling the general mathematical formulation of SVMs. Starting from the linearly separable case, optimal hyperplanes are introduced, and then the classification problem is modified to handle nonlinearly separable data and a brief description of multiclass strategies is given. Finally, kernel methods are presented.

Linear SVMs. For a two-class problem in a n-dimensional space $\mathbb{R}^n$, we assume that N training samples, $\mathbf{x}_i \in \mathbb{R}^n$, are available with their corresponding label $y_i = \pm 1$: $\{(\mathbf{x}_i, y_i) | i \in [1, N]\}$. The SVMs method consists in finding the hyperplane that maximizes the margin (see Figure 12.11)—that is, the distance to the closest training data points in both classes. Noting $\mathbf{w} \in \mathbb{R}^n$ as the vector normal to the hyperplane and $b \in \mathbb{R}$ as the bias, the hyperplane H_p is defined as

$$\mathbf{w} \cdot \mathbf{x} + b = 0, \qquad \forall \mathbf{x} \in H_p \tag{12.26}$$

where $\mathbf{w} \cdot \mathbf{x}$ is the dot product between $\mathbf{w}$ and $\mathbf{x}$. If $\mathbf{x} \notin H_p$, then $f(\mathbf{x}) = \mathbf{w} \cdot \mathbf{x} + b$ is the distance of $\mathbf{x}$ to H_p. According to the previous statement, such a hyperplane has to satisfy

$$y_i(\mathbf{w} \cdot \mathbf{x}_i + b) > 1, \qquad \forall i \in [1, N] \tag{12.27}$$

Finally, the optimal hyperplane has to maximize the margin, $\| \mathrm{w} \|$; this is equivalent to minimize $\| \mathrm{w} \|/2$ and leads to the following quadratic optimization

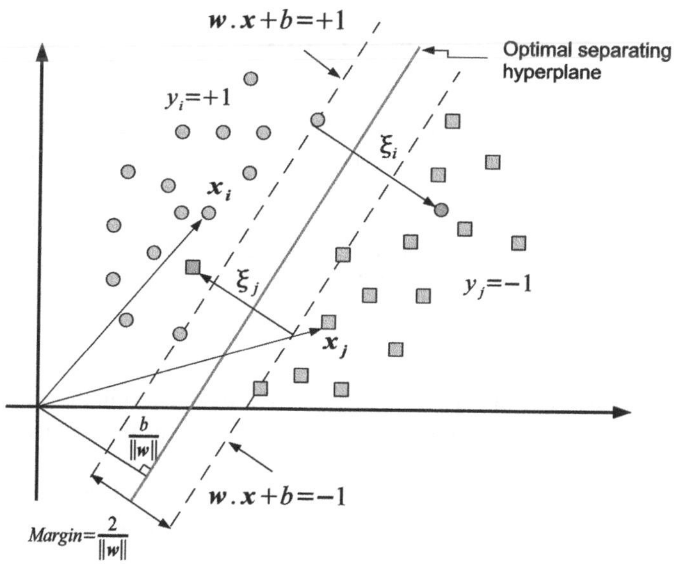

Figure 12.11. Classification of a nonlinearly separable case by SVMs. There is one non separable feature vector in each class.

problem:

$$\min\left[\frac{\| \mathbf{w} \|^2}{2}\right], \qquad \text{subject to (12.27)} \tag{12.28}$$

For nonlinearly separable data, *slack* variables ξ are introduced to deal with misclassified samples (see Figure 12.11). Equation (12.27) becomes

$$y_i(\mathbf{w} \cdot \mathbf{x}_i + b) > 1 - \xi_i, \xi_i \geq 0, \qquad \forall i \in [1, N] \tag{12.29}$$

The final optimization problem becomes

$$\min\left[\frac{\| \mathbf{w} \|^2}{2} + C\sum_{i=1}^{N} \xi_i\right], \qquad \text{subject to (12.29)} \tag{12.30}$$

where the constant C controls the amount of penalty. It can be solved by considering the dual optimization problem using Lagrange multipliers α:

$$\begin{aligned} &\max_{\alpha} \sum_{i=1}^{N} \alpha_i - \frac{1}{2}\sum_{i,j=1}^{N} \alpha_i \alpha_j y_i y_j (\mathbf{x}_i \cdot \mathbf{x}_j) \\ &\text{subject to } 0 \leq \alpha_i \leq C, \qquad \forall_i \in [1, N] \\ &\sum_{i=1}^{N} \alpha_i y_i = 0 \end{aligned} \tag{12.31}$$

Finally,

$$\mathbf{w} = \sum_{i=1}^{N} \alpha_i y_i \mathbf{x}_i \tag{12.32}$$

The solution vector is a linear combination of some samples of the training set, whose α_i is nonzero, called *support vectors.* The hyperplane decision function can thus be written as

$$y_u = \operatorname{sgn}\left(\sum_{i=1}^{N} y_i \alpha_i (\mathbf{x}_u \cdot \mathbf{x}_i) + b\right) \tag{12.33}$$

where $\mathbf{x}_u$ is an unseen sample.

Multiclass SVMs. SVMs are designed to solve binary problems where the class labels can only take two values: ± 1. For a remote sensing application, several classes are usually of interest. Various approaches have been proposed to address this problem. They usually combine a set of binary classifiers. Two main approaches were originally proposed for the m-classes problem [5].

- **One Versus the Rest.** m binary classifiers are applied on each class against the others. Each sample is assigned to the class with the *maximum* output.

- **Pairwise Classification.** $\frac{m(m-1)}{2}$ binary classifiers are applied on each pair of classes. Each sample is assigned to the class getting the highest number of votes. A vote for a given class is defined as a classifier assigning the pattern to that class.

The *pairwise classification* has shown to be more suitable for large problems [44]. Even though the number of classifiers to handle is larger than for the *one versus the rest* approach, the whole classification problem is decomposed into much simpler ones. This second approach is therefore used in this chapter.

Nonlinear SVMs. *Kernel methods* are a generalization of SVMs providing nonlinear decision functions and thus improving classification abilities. Input data are mapped onto a higher-dimensional space $\mathbb{H}$ using a nonlinear function Φ:

$$\begin{aligned} \mathbb{R}^n &\rightarrow \mathbb{H} \\ \mathbf{x} &\rightarrow \Phi(\mathbf{x}) \\ \mathbf{x}_i \cdot \mathbf{x}_j &\rightarrow \Phi(\mathbf{x}_i) \cdot \Phi(\mathbf{x}_j) \end{aligned} \tag{12.34}$$

The expensive computation of $\Phi(\mathbf{x}_i) \cdot \Phi(\mathbf{x}_j)$ in $\mathbb{H}$ is reduced using the *kernel trick* [5]:

$$\Phi(\mathbf{x}_i) \cdot \Phi(\mathbf{x}_j) = K(\mathbf{x}_i, \mathbf{x}_j) \tag{12.35}$$

The kernel K should fulfill Mercer's condition [4]. Using kernels, we never explicitly work in $\mathbb{H}$, and all the computations are done in the original space $\mathbb{R}^n$.

For classification of remote sensing images, two kernels are popular: the inhomogeneous polynomial function and the Gaussian radial basis function (RBF).

$$K_{\text{POLY}}(\mathbf{x}_i, \mathbf{x}_j) = [(\mathbf{x}_i \cdot \mathbf{x}_j) + 1]^p \tag{12.36}$$

$$K_{\text{RBF}}(\mathbf{x}_i, \mathbf{x}_j) = \exp[-\gamma \, \|\mathbf{x}_i - \mathbf{x}_j\|^2] \tag{12.37}$$

Radial basis functions can be written as follows [5]: $K(\mathbf{x}_i, \mathbf{x}_j) = f(d(\mathbf{x}_i, \mathbf{x}_j))$, where d is a metric on $\mathbb{R}^n$ and f is a function on $\mathbb{R}_0^+$. For the Gaussian RBF, $f(t) = \exp(-\gamma t^2)$, $t \in \mathbb{R}_0^+$, and $d(\mathbf{x}_i, \mathbf{x}_j)) = \|\mathbf{x}_i - \mathbf{x}_j\|$, that is, the Euclidean distance. As mentioned in Keshava [43], Euclidean distance is not scale-invariant; however, due to atmospheric attenuation or variation in illumination, spectral energy can be different for two samples even if they belong to the same class. To handle such a problematic case, scale-invariant metrics can be considered. *Spectral angle mapper* (SAM) is a well-known scale-invariant metric, has been widely used in many remote sensing problems, and has been shown to be robust to variations in spectral energy [43]. This metric α focuses on the angle between two vectors:

$$\alpha(\mathbf{x}_i, \mathbf{x}_j) = \arccos\left(\frac{\mathbf{x}_i \cdot \mathbf{x}_j}{\|\mathbf{x}_i\| \cdot \|\mathbf{x}_j\|}\right) \tag{12.38}$$

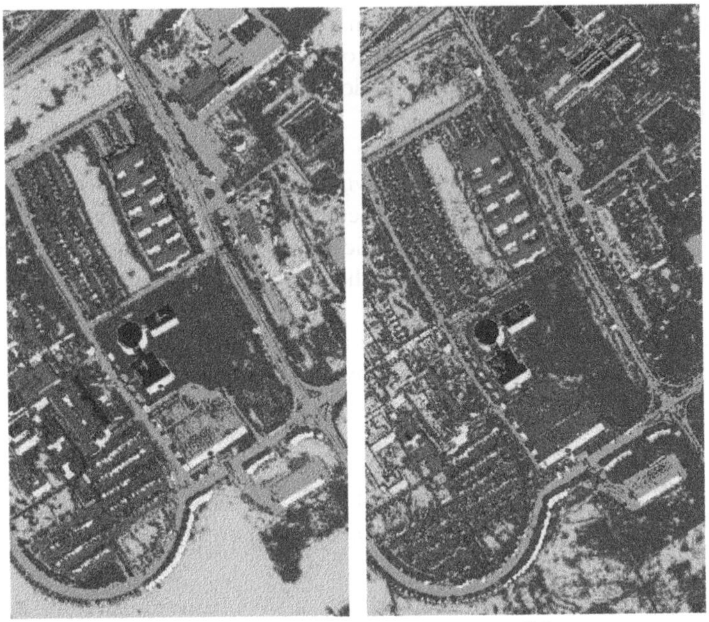

(a) (b)

Figure 12.12. Result of the classification using the neural network (a) and the support vector machine (b).

See Fauvel et al. [42] for a comparison of RBF kernels with the Euclidean distance (12.37) and the spectral angle mapper (12.39).

$$K_{\text{SAM}}(\mathbf{x}_i, \mathbf{x}_j) = \exp[-\gamma\alpha(\mathbf{x}_i, \mathbf{x}_j)^2] \quad (12.39)$$

Both kernels fulfill Mercer's conditions and optimal hyperplanes can therefore be found. In this study, we will only consider RBF kernels with the Euclidean distance which turned out to be better suited for the analysis of urban areas.

Results. Figure 12.12b presents the thematic classification map obtained with this Gaussian kernel SVM classifier. The corresponding confusion matrix is given by Table 12.4. Generally speaking, the results obtained are comparable with the results provided by the neural network. The overall accuracy is lower but the average accuracy is higher since the performances are more uniform among the different classes. In particular, this method performs much better than the neural network for the "gravel" class and significantly better for the "bitumen" class as well.

As a conclusion, the two presented classifiers provide complementary information and a fusion of their results should improve the performances. As a prior knowledge for the fusion algorithm, we can define indices of confidence (global accuracy) for the two algorithms. They are presented in Table 12.5): Both

TABLE 12.4. Confusion Matrix for the SVM-Based Classifier

	Asphalt Ref.	Meadow Ref.	Gravel Ref.	Tree Ref.	Metal Sheet Ref.	Bare Soil Ref.	Bitumen Ref.	Brick Ref.	Shadow Ref.		
Asphalt	5551	0	29	0	0	21	99	35	22		
Meadow	32	13,100	11	28	0	193	0	14	0		
Gravel	113	0	1476	2	0	0	1	186	7		
Tree	21	2037	0	2997	2	55	0	6	0		
Metal sheet	20	0	0	5	1337	97	0	0	0		
Bare soil	16	3493	6	31	0	4639	0	23	0		
Bitumen	346	0	5	0	0	0	1218	0	0		
Brick	515	19	572	1	0	24	12	3409	3		
Shadow	17	0	0	0	6	0	0	0	915	Average Accuracy	Overall Accuracy
%	83.71	70.26	70.32	97.81	99.41	92.25	91.58	92.59	96.62	88.28	80.98

TABLE 12.5. Indices of Confidence

	Neural Network	SVM
Asphalt	1	1
Meadow	1	1
Gravel	0	1
Tree	1	1
Metal sheet	1	1
Bare soil	1	1
Bitumen	1	1
Brick	0	1
Shadow	1	1

algorithms are well-suited for all the classes, except the neural network for the "gravel" and the "brick" classes.

Assessing the point-wise accuracy. Since a *pairwise classification* has been used, $\frac{m(m-1)}{2}$ binary classifiers are evaluated; thus it is not possible to derive directly a fuzzy set $\pi_i(x)$ as with the neural network based algorithm. Probabilities are built following a standard method [45]: For each considered class, the obtained number of votes is divided by the total number of votes, that is, the number of pairwise classifiers (i.e., $\frac{m(m-1)}{2}$). For every pixel, it results in the construction of a fuzzy set where a probability value ranging form 0 to 1 is associated to each class. These probabilities are used as the input to the previously described fusion scheme for the SVM-based classifier.

12.5.4. Decision Fusion

In this section, we present the results obtained with the different fusion operators presented in Section 12.3.

Figures 12.13d and 12.13e present the thematic maps obtained with the min and the max operator, respectively. Corresponding confusion matrices are given in Tables 12.6 and 12.7, respectively. This simple operators do not lead to any improvement, with the class "tree" even being completely lost and confused with "asphalt." This underlines the need for a proper handling of conflictual situations.

Figures 12.13a, 12.13b, and 12.13c present the thematic maps obtained with the operators from Eqs. (12.18), (12.19), and (12.20), respectively. Corresponding confusion matrices are given in Tables 12.8, 12.9, and 12.10, respectively. For these operators, the measure of conflict between the different classifers is given by Eq. 12.17. For the operators (12.19) and (12.20), the neural network classifier is chosen as the priority in case of conflict, since this is the one providing the best overall accuracy. It is striking to note that operators (12.18) and (12.19) actually lead to the same results as operator min (they are almost identical). This is due to the measure of conflict $1 - C$ that is most of the times quite low: With values of C higher than 0.6, operators (12.18) and (12.19) indeed converge toward the min

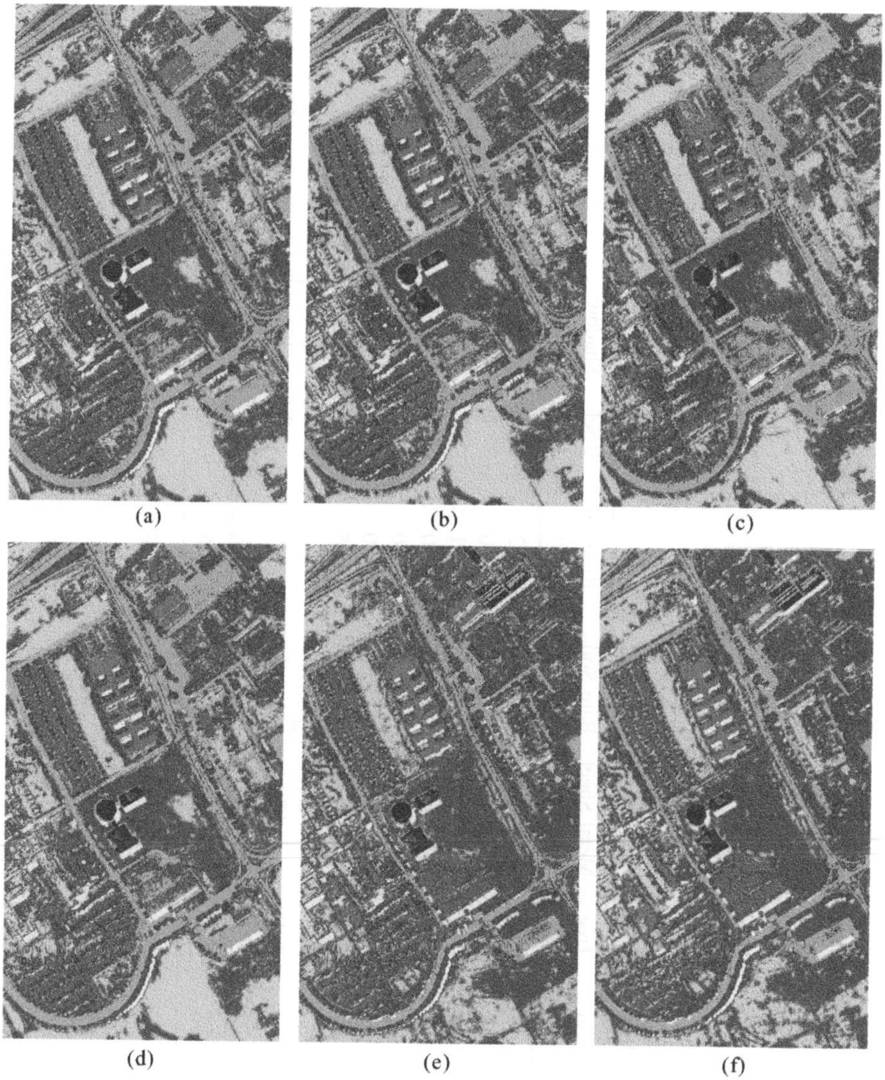

Figure 12.13. Result of the classification with a decision fusion using various operators. (a) Operator (12.18), (b) operator (12.19), (c) operator (12.20), (d) operator min, (e) operator max, and (f) proposed adaptive operator.

operator. None of these operators provide fully satisfactory results. As a matter of fact, more flexibility is needed in the fusion of conflictual situations. This is achieved with the proposed adaptice fusion scheme whose results are presented on Figure 12.13f and in Table 12.11. Though providing quantitative results close to the results obtained with the SVM, the thematic map is less noisy, and no class has bad results, unlike with the neural network ("gravel") or the other fusion operators ("tree").

TABLE 12.6. Confusion Matrix for the Decision Fusion Using the min Operator

	Asphalt Ref.	Meadow Ref.	Gravel Ref.	Tree Ref.	Metal Sheet Ref.	Bare Soil Ref.	Bitumen Ref.	Brick Ref.	Shadow Ref.		
Asphalt	6114	1857	2	2948	23	53	82	26	94		
Meadow	26	16,496	0	59	0	902	0	4	0		
Gravel	25	0	1521	3	59	5	1	256	70		
Tree	0	0	0	0	0	0	0	0	0		
Metal sheet	0	0	0	0	1198	7	0	0	0		
Bare soil	28	189	0	47	0	3959	0	0	0		
Bitumen	158	0	0	0	0	0	1245	0	0		
Brick	278	107	576	6	48	103	2	3396	31		
Shadow	2	0	0	1	17	0	0	0	752	Average Accuracy	Overall Accuracy
%	92.20	88.46	72.46	0	89.07	78.72	93.61	92.23	79.41	76.24	81.08

TABLE 12.7. Confusion Matrix for the Decision Fusion Using the max Operator

	Asphalt Ref.	Meadow Ref.	Gravel Ref.	Tree Ref.	Metal Sheet Ref.	Bare Soil Ref.	Bitumen Ref.	Brick Ref.	Shadow Ref.		
Asphalt	5572	1921	13	3011	2	52	97	38	18		
Meadow	46	13932	9	29	0	193	0	12	0		
Gravel	110	0	1485	0	0	0	1	186	8		
Tree	0	0	0	0	0	0	0	0	0		
Metal sheet	17	2	5	5	1337	98	1	0	1		
Bare soil	19	2775	6	19	0	4665	1	32	2		
Bitumen	352	0	3	0	0	0	1219	9	0		
Brick	500	19	578	0	0	21	11	3405	1		
Shadow	15	0	0	0	6	0	0	0	917	Average Accuracy	Overall Accuracy
%	84.03	74.71	70.75	0	99.41	92.76	91.65	92.48	96.83	78.07	76.05

TABLE 12.8. Confusion Matrix for the Decision Fusion Using Operator (12.18)

	Asphalt Ref.	Meadow Ref.	Gravel Ref.	Tree Ref.	Metal Sheet Ref.	Bare Soil Ref.	Bitumen Ref.	Brick Ref.	Shadow Ref.	Average Accuracy	Overall Accuracy
Asphalt	6114	1857	2	2948	23	53	82	26	94		
Meadow	26	16,496	0	59	0	902	0	4	0		
Gravel	25	0	1521	3	59	5	1	256	70		
Tree	0	0	0	0	0	0	0	0	0		
Metal sheet	0	0	0	0	1198	7	0	0	0		
Bare soil	28	189	0	47	0	3959	0	0	0		
Bitumen	158	0	0	0	0	0	1245	0	0		
Brick	278	107	576	6	48	103	2	3396	31		
Shadow	2	0	0	1	17	0	0	0	752		
%	92.20	88.46	72.46	0	89.07	78.72	93.61	92.23	79.41	76.24	81.08

TABLE 12.9. Confusion Matrix for the Decision Fusion Using Operator (12.19)

	Asphalt Ref.	Meadow Ref.	Gravel Ref.	Tree Ref.	Metal Sheet Ref.	Bare Soil Ref.	Bitumen Ref.	Brick Ref.	Shadow Ref.		
Asphalt	6116	1857	2	2948	23	63	82	26	94		
Meadow	26	16,496	0	59	0	898	0	4	0		
Gravel	23	0	1521	3	59	5	1	256	70		
Tree	0	0	0	0	0	0	0	0	0		
Metal sheet	0	0	0	0	1198	7	0	0	0		
Bare soil	28	189	0	47	0	3959	0	0	0		
Bitumen	158	0	0	0	0	0	1245	0	0		
Brick	278	107	576	6	48	97	2	3396	31		
Shadow	2	0	0	1	17	0	0	0	752	Average Accuracy	Overall Accuracy
%	92.23	88.46	72.46	0	89.07	78.72	93.61	92.23	79.41	76.24	81.08

TABLE 12.10. Confusion Matrix for the Decision Fusion Using Operator (12.20)

	Asphalt Ref.	Meadow Ref.	Gravel Ref.	Tree Ref.	Metal Sheet Ref.	Bare Soil Ref.	Bitumen Ref.	Brick Ref.	Shadow Ref.	Average Accuracy	Overall Accuracy
Asphalt	6023	1688	96	2884	10	72	60	367	107		
Meadow	216	16,164	6	67	123	1170	3	19	8		
Gravel	38	17	1347	15	9	1	2	1028	121		
Tree	0	0	0	0	0	0	0	0	0		
Metal sheet	7	194	379	16	1130	19	4	55	4		
Bare soil	83	490	10	58	18	3676	16	38	137		
Bitumen	167	0	1	2	9	2	1244	0	0		
Brick	94	96	260	12	10	82	1	2175	80		
Shadow	3	0	0	10	36	7	0	0	490		
%	90.83	86.67	64.17	0	84.01	73.10	93.53	59.07	51.74	67.01	75.39

TABLE 12.11. Confusion Matrix for the Decision Fusion Using the Proposed Adaptive Operator

	Asphalt Ref.	Meadow Ref.	Gravel Ref.	Tree Ref.	Metal Sheet Ref.	Bare Soil Ref.	Bitumen Ref.	Brick Ref.	Shadow Ref.	Average Accuracy	Overall Accuracy
Asphalt	6370	3681	356	0	36	229	98	232	80		
Meadow	2	12,272	2	16	0	26	0	8	0		
Gravel	0	0	1350	0	0	0	0	27	1		
Tree	0	159	0	3041	0	0	0	0	0		
Metal sheet	2	0	0	4	1306	78	0	0	0		
Bare soil	1	2535	1	1	0	4692	0	9	0		
Bitumen	35	0	0	0	0	0	1232	3	0		
Brick	221	2	390	0	0	4	0	3403	0		
Shadow	0	0	0	2	3	0	0	0	866		
%	96.06	65.81	64.32	99.25	97.10	93.30	92.63	92.42	91.45	88.04	80.73

12.6. CONCLUSION

In this chapter, we have presented two classifiers dealing with hyperspectral images. The first one uses operators derived from mathematical morphology to extract relevant features. The classification is performed using an artificial neural network. The second one classifies each pixel directly from its spectral value using a kernel support vector machine.

To take advantage of these two classifiers and their complementary properties, a general decision fusion framework, based on a fuzzy combination rule, is presented. The proposed method is adaptive and enables a satisfactory handling of the conflictual situations where the different classifiers disagree. Two measures of accuracy are used in the combination rule: The first one, based on prior knowledge, defines global reliabilities, both for each classifier and each class. The second one automatically estimates the pointwise reliability of the results provided by each classifier and, thus, enables the adaptation of the fusion rule to local context. The proposed approach does not need any training, and the computational load remains low. As a result, a better classification is obtained, providing less noisy thematic maps.

One should underline that no prior assumption is needed regarding the modeling of the data (e.g., Bayes theory, possibility theory, etc.) before the data are fused. A key point lies in the generality of the presented framework for *decision level fusion*. Though only two classifiers were used in this study, additional algorithms could easily be added to the process. For instance, dedicated algorithms such as street trackers could be used without increasing errors in the other classes.

In this chapter, the α-quadratic entropy was chosen for the fuzziness evaluation because the sensibility of that measure can be modified with the value of α. Note that several other measures can be used—for example, the *fuzzy entropy* [29]. One limitation of the proposed approach is the use of binary values for the global confidence. With fuzzy confidence, the combination rule could be rewritten with *T-conorm* and *T-norm*, both of which are less indulgent and less severe than max and min. Moreover, the use of the *T-conorm* and *T-norm* would allow a finer definition of global accuracy.

ACKNOWLEDGMENT

This research was supported in part by the Research Fund of the University of Iceland and the Jules Verne Program of the French and Icelandic governments (PAI EGIDE).

REFERENCES

1. C. Lee and D. A. Landgrebe, Analysing high dimensional multispectral data, *IEEE Transactions on Geoscience and Remote Sensing*, vol. 31, pp. 792–800, 1993.
2. G. F. Hughes, On the mean accuracy of statistical pattern recognizers, *IEEE Transactions on Information Theory*, vol. IT-14, pp. 55–63, 1968.

3. V. Vapnik, *Statistical Learning Theory*, John Wiley & Sons, New York, 1998.
4. C. Burges, A tutorial on support vector machines for pattern recognition, *Data Mining and Knowledge Discovery*, vol. 2, pp. 121–167, 1998.
5. B. Scholkopf and A. J. Smola, *Learning with Kernels*, MIT Press, Cambridge, MA, 2002.
6. J. A. Gualtieri and S. Chettri, Support vector machines for classification of hyperspectral data, in *Geoscience and Remote Sensing Symposium*, Vol. 2. IGARSS '00 Proceedings, July 2000.
7. G. H. Halldorsson, J. A. Benediktsson, and J. R. Sveinsson, Support vector machines in multisource classification, in *Geoscience and Remote Sensing Symposium*, Vol. 3. IGARSS '03. Proceedings, July 2003.
8. F. Melgani and L. Bruzzone, Classification of hyperspectral remote sensing images with support vector machines, *IEEE Transactions on Geoscience and Remote Sensing*, vol. 42, pp. 1778–1790, 2004.
9. G. F. Foody and A. Mathur, A relative evaluation of multiclass image classification by support vector machines, *IEEE Transactions on Geoscience and Remote Sensing*, vol. 42, pp. 1335–1343, 2004.
10. G. Mercier and M. Lennon, Support vector machines for hyperspectral image classification with spectral-based kernels, in *Geoscience and Remote Sensing Symposium*, vol. 1. IGARSS '03. Proceedings, July 2003.
11. M. Fauvel, J. Chanussot, and J. A. Benediktsson, Decision fusion for the classification of urban remote sensing images, *IEEE Transactions on Geoscience and Remote Sensing*, vol. 44, no. 10, pp. 2828–2838, Oct. 2006.
12. P. Soille and M. Pesaresi, Advances in mathematical morphology applied to geoscience and remote sensing, *IEEE Transactions on Geoscience and Remote Sensing*, vol. 40, no. 9, pp. 2042–2055, 2002.
13. P. Sollie, *Morphological Image Analysis, Principles and Applications*, 2nd edition, Springer, Berlin, 2003.
14. M. Pesaresi and J. A. Benediktsson, A new approach for the morphological segmentation of high resolution satellite imagery, *IEEE Transactions on Geoscience and Remote Sensing*, vol. 39, no. 2, pp. 309–320, 2001.
15. J. A. Benediktsson, M. Pesaresi, and K. Arnason, Classification and feature extraction for remote sensing images from urban areas based on morphological transformations, *IEEE Transactions on Geoscience and Remote Sensing*, vol. 41, no. 9, pp. 1940–1949, 2003.
16. F. Dell'Acqua, P. Gamba, A. Ferrari, J. Palmason, J. Benediktsson, and K. Arnason, Exploiting spectral and spatial information in hyperspectral urban data with high resolution, *IEEE Geoscience and Remote Sensing Letters*, vol. 1, no. 4, pp. 322–326, 2004.
17. J. A. Benediktsson, J. A. Palmason, and J. R. Sveinsson, Classification of hyperspectral data from urban areas based on extended morphological profiles, *IEEE Transactions on Geoscience and Remote Sensing*, vol. 43, no. 3, pp. 480–491, 2005.
18. A. Plaza, P. Martinez, J. Plaza, and R. Perez, Dimensionality reduction and classification of hyperspectral image data using sequences of extended morphological transformations, *IEEE Transactions on Geoscience and Remote Sensing*, vol. 43, no. 3, pp. 466–479, 2005.
19. J. A. Benediktsson and I. Kanellopoulos, Classification of multisource and hyperspectral data based on decision fusion, *IEEE Transactions on Geoscience and Remote Sensing*, vol. 37, pp. 1367–1377, 1999.

20. B. Jeon and D. A. Landgrebe, Decision fusion approach for multitemporal classification, *IEEE Transactions on Geoscience and Remote Sensing*, vol. 37, pp. 1227–1233, 1999.
21. G. Lisini, F. Dell'Acqua, G. Triani, and P. Gamba, Comparison and combination of multiband classifiers for landsat urban land cover mapping, in *Geoscience and Remote Sensing Symposium*, vol. CD ROM. IGARSS '05. Proceedings, August 2005.
22. F. Tupin, I. Bloch, and H. Maitre, A first step toward automatic interpretation of SAR images using evidential fusion of several structure detectors, *IEEE Transactions on Geoscience and Remote Sensing*, vol. 37, no. 3, pp. 1327–1343, 1999.
23. F. Tupin and M. Roux, Detection of building outlines based on the fusion of SAR and optical features, *ISPRS Journal of Photogrammetry and Remote Sensing*, vol. 58, pp. 71–82, 2003.
24. J. Chanussot, G. Mauris, and P. Lambert, Fuzzy fusion techniques for linear features detection in multitemporal sar images, *IEEE Transactions on Geoscience and Remote Sensing*, vol. 37, no. 3, pp. 1292–1305, 1999.
25. R. R. Yager, A general approach to the fusion of imprecise information, *International Journal of Intelligent Systems*, vol. 12, pp. 1–29, 1997.
26. L. A. Zadeh, Fuzzy sets, *Information and Control*, pp. 338–353, 1965.
27. G. J. Klir and B. Yuan, *Fuzzy Set and Fuzzy Logic: Theory and Applications*, Prentice-Hall PTR, Englewood Cliffs, NJ, 1995.
28. K. Tanaka, *An Introduction to Fuzzy Logic for Practical Application*, Springer, Berlin, 1996.
29. A. D. Luca and S. Termini, A definition of non-probabilistic entropy in the setting of fuzzy set theory, *Information and Control*, pp. 301–312, 1972.
30. L. A. Zadeh, Probability measures of fuzzy events, *Journal of Mathematical Analysis and Application*, vol. 23, pp. 421–427, 1968.
31. B. R. Ebanks, On measures of fuzziness and their representations, *Journal of Mathematics and Analysis and Application*, vol. 94, pp. 421–427, 1983.
32. C. Bezdeck, Measuring fuzzy uncertainty, *IEEE Transactions on Fuzzy Systems*, pp. 107–118, 1994.
33. I. Bloch, *Fusion d'informations en traitement du signal et des images*, edited by H. Sciences, Germes Lavoisier, Paris, 2003.
34. M. Oussalah, Study of some algebraical properties of adaptive combination rules, *Fuzzy Set and Systems*, vol. 114, pp. 391–409, 2000.
35. H. Prade and D. Dubois, Possibility theory in information fusion, *Proceedings of the Third International Conference on Information Fusion*, 2000.
36. I. Bloch, Information combination operators for data fusion: A comparative review with classification, *IEEE Transactions on Systems, Man, and Cybernetics—Part A: Systems and Humans*, vol. 26, no. 1, pp. 52–67, 1996.
37. D. Dubois and H. Prade, *Combination of Information in the Framework of Possibility Theory*, edited by M. A. A. et al., Academic, New York, 1992.
38. R. Gonzalez and R. Woods, *Digital Image Processing*, 2nd edition, Prentice Hall, Upper Saddle River, NJ, 2002.
39. J. Serra, *Mathematical Morphology, Theoretical Advances*, vol. 2, Academic Press, New York, 1988.

40. E. Dougherty, *Mathematical Morphology in Image Processing*, Marcel Dekker, New York, 1993.

41. J. Crespo, J. Serra, and R. Schafer, Theoretical aspects of morphological filters by reconstruction, *Signal Processing*, vol. 47, pp. 201–225, 1995.

42. M. Fauvel, J. Chanussot, and J. A. Benediktsson, Kernels Evaluation of multiclass classification of hyperspectral remote sensing data, in *Proceedings of IEEE International Conference on Acoustics, Speech, and Signal Processing*, Toulouse, France, 2006.

43. N. Keshava, Distance metrices and band selection in hyperspectral processing with application to material identification and spectral libraries, *IEEE Transactions on Geoscience and Remote Sensing*, vol. 42, pp. 1552–1565, 2004.

44. C. W. Hsu and C. J. Lin, A comparison of methods for multiclass support vector machines, *IEEE Transactions on Neural Networks*, vol. 13, pp. 415–425, March 2002.

45. T. Wu, C. Lin, and R. Weng, Probability estimates for multi-class classification by pairwise coupling, *Journal of Machine Learning Research*, vol. 5, pp. 975–1005, 2004.

CHAPTER 13

MORPHOLOGICAL HYPERSPECTRAL IMAGE CLASSIFICATION: A PARALLEL PROCESSING PERSPECTIVE

ANTONIO J. PLAZA
Department of Computer Science, University of Extremadura, E-10071 Caceres, Spain

13.1. INTRODUCTION

Mathematical morphology (MM) is a theory for spatial structure analysis that was established by introducing fundamental operators applied to two sets [1]. A set is processed by another one having a carefully selected shape and size, known as the structuring element (SE). In the context of image processing, the SE acts as a probe for extracting or suppressing specific structures of the image objects, checking that each position of the SE fits within those objects. Based on these ideas, two fundamental operators are defined in MM, namely *erosion* and *dilation*. The application of the erosion operator to an image yields an output image, which shows where the SE *fits* the objects in the image. On the other hand, the application of the dilation operator to an image produces an output image, which shows where the SE *hits* the objects in the image. All other MM operations can be expressed in terms of erosion and dilation [2]. For instance, the notion behind the *opening* operator is to dilate an eroded image in order to recover as much as possible of the eroded image. In contrast, the *closing* operator erodes a dilated image so as to recover the initial shape of image structures that have been dilated. The filtering properties of the opening and closing are based on the fact that, depending on the size and shape of the considered SE, not all structures from the original image will be recovered when these operators are applied. Because of the nonlinear properties of MM filters, their application generally results in an irreversible, though controlled, loss of information.

Although MM operators were originally defined for binary images, they were soon extended to gray-tone (mono-channel) images by viewing these data as an imaginary topographic relief in which the brighter the gray tone, the higher the corresponding elevation [3]. Here, morphological operations can be graphically

Hyperspectral Data Exploitation: Theory and Applications, Edited by Chein-I Chang

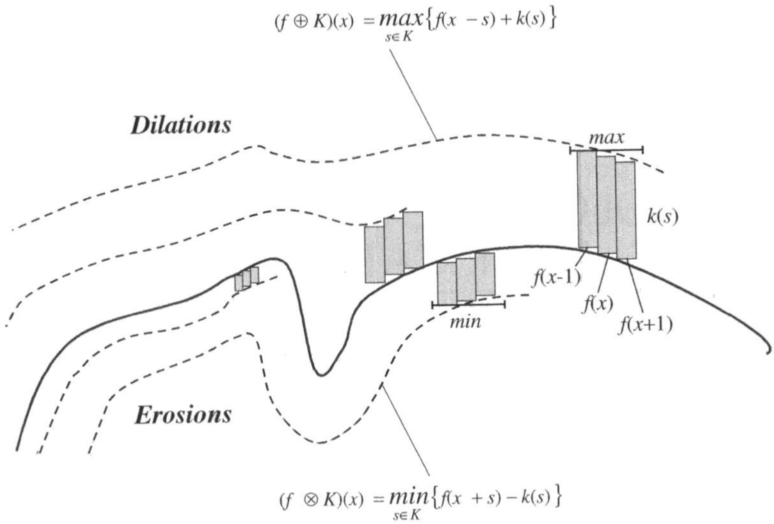

Figure 13.1. Graphical interpretation of grayscale morphological erosion and dilation operations.

interpreted as the result of sliding a flat SE over the topographical relief, so that the SE defines the new (dilated or eroded) scene based on its spatial properties such as height or width (see Figure 13.1). However, extension of MM operators to multichannel data such as hyperspectral imagery with hundreds of spectral channels is not straightforward. A simple approach consists in applying grayscale MM techniques to each channel separately, an approach that has been called marginal MM in the literature [4]. However, the marginal approach is often unacceptable in remote sensing applications because, when MM techniques are applied independently to each image channel, analysis techniques are subject to the well-known problem of "false colors"; that is, it is very likely that new spectral constituents not present in the original image may be created as a result of processing the channels separately [5]. An alternative way to approach the problem of multichannel MM is to treat the data at each pixel as a vector. Unfortunately, there is no unambiguous means of defining the minimum and maximum values between two vectors of more than one dimension, and thus it is important to define an appropriate arrangement of vectors in the selected vector space.

To be able to define appropriate vector ordering schemes for remote sensing-driven applications, it is important to take into account the requirements of available techniques for analyzing data of interest. In particular, the special characteristics of hyperspectral images pose different processing problems, which must be necessarily tackled under specific mathematical formalisms, such as classification, segmentation, spectral unmixing, and so on. A diverse array of techniques has been applied to extract information from hyperspectral data during the last decade. They are inherently either full pixel techniques or mixed pixel techniques, where each pixel vector in a hyperspectral scene provides a "spectral signature" that uniquely characterizes the underlying materials at each site in a scene. The underlying assumption governing full pixel techniques is that each pixel vector measures the response of one single material. In contrast, the underlying assumption governing mixed pixel techniques (also called "spectral unmixing" approaches) is that each pixel vector measures the response of multiple underlying materials at each site. A hyperspectral scene (sometimes referred to as "data cube") is often a combination of the two situations, where a few sites in a scene are pure materials, but many other are mixtures of materials. Most available techniques for hyperspectral data processing focus on analyzing the data without incorporating information on the spatially adjacent data; that is, the hyperspectral data are treated not as an image but as an unordered listing of spectral measurements. It is worth noting that such spectral-based techniques would yield the same result for a data cube, as well as for the same data cube where the spatial positions have been randomly permuted. However, one of the distinguishing properties of hyperspectral data is the multivariate information coupled with a two-dimensional (pictorial) representation amenable to image interpretation. Subsequently, there is a need to incorporate the spatial component of the data in the development of techniques for hyperspectral data exploitation. As will be shown in this chapter, mathematical morphology offers a remarkable framework to achieve the desired integration of spatial and spectral information in hyperspectral data analysis.

While integrated spatial/spectral developments hold great promise for hyperspectral image analysis, they also introduce new processing challenges. In particular, the price paid for the wealth of spatial and spectral information available from hyperspectral sensors is the enormous amounts of data that they generate. Several applications exist, however, where having the desired information calculated in near real time is a requirement. Such is the case of automatic target recognition for military and defense/security deployment. Other relevant examples include (a) environmental monitoring and assessment, (b) urban planning and management studies, and (c) risk/hazard prevention and response including wild-land fire tracking, biological threat detection, and monitoring of oil spills and other types of chemical contamination. With the recent explosion in the amount and complexity of hyperspectral imagery, parallel processing has soon become a tool of choice in many remote sensing missions, especially with the advent of low-cost systems such as commodity clusters and distributed networks of workstations.

The main goal of this chapter is to provide a seminal view on recent, consolidated advances in morphological techniques for efficient processing of hyperspectral imagery. Our main focus is on the development of joint spatial/spectral developments, able to exploit knowledge about the spatial arrangement of objects in the scene and to take advantage of the wealth of spectral information present in the data. Two new algorithms are developed, based on different strategies for ordering pixel vectors in spectral space. The considered approaches are (i) a supervised mixed pixel classification algorithm that naturally integrates both spatial and spectral information simultaneously and (ii) a morphological watershed-based algorithm that segments a hyperspectral scene in fully unsupervised fashion by first exploiting the spectral information and then making use of spatial context.

The chapter is structured as follows. Section 13.2 provides several vector ordering strategies used in this work to extend morphological operations to high-dimensional spaces. Section 13.3 develops two new algorithms for morphological classification of hyperspectral image data sets. Section 13.4 provides parallel processing support for the two algorithms above. An evaluation of the proposed techniques is then provided, along with a comparison to other existing approaches, using real hyperspectral data sets collected by the 224-channel NASA's Jet Propulsion Laboratory Airborne Visible-Infrared Imaging Spectrometer (AVIRIS) system. Parallel performance results are given in a massively parallel Beowulf cluster. Our final section concludes with some summarizing points and hints at plausible future research.

13.2. VECTOR ORDERING STRATEGIES FOR MULTIDIMENSIONAL MORPHOLOGICAL OPERATIONS

This section first provides an overview of available approaches for vector ordering in different applications, along with a detailed discussion on the appropriateness of using such strategies in the context of hyperspectral imaging applications. Then, it describes our adopted vector ordering strategy and briefly illustrates its advantages and disadvantages.

13.2.1. Available Approaches

Several vector ordering schemes have been discussed in the literature [4]. The choice between the different available options is generally application-driven. Four classes of ordering methods (marginal, reduced, partial, and conditional ordering) will be shortly illustrated here in the context of hyperspectral imaging applications. Let us first consider a hyperspectral image $\boldsymbol{f}$, defined on the N-dimensional space, where N is the number of spectral channels. Let $\boldsymbol{f}(x,y)$ and $\boldsymbol{f}(x',y')$ denote two pixels vectors at spatial locations (x,y) and (x',y'), respectively, with $\boldsymbol{f}(x,y) = [f_1(x,y),\dots,f_N(x,y)]^T$ and $\boldsymbol{f}(x',y') = [f_1(x',y'),\dots,f_N(x',y')]^T$. In marginal ordering (M-ordering), each pair of observations $f_i(x,y)$ and $f_i(x',y')$ would be ordered independently along each of the N channels [4]. In reduced ordering (R-ordering), a scalar parameter function g would be computed for each pixel of the image, and the ordering is performed according to the resulting scalar values. The ordered vectors would satisfy the relationship $\boldsymbol{f}(x,y) \leq \boldsymbol{f}(x',y') \Rightarrow g[\boldsymbol{f}(x,y)] \leq g[\boldsymbol{f}(x',y')]$. In partial ordering (P-ordering), the input multivariate samples would be partitioned into smaller groups, which would then be ordered. Both R-ordering and P-ordering may lead to the existence of more than one suprema (or infima) and, thus, introduce ambiguity in the resulting data. In conditional ordering (C-ordering), the pixel vectors would be initially ordered according to the ordered values of one of their components—for example, the first component, $f_1(x,y)$ and $f_1(x',y')$. As a second step, vectors with the same value for the first component would be ordered according to the ordered values of another component—for example, the second component, $f_2(x,y)$ and $f_2(x',y')$, and so on. This type of ordering is not generally appropriate for hyperspectral data, where each spectral feature *as a whole* contains relevant information about the optical and physical properties of the observed land cover. In addition, pixel vectors in remote sensing are usually affected by atmospheric and illumination interferers, which may introduce fluctuations in the amount of energy collected by the sensor at the different wavelength channels. The incident signal is electromagnetic radiation that originates from the sun and is measured by the sensor after it has been reflected upwards by materials on the surface of the Earth. As a result, two differently illuminated pixels that belong to the same spectral constituent may be ordered inconsistently by the C-ordering and M-ordering schemes.

13.2.2. Proposed Vector-Ordering Strategy

In this subsection, we develop an application-driven vector ordering technique based on a spectral purity-based criterion, where each pixel vector is ordered according to its spectral distance to other neighboring pixel vectors in the data. This type of ordering, which can be seen as a modification of the D-ordering available in the literature [5], is based on the definition of a cumulative distance $D_B[\boldsymbol{f}(x,y)]$ between one particular pixel $\boldsymbol{f}(x,y)$, where $\boldsymbol{f}(x,y)$ denotes the N-dimensional vector at discrete spatial coordinates $(x,y) \in Z^2$, and all the pixel vectors in the spatial neighborhood given by B (B-neighborhood):

$$D_B[\boldsymbol{f}(x,y)] = \sum_s \sum_t \text{Dist}[\boldsymbol{f}(x,y),\boldsymbol{f}(s,t)], \qquad \forall (s,t) \in Z^2(B) \tag{13.1}$$

where Dist is a linear pointwise distance measure between two N-dimensional vectors. As a result, $D_B[\boldsymbol{f}(x,y)]$ is given by the sum of Dist scores between $\boldsymbol{f}(x,y)$ and every other pixel vector in the B-neighborhood. To be able to define the usual MM operators in a complete lattice framework, we need to be able to define a supremum and an infimum given an arbitrary set of vectors $S = \{\boldsymbol{s}_1, \boldsymbol{s}_2, \ldots, \boldsymbol{s}_n\}$, where n is the number of vectors in the set. This can be achieved by computing $D_B(S)$ and selecting $\boldsymbol{s}_p$ such that $D_B[\boldsymbol{s}_p]$ is the minimum of that set, with $1 \leq p \leq n$. In similar fashion, we can select $\boldsymbol{s}_k$ such that $D_B[\boldsymbol{s}_k]$ is the maximum of that set, with $1 \leq k \leq n$. The selection of a maximum pixel vector making use of a pointwise distance measure Dist is graphically illustrated in Figure 13.2, where four pixels (vectors) respectively denoted by ***a***, ***b***, ***c***, and ***d*** are ordered by calculating their respective D_B-based scores and selecting the one that results in the maximum score. In the figure, we assume that the four pixel vectors above belong to the same spatial neighborhood given by a SE denoted by B.

Based on the simple definitions above, the extended erosion of $\boldsymbol{f}$ by B is based on the selection of the B-neighborhood pixel vector that produces the minimum value for D_B:

$$\begin{aligned}(\boldsymbol{f}\Theta B)(x,y) &= \{\boldsymbol{f}(x+s',y+t'), (s',t') \\ &= \arg\min_{(s,t)\in Z^2(B)} \{D_B[\boldsymbol{f}(x+s,y+t)]\}\}, \qquad (x,y) \in Z^2\end{aligned} \tag{13.2}$$

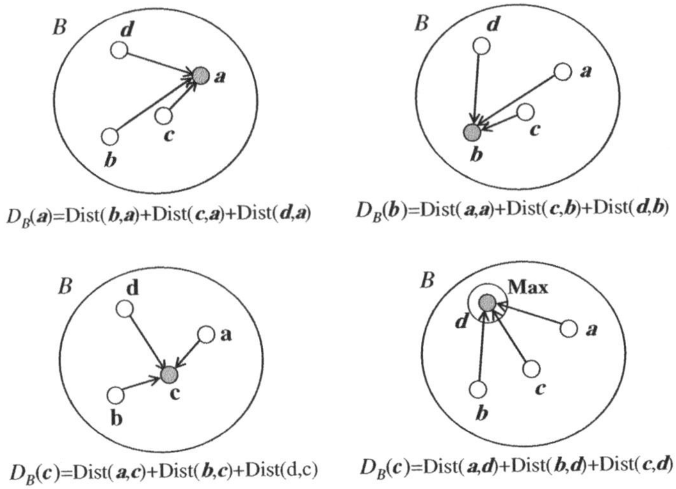

Figure 13.2. Example illustrating the proposed distance-based vector ordering strategy.

where the arg min operator selects the pixel vector which is most highly similar, according to the distance Dist, to all the other pixels in the in the B-neighborhood. On other hand, the flat extended dilation of $\boldsymbol{f}$ by B selects the B-neighborhood pixel vector that produces the maximum value for D_B:

$$(\boldsymbol{f} \oplus B)(x,y) = \{\boldsymbol{f}(x+s', y+t'), (s',t') = \arg\max_{(s,t)\in Z^2(B)} \{D_B[\boldsymbol{f}(x+s, y+t)]\}\}, \qquad (x,y) \in Z^2 \tag{13.3}$$

where the arg max operator selects the pixel vector that is most highly different, according to Dist, to all the other pixels in the B-neighborhood. Using the above notation, the multichannel morphological gradient at pixel $\boldsymbol{f}(x,y)$ using B can be simply defined as follows:

$$G_B(\boldsymbol{f}(x,y)) = \text{Dist}((\boldsymbol{f} \oplus B)(x,y), (\boldsymbol{f} \otimes B)(x,y)) \tag{13.4}$$

In this chapter, we assume that Dist is the spectral angle mapper (SAM), a standard metric in hyperspectral analysis. The SAM between two pixel vectors $\boldsymbol{f}(x,y)$ and $\boldsymbol{f}(x',y')$ is given by the following expression: $\text{SAM}[\boldsymbol{f}(x,y), \boldsymbol{f}(x',y')] = \cos^{-1}[\boldsymbol{f}(x,y) \cdot \boldsymbol{f}(x',y') / \| \boldsymbol{f}(x,y) \| \cdot \| \boldsymbol{f}(x',y') \|]$. It should be noted that the proposed extended operators are vector-preserving in the sense that no vector (spectral constituent) that is not present in the input data is generated as a result of the extension process. An important ambiguity not sufficiently explored in previous work has to do with the fact that the ordering imposed above is not injective in general; that is, two or more distinct vectors may output the same minimum or maximum distance. A solution suggested in the literature is to break the tie by using a space-filling curve such as a Peano curve [5, 6]. However, we believe that the total ordering so created is rather artificial and lacks physical interpretation in remote sensing applications. A further approach is to apply component transformations such as the principal component analysis or maximum noise transform [7] and then consider the first component only [8]. This approach discards significant information that can be very useful for the discrimination of different materials. In this work, we explore the effectiveness of a more physically meaningful approach, based on applying component transformations to bring the data from a high order dimension to a low order dimension, and then exploit all the information available in the reduced feature space to separate the different land cover classes. Our proposed partial solution to alleviate this problem is given by the following tiebreak approach, which is defined in the context of remote sensing applications:

1. Apply a component transformation h to the original N-dimensional data $\boldsymbol{f}$, aimed at increasing the spectral separability between the observed land-cover surfaces in the reduced feature space. Two widely used techniques in remote sensing are considered: principal component analysis (PCA) and maximum noise fraction (MNF). In the two cases, a new M-dimensional data set $\boldsymbol{g} = h(\boldsymbol{f})$ is obtained, with $M \leq N$ and the resulting M channels ordered in terms of decreasing information content. By virtue of the considered transformations, features not discernible in the original data may be evident in the reduced feature space.
2. Order the pixels by according to their spectral properties in the reduced feature space. If we assume that $\boldsymbol{g}(x',y') = [g_1(x',y'),\ldots,g_M(x',y')]^T$ and $\boldsymbol{g}(x'',y'') = [g_1(x'',y''),\ldots,g_M(x'',y'')]^T$ are the correspondent representations of $\boldsymbol{g}(x',y')$ and $\boldsymbol{g}(x'',y'')$ in the M-dimensional space, a partial solution to address multiple suprema or infima in the original N-dimensional space is to order $\boldsymbol{g}(x',y')$ and $\boldsymbol{g}(x'',y'')$ instead of $\boldsymbol{f}(x',y')$ and $\boldsymbol{f}(x'',y'')$, respectively. Following available approaches in the literature [9], we explore two different alternatives to accomplish the above goal:

   (a) *D-ordering.* This approach is based on the application of function D_B to the two M-dimensional representations of the original pixels, that is, $D_B[\boldsymbol{g}(x',y')]$ and $D_B[\boldsymbol{g}(x'',y'')]$. The pixels may now be ordered according to the ordered values of their cumulative distances in the reduced feature space.

   (b) *R-ordering about the centroid.* This approach is based on ordering of the multivariate reduced sample pixels $\boldsymbol{g}(x',y')$ and $\boldsymbol{g}(x'',y'')$ according to their spectral similarity (either SAM or SID) to a preselected location $\boldsymbol{c}_B(x,y)$, that is, the centroid of all the reduced multivariate samples which are located in the spatial neighborhood of $\boldsymbol{g}(x,y)$ defined by B. According to our interpretation in terms of spectral purity, the centroid is the most highly mixed pixel in the B-neighborhood [10]. It can be simply calculated by taking advantage of our multichannel erosion operation as follows: $\boldsymbol{c}_B(x,y) = (\boldsymbol{g}\Theta B)(x,y)$.

To conclude this subsection, we emphasize that the proposed approach represents only a partial solution to the problem since ties in the reduced M-dimensional feature space after the component transformation may still occur. However, we have experimentally tested that using both PCA and MNF in conjunction with the proposed ordering schemes can significantly reduce the percentage of ties found in the original N-dimensional space, as will be demonstrated by experiments in Section 13.5.

13.3. MORPHOLOGICAL PROCESSING OF HYPERSPECTRAL IMAGE SCENES

This section develops two innovative algorithms for processing hyperspectral images using mathematical morphology concepts. The first one makes use of the concept of morphological profile to produce feature vectors for supervised classification using neural networks. The second one is a fully unsupervised approach that expands the morphological waterhsed transformation to hyperspectral analysis. Parallel processing support for these two algorithms is provided in the following section.

13.3.1. Morphological Profile-Based Classification Algorithm

In this subsection, we propose a supervised classification approach that systematically analyzes spatial and spectral patterns in simultaneous fashion. The classifier makes use of sequences of multichannel transformations with SEs of varying width. A similar approach was applied by Pesaresi and Benediktsson [11], who used a composition of mono-channel morphological operations based on SEs of different sizes in order to characterize image structures in high-resolution grayscale urban satellite data. In this work, we use the concept of multichannel morphological profile, defined as a vector where a measure of the spectral variation of the result of pseudo-morphological transformations is stored for every step of an increasing SE series [12]. Following previous work by Benediktsson et al. [13], we use an artificial neural network-based approach for the classification of the resulting morphological features. Below, we describe the proposed algorithm in two stages: (i) morphological feature extraction and (ii) neural network-based classification.

13.3.1.1. Morphological Feature Extraction. The concept of morphological profile relies on opening and closing by reconstruction [2] a special class of morphological transformations that do not introduce discontinuities and which therefore preserve the shapes observed in input images. The basic contrast imposed by conventional opening and closing versus reconstruction-based opening and closing can be described as follows: Conventional opening and closing remove the parts of the objects that are smaller than the SE, whereas opening and closing by reconstruction either completely removes the features or retains them as a whole. In order to define the concept of multichannel morphological profiles using a simple notation, we have adapted the terminology used by Soille [2], where the spatial coordinates of pixel vectors have been ommitted from the formulation for simplicity.

It should be noted, however, that multichannel morphological profiles defined below are calculated for each pixel vector in the input data. First, we define the geodesic dilation operator $\delta_B^{(1)}$ of $\boldsymbol{f}$ under $\boldsymbol{g}$ as

$$\delta_B^{(1)}(\boldsymbol{f},\boldsymbol{g}) = \max\{\delta_B(\boldsymbol{f}),\boldsymbol{g}\} \tag{13.5}$$

where $\boldsymbol{f}$ and $\boldsymbol{g}$ are multichannel images and $\delta_B(\boldsymbol{f})$ is the elementary pseudo-dilation. Similarly, we define the geodesic erosion $\varepsilon_B^{(1)}$ as

$$\varepsilon_B^{(1)}(\boldsymbol{f},\boldsymbol{g}) = \min\{\varepsilon_B(\boldsymbol{f}),\boldsymbol{g}\} \tag{13.6}$$

where $\varepsilon_B(\boldsymbol{f})$ is the elementary erosion. Then, successive geodesic dilations and erosions can respectively be obtained by

$$\begin{aligned} \delta_B^{(k)}(\boldsymbol{f},\boldsymbol{g}) &= \underbrace{\delta_B^{(1)}\left(\delta_B^{(1)}\left(\cdots\delta_B^{(1)}(\boldsymbol{f},\boldsymbol{g}),\boldsymbol{g}\right),\boldsymbol{g}\right)}_{k\text{ times}} \quad \text{and} \\ \varepsilon_B^{(k)}(\boldsymbol{f},\boldsymbol{g}) &= \underbrace{\varepsilon_B^{(1)}\left(\varepsilon_B^{(1)}\left(\cdots\varepsilon_B^{(1)}(\boldsymbol{f},\boldsymbol{g}),\boldsymbol{g}\right),\boldsymbol{g}\right)}_{k\text{ times}} \end{aligned} \tag{13.7}$$

The reconstruction by dilation of $\boldsymbol{f}$ under $\boldsymbol{g}$ is then given by $\rho_B^{\delta}(\boldsymbol{f},\boldsymbol{g}) = \delta_B^{(\infty)}(\boldsymbol{f},\boldsymbol{g})$, i.e., until idempotence [2]. Similarly, the reconstruction by erosion of $\boldsymbol{f}$ under $\boldsymbol{g}$ is given by $\rho_B^{\varepsilon}(\boldsymbol{f},\boldsymbol{g}) = \varepsilon_B^{(\infty)}(\boldsymbol{f},\boldsymbol{g})$. With the above definitions in mind, the opening by reconstruction of size k of an image $\boldsymbol{f}$ can be simply defined as the reconstruction of $\boldsymbol{f}$ from the erosion of size k of $\boldsymbol{f}$:

$$\gamma_B^{(k)}(\boldsymbol{f}) = \rho_B^{\delta}\left(\varepsilon_B^{(k)}(\boldsymbol{f}),\boldsymbol{f}\right) \tag{13.8}$$

and the closing by reconstruction is defined by duality:

$$\phi_B^{(k)}(\boldsymbol{f}) = \rho_B^{\varepsilon}\left(\delta_B^{(k)}(\boldsymbol{f}),\boldsymbol{f}\right) \tag{13.9}$$

Using (13.8) and (13.9), multichannel profiles are defined as follows. The opening profile is defined as the vector $\boldsymbol{p}_i^{\gamma}(\boldsymbol{f}) = \left\{\gamma_B^{(i)}(\boldsymbol{f})\right\}$, while the closing profile is given by $\boldsymbol{p}_i^{\phi}(\boldsymbol{f}) = \left\{\phi_B^{(i)}(\boldsymbol{f})\right\}$, with $i = \{0, 1, \ldots, k\}$. Here, $\phi_B^{(i)}(\boldsymbol{f}) = \gamma_B^{(i)}(\boldsymbol{f}) = \boldsymbol{f}$ by the definition of opening and closing by reconstruction [2]. We define the combined derivative profile $\Delta\boldsymbol{p}_i$ as the vector:

$$\Delta\boldsymbol{p}_i = \left\{\text{Dist}\left[\gamma_B^{(i)}(\boldsymbol{f}),\gamma_B^{(i-1)}(\boldsymbol{f})\right]\right\} \cup \left\{\text{Dist}\left[\phi_B^{(i)}(\boldsymbol{f}),\phi_B^{(i-1)}(\boldsymbol{f})\right]\right\},$$
$$\text{with } i = \{1, 2, \ldots, \text{k}\} \tag{13.10}$$

13.3.1.2. Neural Network-Based Classification. Although the resulting morphological feature vectors can be usually regarded as low-dimensional when compared to the original hyperspectral signals, much redundancy may still be present in the feature vectors. Therefore, the application of feature extraction techniques to the resulting feature set after applying MM sequences is of great interest in order to select the most relevant features for class discrimination. In this work, we use decision boundary feature extraction (DBFE), introduced by Lee and Landgrebe [14]. DBFE has been demonstrated to be a very powerful approach for extracting all the necessary features for classification using neural networks [15]. The resulting features after DBFE-based feature extraction were then used to train a back-propagation neural-network-based classifier with one hidden layer, where the number of hidden neurons was empirically set to twice the number of input features and information classes. Different training-testing sets were used in experiments, as shown by experimental results in Section 13.4.

13.3.2. Morphological Watershed-Based Classification Algorithm

The morphological watershed transformation [16] consists of a combination of seeded region growing [17, 18] and edge detection. It relies on a marker-controlled approach that considers the image data as imaginary topographic relief. Let us assume that a drop of water falls on such a topographic surface. The drop will flow down along the steepest slope path until it reaches a minimum. The set of points of the surface whose steepest slope path reach a given minimum constitutes the *catchment basin* associated with that minimum, while the watersheds are the zones dividing adjacent catchment basins. Another way of visualizing the watershed concept is by analogy to immersion [2]. Starting from every minimum, the surface is progressively flooded until water coming from two different minima meet. At this point, a watershed line is erected. In order to extend the above algorithm to hyperspectral images, we propose a multichannel watershed-based algorithm with two stages: (i) minima selection and (ii) flooding.

13.3.2.1. Minima Selection. The key of an accurate watershed-based segmentation resides in the initialization—that is, the selection of "markers" or minima from which the transform is started. Following a recent work [19], we hierarchically order all minima according to their *deepness* and then select only those above a threshold. This concept can be easily explained using the immersion simulation. The deepness of a basin would be the level the water would reach, coming in through the minimum of the basin, before the water would overflow into a neighbor basin—that is, the height from the minimum to the lowest point in the watershed line of the basin. Deepness can be computed using morphological reconstruction applied to the multichannel gradient in Eq. (13.4). Given a "flat" SE of minimal size, designed by B, and the multichannel gradient $G_B(f)$ of an n-dimensional image f, morphological reconstruction by erosion of $G_B(f)$ using B can be defined as follows:

$$(G_B(f) \otimes B)^t(x,y) = \bigvee_{k\geq 1} \left[\delta_B^t(G_B(f) \otimes B|G_B(f))\right](x,y) \tag{13.11}$$

where

$$[\delta_B^t(G_B(f) \otimes B|G_B(f))](x,y) = \left[\overbrace{\delta_B\delta_B\cdots\delta_B}^{t\text{times}}(G_B(f) \otimes B|G_B(f))\right](x,y) \quad (13.12)$$

and

$$[\delta_B(G_B(f) \otimes B|G_B(f))](x,y) = \vee\{(G_B(f) \otimes B)(x,y), G_B(f(x,y))\} \quad (13.13)$$

In the above operation, $G_B(f) \otimes B$ is the standard erosion of the multichannel gradient image, which acts as a "marker" image for the reconstruction, while $G_B(f)$ acts as a "mask" image. Reconstruction transformations always converge after a finite number of iterations, t—that is, until the propagation of the marker image is totally impeded by the mask image. It can be proven that the morphological reconstruction $(G_B(f) \otimes B)^t$ of $G_B(f)$ from $G_B(f) \otimes B$ will have a watershed transform in which the regions with deepness lower than a certain value v have been joined to the neighbor region with closer spectral properties; that is, parameter v is a minima selection threshold.

13.3.2.2. Flooding. In this section, we formalize the flooding process following a standard notation [2]. Let the set $P = \{\mathbf{p}_1, \mathbf{p}_2, \ldots, \mathbf{p}_k\}$ denote the set of k minimum pixel vectors selected after multidimensional minima selection. Similarly, let the catchment basin associated with a minimum pixel $\mathbf{p}_i$ be denoted by $\text{CB}(\mathbf{p}_i)$. The points of this catchment basin which have an altitude less than or equal to a certain deepness score d [19] are denoted by

$$\text{CB}_d(\mathbf{p}_i) = \{f(x,y) \in CB(\mathbf{p}_i)|\text{Dist}(\mathbf{p}_i, f(x,y)) \leq d\} \quad (13.14)$$

We also denote by $\mathbf{X}_d = \cup_{i=1}^{k} CB_d(\mathbf{p}_i)$ the subset of all catchment basins which contain a pixel vector with a deepness value less than or equal to d. Finally, the set of points belonging to the regional minima of deepness d are denoted by $\text{RMIN}_d(f(x,y))$. The catchment basins are now progressively created by simulating the flooding process. The first pixel vectors reached by water are the points of highest deepness score. These points belong to $\text{RMIN}_{\mathbf{p}_j}(f(x,y)) = \mathbf{X}_{D_B(\mathbf{p}_j)}$, where $\mathbf{p}_j$ is the deepest pixel in P; that is, $D_B(\mathbf{p}_j)$ is the minimum, with $1 \leq j \leq k$. From now on, the water either (a) expands the region of the catchment basin already reached by water or (b) starts to flood the catchment basin whose minima have a deepness equal to $D_B(\mathbf{p}_l)$, where $\mathbf{p}_l$ is the deepest pixel in the set of $P - \{\mathbf{p}_j\}$. This operation is repeated until $P = \emptyset$. At each iteration, there are three possible relations between a connected component $\mathbf{Y}$ and $\mathbf{Y} \cap \mathbf{X}_{D_B(\mathbf{p}_j)}$:

1. If $\mathbf{Y} \cap \mathbf{X}_{D_B(\mathbf{p}_j)} = \emptyset$, then a new minimum $\mathbf{Y}$ has been discovered at level $D_B(\mathbf{p}_l)$. In this case, the set of all minima at level $D_B(\mathbf{p}_l)$—that is, $\text{RMIN}_{\mathbf{p}_l}(f(x,y))$ —will be used for defining $\mathbf{X}_{D_B(\mathbf{p}_l)}$.

2. If $Y \cap X_{D_B(p_j)} \neq \emptyset$ and is connected, then the flooded region is expanding and Y corresponds to the pixels belonging to the catchment basin associated with the minimum and having a deepness score less than or equal to $D_B(p_l)$, i.e. $Y = CB_{D_B(p_l)}(Y \cap X_{D_B(p_j)})$.
3. Finally, if $Y \cap X_{D_B(p_j)} \neq \emptyset$ and is not connected, then the flooded regions of the catchment basins of two distinct minima at level $D_B(p_j)$ are expanding and merged together.

Once all levels have been flooded, the set of catchment basins of a multidimensional image f is equal to the set $X_{D_B(p_m)}$, where p_m is the least deep pixel in P; that is, $D_B(p_m)$ is the maximum, with $1 \leq m \leq k$. The set of catchment basins following multidimensional watershed can be represented as a set $\{\mathrm{CB}(p_i)\}_{i=1}^{k}$, where each element corresponds to the catchment basin of a regional minimum of the input image f. This is the final segmentation output for the algorithm.

13.4. PARALLEL IMPLEMENTATIONS

In this section, we develop parallel implementations for the two mathematical morphology-based algorithms addressed in the previous section. To reduce code redundancy and enhance reusability, our goal was to reuse much of the code for the sequential algorithms in the parallel implementations. The kernel (SE)-based nature of morphological algorithms introduces some border-handling and overlapping issues that are also carefully addressed below. Before describing our parallel implementations, we first discuss data partitioning schemes for hyperspectral data.

13.4.1. Data Partitioning Strategy

Two types of partitioning can be exploited in the considered application: spectral-domain partitioning and spatial-domain partitioning. Spectral-domain partitioning subdivides the data volume into small cells or subvolumes made up of contiguous spectral bands and assigns one or more subvolumes to each processor. It should be noted that most information extraction techniques for hyperspectral imaging focus on analyzing the data based on the properties of spectral signatures; that is, they utilize with the information provided by each pixel vector *as a whole*. With a spectral-domain partitioning model, each pixel vector may be split amongst several processors, and the calculations made for each spectral signature would need to originate from several processors. Although such spectral-domain partitioning strategy has been used in very low-dimensional remote sensing applications (e.g., those based on color images), it was soon discarded in our framework due to several reasons [20]. First, spatial-domain partitioning is a more natural approach for neighbor-based image processing, because many image processing operations require the same function to be applied to a small set of elements around each pixel vector in the input data volume. A second major reason has to do with the overhead introduced by inter-processor communications. Since most hyperspectral algorithms extract information based on spectral signatures *as a whole*, partitioning the data in the spectral domain would linearly increase communications with the increase in the number of processing elements, thus complicating the design of parallel algorithms. A final issue is code reusability; to reduce code redundancy and enhance portability, it is desirable to reuse much of the sequential code in the design of the parallel version.

13.4.2. Parallel Morphological Profile-Based Classification Algorithm

Multichannel morphological erosion and dilation operations are characterized by their regularity. Therefore, in order to paralellelize the construction of morphological profiles, a desirable goal is to be able to provide parallel support for standard morphological erosion and dilation since these two operations are used as a baseline to construct morphological profiles in iterative fashion. In this subsection, we provide a simple parallel strategy for optimization of standard multichannel erosions and dilations which has been used to improve computational complexity of the proposed morphological profile-based classification algorithm.

To achieve this goal, we first partition the data in the spatial domain (a pixel vector is never partitioned) so that the size of partitions assigned to the different processors is proportional to their speeds [21]. Then, morphological operations (erosion and dilation) are carried out in parallel by the different processing nodes so that each node works independently, making use of its local data portion only, applying the formulation given in Section 13.2.2. An important issue in SE-based morphological image processing operations of this kind is that accesses to pixels outside the spatial domain of the input image are possible. This is particularly so when the SE is centered on a pixel located in the border of the original image. In sequential implementations, it is common practice to redirect such accesses according to a predefined border handling strategy. In our application, a border handling strategy is adopted when the location of the SE is such that some of the pixel positions in the SE are outside the input image domain (see Figure 13.3a).

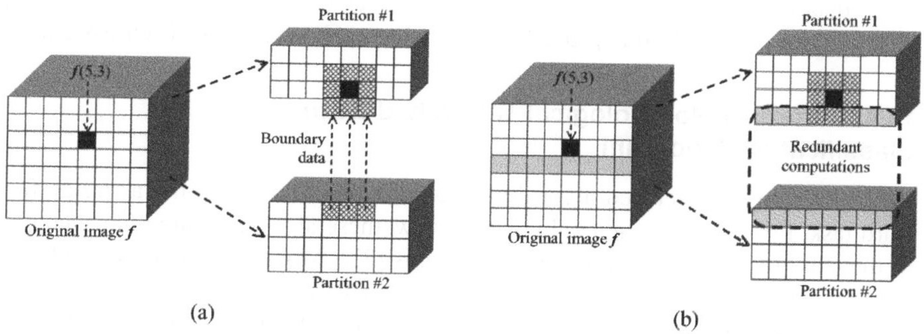

Figure 13.3. (a) 3 × 3-pixel structuring element computation split among two processing nodes. (b) Introduction of redundant computations to minimize inter-processor communication in a 3 × 3-pixel structuring element computation.

In this situation, only those pixels inside the image domain are read for the morphological calculation. This strategy is equivalent to the common mirroring technique used in digital image processing applications.

Apart from the border handling strategy above, a function to update overlapping parts of partial data structures has been implemented in order to avoid inter-processor communication when the SE computation is split amongst several different processing nodes (see Figure 13.3b). Here, we use an *overlap mapping* strategy based on replicating the pixel vectors at the border of a partition so that each workstation can perform all the calculations independently, with no need to know which other workstations have pixels close to the boundary of the partition. It should be noted that Figure 13.3b gives a simplified view, because some steps of the operation are not shown. For example, depending on how many adjacent spatial-domain partitions are involved in the parallel computation of a SE, it may be necessary to place a *scratch border* around each partition to completely avoid inter-processor communication. It is also important to note that the amount of redundant information introduced by the overlapping scatter depends on the size of the SE used in the computation. In order to perform the final data partitioning in both cases, we adopt a simple hybrid methodology that consists of two main steps:

1. Partition the hyperspectral data set so that the number of rows in each partition is proportional to the speed of the processor, assuming that no upper bound exists on the number of pixel vectors that can be stored by the processor (at the same time, replicate necessary information for the overlap partitioning strategy). If the number of pixel vectors in each partition assigned to each processor is less than the upper bound on the number of elements that can be stored by the processor, we have an optimal distribution.
2. For each processor, check if the number of pixel vectors assigned to it is greater than the upper bound on the number of elements that it can store. For all the processors whose upper bounds are exceeded, assign them a number of pixels equal to their upper bounds. Now, we solve the partitioning problem of a set with remaining pixel vectors (including those resulting from data replication in the overlap strategy) over the remaining processors. We recursively apply this procedure until all the elements have been assigned.

13.4.3. Parallel Morphological Watershed-Based Classification Algorithm

Parallelization of watershed algorithms that simulate flooding is not a straightforward task. From a computational point of view, these algorithms are representative of the class of irregular and dynamic problems [22]. Moreover, the watershed process has a very volatile behavior, starting with a high degree of parallelism that very rapidly diminishes to a much lower degree of parallelism. Our proposed parallel implementation uses a simple master–slave model. The master processor divides the multichannel image f into a set of subimages f_i which are sent to different processors, so that the domain of each subimage is an extended subdomain given by $D^e_{f_i}$. The slave processors run the classification algorithm on the respective subimages and also exchange data among themselves for uniform segmentation. After the segmented regions become stable, the slaves send the output to the master, which combines all of them in a proper way and provides the final segmentation. If we assume that the parallel system has p processors available, then one of the processors is reserved to act as the master, while each of the remaining $p - 1$ processors create a local queue Q_i with $1 \leq i \leq p - 1$.

The minima selection algorithm is run locally at each processor to obtain a set of minima pixels surrounded by nonminima, which are then used to initialize each queue Q_i. Flooding is then performed locally in each processor as in the serial algorithm. However, due to the image division, flooding is confined only to the local subdomain. Therefore, there may exist parts of the subimage that cannot be reached by flooding since they are contained in other subimages. Our approach to deal with this problem is to first flood locally at every deepness score in the subimage. Once the local flooding is finished, each processor exchanges segmentation labels of pixels in the boundary with appropriate neighboring processors. Subsequently, a processor can receive segmentation labels corresponding to pixels in the extended subdomain. The processor must now "reflood" the local subdomain from those pixels, a procedure that may introduce changes in segmentation labels of the local subdomain. Communication and reflooding are again repeated until stabilization (i.e., no more changes occur). When the flood–reflood process is finished, each slave processor sends the final segmentation labels to the master processor, which combines them together and performs region merging to produce final set of segmentation labels.

To conclude this subsection, we emphasize that although the flooding–reflooding scheme above seems complex, we have experimentally tested that this parallelization strategy is more effective than other approaches that exchange segmentation labels without first flooding locally at every deepness score. The proposed parallel framework guarantees that the processors are not tightly synchronized [23]. In addition, the processors execute a similar amount of work at approximately the same time, thus achieving load balance.

13.4. EXPERIMENTAL RESULTS

This section describes the experimental analyses conducted to validate the performance of the proposed morphological hyperspectral processing algorithms. The section is structured as follows. First, we provide a description of the hyperspectral data sets used in experiments. Then, we conduct a thorough quantitative and comparative assessment of the two novel morphological classification algorithms introduced in this chapter, along with a study of the impact of using different vector ordering strategies for the definition of morphological operations. The section concludes with an assessment of the parallel performance of the proposed parallel versions, along with a detailed description of the parallel computing architectures used for evaluation purposes.

13.4.1. Hyperspectral Image Data Set

The data set used in experiments is a hyperspectral scene collected by the NASA/Jet Propulsion Laboratory's Airborne Visible Infra-Red Imaging Spectrometer (AVIRIS) system [24] in 1998 over the Salinas Valley, California. The full scene consists of 512 lines by 217 samples with 224 spectral bands from 0.4 μm to 2.5 μm, nominal spectral resolution of 10 nm, and 16-bit radiometric resolution. It was taken at low altitude with a pixel size of 3.7 m. The data include vegetables, bare soils, and vineyard fields. Figure 13.4a shows the entire scene and a subscene of the data set (called hereinafter Salinas A), outlined by a white rectangle, which comprises 83 × 86 pixels. Figure 13.4b shows available ground-truth regions. As shown in Figure 13.4b, ground-truth is available for about two-thirds of the entire Salinas scene. The data set represents a challenging classification problem due to the early growth stage of most of the crops in the area, which resulted in a lot of mixed pixels (in particular, in the subscene labeled as Salinas A, which comprises lettuce romaine classes at different weeks since planting).

13.4.2. Quantitative Assessment of the Morphological Profile-Based Classifier

In this subsection, we test the performance of the proposed morphological profile-based supervised classification approach. To do so, we extracted a random sample

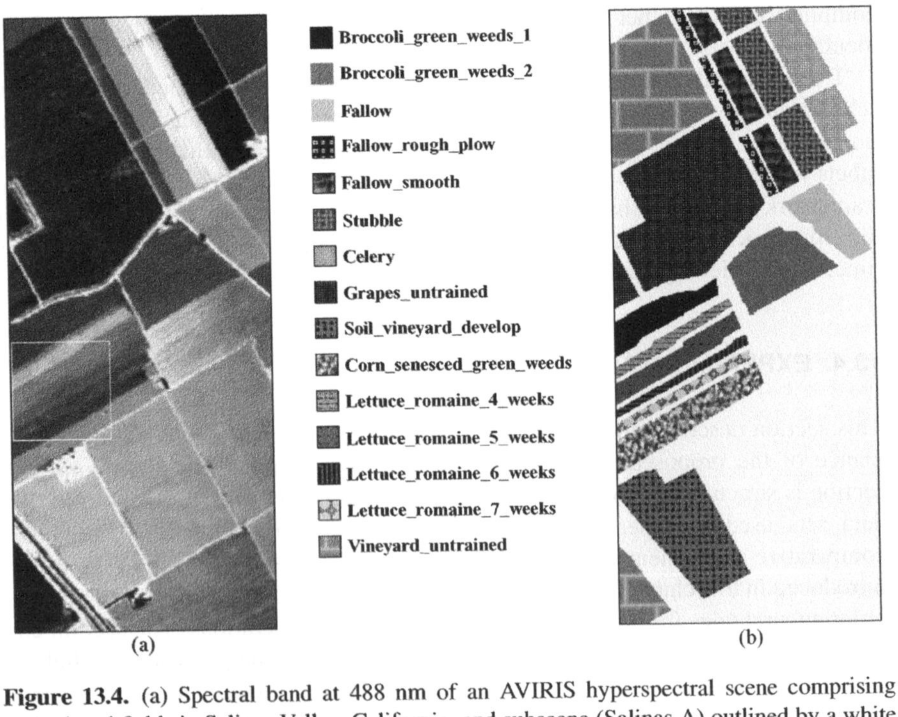

Figure 13.4. (a) Spectral band at 488 nm of an AVIRIS hyperspectral scene comprising agricultural fields in Salinas Valley, California, and subscene (Salinas A) outlined by a white rectangle. (b) Land-cover ground truth classes.

of only 2% of the pixels from the known ground-truth of the 15 ground-truth classes in Figure 13.4b; the number of training pixels selected in each class was proportional to the size of the class. Multichannel morphological profiles were then constructed for the selected training samples, thus allowing for the application of our proposed supervised framework. Finally, we also used the original spectral information contained in the hyperspectral image as an input to our classification system. The resulting features were used to train a back-propagation neural network-based classifier with one hidden layer, where the number of hidden neurons was empirically set to twice the number of input features and information classes. The trained classifier was then applied to the remaining 98% of the known ground pixels in the scene.

Figure 13.5 displays the overall test classification accuracies obtained after applying our classification system to multichannel and mono-channel morphological profiles as a function of the number of opening/closing operations used in the process. Four different approaches were tested in the construction of multichannel morphological operations, given by all possible combinations of the dimensionality reduction transformation (MNF or PCA) plus the reduced ordering strategy (D-ordering or R-ordering). Similarly, two different approaches were considered in the construction of mono-channel profiles based on processing of the first MNF or PCA component. As demonstrated by Figure 13.5, the best overall accuracies were achieved when MNF+D-ordering multichannel morphological profiles were used for feature extraction, followed by PCA+D-ordering. This fact revealed that D-ordering is more appropriate than R-ordering for this application. In all cases, multichannel profiles produced classification results which are clearly better than those found using mono-channel profiles. From Figure 13.5, it is also evident

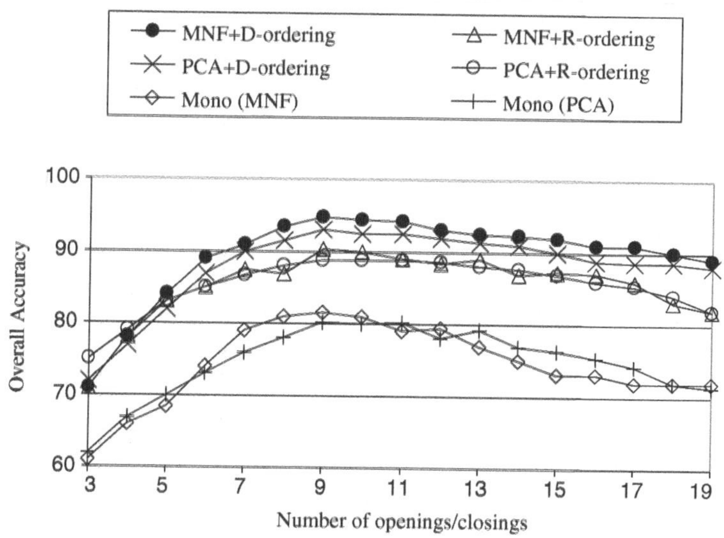

Figure 13.5. Overall classification accuracies for the proposed supervised classifier using different morphological features.

that the width in pixels of patterns of interest in the Salinas AVIRIS scene makes nine opening/closing iterations a reasonable parameter selection for most of the methods tested in this experiment. The construction of morphological feature vectors with larger data dimensions generally causes a loss in the classification performance.

Table 13.1 reports overall (OA), average (AVE), and individual test accuracies for each of the classes in the Salinas data set, using nine iterations. In the table, OA represents the accuracy for all samples whereas AVE represents the average classification accuracy for the samples. The results obtained by using the original spectral information in the hyperspectral scene are also shown for comparison. As can be examined, the OAs exhibited by the D-ordering-based multichannel classifiers are

TABLE 13.1. Overall (OA), Average (AVE), and Individual Test Accuracies in Percentage, Obtained after Applying the Proposed Classification System with Monochannel and Multichannel (MNF- and PCA-based) Profiles with Nine Iterations to the Salinas Scene[a]

Class	Original Spectral Information	Mono (MNF)	Mono (PCA)	MNF+D-Ordering	MNF+R-Ordering	PCA+D-Ordering	PCA+R-Ordering
Broccoli_green_weeds_1	78.42	76.21	75.53	82.64	79.36	81.25	79.01
Broccoli_green_weeds_2	80.13	74.58	75.86	86.31	81.26	83.02	81.17
Fallow	92.98	88.51	86.43	98.15	97.54	96.59	95.40
Fallow_rough_plow	96.51	86.77	85.23	96.51	95.30	94.52	92.37
Fallow_smooth	93.72	89.35	86.24	97.63	95.89	95.01	92.89
Stubble	94.71	85.19	86.02	98.96	95.48	98.02	95.17
Celery	89.34	88.40	85.56	98.03	96.75	99.05	93.67
Grapes_untrained	88.02	83.07	82.43	95.34	92.31	93.78	90.67
Soil_vineyard_develop	88.55	78.13	79.63	90.45	87.32	89.13	88.34
Corn_senesced_green_weeds	87.46	70.28	69.83	82.54	80.46	83.90	84.02
Lettuce_romaine_4_weeks	78.86	73.10	72.64	83.21	81.42	82.28	81.49
Lettuce_romaine_5_weeks	91.35	72.57	73.22	82.14	77.43	79.28	78.09
Lettuce_romaine_6_weeks	88.53	74.25	75.39	84.56	80.76	81.81	79.15
Lettuce_romaine_7_weeks	84.85	76.21	77.05	86.57	84.76	84.23	81.47
Vineyard_untrained	87.14	80.04	78.98	92.93	89.23	91.27	87.81
OA	87.25	81.43	80.28	94.82	90.45	93.12	89.03
AVE	87.93	79.73	79.29	90.66	87.96	88.93	87.98

[a]Results using the full spectral information in the original scene are also displayed for comparison.

higher than the OAs provided by R-ordering-based approaches. This confirms the effectiveness of D-ordering with regards to R-ordering in this example. It is also clear from Table 13.1 that the proposed multichannel classifiers outperform single-channel-based approaches in terms of classification accuracies. Most importantly, it should be noted that both D-ordering and R-ordering, when combined with PCA and MNF transformations, produce results that outperform those found using the original spectral information in terms of both OA and AVE. Interestingly enough, however, a deeper analysis of the results reveals some limitations in the proposed techniques. For example, the individual test accuracies exhibited by the MNF+D-ordering, MNF+R-ordering, PCA+D-ordering, and PCA+R-ordering classifiers on the broccoli_green_weeds_1, corn_senesced_green_weeds and four lettuce_romaine (at different weeks since planting) classes are similar to the accuracies produced by the MNF+C-ordering and PCA+C-ordering classifiers on the same classes, and only slightly better than those found by either mono-channel-based morphological classifiers or the original spectral information. It should be noted that the above six classes, all of them contained in the Salinas A subscene (see Figure 13.3a), are dominated by highly mixed pixels such as broccoli plus green weeds, corn senesced plus green weeds, and lettuce romaine plus soil.

13.4.3. Quantitative Assessment of the Morphological Watershed-Based Classifier

In order to test the fully unsupervised classifier, we first used the virtual dimensionality (VD) concept [25] to estimate the number of different classes in the AVIRIS Salinas scene. Here, we masked out the unlabeled pixels in Figure 13.4b in order to assess the algorithm using the full set of ground-truth information available. Interestingly, the VD concept estimated 15 classes which is exactly the number of classes available in Figure 13.4b. It is interesting to note that the VD was able to separate among the four lettuce_romaine classes, which is a significant achievement given the high spectral similarity among those classes. Apart from the number of classes to be extracted, which defined the number of minima ($k = 15$) to be extracted by the minima selection step of the algorithm, parameter v was set automatically using the multilevel Otsu thresholding method as explained in Plaza et al. [26].

Table 13.2 reports the overall (OA), average (AVE), and individual test accuracies for each of the classes in the Salinas data set after applying the proposed multichannel segmentation algorithm using a disk-shaped SE with radius of nine pixels (this size was set empirically after extensive experiments with larger and smaller SEs). For illustrative purposes, results by other two standard unsupervised classification algorithms: The ISODATA [7] and Soille's watershed-based clustering [27] are also reported. As shown by Table 13.2, the use of appropriate SE sizes in the proposed method produced segmentation results that were superior to those found by ISODATA and Soille's watershed-based clustering algorithm. In particular, the best results were obtained when a disk-shaped SE with a radius of nine pixels was used. This is mainly due to the relation between the SE and the spatial properties of

TABLE 13.2. Overall (OA), Average (AVE), and Individual Test Accuracies in Percentage, Obtained After Applying the Proposed Unsupervised Classification System with Mono-channel and Multichannel (MNF- and PCA-based) Profiles with Nine Iterations to the Salinas Scene[a]

Class	ISODATA	Soille's	MNF+D-Ordering	MNF+R-Ordering	PCA+D-Ordering	PCA+R-Ordering
Broccoli_green_weeds_1	69.05	70.81	80.45	77.01	78.65	76.23
Broccoli_green_weeds_2	67.01	70.11	84.03	79.44	80.43	78.45
Fallow	82.23	83.28	95.03	94.81	93.25	92.24
Fallow_rough_plow	79.48	80.44	93.28	92.54	91.03	89.00
Fallow_smooth	81.44	81.99	93.99	93.01	92.67	89.34
Stubble	78.00	81.04	95.03	93.24	96.13	92.06
Celery	81.98	80.99	94.28	94.03	96.01	90.45
Grapes_untrained	76.42	77.93	91.23	89.93	91.23	87.79
Soil_vineyard_develop	71.05	76.79	87.34	85.12	87.41	86.19
Corn_senesced_green_weeds	63.92	64.02	80.23	77.76	80.03	82.23
Lettuce_romaine_4_weeks	66.21	68.08	81.34	79.23	79.14	79.44
Lettuce_romaine_5_weeks	65.05	68.72	80.02	75.12	77.81	76.35
Lettuce_romaine_6_weeks	67.57	70.97	81.23	78.21	79.70	76.54
Lettuce_romaine_7_weeks	69.78	72.45	84.16	81.94	82.04	78.23
Vineyard_untrained	73.28	74.25	89.00	87.42	88.76	84.49
OA	74.03	76.17	91.95	89.04	90.16	86.55
AVE	72.24	75.03	88.78	85.12	86.88	83.80

[a]Results using the full spectral information in the original scene are also displayed for comparison.

regions of interest in the scene. Interestingly, as observed in the previous subsection, both D-ordering and R-ordering strategies performed similarly, but MNF-based dimensional reduction proved to be much more effective than PCA-based dimensional reduction. This effect was already observed and thoroughly analyzed in Plaza et al. [28]. Overall, the results shown in Table 13.2 reveal that the proposed algorithm can achieve classification results that are comparable to those reported by the supervised approach in a complex analysis scenario given by agricultural classes with very similar spectral features, in fully unsupervised fashion (complemented with the use of VD to estimate the number of classes). It should also be noted that the two proposed algorithms required more than one hour of computation in a last-generation PC with Intel Centrino 3-GHz processor and 2 GB of RAM, which creates the need for parallel implementations. Performance data for the parallel versions of the two algorithms above are given in the following subsection.

13.4.4. Parallel Performance Evaluation

This subsection provides an assessment of parallel hyperspectral algorithms in providing significant performance gains (without loss of accuracy) with regards to their serial versions. The section is organized as follows. First, we provide an overview of the parallel computing architectures used for evaluation purposes. Second, a quantitative assessment of the two proposed parallel approaches, in comparison with their respective sequential counterparts, is provided.

The parallel algorithms were coded using the C++ programming language with calls to message passing interface (MPI). They were tested on the Thunderhead Beowulf cluster at NASA's Goddard Space Flight Center (NASA/GSFC). From the early 1990, the overwhelming computational needs of Earth and space scientists have driven NASA/GSFC to be one of the leaders in the application of low cost high-performance computing [29]. In 1997, the HIVE (Highly Parallel Virtual Environment) project was started to build a commodity cluster intended to be exploited by different users in a wide range of scientific applications. The Thunderhead system can be seen as an evolution of the HIVE project (see http://thunderhead.nasa.gov). It is composed of 256 dual 2.4-GHz Intel Xeon nodes, each with 1 Gb of memory and 80 GB of main memory. The total peak performance of the system is 2457.6 Gflops. Along with the 512-processor computer core, Thunderhead has several nodes attached to the core with 2-GHz optical fiber Myrinet. The parallel algorithms tested in this work were run from one of such nodes, called thunder1. The operating system used at the time of experiments was Linux RedHat 8.0, and MPICH was the message-passing library used.

To empirically investigate the scaling properties of the considered parallel algorithms, Figure 13.6 plots the speedup factors (i.e., the ratio of increase in algorithm performance with regards to a single-processor run of the algorithm in a single Thunderhead node), as a function of the number of available processors. Since the use of PCA/MNF or D-ordering/R-ordering seemed to be irrelevant for parallel performance results, Figure 13.6 only displays a general case study for the two considered algorithms. Results in Figure 13.6 reveal that the performance drop from linear speedup in the watershed-based algorithm was more significant than that

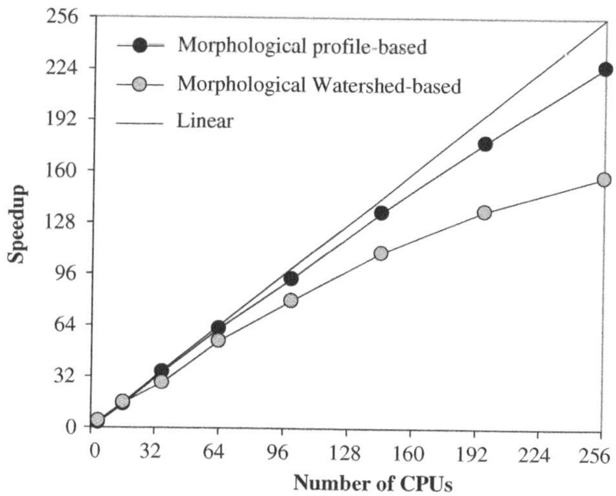

Figure 13.6. Speedup factors achieved by our parallel implementations on Thunderhead.

TABLE 13.3. Load-Balancing Rates and Execution Processing Times for the Parallel Algorithms on Thunderhead.

	Morphological Profile-Based			Watershed-Based		
# CPUs	Time	D_{All}	D_{Minus}	Time	D_{All}	D_{Minus}
1	1874	1.07	1.03	4095	1.35	1.24
4	580	1.05	1.02	1170	1.41	1.30
16	132	1.08	1.04	269	1.42	1.29
36	53	1.07	1.03	144	1.37	1.32
64	30	1.06	1.02	75	1.44	1.31
100	20	1.08	1.04	52	1.45	1.33
144	14	1.07	1.03	37	1.39	1.30
196	11	1.09	1.05	30	1.40	1.28
256	8	1.08	1.03	26	1.45	1.33

observed for the morphological profile-based algorithm as the number of processors increase. This is due to the irregular parallel nature of the flooding stage of the Watershed-based algorithm, which prevents the worker nodes from completing their calculations at exactly the same time. Quite opposite, load balance in the morphological profile-based algorithm was much better due to the regularity in the computations.

In order to fully validate the above remarks, Table 13.3 shows the execution times along with the imbalance scores [30] achieved by the considered parallel algorithms on Thunderhead. The imbalance is defined as $D = R_{max}/R_{min}$, where R_{max} and R_{min} are the maxima and minima processor run times, respectively. Therefore, perfect balance is achieved when $D = 1$. In the table, we display the imbalance considering all processors, D_{All}, and also considering all processors but the root, D_{Minus}. As we can see from Table 13.3, the parallel version of the morphological profile-based parallel algorithm was able to provide values of D_{All} close to 1 in all cases. Furthermore, the above algorithm provided almost the same results for both D_{All} and D_{Minus} while, for the watershed-based parallel algorithm, load balance was much better when the root processor was not included. Also, the high scores reported for D_{Minus} in this case indicate that the workload is not appropriately balanced among the different processors. The problem of finding an effective workload distribution for region growing algorithms has been identified as a very challenging one in the literature [31], and further work is required to address this issue for the proposed parallel watershed-based algorithm.

For the sake of quantitative comparison, execution times in Table 13.3 reveal that the tested algorithms were able to obtain relevant information from the considered hyperspectral data sets (in light of results in Tables 13.1 and 13.2), but also quickly enough for practical use. For instance, using 256 processors, the morphological profile-based parallel algorithm provided a highly accurate classification result of the Salinas AVIRIS scene in only 8 seconds, while the watershed-based parallel algorithm was able to provide a comparable classification result in 26 seconds, using the same number of processors but in fully unsupervised fashion.

The above results indicate significant improvements over the single-processor runs of the same algorithms, which can take up to more than one hour of computation for the considered problem size, as indicated by Table 13.3. Contrary to common perception that spatial/spectral information extraction algorithms are too computationally demanding for practical use, our results demonstrate that such combined approaches may indeed be very appealing for parallel design and implementation, not only due to the window-based nature of such algorithms but also because they can efficiently distribute the workload among the different processors and reduce sequential computations at the master node, thus enhancing parallel performance.

13.5. CONCLUSIONS AND FUTURE LINES

The recent application of mathematical morphology theory to hyperspectral image data has opened ground-breaking perspectives—in particular, from the viewpoint of naturally integrating the wealth of different sources of information present in the input data. This chapter has described new trends in multichannel processing of hyperspectral imagery by simultaneously considering both spatial and spectral information. A physically meaningful vector organization scheme has been introduced to extend classic morphological operations to high-dimensional spaces, and two different approaches (D-ordering and R-ordering) were used to propose two innovative algorithms for classification of hyperspectral imagery. In the first one, multichannel profiles were used to extract relevant features for classification using neural networks. The second proposed algorithm is a fully unsupervised one which takes advantage of the concept of morphological watershed, originally proposed for grayscale imagery and extended here to the case of multichannel image data. In both cases, component transformations such as PCA or MNF are used to address the issue of partial ordering of pixel vectors. These methods offer a highly representative sample of available and new techniques in morphological hyperspectral analysis research. Our experimental assessment of these two algorithms demonstrated that the first one provides results that are better than those found using both the entire spectral information in the original hyperspectral image and standard, mono-channel morphological processing techniques. The second algorithm provided comparable results (but in fully unsupervised fashion), and slightly better ones than those provided by other similar unsupervised classification approaches. Overall, results in this chapter demonstrated that the combined use of spatial and spectral information in hyperspectral data analysis, achieved via mathematical morphology concepts, can greatly improve the results found by available techniques that consider the spectral information alone—in particular, when adequate vector ordering strategies are employed in the proposed morphological-oriented data processing framework.

A drawback in the proposed approaches has to do with the need to heed a range of filters with increasing SE sizes, a labor that results in a heavy computational burden when processing high-dimensional data. This phenomenon is particularly relevant for the case of images with large and spectrally homogeneous regions. For that purpose, this chapter has also developed parallel processing support for the two morphological algorithms above. An interesting finding by experiments in this chapter is that spatial/spectral parallel implementations offer a surprisingly simple, yet effective and highly scalable, alternative to standard, spectral-based algorithms for hyperspectral image analysis. Combining the readily available computational power offered by commodity cluster-based parallel architectures such as NASA's Thunderhead system with last-generation sensor and parallel processing technology may introduce substantial changes in the systems currently used by NASA and other agencies for exploiting the sheer volume of Earth and planetary remotely sensed data collected on a daily basis. Although the parallel processing times reported in this chapter are very encouraging, close to (near) real-time processing, we continue our exploration of parallel processing strategies for morphological analysis of hyperspectral image data in real time using hardware-based parallel computing platforms such as field programmable gate arrays (FPGAs). Future work will also include the study of alternative approaches to be used in the extension of morphological operations and an investigation of additional strategies to reduce the impact of partial ordering in the vector space, without losing the physical interpretation inherent to the multichannel remotely sensed data.

ACKNOWLEDGMENTS

The author would like to gratefully thank Professor Chein-I Chang for his guidance, encouragement, and support during a research visit to his laboratory, Remote Sensing Signal and Image Processing Laboratory (RSSIPL). He would also like to thank Drs. John Dorband, Anthony Gualtieri, and James C. Tilton for their collaboration in experiments on the Thunderhead Beowulf cluster. The author would like to acknowledge support received from the Spanish Ministry of Education and Science (Fellowship PR2003-0360), which allowed him to conduct postdoctoral research at University of Maryland Baltimore County and NASA's Goddard Space Flight Center in 2004.

REFERENCES

1. J. Serra, *Image Analysis and Mathematical Morphology*, Academic, New York, 1982.
2. P. Soille, *Morphological Image Analysis: Principles and Applications*, 2nd edition, Springer, Berlin, 2003.
3. S. R. Sternberg, Grayscale morphology, *Computer Graphics and Image Processing.*, vol. 35, pp. 333–355, 1986.
4. I. Pitas and C. Kotropoulos, Multichannel L filters based on marginal data ordering, *IEEE Transactions on Signal Processing*, vol. 42, pp. 2581–2595, 1994.
5. J. Chanussot and P. Lambert, Total ordering based on space filling curves for multivalued morphology, in *Mathematical Morphology and Its Applications to Image and Signal*

Processing, edited by H. Heijmans and J. Roerdink, pp. 51–58, Kluwer Academic Publishers, Dordrecht, 1998.

6. C. Regazzoni and A. Teschioni, A new approach to vector median filtering based on space filling curves, *IEEE Transactions on Image Processing*, vol. 6, pp. 1025–1037, 1997.
7. X. Jia, J. A. Richards and D. E. Ricken, *Remote Sensing Digital Image Analysis: An Introduction*, Springer: Berlin, 1999.
8. J. Goutsias, H. Heijmans, and K. Sivakumar, Morphological operators for image sequences, *Computer Vision and Image Understanding*, vol. 62, pp. 326–346, 1995.
9. J. Astola, P. Haavisto, and Y. Neuvo, Vector median filters, *Proceedings of IEEE*, vol. 78, pp. 678–689, 1990.
10. A. Plaza, P. Martinez, R. Perez, and J. Plaza, Spatial/spectral endmember extraction by multidimensional morphological operations, *IEEE Transactions on Geoscience and Remote Sensing*, vol. 40, pp. 2025–2041, 2002.
11. M. Pesaresi and J. A. Benediktsson, A new approach for the morphological segmentation of high resolution satellite imagery, *IEEE Transactions on Geoscience and Remote Sensing*, vol. 39, pp. 309–320, 2001.
12. A. Plaza, P. Martinez, R. Perez, and J. Plaza, A new approach to mixed pixel classification in hyperspectral imagery based on extended morphological profiles, *Pattern Recognition*, vol. 37, pp. 1097–1116, 2004.
13. J. A. Benediktsson, M. Pesaresi, and K. Arnason, Classification and feature extraction for remote sensing images from urban areas based on morphological transformations, *IEEE Transactions on Geoscience and Remote Sensing*, vol. 41, pp. 1940–1949, 2003.
14. C. Lee and D. Landgrebe, Decision boundary feature extraction for neural networks, *IEEE Transactions on Neural Networks*, vol. 8, pp. 75–83, 1997.
15. J. A. Benediktsson, J. A. Palmason, and J. R. Sveinsson, Classification of hyperspectral data from urban areas based on extended morphological profiles, *IEEE Transactions on Geoscience and Remote Sensing*, vol. 43, pp. 480–491, 2005.
16. S. Beucher, Watershed, hierarchical segmentation and waterfall algorithm, in *Mathematical Morphology and Its Applications to Image Processing*, edited by E. Dougherty, Kluwer, Boston, 1994.
17. R. Adams and L. Bischof, Seeded region growing, *IEEE Transactions on Pattern Analysis and Machine Intelligence*, vol. 16, pp. 641–647, 1994.
18. A. Mehnert and P. Jackway, An improved seeded region growing algorithm, *Pattern Recognition Letters*, vol. 18, pp. 1065–1071, 1997.
19. N. Malpica, J. E. Ortuño, and A. Santos, A multichannel watershed-based algorithm for supervised texture segmentation, *Pattern Recognition Letters*, vol. 24, pp. 1545–1554, 2003.
20. A. Plaza, D. Valencia, P. Martínez, and J. Plaza, Commodity cluster-based parallel processing of hyperspectral imagery, *Journal of Parallel and Distributed Computing*, vol., pp. 2006.
21. D. Valencia, A. Plaza, P. Martínez, and J. Plaza, On the use of cluster computing architectures to process hyperspectral imagery, in *Proceedings of the IEEE Symposium on Computers and Communications*, Cartagena, Spain, pp. 995–1000, 2005.
22. A. N. Moga and M. Gabbouj, Parallel image component labeling with watershed transformation, *IEEE Trans. Pattern Analysis and Machine Intelligence*, vol. 19, pp. 441–450, 1997.

23. A. N. Moga and M. Gabbouj, Parallel marker-based image segmentation with watershed transformation, *Journal of Parallel and Distributed Computing*, vol. 51, pp. 27–45, 1998.
24. R. O. Green et al., Imaging spectroscopy and the airborne visible/infrared imaging spectrometer (AVIRIS), *Remote Sensing of Environment*, vol. 65, pp. 227–248, 1998.
25. C.-I Chang, *Hyperspectral Imaging: Techniques for Spectral Detection and Classification*, Kluwer, New York, 2003.
26. A. Plaza, P. Martinez, R. Perez, and J. Plaza, A quantitative and comparative analysis of endmember extraction algorithms from hyperspectral data, *IEEE Transactions on Geoscience and Remote Sensing*, vol. 42, pp. 650–663, 2004.
27. P. Soille, Morphological partitioning of multispectral images, *Journal of Electronic Imaging*, vol. 5, pp. 252–265, 2001.
28. A. Plaza, P. Martinez, J. Plaza and R.M. Perez, Dimensionality reduction and classification of hyperspectral image data using sequences of extended morphological transformations, *IEEE Transactions on Geoscience and Remote Sensing*, vol. 43, no. 3, pp. 466–479, 2005.
29. J. Dorband, J. Palencia, and U. Ranawake, Commodity computing clusters at Goddard Space Flight Center, *Journal of Space Communication*, vol. 1, no. 3, 2003.
30. M. J. Martín, D. E. Singh, J. C. Mouriño, F. F. Rivera, R. Doallo, and J. D. Bruguera, High performance air pollution modeling for a power plan environment, *Parallel Computing*, vol. 29, pp. 1763–1790, 2003.
31. M. G. Montoya, C. Gil, and I. García, The load unbalancing problem for region growing image segmentation algorithms, *Journal of Parallel and Distributed Computing*, vol. 63, pp. 387–395, 2003.

CHAPTER 14

THREE-DIMENSIONAL WAVELET-BASED COMPRESSION OF HYPERSPECTRAL IMAGERY

JAMES E. FOWLER AND JUSTIN T. RUCKER
Department of Electrical and Computer Engineering, GeoResources Institute, Mississippi State University, Mississippi State, MS 39762

14.1. INTRODUCTION

Since hyperspectral imagery is generated by collecting hundreds of contiguous bands, uncompressed hyperspectral imagery can be very large, with a single image potentially occupying hundreds of megabytes. For instance, the Airborne Visible Infrared Imaging Spectrometer (AVIRIS) sensor is capable of collecting several gigabytes of data per day. Compression is thus necessary to facilitate both the storage and the transmission of hyperspectral images. Since hyperspectral imagery is typically collected on remote acquisition platforms, such as satellites, the transmission of such data to central, often terrestrial, reception sites can be a critical issue. Thus, compression schemes oriented to the task of remote transmission are becoming increasingly of interest in hyperspectral applications.

Although there have been a number of approaches to the compression of hyperspectral imagery proposed in recent years—prominent techniques would include vector quantization (VQ) (e.g., [1, 2]) or principal component analysis (PCA) (e.g., [3, 4]) applied to spectral pixel vectors, as well as three-dimensional (3D) extensions of common image-compression methods such as the discrete cosine transform (DCT) (e.g., [5])—most of the approaches as proposed are not particularly well-suited to the image-transmission task. That is, in many applications involving the communication of images, *progressive transmission* is desired in that successive reconstructions of the image are possible. In such a scenario, the receiver can produce a low-quality representation of the image after having received only a small portion of the transmitted bitstream, and this "preview" representation can be successively refined in quality as more and more of the bitstream is received.

Hyperspectral Data Exploitation: Theory and Applications, Edited by Chein-I Chang

In many hyperspectral compression techniques, such progressive transmission is not supported, and, if the bitstream is not received in its entirety, no data set can be reconstructed. In this case, the bits that are received are generally useless in the application. It is anticipated that progressive-transmission capabilities will be of increasing interest, particularly for hyperspectral applications involving satellite-to-ground communications that are inherently susceptible to transmission failure due to high noise levels and limited bandwidth.

Wavelet-based compression schemes have garnered significant attention in recent years, in part due to their widespread support for progressive transmission. Wavelet-based compression techniques typically implement progressive transmission through the use of embedded coding. An *embedded coding* of a data set can be defined as any coding such that (1) any prefix of length N bits of an M-bit coding is also a valid coding of the entire data set, $0 < N \leq M$, and (2) if $N' > N$, then the quality upon reconstructing from the length-N' prefix is greater than or equal to that associated with the length-N prefix. Figures 14.1 and 14.2 illustrate the difference between transmission of typical nonembedded and embedded codings. With an embedded coding, applications may be able to process partially reconstructed data sets—for example, in the case of a bitstream being truncated prematurely due to a communication failure—whereas the nonembedded bitstream is generally of little use unless received in its entirety.

In this chapter, we overview embedded wavelet-based algorithms as applied to the compression of hyperspectral imagery. First, we review the major components of which modern wavelet-based coders are composed in Section 14.2 as well as various measures of compression performance in Section 14.3. We then overview specific compression algorithms in Section 14.4. In Section 14.5, we consider several issues concerning encoder design for JPEG2000 [6–8], perhaps the most

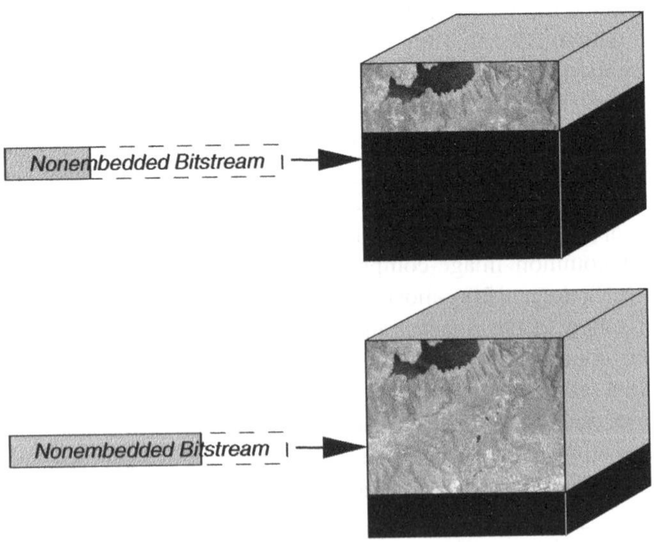

Figure 14.1. Transmission of a nonembedded coding.

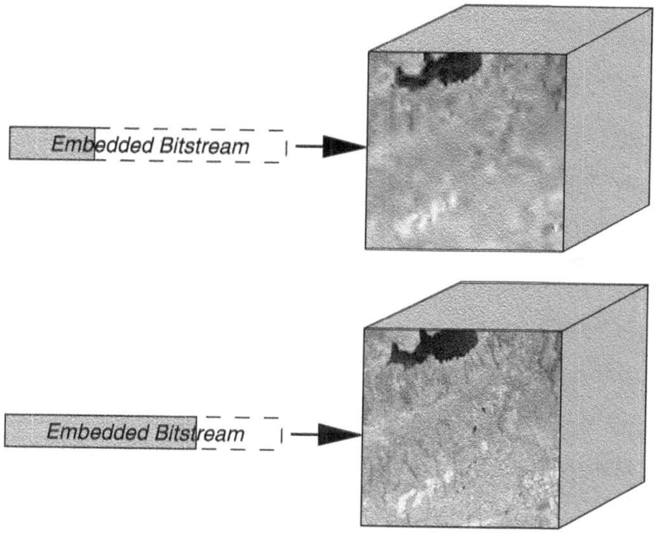

Figure 14.2. Transmission of an embedded coding.

prominent wavelet-based coder used for hyperspectral compression. We follow with a body of experimental results in Section 14.6 that compares the relative compression performance of the various wavelet-based approaches considered. Finally, we make some concluding observations in Section 14.7.

14.2. EMBEDDED WAVELET-BASED COMPRESSION OF 3D IMAGERY

The general philosophy behind embedded coding lies in the recognition that each successive bit of the bitstream that is received improves the quality of the reconstructed image by a certain amount. Consequently, in order to achieve an embedded coding, we must organize information in the bitstream in decreasing order of importance, where the most important information is defined to be that which produces the greatest increase in quality upon reconstruction. Although it is usually not possible to exactly achieve this ordering in practice, modern embedded compression algorithms do come close to approximating this optimal embedded ordering. Embedded wavelet-based coders are based upon four major precepts: a wavelet transform; significance-map encoding; successive-approximation coding (i.e., bitplane coding); and some form of entropy coding, most often arithmetic coding. These components are described in detail below.

14.2.1. Discrete Wavelet Transform (DWT)

Transforms aid the establishment of an embedded coding in that low-frequency components typically contain the majority of signal energy and are thus more

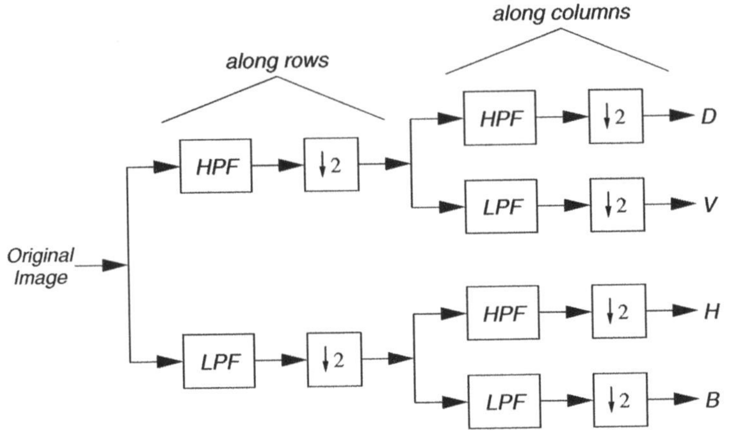

Figure 14.3. One stage of 2D DWT decomposition composed of lowpass (LPF) and highpass (HPF) filters applied to the columns and rows independently.

important than high-frequency components to reconstruction. Wavelet transforms are currently the transform of choice for modern two-dimensional (2D) image coders, since they not only provide this partitioning of information in terms of frequency but also retain much of the spatial structure of the original image. Wavelet-based coders for hyperspectral imagery extend the 2D transform structure into three dimensions.

A 2D *discrete wavelet transform* (DWT) can be implemented as a filter bank as illustrated in Figure 14.3. This filter bank decomposes the original image into horizontal (H), vertical (V), diagonal (D), and baseband (B) subbands, each being one-fourth the size of the original image. Wavelet theory provides filter-design methods such that the filter bank is perfectly reconstructing (i.e., there exists a reconstruction filter bank that will generate exactly the original image from the decomposed subbands H, V, D, and B) and such that the lowpass and highpass filters have finite impulse responses (which aids practical implementation). Multiple stages of decomposition can be cascaded together by recursively decomposing the baseband; the subbands in this case are usually arranged in a pyramidal form as illustrated in Figure 14.4.

For hyperspectral imagery, the 2D-transform decomposition of Figure 14.4 is extended to three dimensions to accommodate the addition of the spectral dimension. A 3D wavelet transform, like the 2D transform, is implemented in separable fashion, employing one-dimensional (1D) transforms separately in the spatial-row, spatial-column, and spectral-slice directions. However, the addition of a third dimension permits several options for the order of decomposition. For instance, we can perform one scale of decomposition along each direction, then further decompose the lowpass subband, leading to the dyadic decomposition, as is illustrated in Figure 14.5. This dyadic decomposition structure is the most straightforward 3D generalization of the 2D dyadic decomposition of Figure 14.4. However,

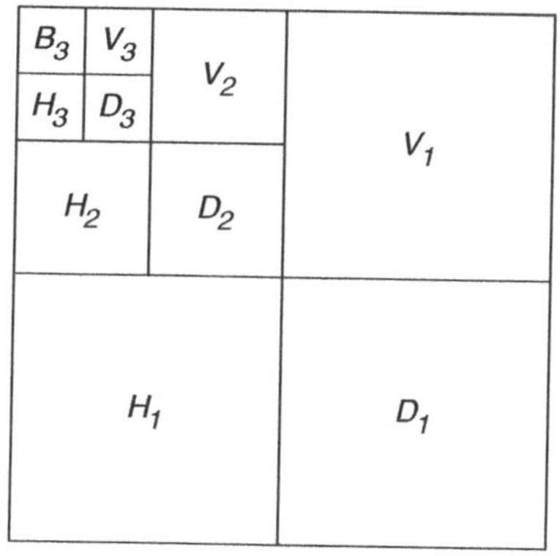

Figure 14.4. A three-scale, 2D DWT pyramid arrangement of subbands.

in 3D, we can alternatively use a so-called wavelet-packet transform, in which we first decompose each spectral slice using a separable 2D transform and then follow with a 1D decomposition in the spectral direction. With this approach, we employ an m-scale decomposition spatially, followed by an n-scale decomposition spectrally, where it is possible for $m \neq n$. For example, the wavelet-packet transform depicted in Figure 14.6 uses a three-scale decomposition ($m = n = 3$) in all directions. In comparing the two decomposition structures, the wavelet-packet transform is more flexible, because the spectral decomposition can be better tailored to the data at hand than in the dyadic transform. In Section 14.6.1, we will see that this wavelet-packet decomposition typically yields more efficient coding for hyperspectral datasets than does the dyadic decomposition. Additionally, it has been shown [9] that the particular wavelet-packet decomposition of Figure 14.6 is typically

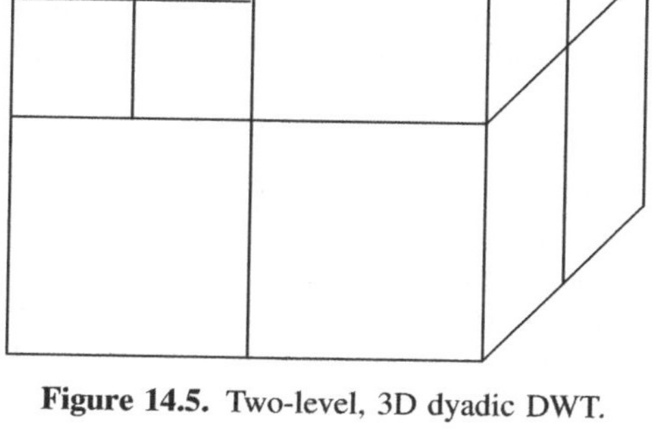

Figure 14.5. Two-level, 3D dyadic DWT.

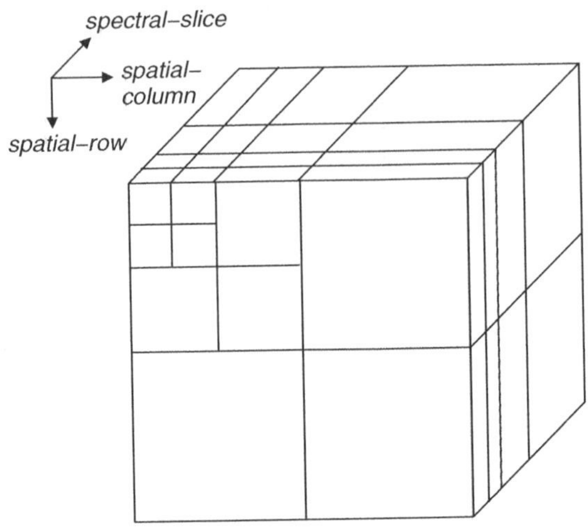

Figure 14.6. Three-dimensional packet DWT, with $m = 3$ spatial decompositions and $n = 3$ spectral decompositions.

very close in performance to the optimal 3D transform structure selected in a best-basis sense for the data set at hand.

Wavelet-based coders, 2D or 3D, base their operation on the following observations about the DWT: (1) Since most images are lowpass in nature, most signal energy is compacted into the baseband and lower-frequency subbands; (2) most coefficients are zero in the higher-frequency subbands; (3) small- or zero-valued coefficients tend to be clustered together within a given subband; and (4) clusters of small- or zero-valued coefficients in one subband tend to be located in the same relative spatial/spectral position as similar clusters in subbands of the next decomposition scale. The techniques we describe in Section 14.4 exploit one or more of these DWT properties to achieve efficient coding performance.

14.2.2. Bitplane Coding

The partitioning of information into DWT subbands somewhat inherently supports embedded coding in that transmitting coefficients by ordering the subbands from the low-resolution baseband subband toward the high-resolution highpass subbands implements a decreasing order of importance. However, more is needed to produce a truly embedded bitstream; even if some coefficient is more important than some other coefficient, not every bit of the first coefficient is necessarily more important than every bit of second. That is, not only should the coefficients be transmitted in decreasing order of importance, but also the individual bits that constitute the coefficients should be ordered as well.

Specifically, to effectuate an embedded coding of a set of coefficients, we represent the coefficients in sign-magnitude form as illustrated in Figure 14.7 and code the sign and magnitude of the coefficients separately. For coefficient-magnitude

		Coefficients			
		11	2	−3	6
	Sign	0	0	1	0
Bitplane	MSB 3	1	0	0	0
	2	0	0	0	1
	1	1	1	1	1
	LSB 0	1	0	1	0

Figure 14.7. Coefficients in sign-magnitude bitplane representation.

coding, we transmit the most significant bit (MSB) of *all* coefficient magnitudes, then the next-most significant bit of all coefficient magnitudes, and so on, such that each coefficient is successively approximated. This *bitplane-coding* scheme is contrary to the usual binary representation that would output all bits of a coefficient at once. The net effect of bitplane coding is that each coefficient magnitude is successively quantized by dividing the interval in which it is known to reside in half and outputting a bit to designate the appropriate subinterval, as illustrated in Figure 14.8.

In practice, bitplane coding is usually implemented by performing two passes through the set of coefficients for each bitplane: the significance pass and the refinement pass. Suppose the coefficient located at position $[x_1, x_2, x_3]$ in the 3D hyperspectral volume is $c[x_1, x_2, x_3]$. We define the significance state with respect to threshold t of the coefficient as

$$s[x_1, x_2, x_3] = \begin{cases} 1, & |c[x_1, x_2, x_3]| \geq t \\ 0, & \text{otherwise} \end{cases} \tag{14.1}$$

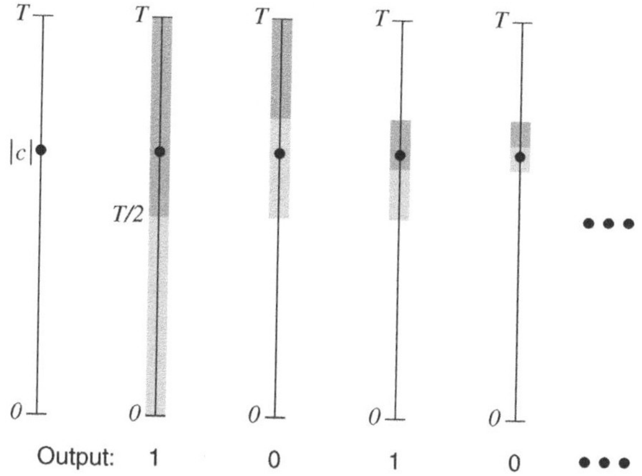

Figure 14.8. Successive-approximation quantization of a coefficient magnitude $|c|$ in interval $[0, T]$ where T is an integer power of 2.

We say that $c[x_1, x_2, x_3]$ is a *significant* coefficient when $s[x_1, x_2, x_3] = 1$; otherwise, $c[x_1, x_2, x_3]$ is *insignificant*. The significance pass describes $s[x_1, x_2, x_3]$ for all the coefficients in the DWT that are currently known to be insignificant but may become significant for the current threshold. On the other hand, the refinement pass produces a successive approximation to those coefficients that are already known to be significant by coding the current coefficient-magnitude bitplane for those significant coefficients. After each iteration of the significance and refinement passes, the significance threshold is divided in half, and the process is repeated for the next bitplane.

14.2.3 Significance-Map Coding

The collection of $s[x_1, x_2, x_3]$ values for all the coefficients in the DWT of an image is called the *significance map* for a particular threshold value. Given our observations in Section 14.2.1 of the nature of DWT coefficients, we see that for most of the bitplanes (particularly for large t), the significance map will be only sparsely populated with nonzero values. Consequently, the task of the significance pass is to create an efficient coding of this sparse significance map at each bitplane; the efficiency of this coding will be crucial to the overall compression efficiency of the coder. Section 14.4 is devoted to reviewing approaches that prominent algorithms have taken for the efficient coding of significance-map information. These algorithms are largely 2D image coders which have been extended to 3D and modified to accommodate the addition of spectral information.

14.2.4. Refinement and Sign Coding

In most embedded image coders, after the significance map is coded for a particular bitplane, a refinement pass proceeds through the coefficients, coding the current bitplane value of each coefficient that is already known to be significant but did not become significant in the immediately preceding significance pass. These *refinement bits* permit the reconstruction of the significant coefficients with progressively greater accuracy. It is usually assumed that the occurrence of a 0 or 1 is equally likely in bitplanes other than the MSB for a particular coefficient; consequently, most algorithms take little effort to code the refinement bits and may simply output them unencoded into the bitstream. Recently, it has been recognized that the refinement bits typically possess some correlation to their neighboring coefficients [10], particularly for the more significant bitplanes; consequently, some coders (e.g., JPEG2000) employ entropy coding for refinement bits.

The significance and refinement passes encode the coefficient magnitudes; to reconstruct the wavelet coefficients, the coefficient signs must also be encoded. As with the refinement bits, most algorithms assume that any given coefficient is equally likely to be positive or negative; however, recent work [10–12] has shown that there is some structure to the sign information that can be exploited to improve coding efficiency. Thus, certain coders (e.g., JPEG2000) also employ entropy coding for coefficient signs as well as for refinement bits.

14.2.5. Arithmetic Coding

Most wavelet-based coders incorporate some form of lossless entropy coding at the final stage before producing the compressed bitstream. In essence, such entropy coders assign shorter bitstream codewords to more frequently occurring symbols in order to maximize the compactness of the bitstream representation.

Most wavelet-based coders use *adaptive arithmetic coding* (AAC) [13] for lossless entropy coding. AAC codes a stream of symbols into a bitstream with length very close to its theoretical minimum limit. Suppose source X produces symbol i with probability p_i. The *entropy* of source X is defined to be

$$H(X) = -\sum_i p_i \log_2 p_i \tag{14.2}$$

where $H(X)$ has units of bits per symbol (bps). One of the fundamental tenets of information theory is that the average bit rate in bps of the most efficient lossless (i.e., invertible) compression of source X cannot be less than $H(X)$. In practice, AAC often produces compression quite close to $H(X)$ by estimating the probabilities of the source symbols with frequencies of occurrence as it codes the symbol stream. Essentially, the better able AAC can estimate p_i, the closer it will come to the $H(X)$ lower bound on compression efficiency. Oftentimes, the efficiency of AAC can be improved by *conditioning* the coder with known *context* information and maintaining separate symbol-probability estimates for each context. That is, limiting attention of AAC to a specific context usually reduces the variety of symbols, thus permitting better estimation of the probabilities within that context and producing greater compression efficiency. From a mathematical standpoint, the *conditional entropy* of source X with known information Y is $H(X|Y)$. Since it is well known from information theory that

$$H(X|Y) \leq H(X) \tag{14.3}$$

conditioning AAC with Y as the context will (usually) produce a bitstream with a smaller bit rate.

14.3. PERFORMANCE MEASURES FOR HYPERSPECTRAL COMPRESSION

Traditionally, performance for lossy compression is determined by simultaneously measuring both *distortion* and *rate*. Distortion measures the fidelity of the reconstructed data to the original data, while rate essentially measures the amount of compression incurred. Distortion is commonly measured via a *signal-to-noise ratio* (SNR) between the original and reconstructed data. Let $c[x_1, x_2, x_3]$ be an $N_1 \times N_2 \times N_3$ hyperspectral dataset with variance of σ^2. Let $\widehat{c}[x_1, x_2, x_3]$ be the dataset as reconstructed from the compressed bitstream. The *mean squared error* (MSE) is defined as

$$\text{MSE} = \frac{1}{N_1 N_2 N_3} \sum_{x_1,x_2,x_3} (c[x_1,x_2,x_3] - \widehat{c}[x_1,x_2,x_3])^2 \tag{14.4}$$

while the SNR in decibels (dB) is defined in terms of the MSE as

$$\text{SNR} = 10 \log_{10} \frac{\sigma^2}{\text{MSE}} \tag{14.5}$$

Both the MSE and SNR provide a measure of the performance of a coder in an *average* sense over the entire volume. Such an average measure may or may not be of the greatest use, depending on the application to be made of the reconstructed data. Hyperspectral imagery is often used in applications involving extensive analysis; consequently, it is paramount that the compression of hyperspectral data does not alter the outcome of such analysis. As an alternative to the SNR measure for distortion, one can examine the difference between performance of application-specific analysis as applied to the original data and the reconstructed data. As an example, unsupervised classification of hyperspectral pixel vectors is representative of methods that segment an image into multiple constituent classes. To form a distortion measure, we can apply unsupervised classification on the original hyperspectral image as well as on the reconstructed image, counting the number of pixels that change assigned class as a result of the compression. We call the resulting distortion measure *preservation of classification* (POC), which is measured as the percentage of pixels that do not change class due to compression.

In the subsequent experimental results reported in Section 14.6, all POC results are calculated using the ISODATA and k-means unsupervised classification as implemented in ENVI Version 4.0. A maximum of 10 classes are used, and POC performance is determined by applying the classification to the original dataset as well as to the reconstructed volume and comparing the classification map produced for reconstructed volume to that of the original dataset. In this manner, the classification map of the original dataset is effectively used as "ground truth." Figure 14.9 depicts typical classification maps generated in this manner.

In addition to distortion, it is necessary to gauge compression techniques according to the amount of compression incurred, due to the inherent trade-off between distortion and compression: The more highly compressed a reconstructed data set is, the greater the expected distortion between the original and reconstructed data. Typically, for hyperspectral imagery, one measures the rate as the number of *bits per pixel per band* (bpppb) which gives the average number of bits to represent a single sample of the hyperspectral data set. A *compression ratio* can then be determined as the ratio of the bpppb of the original data set (usually 16 bpppb) to the bpppb of the compressed dataset.

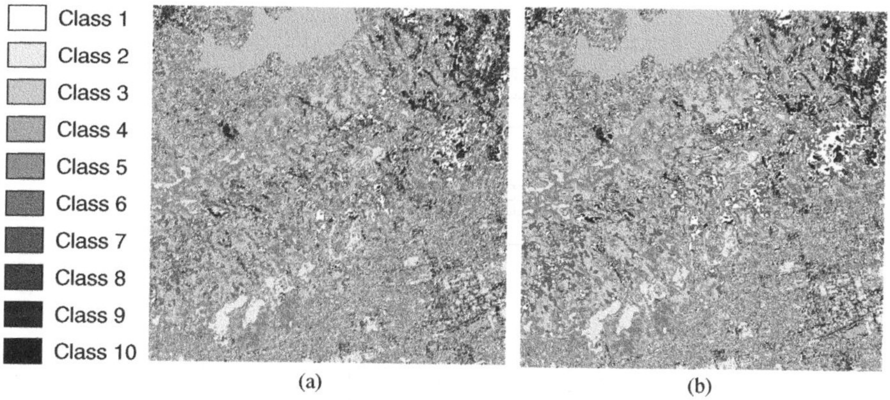

Figure 14.9. Classification map for the Moffett image using k-means classification. (a) Map for original image. (b) Map after JPEG2000 compression.

14.4. PROMINENT TECHNIQUES FOR SIGNIFICANCE-MAP CODING

The primary difference between wavelet-based coding algorithms is how coding of the significance map is performed. Several techniques for significance-map coding that have been used for hyperspectral imagery are discussed below. These techniques were typically developed originally for 2D images and then subsequently extended and modified for 3D coding. As a consequence, we briefly overview the original 2D algorithm—which is usually more easily conceptualized—before discussing its 3D extension for each of the techniques considered below.

14.4.1. Zerotrees

Zerotrees are one of the most widely used techniques for coding significance maps in wavelet-based coders. Zerotrees capitalize on the fact that insignificant coefficients tend to cluster together within a subband, and clusters of insignificant coefficients tend to be located in the same location within subbands of different scales. As illustrated for a 2D DWT in Figure 14.10, "parent" coefficients in a subband can be related to *four* "children" coefficients in the same relative spatial location in a subband at the next scale. A *zerotree* is formed when a coefficient and all of its descendants are insignificant with respect to the current threshold, while a *zerotree root* is defined to be a coefficient that is part of a zerotree but is not the descendant of another zerotree root.

The Embedded Zerotree Wavelet (EZW) algorithm [14] was the first 2D image coder to make use of zerotrees for the coding of significance-map information. This coder is based on the observation that if a coefficient is found to be insignificant, it is likely that its descendants are also insignificant. Consequently, the occurrence of a zerotree root in the baseband or in the lower-frequency subbands can lead to substantial coding efficiency since we can denote the zerotree root as a special "Z"

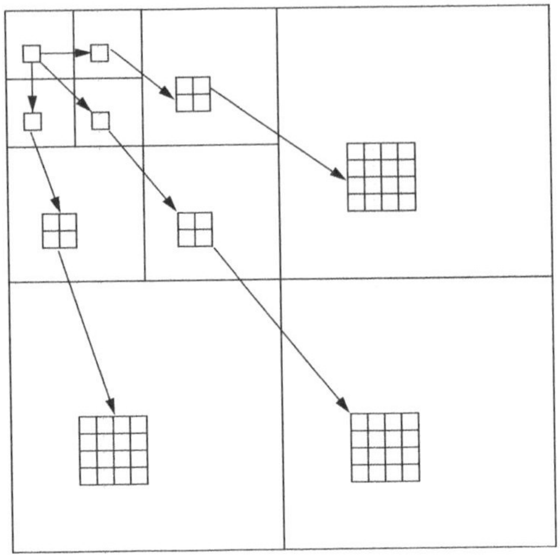

Figure 14.10. Parent–child relationships between subbands of a 2D DWT.

symbol in the significance map, and not code all of the descendants which are known then to be insignificant by definition. The EZW algorithm then proceeds to code the significance map in a raster scan within each subband, starting with the baseband and progressing to the high-frequency subbands. A lossless entropy coding of symbols from this raster scan then produces a compact representation of the significance map.

The Set Partitioning in Hierarchical Trees (SPIHT) algorithm [15] improves upon the zerotree concept by replacing the raster scan with a number of sorted lists that contain sets of coefficients (i.e., zerotrees) and individual coefficients. These lists are illustrated in Figure 14.11. In the significance pass of the SPIHT algorithm, the list of insignificant sets (LIS) is examined in regard to the current threshold; any set in the list that is no longer a zerotree with respect to the current threshold is then partitioned into one or more smaller zerotree sets, isolated insignificant coefficients,

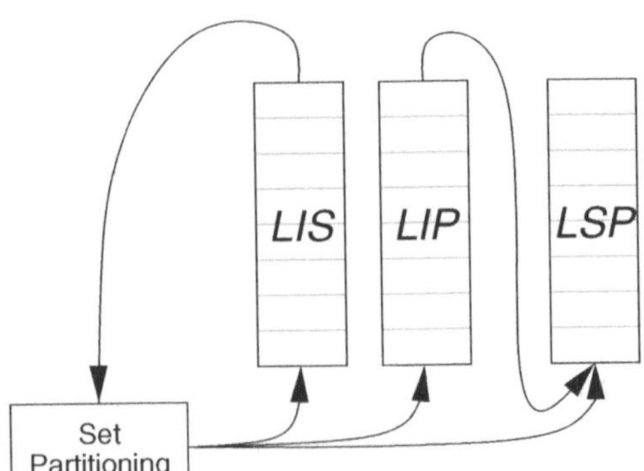

Figure 14.11. Processing of sorted lists in SPIHT.

or significant coefficients. Isolated insignificant coefficients are appended to the list of insignificant pixels (LIP), while significant coefficients are appended to the list of significant pixels (LSP). The LIP is also examined, and, as coefficients become significant with respect to the current threshold, they are appended to the LSP. Binary symbols are encoded to describe motion of sets and coefficients between the three lists. Since the lists remain implicitly sorted in an importance ordering, SPIHT achieves a high degree of embedding and compression efficiency.

Originally developed for 2D images, SPIHT has been extended to 3D in several contexts [16–21]. In the case of a dyadic transform such as in Figure 14.5, the 3D zerotree is a straightforward extension of the parent–child relationship of 2D zerotrees; that is, one coefficient is the parent to a $2 \times 2 \times 2$ cube of eight offspring coefficients in the next scale. However, in the case of a wavelet-packet transform, there are several approaches to fitting a zerotree structure to the wavelet coefficients. The first, proposed in Kim et al. [18], recognizes that wavelet-packet subbands appear as "split" versions of their dyadic counterparts; consequently, one should "split" the $2 \times 2 \times 2$ offspring nodes of the dyadic zerotree structure appropriately. An alternative zerotree structure for packet transforms was proposed in reference 19 and used subsequently in references 20 and 21. In essence, this zerotree structure consists of 2D zerotrees within each "slice" of the subband-pyramid volume, with parent–child relationships set up between the tree-root coefficients of the 2D trees. Cho and Pearlman [20] called this alternative structure an *asymmetric packet zerotree*, with the original splitting-based packet structure of Kim et al. [18] then being a *symmetric packet zerotree*. The asymmetric structure, which is depicted in Figure 14.12, usually offers somewhat more efficient compression performance than the symmetric packet structure [19–21]. Additionally, the wavelet-packet transform can have the number of spectral decomposition levels different from the number of spatial decomposition levels when the asymmetric tree is used, whereas the number of spatial and spectral decompositions must be the same in order to use the symmetric packet zerotree.

14.4.2. Spatial-Spectral Partitioning

Another approach to significance-map coding is *spatial-spectral partitioning*. The Set-Partitioning Embedded Block Coder (SPECK) [22, 23], originally developed as a 2D image coder, employs quadtree partitioning (see Figure 14.13) in which the significance state of an entire block of coefficients is tested and coded. Then, if the block contains at least one significant coefficient, the block is subdivided into four subblocks of approximately equal size, and the significance-coding process is repeated recursively on each of the subblocks.

In 2D-SPECK, there are two types of sets: S sets and I sets. The first S set is the baseband, and the first I set contains everything that remains. There are also two linked lists in SPECK: the List of Insignificant Sets (LIS), which contains sorted lists of decreasing sizes that have not been found to contain a significant pixel as compared with the current threshold, and the List of Significant Pixels (LSP), which contains single pixels that have been found to be significant through sorting and

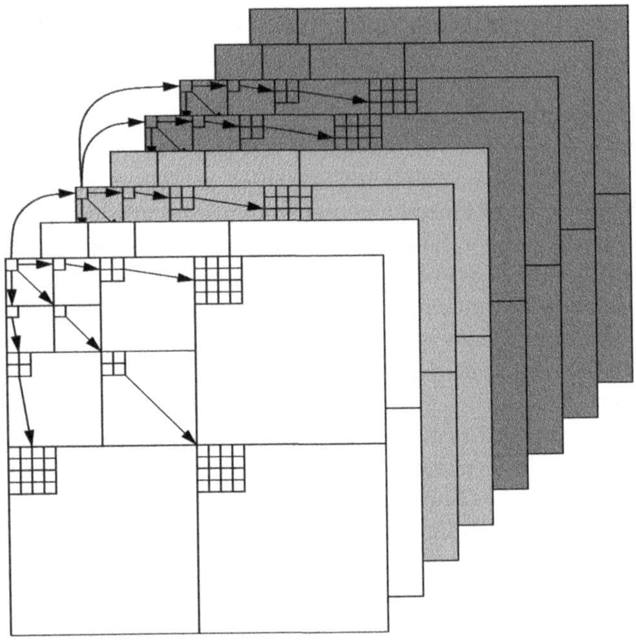

Figure 14.12. The asymmetric packet zerotree in a 3D packet DWT of $m = 3$ spatial decompositions and $n = 2$ spectral decompositions (adapted from Cho and Pearlman [20]). The spectral subbands are indicated by different shades of gray.

refinement passes. An S set remains in the LIS until it is found to be significant against the current threshold. The set is then divided into four approximately equal-sized sets, and the significance of each of the resulting four sets is tested. If the set is not significant, then it is placed in its appropriate place in the LIS. If the set is significant and contains a single pixel, it is appended to the LSP; otherwise, the set is recursively split into four subsets. Following the significant pass, the coefficients in the LSP go through a refinement pass in which coefficients that have been previously found to be significant are refined.

The SPECK algorithm was extended to 3D in references 24 and 25 by replacing quadtrees with octrees as illustrated in Figure 14.14. Unlike the original 2D-SPECK algorithm, the 3D-SPECK algorithm uses only one type of set, rather than having S and I sets as in 2D-SPECK. Consequently, each subband in the DWT decomposition is added to an LIS at the start of the 3D-SPECK algorithm, whereas the 2D

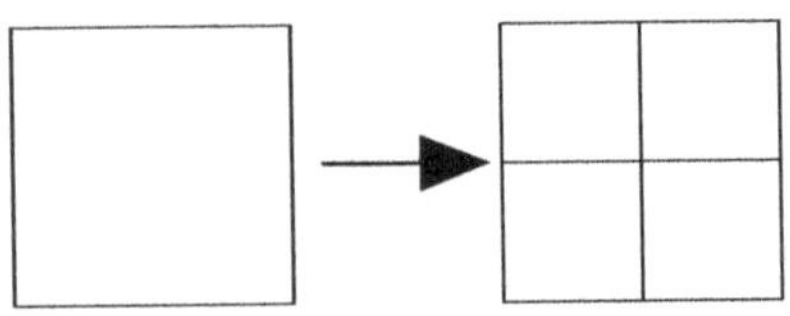

Figure 14.13. Two-dimensional quadtree block partitioning as performed in 2D SPECK.

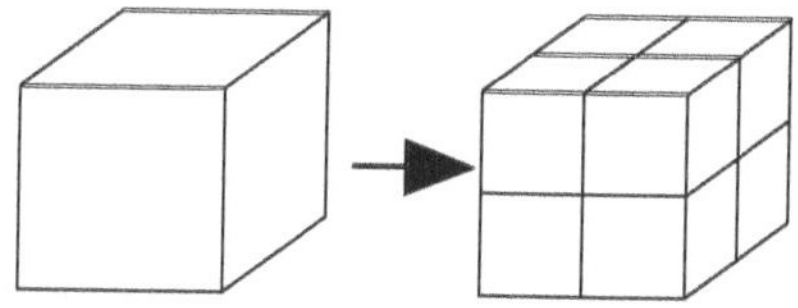

Figure 14.14. Three-dimensional octree cube partitioning as performed in 3D SPECK.

algorithm initializes with only the baseband subband in an LIS. An advantage of the set-partitioning processing of 3D-SPECK is that sets are confined to reside within a single subband at all times throughout the algorithm, whereas sets in SPIHT (i.e., the zerotrees) span across scales. This is beneficial from a computational standpoint because the coder need buffer only a single subband at a given time, leading to reduced dynamic memory needed [23]. Furthermore, 3D-SPECK is easily applied to both the dyadic and packet transform structures of Figures 14.5 and 14.6 with no algorithmic differences.

14.4.3. Conditional Coding

Recent work [26] has indicated that typically the ability to predict the insignificance of a coefficient through parent–child relationships, such as those employed by zerotree algorithms, is somewhat limited compared to the predictive ability of neighboring coefficients within the same subband. Consequently, recent algorithms, such as SPECK above, have focused on coding significance-map information using only within-subband information. Another approach to within-subband coding is to employ extensively conditioned, multiple-context AAC to capitalize on the theoretical advantages that conditioning provides for entropy coding as discussed in Section 14.2.5.

The usual approach to employing AAC with context conditioning for the significance-map coding of an image is to use the known significance states of neighboring coefficients to provide the context for the coding of the significance state of the current coefficient. Assuming a 2D image, the eight neighboring significance states to x_i are shown in Figure 14.15. Given that each neighbor takes on a binary value, there are $2^8 = 256$ possible contexts.

JPEG2000 [6–8], the most prominent *conditional-coding* technique, uses contexts derived from the neighbors depicted in Figure 14.15, but reduces the number of distinct contexts to nine, since not all possible contexts were found to be useful. The context definitions, which vary from subband to subband, are shown in

d_0	v_0	d_1
h_0	x_i	h_1
d_2	v_1	d_3

Figure 14.15. Significance-state neighbors to x_i.

B_j and V_j subbands			H_j subbands			D_j subbands		
$\sum h_i$	$\sum v_i$	$\sum d_i$	$\sum h_i$	$\sum v_i$	$\sum d_i$	$\sum(h_i + v_i)$	$\sum d_i$	Context
2				2			≥ 3	8
1	≥ 1		≥ 1	1		≥ 1	2	7
1	0	≥ 1	0	1	≥ 1	0	2	6
1	0	0	0	1	0	≥ 2	1	5
0	2		2	0		1	1	4
0	1		1	0		0	1	3
0	0	≥ 2	0	0	≥ 2	≥ 2	0	2
0	0	1	0	0	1	1	0	1
0	0	0	0	0	0	0	0	0

Figure 14.16. The AAC contexts for JPEG2000.

Figure 14.16. To further improve the context conditioning, as well as to increase the degree of embedding, JPEG2000 splits the coding of the significance map into two separate passes rather than employing one significance pass as do most other algorithms. Specifically, JPEG2000 uses a significance-propagation pass that codes those coefficients that are currently insignificant but have at least one neighbor that is already significant. This pass accounts for all coefficients that are likely to become significant in the current bitplane. The remaining insignificant coefficients are coded in the cleanup pass; these coefficients, which are surrounded by insignificant coefficients, are likely to remain insignificant. Both passes use the same nine contexts depicted in Figure 14.16. In addition, the cleanup pass includes one additional context used to encode four successive insignificant coefficients together with a single "insignificant run" symbol.

To code a single-band (i.e., 2D) image, a JPEG2000 encoder first performs a 2D wavelet transform on the image and then partitions each transform subband into small, 2D rectangular blocks called codeblocks, which are typically of size 32×32 or 64×64 pixels. Subsequently, the JPEG2000 encoder independently generates an embedded bitstream for each codeblock. To assemble the individual codeblock bitstreams into a single, final bitstream, each codeblock bitstream is truncated in some fashion, and the truncated bitstreams are concatenated together to form the final bitstream. The method for codeblock–bitstream truncation is an implementation issue concerning only the encoder because codeblock–bitstream lengths are conveyed to the decoder as header information. Consequently, this truncation process is not covered by the JPEG2000 standard.

It is highly likely that, for codeblocks residing in a single spectral band, any given JPEG2000 encoder with perform a Lagrangian rate-distortion optimal truncation as described as part of Taubman's EBCOT algorithm [8, 10]. This optimal truncation technique, *post-compression rate-distortion (PCRD) optimization*, is a primary factor in the excellent rate-distortion performance of the EBCOT algorithm. PCRD optimization is performed simultaneously across all of the codeblocks from the image, producing an optimal truncation point for each codeblock. The truncated codeblocks are then concatenated together to form a single bitstream. The PCRD optimization, in effect, distributes the total rate for the image spatially across the codeblocks in a rate-distortion-optimal fashion such that codeblocks with higher energy, which tend to more heavily influence the distortion measure, tend to receive greater rate.

As described in the standard, JPEG2000 is, in essence, a 2D image coder. Although the standard does make a few provisions for multiband imagery such as hyperspectral data, the core coding procedure is based on within-band coding of 2D blocks as described above. Furthermore, the exact procedure employed for 3D imagery (e.g., the 3D wavelet transform and PCRD optimization across multiple bands) largely entails design issues for the encoder and thus lies outside the realm of the JPEG2000 standard, which covers only the decoder. Given the increasing prominence that JPEG2000 is garnering for the coding of hyperspectral imagery, we return to consider these encoder-centric issues in depth in Section 14.5. Finally, we note that JPEG2000 with truly 3D coding, consisting of AAC coding of 3D codeblocks as in Schelkens et al. [27], has been proposed as JPEG2000 Part 10 (JP3D), an extension to the core JPEG2000 standard. However, at the time of this writing, this proposed extension is in the preliminary stages of development, and currently, JPEG2000 for hyperspectral imagery is employed as discussed in Section 14.5.

14.4.4. Runlength Coding

Since, for a given significance threshold, the significance map is essentially a binary image, techniques that have long been employed for the coding of bilevel images are applicable. Specifically, *runlength coding* is the fundamental compression algorithm behind the Group 3 fax standard; the Wavelet Difference Reduction (WDR) [28] algorithm combines runlength coding of the significance map with an efficient lossless representation of the runlength symbols to produce an embedded image coder. Originally developed for 2D imagery in [28], WDR was extended to 3D as an implementation in QccPack [29]; this 3D extension merely deploys the runlength scanning as a 3D raster scan of each subband of the 3D DWT, which is easily accomplished in either dyadic or packet DWT decompositions.

14.4.5. Density Estimation

An all-together different approach to significance-map coding was proposed in Simard et al. [30] wherein an explicit estimate of the probability of significance of wavelet coefficients is used to code the significance map. Specifically, the significance state of a set of coefficients for a given threshold is coded via a raster scan through the coefficients. For coding efficiency, an entropy coder codes the significance state for each coefficient, using the probability that the coefficient is significant as determined by the density-estimation procedure. The density estimate is in the form of a multidimensional convolution implemented as a sequence of 1D filtering operations coined *tarp filtering*. In Simard et al. [30], the tarp filtering procedure was originally developed for 2D image coding; 3D tarp, with the tarp-filtering procedure suitably extended to three dimensions, was proposed in Wang et al. [31, 32].

Of the various significance-map coding techniques considered in this section, conditional coding in the form of JPEG2000 has achieved the most widespread prominence for the coding of hyperspectral imagery. In the next section, we explore several issues concerning JPEG2000 encoding that lie outside the scope of the JPEG2000 standard but yield significant impact on compression performance.

14.5. JPEG2000 ENCODING STRATEGIES

JPEG2000 is increasingly being considered for the coding of hyperspectral imagery as well as other types of volumetric data, such as medical imagery. JPEG2000 is attractive because of its proven state-of-the-art performance for the compression of grayscale and color photographic imagery. However, its performance for hyperspectral compression can vary greatly, depending on how the JPEG2000 encoder handles multiple-band images—that is, images with multiple spectral bands. In effect, the JPEG2000 standard specifies the syntax and semantics of the compressed bitstream and, consequently, the operation of the *decoder*. The exact architecture of the *encoder*, on the other hand, is left largely to the designer of the compression system.

In deploying JPEG2000 on hyperspectral imagery, there are two primary issues that must be considered in the implementation of the JPEG2000 encoder: (1) spectral decorrelation, and (2) rate allocation between spectral bands. The first issue arises because there tends to exist significant correlation between consecutive bands in a hyperspectral image. As a consequence, spectral decorrelation, via a wavelet transform, yields significant performance improvement.

The second encoder-design issue—rate allocation between spectral bands—arises from the fact that, essentially, JPEG2000 is a 2D compression algorithm. Consequently, given a specific target rate of R bpppb, the JPEG2000 encoder must determine how to allocate this total rate appropriately between spectral bands. It is usually the case that certain bands have significantly higher energy than other bands and thus will weigh more heavily in distortion measures than the other, weaker-energy bands. Consequently, it is likely that the JPEG2000 encoder will need to allocate proportionally greater rate to the higher-energy bands in order to maximize distortion performance for a given total rate R. Below, we explore several rate-allocation strategies; we will find significant performance difference between these strategies later in experimental results in Section 14.6.2.

14.5.1. Spectral Decorrelation for Multiple-Component Images

The JPEG2000 standard allows for images with up to 16,385 spectral bands to be included in a single bitstream; however, the standard does not specify how these spectral bands should be encoded for best performance. Whereas Part I of the JPEG2000 standard [6] permits spectral decorrelation only in the case of three-band images (i.e., red–green–blue), Annexes I and N of Part II of the standard [7] make provisions for arbitrary spectral decorrelation, including wavelet transforms.

By applying a 1D wavelet transform spectrally and then subsequently employing a 2D wavelet transform spatially, we effectively implement the wavelet-packet transform of Figure 14.6. We note that many JPEG2000 implementations are not yet fully compliant with Part II of the standard. In this case, we can "simulate" the spectral decorrelation permitted under Part II by employing a 1D wavelet transform spectrally on each pixel in the scene before the image cube is sent to the Part-I-compliant JPEG2000 encoder. Such an external spectral transform as been used previously [32, 33] to implement a "2D spatial + 1D spectral" wavelet-packet transform with Part-I-compliant coders.

14.5.2. Rate-Allocation Strategies Across Multiple Image Components

The PCRD optimization procedure of EBCOT described in Section 14.4.3 produces a rate-distortion-optimal bitstream for a single-band image by optimally truncating the independent codeblock bitstreams from the spectral band. However, there are several ways that this single-band truncation procedure can be extended to the multiband case, and the resulting multiband truncation procedure, in effect, dictates how the total rate available for coding the hyperspectral image is allocated between the individual spectral bands.

That is, for a multiple-band image, a JPEG2000 encoder will partition each spectral band into 2D codeblocks that are coded into independent bitstreams identically to the process used for single-band imagery. To assemble a final bitstream, these individual codeblock bitstreams are truncated and concatenated together. Although the method for codeblock-bitstream truncation is an implementation issue concerning only the encoder and is thus not covered by the JPEG2000 standard, it is highly likely that any given multiband JPEG2000 encoder will perform PCRD optimization for *at least* the codeblocks originating from a single spectral band. How this truncation process is extended across the multiple bands may vary with encoder implementation. Below, we describe three possible multiband rate-allocation strategies. In the following, let a hyperspectral image volume X be composed of N bands X_i, that is, $X = \{X_1, X_2, \ldots, X_N\}$. We code X with a total rate of R bpppb. Assume that $B_i = \texttt{JPEG2000\_Encode}(R_i, X_i)$ is a single-band JPEG2000 encoder that encodes spectral band X_i with rate R_i using PCRD optimization, producing a bitstream B_i.

JPEG2000-BIFR. The most straightforward method of allocating rate between multiple spectral bands is to simply code each band independently and assign to each an identical rate. This JPEG2000 band-independent fixed-rate (JPEG2000-BIFR) strategy operates as follows, where the "$\circ$" operator denotes bitstream concatenation:

```
JPEG2000_BIFR (R, {X_1, ..., X_N})
  B = ∅
  for i = 1, 2, ..., N
```

$B_i = \texttt{JPEG2000\_Encode}\ (R, X_i)$

$B = B \circ B_i$

return B

JPEG2000-BIRA. JPEG2000 band-independent rate allocation (JPEG2000-BIRA) also codes each band independently; however, rates are allocated explicitly so that more important bands are coded with higher rate, and less important bands are coded at a lower rate.

JPEG2000_BIRA $(R, \{X_1, \ldots, X_N\})$

$B = \emptyset$

for $i = 1, 2, \ldots, N$

$\sigma_i^2 = \texttt{variance}\ [X_i]$

for $i = 1, 2, \ldots, N$

$$R_i = \frac{\log_2 \sigma_i}{\sum_{j=1}^{N} \log_2 \sigma_j} \cdot RN$$

$B_i = \texttt{JPEG2000\_Encode}\ (R_i, X_i)$

$B = B \circ B_i$

return B

The rates, R_i, are determined so that bands with larger variances (i.e., higher energy) are coded at a higher rate than those with lower variances, while the total rate for the entire volume is R. This approach is, in essence, an *ad hoc* variant of classical optimal rate allocation for a set of quantizers based on log variances [34, Chapter 8; 35].

JPEG2000-MC. The final approach, JPEG2000 multicomponent (JPEG2000-MC), can be employed when the JPEG2000 encoder is capable of performing PCRD optimization across multiple bands. That is, all of the spectral bands are input to the encoder, which produces codeblock bitstreams for every codeblock in every subband of every spectral band. Then, PCRD optimal truncation is applied to all codeblock bitstreams from all bands simultaneously, rather than simply the codeblock bitstreams for a single band. In this way, the PCRD optimization performs to the maximum of its potential, implicitly allocating rate in a rate-distortion fashion, not only spatially within each spectral band, but also spectrally across the multiple bands.

14.6. COMPRESSION PERFORMANCE

All the data sets used in the experiments were collected by AVIRIS, an airborne hyperspectral sensor with data in 224 contiguous bands from 400 nm to 2500 nm.

For the results here, we crop the first scene in each data set to produce image cubes with dimensions of 512 × 512 × 224. In all cases, unprocessed radiance data were used. All coders use the popular biorthogonal 9-7 wavelet [36] with symmetric extension as used extensively in image-compression applications, and a transform decomposition of four spatial and spectral levels is employed. All rate measurements are expressed in bits per pixel per band (bpppb). All JPEG2000 coding uses Kakadu* Version 4.3 with a quantization step size of 0.0000001. Since Kakadu is not yet fully compliant with Part II of the JPEG2000 standard, the spectral transform is applied externally as described in Section 14.2.1 and in [32, 33]. We note that the results below are selected from extensive empirical evaluations we have conducted; a more complete presentation of results is available in Rucker [37].

14.6.1. Performance of Dyadic and Packet Transforms

As was discussed in Section 14.2.1, there are two contending transform arrangements for the 3D DWT. The 3D dyadic transform (Figure 14.5) is a direct extension of the 2D dyadic transform in which we transform once in each direction and then further decompose the baseband. In the case of the 3D packet transform (Figure 14.6), the coefficients in each spectral slice are transformed with a 2D dyadic transform, which is then followed by a spectral transform. Figure 14.17 depicts the typical rate-distortion performance achieved by a coder using these

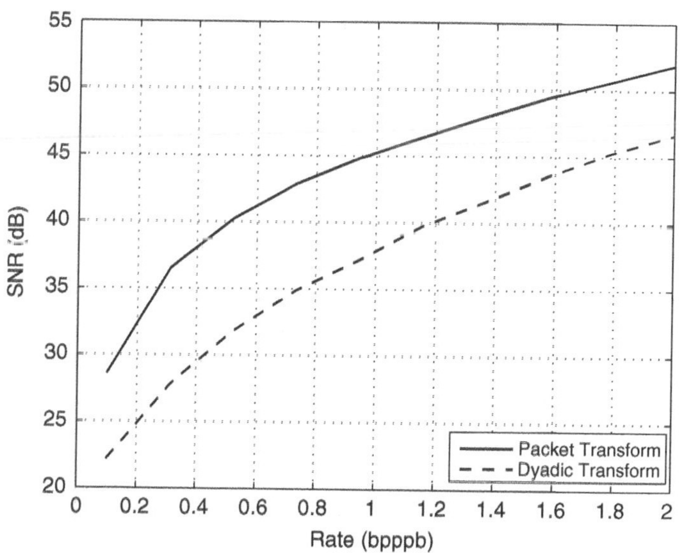

Figure 14.17. Comparison of the typical rate-distortion performance for the dyadic transform of Figure 14.5 versus that of the packet transform of Figure 14.6. This plot is for the Moffett image using 3D-SPIHT.

*http://www.kakadusoftware.com

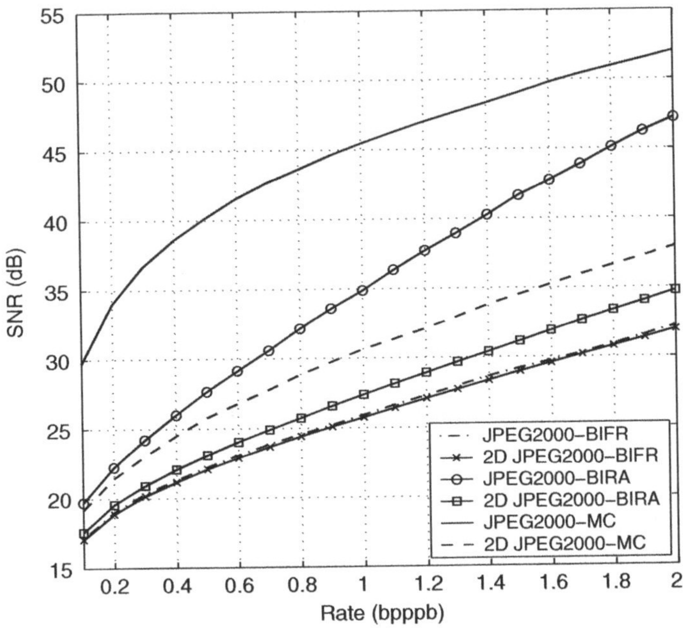

Figure 14.18. Rate-distortion performance for Moffett for the JPEG2000 encoding strategies.

two transform structures. We see that performance for the packet transform is greatly superior to that for the dyadic transform. As we have observed similar results for other coders and other datasets, we use the packet transform exclusively for all subsequent results.

14.6.2. Performance of JPEG2000 Encoding Strategies

Section 14.5 presents several strategies for the design of a JPEG2000 encoder. We evaluate these strategies now, focusing first on rate-distortion performance before considering POC performance as described in Section 14.3. In Figure 14.18, we plot the rate-distortion performance of JPEG2000 for a range of rates, while in Table 14.1, distortion performance at a single rate is tabulated. In these results,

TABLE 14.1. SNR Performance in dB at 1.0 bpppb for the JPEG2000 Encoding Strategies

Data Set	2D BIFR	BIFR	2D BIRA	BIRA	2D MC	MC
Moffett	25.8	25.9	27.4	34.9	30.6	45.5
Jasper Ridge	24.0	23.8	25.7	33.4	29.8	44.8
Cuprite	32.9	32.8	34.9	42.6	38.3	51.0

TABLE 14.2. POC Performance at 1.0 bpppb for the JPEG2000 Encoding Strategies

Data Set	2D BIFR	BIFR	2D BIRA	BIRA	2D MC	MC
	ISODATA POC (%)					
Moffett	83.4	94.5	86.6	94.5	93.2	99.7
Jasper Ridge	77.3	75.5	82.2	93.7	93.9	99.7
Cuprite	80.3	78.1	85.1	94.7	94.7	99.8
	k-means POC (%)					
Moffett	75.4	73.2	79.9	91.7	89.8	99.6
Jasper Ridge	67.2	64.7	73.9	90.4	91.0	99.5
Cuprite	71.3	68.3	77.6	92.2	92.1	99.6

techniques labeled as "2D" do not use any spectral transform (i.e., only 2D wavelet transforms are applied spatially), while the other techniques use the 3D wavelet-packet transform which includes a spectral transform. For each data set, we present performance for the three rate-allocation techniques described in Section 14.5.2, both with and without the spectral-decorrelation transform. With the exception of JPEG2000-BIFR, all the rate-allocation techniques perform significantly better when a spectral transform is performed. We see that JPEG2000-MC substantially outperforms the other techniques by at least 5–10 dB.

We now turn our attention to POC performance to gauge the preservation of unsupervised-classification performance. We see that the POC performances in Table 14.2 correlate well with SNR figures of Table 14.1 in that, if one technique outperforms another in the rate-distortion realm, then it will mostly likely have higher POC performance as well. As expected, JPEG2000-MC performs substantially better than the other techniques in terms of POC.

We note that both Kakadu Version 4.3 and the JPEG2000 encoder in ENVI Version 4.1 (which uses the Kakadu coder) implement JPEG2000-MC rate allocation, yet neither one of these supports the use of a spectral transform since they are not fully compliant with Part II of the JPEG2000 standard. Thus, the performance of these coders is equivalent to that of the 2D JPEG2000-MC approach considered here. As our results indicate, adding a spectral transform would significantly enhance the performance of these coders.

14.6.3. Algorithm Performance

Rate-distortion performance for a variety of the algorithms described in Section 14.4 (3D-WDR, 3D-tarp, 3D-SPECK, 3D-SPIHT, and JPEG2000-MC) is shown in Figures 14.19–14.21 as well as in Table 14.3. In these results, we see that all five techniques provide largely similar rate-distortion performance for the datasets considered, with JPEG2000-MC usually slightly outperforming the others. Similar conclusions are drawn from the POC results of Table 14.4.

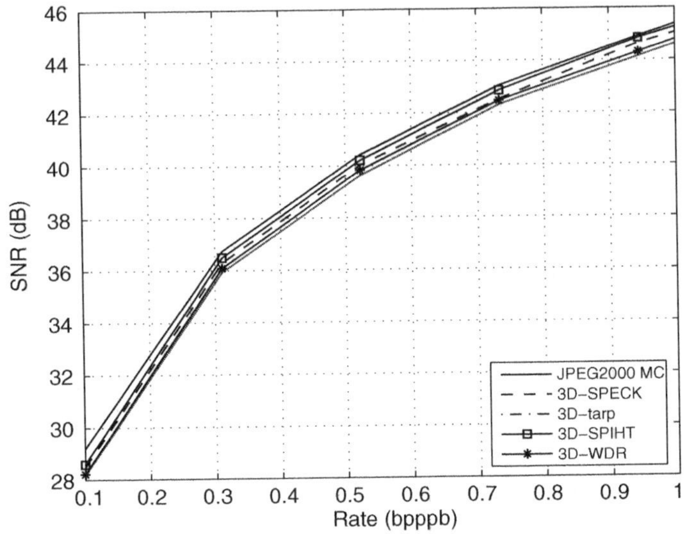

Figure 14.19. Rate-distortion performance for Moffett.

14.7. SUMMARY

In this chapter, we overviewed the major concepts in 3D embedded wavelet-based compression for hyperspectral imagery. We reviewed several popular compression techniques that have been considered for the coding of hyperspectral imagery,

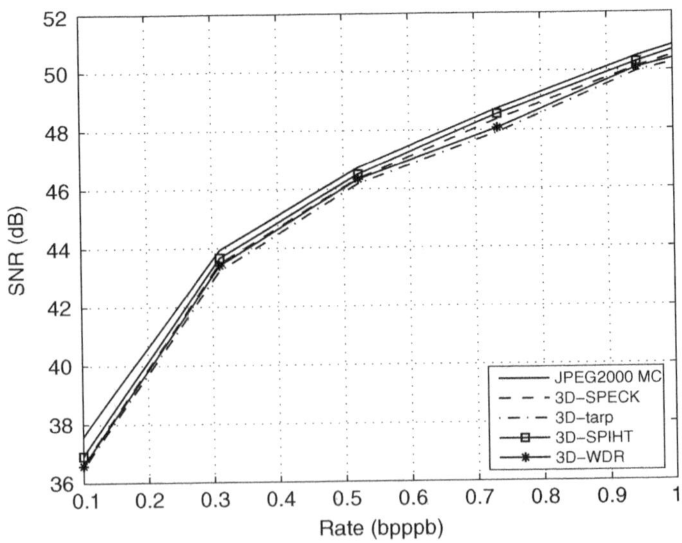

Figure 14.20. Rate-distortion performance for Cuprite.

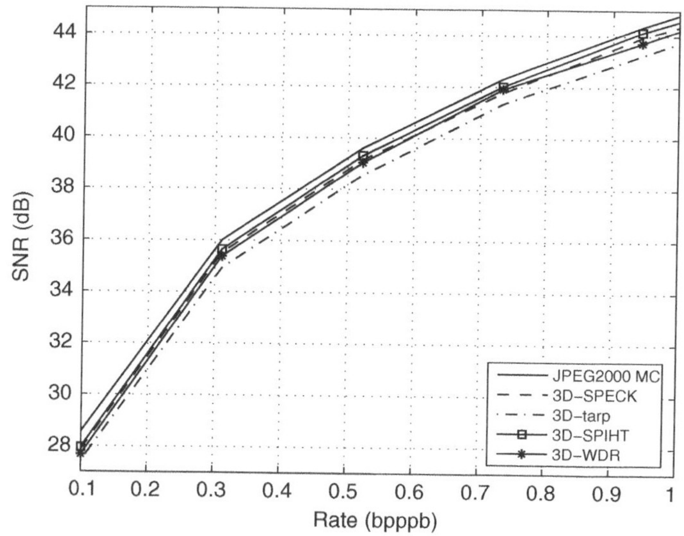

Figure 14.21. Rate-distortion performance for Jasper Ridge.

focusing on the primary difference between techniques—how the significance map is coded. We found that the different techniques offered performance that is roughly similar, both in terms of rate-distortion performance and a more application-specific preservation of performance at unsupervised classification. We discussed that the most prominent of the algorithms considered, JPEG2000, is subject to an international standard that covers only the decoder, leaving many design details regarding the encoder unspecified, particularly as pertaining to the coding of multiband imagery. We presented experimental results that demonstrate how a JPEG2000 encoder allocates rate between spectral bands substantially affects performance. Additionally, we saw that JPEG2000 performance almost always benefits greatly from the application of a 1D spectral wavelet transform to remove correlation in the spectral direction.

As a final note, we observe that, in many situations, it may be necessary to store hyperspectral datasets in their original state—that is, without any compression loss.

TABLE 14.3. SNR at 1.0 bpppb

	SNR (dB)				
Data Set	JPEG2000	SPECK	SPIHT	TARP	WDR
Moffett	45.4	45.1	45.3	44.5	44.7
Jasper Ridge	44.9	44.4	44.7	43.7	44.2
Cuprite	50.8	50.5	50.7	50.3	50.4
Low altitude	27.6	27.3	27.4	25.2	27.1
Lunar Lake	46.4	45.9	46.1	43.7	45.9

TABLE 14.4. POC Performance at 1.0 bpppb

	ISODATA POC (%)				
Data Set	JPEG2000	SPECK	SPIHT	TARP	WDR
Moffett	99.8[a]	99.7	99.7	99.7	99.7
Jasper Ridge	99.8	99.7	99.7	99.7	99.7
Cuprite	99.8	99.8	99.8	99.8	99.8
Low altitude	97.9	98.1	97.9	97.3	97.9
Lunar Lake	99.7	99.5	99.7	99.7	99.7
	k-Means POC (%)				
Moffett	99.7	99.6	99.6	99.6	99.6
Jasper Ridge	99.6	99.5	99.5	99.5	99.5
Cuprite	99.7	99.7	99.6	99.7	99.7
Low altitude	96.6	96.8	96.6	96.1	96.7
Lunar Lake	99.5	99.6	99.5	99.2	99.6

Such archival applications may necessitate the *lossless* compression of hyperspectral imagery, whereas the discussion in this chapter has focused exclusively on *lossy* compression algorithms. However, it is fairly straightforward to modify the lossy algorithms considered here to render them lossless, while still preserving their progressive-transmission capability. Such embedded wavelet-based coders then provide "lossy-to-lossless" performance, such that any truncation of the bitstream can be reconstructed to a lossy representation, yet, if the entire bitstream is decoded, a lossless reconstruction of the original data set is obtained. Such lossy-to-lossless coding has been proposed in several contexts [38, 39], including hyperspectral-image compression [40], by adding an integer-to-integer wavelet transform [41] to a lossy technique. JPEG2000 supports such lossy-to-lossless coding in Part I of the standard in exactly this way.

REFERENCES

1. S. Gupta and A. Gersho, Feature predictive vector quantization of multispectral images, *IEEE Transactions on Geoscience and Remote Sensing*, vol. 30, no. 3, pp. 491–501, 1992.
2. S.-E. Qian, A. B. Hollinger, S. Williams, and D. Manak, Vector quantization using spectral index-based multiple subcodebooks for hyperspectral data compression, *IEEE Transactions on Geoscience and Remote Sensing,* vol. 38, no. 3, pp. 1183–1190, 2000.
3. B. R. Epstein, R. Hingorani, J. M. Shapiro, and M. Czigler, Multispectral KLT-wavelet data compression for Landsat thematic mapper images, in *Proceedings of the IEEE Data Compression Conference*, edited by J. A. Storer and M. Cohn, Snowbird, UT, pp. 200–208, 1992.

4. J. A. Saghri, A. G. Tescher, and J. T. Reagan, Practical transform coding of multispectral imagery, *IEEE Signal Processing Magazine*, vol. 12, no. 1, pp. 32–43, 1995.

5. G. P. Abousleman, M. W. Marcellin, and B. R. Hunt, Compression of hyperspectral imagery using the 3-D DCT and hybird DPCM/DCT, *IEEE Transactions on Geoscience and Remote Sensing*, vol. 33, no. 1, pp. 26–34, 1995.

6. *Information Technology—JPEG 2000 Image Coding System—Part 1: Core Coding System*, ISO/IEC 15444-1, 2000.

7. *Information Technology—JPEG 2000 Image Coding System—Part 2: Extensions*, ISO/IEC 15444-2, 2004.

8. D. S. Taubman and M. W. Marcellin, *JPEG2000: Image Compression Fundamentals, Standards and Practice*, Kluwer Academic Publishers, Boston, MA, 2002.

9. B. Penna, T. Tillo, E. Magli, and G. Olmo, Progressive 3-D coding of hyperspectral images based on JPEG 2000, *IEEE Geoscience and Remote Sensing Letters*, vol. 3, no. 1, pp. 125–129, 2006.

10. D. Taubman, High performance scalable image compression with EBCOT, *IEEE Transactions on Image Processing*, vol. 9, no. 7, pp. 1158–1170, 2000.

11. A. Deever and S. S. Hemami, What's your sign?: Efficient sign coding for embedded wavelet image coding, in *Proceedings of the IEEE Data Compression Conference*, edited by J. A. Storer and M. Cohn, Snowbird, UT, pp. 273–282, 2000.

12. A. T. Deever and S. S. Hemami, Efficient sign coding and estimation of zero-quantized coefficients in embedded wavelet image codecs, *IEEE Transactions on Image Processing*, vol. 12, no. 4, pp. 420–430, April 2003.

13. I. H. Witten, R. M. Neal, and J. G. Cleary, Arithmetic coding for data compression, *Communications of the ACM*, vol. 30, no. 6, pp. 520–540, 1987.

14. J. M. Shapiro, Embedded image coding using zerotrees of wavelet coefficients, *IEEE Transactions on Signal Processing*, vol. 41, no. 12, pp. 3445–3462, 1993.

15. A. Said and W. A. Pearlman, A new, fast, and efficient image codec based on set partitioning in hierarchical trees, *IEEE Transactions on Circuits and Systems for Video Technology*, vol. 6, no. 3, pp. 243–250, 1996.

16. B.-J. Kim and W. A. Pearlman, An embedded wavelet video coder using three-dimensional set partitioning in hierarchical trees (SPIHT), in *Proceedings of the IEEE Data Compression Conference*, edited by J. A. Storer and M. Cohn, Snowbird, UT, pp. 251–257, 1997.

17. P. L. Dragotti, G. Poggi, and A. R. P. Ragozini, Compression of multispectral images by three-dimensional SPIHT algorithm, *IEEE Transactions on Geoscience and Remote Sensing*, vol. 38, no. 1, pp. 416–428, 2000.

18. B.-J. Kim, Z. Xiong, and W. A. Pearlman, Low bit-rate scalable video coding with 3-D set partitioning in hierarchical trees (3-D SPIHT), *IEEE Transactions on Circuits and Systems for Video Technology*, vol. 10, no. 8, pp. 1374–1387, 2000.

19. C. He, J. Dong, Y. F. Zheng, and Z. Gao, Optimal 3-D coefficient tree structure for 3-D wavelet video coding, *IEEE Transactions on Circuits and Systems for Video Technology*, vol. 13, no. 10, pp. 961–972, 2003.

20. S. Cho and W. A. Pearlman, Error resilient video coding with improved 3-D SPIHT and error concealment, in *Image and Video Communications and Processing*, edited by B. Vasudev, T. R. Hsing, and A. G. Tescher, Santa Clara, CA, *Proceedings of SPIE* 5022, pp. 125–136, 2003.

21. X. Tang, S. Cho, and W. A. Pearlman, 3D set partitioning coding methods in hyperspectral image compression, in *Proceedings of the International Conference on Image Processing*, Vol. 2, Barcelona, Spain, September 2003, pp. 239–242.
22. A. Islam and W. A. Pearlman, An embedded and efficient low-complexity hierarchical image coder, in *Visual Communications and Image Processing*, edited by K. Aizawa, R. L. Stevenson, and Y.-Q. Zhang, San Jose, CA, *Proceedings of SPIE* 3653, pp. 294–305, 1999.
23. W. A. Pearlman, A. Islam, N. Nagaraj, and A. Said, Efficient, low-complexity image coding with a set-partitioning embedded block coder, *IEEE Transactions on Circuits and Systems for Video Technology*, vol. 14, no. 11, pp. 1219–1235, 2004.
24. X. Tang, W. A. Pearlman, and J. W. Modestino, Hyperspectral image compression using three-dimensional wavelet coding, in *Image and Video Communications and Processing*, edited by B. Vasudev, T. R. Hsing, A. G. Tescher, and T. Ebrahimi, Santa Clara, CA, *Proceedings of SPIE* 5022, 1037–1047, 2003.
25. X. Tang and W. A. Pearlman, Three-dimensional wavelet-based compression of hyperspectral images, in *Hyperspectral Data Compression*, edited by G. Motta, F. Rizzo, and J. A. Storer, Kluwer Academic Publishers, Norwell, MA, 2006. pp. 273–308.
26. M. W. Marcellin and A. Bilgin, Quantifying the parent-child coding gain in zero-tree-based coders, *IEEE Signal Processing Letters*, vol. 8, no. 3, pp. 67–69, 2001.
27. P. Schelkens, J. Barbarien, and J. Cornells, Compression of volumetric medical data based on cube-splitting, in *Applications of Digital Image Processing XXII*, San Diego, CA, *Proceedings of SPIE* 4115, pp. 91–101, 2000.
28. J. Tian and R. Wells, Jr., Embedded image coding using wavelet difference reduction, in *Wavelet Image and Video Compression*, edited by P. N. Topiwala, Kluwer Academic Publishers, Boston, MA, pp. 289–301, 1998.
29. J. E. Fowler, QccPack: An open-source software library for quantization, compression, and coding, in *Applications of Digital Image Processing XXIII*, edited by A. G. Tescher, San Diego, CA, *Proceedings of SPIE* 4115, August 2000, pp. 294–301.
30. P. Simard, D. Steinkraus, and H. Malvar, On-line adaptation in image coding with a 2-D tarp filter, in *Proceedings of the IEEE Data Compression Conference*, edited by J. A. Storer and M. Conn, Snowbird, UT, pp. 23–32, 2002.
31. Y. Wang, J. T. Rucker, and J. E. Fowler, Embedded wavelet-based compression of hyperspectral imagery using tarp coding, in *Proceedings of the International Geoscience and Remote Sensing Symposium*, vol. 3, Toulouse, France, pp. 2027–2029, 2003.
32. Y. Wang, J. T. Rucker, and J. E. Fowler, 3D tarp coding for the compression of hyperspectral images, *IEEE Geoscience and Remote Sensing Letters*, vol. 1, no. 2, pp. 136–140, 2004.
33. H. S. Lee, N. H. Younan, and R. L. King, Hyperspectral image cube compression combining JPEG-2000 and spectral decorrelation, in *Proceedings of the International Geoscience and Remote Sensing Symposium*, Vol. 6, Toronto, Canada, pp. 3317–3319, 2002.
34. A. Gersho and R. M. Gray, *Vector Quantization and Signal Compression*, Kluwer Academic Publishers, Norwell, MA, 1992.
35. J. J. Y. Huang and P. M. Schultheiss, Block quantization of correlated Gaussian random vectors, *IEEE Transactions on Communications*, vol. 11, no. 3, pp. 289–296, 1963.
36. M. Antonini, M. Barlaud, P. Mathieu, and I. Daubechies, Image coding using wavelet transform, *IEEE Transactions on Image Processing*, vol. 1, no. 2, pp. 205–220, 1992.

37. J. T. Rucker, 3D wavelet-based algorithms for the compression of geoscience data, Master's thesis, Mississippi State University, 2005.

38. A. Bilgin, G. Zweig, and M. W. Marcellin, Three-dimensional image compression with integer wavelet transforms, *Applied Optics*, vol. 39, no. 11, pp. 1799–1814, 2000.

39. Z. Xiong, X. Wu, S. Cheng, and J. Hua, Lossy-to-lossless compression of medical volumetric data using three-dimensional integer wavelet transforms, *IEEE Transactions on Medical Imaging*, vol. 22, no. 3, pp. 459–470, 2003.

40. X. Tang and W. A. Pearlman, Lossy-to-lossless block-based compression of hyperspectral volumetric data, in *Proceedings of the International Conference on Image Processing*, Vol. 3, Singapore, pp. 3283–3286, 2004.

41. A. R. Calderbank, I. Daubechies, W. Sweldens, and B.-L. Yeo, Lossless image compression using integer to integer wavelet transforms, in *Proceedings of the International Conference on Image Processing*, Vol. 1, Lausanne, Switzerland, pp. 596–599, 1997.

INDEX

Page references followed by t indicate material in tables.

Hyperspectral Data Exploitation: Theory and Applications, Edited by Chein-I Chang

Printed and bound by CPI Group (UK) Ltd, Croydon, CR0 4YY
16/09/2026
14963443-0004